Classics from IJGIS

Twenty years of the ***International Journal of Geographical Information Science*** and ***Systems***

Classics from IJGIS

Twenty years of the *International Journal of Geographical Information Science* and *Systems*

Edited by
Peter Fisher

CRC Press
Taylor & Francis Group
Boca Raton London New York

CRC Press is an imprint of the
Taylor & Francis Group, an **informa** business
A TAYLOR & FRANCIS BOOK

CRC Press
Taylor & Francis Group
6000 Broken Sound Parkway NW, Suite 300
Boca Raton, FL 33487-2742

First issued in paperback 2019

ISBN-13: 978-0-8493-7042-7 (hbk)
ISBN-13: 978-0-367-39058-7 (hbk)

Library of Congress Cataloging-in-Publication Data

Classics from IJGIS : twenty years of the International journal of geographical information science and systems / editor, Peter Fisher.
p. cm.
Includes bibliographical references and index.
ISBN-13: 978-0-8493-7042-7
ISBN-10: 0-8493-7042-6
1. Geography--Data processing. 2. Geographic information systems. 3. International journal of geographical information science. I. Fisher, Peter F., 1955-

G70.2C53 2006
910.285--dc22 2006004257

Visit the Taylor & Francis Web site at
http://www.taylorandfrancis.com

and the CRC Press Web site at

Dedication

For Jill, Beth, Kate, and Ian.

Preface

Publishing an academic journal really *is* a collaborative venture. It relies on authors to willingly submit their work for review and critical appraisal. The present model of double-blind peer review relies on a pool of individuals who generously give of their time to provide those reviews. Those authors who successfully emerge from the review process are evident in the pages and the list of contents of the journal, and the reviewers are now included in occasional lists by way of acknowledgment. The unsuccessful authors are the poor relations of this process, because no one except they and the editors know their names or the work they have gone to in executing a research programme, and writing the results only to receive a letter of rejection. Sometimes rejected papers are successfully rewritten and resubmitted to the *International Journal of Geographical Information Systems* and subsequently have success in the review process. Others appear in the increasing number of competitor journals which, personally, I welcome, and which are essential in providing alternative outlets for authors. Many do not. I would like to thank all these players for their contributions to the journal, whether that contribution receives the accolade of publication, a mere acknowledgment, or passes unrecognised.

Some people we can name, however, and paramount among those is the founding editor, Terry Coppock, who became well known to geographical information system (as he would have thought of them) researchers of the late 1980s as he roamed the world attending conferences and inviting submissions. While Terry handled manuscripts submitted by authors from much of the world, he was assisted by Eric Anderson and then Steven Guptill as North American editors. During my tenure as editor, I have had the pleasure of working with Keith Clarke, Marc Armstrong, Harvey Miller, and currently Mark Gahegan as North American editors (now editor for the Americas), and with Dave Abel and now Brian Lees as editors for the Western Pacific (now Australasia and Eastern Asia). Over the years, Sandy Crosby, Neil Stuart, and Nick Tate have managed a sporadic but increasing flow of book reviews for the journal.

The editors are supported in their work by an editorial board. Members of the board over the years are too many to number here. In the early years of any journal this board is especially important, encouraging submissions, reviewing, and even writing. They help to set the tone of the journal. For an established journal their importance is not diminished, and they continue to be active in the same roles, especially reviewing. There are also occasional meetings of the board at conferences.

The publisher also has a team of professional staff members. Throughout my period with the *IJGIS,* Richard Steele has been a constant presence, although in recent years his influence has been somewhat removed as his responsibilities within Taylor & Francis have grown. In that later period, the managing contacts for the journal have been Melanie Bartlett and Rachel Sangster, the latter assisted by Virginia

Klaessen. On the production side, managing the flow of manuscripts and the people most authors will have communicated with over proofs (whether they know it or not), are David Chapman, Sophie Middleton, Heidi Cormode, and currently James Baldock. These people have managed the work of typesetters and copy editors who are anonymous to me. I have personally received continuing assistance in communicating with authors and reviewers from Jill Fisher.

As editor of the *IJGIS*, I am probably the only person who realises the full extent of the contributions of all these different people. On behalf of the scientific community, which has seen its work acknowledged and recognised through its publication in the journal, I would like to thank them all.

Finally, I would like to thank Taisuke Soda, Jessica Vakili, Linda Manis, and Jacqueline Callahan who, between them, have been responsible for the production of this commemorative volume.

Peter Fisher

CONTENTS

1 Introduction Twenty years of *IJGIS*: Choosing the Classics

Peter F. Fisher

Celebration

The *International Journal of Geographical Information Systems* (*IJGIS*) started publication in 1986. At the time, only two of the journals that are currently direct competitors of *IJGIS* were being published (Table 1.1), and of those two, *American Cartographer* has quite recently refocused to recognise the importance of first Geographical Information Systems (GISystems) and then Geographical Information Science (GIScience). *Geoprocessing* was short-lived, and appeared sporadically for only a few issues. In that time, however, it did manage to include some landmark papers, which are well cited elsewhere. The real burgeoning of journals in the field occurred in the mid 1990s with the introduction of three direct competitors. Since then a number of other journals have started or are about to start, including two electronic journals (identifiable in Table 1.1 from their Web addresses). Some journals have been renamed to reflect the importance of GIScience, just as the *International Journal of Geographical Information Systems* was renamed the *International Journal of Geographical Information Science* for its second decade of publication (Fisher, 1997). A number of journals have been publishing on topics related to GIScience for much longer; Table 1.2 lists a sample of these journals, and in some, such as *Computers & Geosciences* and *Environment & Planning B*, a substantial proportion of the papers are now of interest to the GIScience community. The remote sensing journals publish a particularly large volume of literature, much of which is related to that area of GIScience.

IJGIS is widely recognised as the primary academic journal for those concerned with how we use computers to manipulate information about the surface of the earth. This book marks the completion of its second decade of publication, celebrating by reprinting a number of papers spanning those two decades which show some of the innovations that have been reported in the pages of the journal. The book, however, has more to offer than simply reprinted work because for most papers, a commentary is also included, written by at least one of the original authors and reflecting on the research.

TABLE 1.1
A chronology of publication of Geographical Information Systems and Science *journals.*

Title	Start date (in its form to end of publication in that form)[a]	Current issues per year
Computers, Environment and Urban Systems	1977	6
Geoprocessing	1978–1985	—
International Journal of Geographical Information Science	1997	10
(Formerly *International Journal of Geographical Information Systems*)	1987–1996	
URISA Journal	1989	2
Cartography and Geographical Information Science	1999	4
(Formerly *Cartography and Geographical Information System;*	1992–1998	—
Formerly *American Cartographer*)	1974	—
Geographical Systems	1994–1997	—
Journal of Geographic Information Sciences	1995	2
Transactions in GIS	1996	4
GeoInformatica	1997	4
Journal of Geographic Information and Decision Analysis (www.geodec.org)	1997	2
Geographical and Environmental Modelling	1997–2002	—
Spatial Cognition and Computation	1999	4
International Journal of Applied Earth Observation and Geoinformation (Formerly the *ITC Journal*)	1999	4
Journal of Geographical Systems	1999	4
GIScience and Remote Sensing (Formerly *Mapping Sciences and Remote Sensing*)	2004	4
*Applied GIS (*http://www.epress.monash.edu/ag/about.html)	2005	3

Note: This list is intended to be indicative, not exhaustive. The ordering reflects the inclusion of computers or information-related issues indicated by the title.

— Indicates that publication *under that title* has ceased.

[a] Some start dates are inferred from the current volume number, assuming an annual volume.

Selection of the articles

Choosing any anthology of writing will rarely satisfy everyone (except possibly the editor of the anthology and the chosen contributors). In this introduction I therefore set out my reasoning, without necessarily justifying the inclusion and exclusion of every single paper ever published in *IJGIS*.

TABLE 1.2
A chronology of publication of some journals associated with Geographical Information Systems and Science.

Title	Start date (in its present form)[a]	Current issues per year
Photogrammetric Engineering and Remote Sensing	1935	12
Geomatica	1947	4
Remote Sensing of Environment	1969	20
Geographical Analysis	1969	4
Environment & Planning A	1969	12
Environment & Planning B	1974	6
Computers & Geosciences	1975	10
International Journal of Remote Sensing	1979	24
Environment & Planning C	1983	6
The Photogrammetric Record	1985	4

Note: This list is intended to be indicative, not exhaustive.

[a] Some start dates are inferred from the current volume number, assuming an annual volume.

Following my analysis of the citation rankings of papers in *IJGIS* (Fisher, 2001), I was aware that there are a few papers that have been published in the journal that have been very well received by the research community, are widely cited (for the right reasons), and could be described as influential on the progress and development of research in the field. My intention is that papers in the collection should have added to our understanding of GIScience in its wider sense.

In choosing the papers I used a number of criteria:

- The first author of each paper should only have one paper included in the collection.
- There should be at least one paper from most years (but not necessarily all).
- Articles concerned with applications were excluded because I wished the papers to focus on the science of information.
- Papers that describe the state-of-the-art of GISystems in particular countries or that review particular research initiatives, were likewise excluded; they tend to have been overtaken by time.
- Review papers on more academic subjects were also excluded (although there have been few such papers).
- Finally, I endeavoured to choose papers from the topics of GIScience that have been covered in the pages of *IJGIS*, but only those that have a level of recognition.

Against these criteria, I used data from the ISI (Institut for Scientific Information) Web of Knowledge on the rates of citations for papers published in *IJGIS*, a request to members of the editorial board for recommendations, and my own opinion. I therefore take full responsibility for the papers that are included, while at the same time being aware that there are undoubtedly papers excluded which another editor would have included, and papers included for which others may not see the reason.

The articles

The articles are presented in chronological order of original publication.

Probably the most important paper ever published in *IJGIS* in relation to the academic position of the science of geographical information is Goodchild's paper included here as Chapter 9. He articulates the justification for a GIScience, and outlines the areas of activity. Interestingly, in the associated commentary he not only discusses the gestation of the paper, but identifies the areas that he failed to point out in the original paper.

As I noted in my review of citations to *IJGIS* (Fisher, 2001), the most cited paper published in *IJGIS* (and still the most cited), is that by Egenhofer and Franzosa (Chapter 7 in this volume). It is the first in a continuing literature of articles that examine the ways in which different spatial object types can interact. The literature has extended from the procedural definitions to cognitive recognition. This is related to the type of research on qualitative spatial reasoning reported by Frank (Chapter 14).

The very first volume of *IJGIS* contained a paper by Openshaw, Charlton, Wymer, and Craft (Chapter 2), which presents an approach to searching for clusters in the occurrence of childhood disease. I identified this as the second most cited paper published in *IJGIS* up to 2001 (Fisher, 2001), although finding records of its citation are time-consuming. Spatial statistical analysis has been a continuing concern in the pages of *IJGIS*, and a particularly significant paper published in *IJGIS* was one of the harbingers of the novel method of geographically weighted regression (Fotheringham, Charlton, and Brunsdon, Chapter 13), which has been a major contribution to that field.

Over the period of publication of *IJGIS*, one of the main developments in computer science generally, and in databases in particular, has been the introduction of object-orientation. Worboys, Hearnshaw, and Maguire (Chapter 6) present a comprehensive introduction to this topic, while Raper and Livingstone (Chapter 11) extend it to the conceptualising of process models in an object-oriented framework.

The analysis of error and uncertainty has always been a major issue for authors in *IJGIS*, and I have previously identified the paper by Heuvelink, Burrough. and Stein (Chapter 4) as the third most cited paper in *IJGIS* (Fisher, 2001). In that paper they are concerned with the analysis of errors in continuous measurements, while the paper by Kiiveri (Chapter 15) is one of the earliest concerned with the analysis of positional uncertainty. Concern with another aspect of uncertainty has been witnessed in the pages of *IJGIS*, namely the stability of analytical results. Skidmore (Chapter 5) looked at the different methods for determining slope and aspect from digital elevations models, while Fisher (Chapter 10) looks not only at algorithms for determining visibility, but also at a variety of results from different implementations.

Temporal analysis has been a long-standing concern among geographers, and many GIS textbooks comment on the lack of temporal analysis capability. One of the outstanding contributions in this area has been by Miller (Chapter 8), where he discusses the space-time prism. A huge volume of subsequent work has developed this theme further, and it is particularly important for those involved in the development of location-based services.

The discussion of generalisation by Brassel and Weibel (Chapter 3) reviews a theme from cartography, but places it in a more holistic framework, particularly bringing out the significance of generalisation in modelling. The field of interactive visualisation and dynamic cartography has been one of the significant developments in the 20 years of publication, and the paper by Andrienko and Andrienko (Chapter 18) is representative of this area.

Jankowski (Chapter 12) discusses the developments in geographical decision making and multicriteria analysis, which has been the subject of a variety of contributions to *IJGIS*. Related to decision making is the area of modelling, and papers in *IJGIS*, particularly on urban modelling with automata, have reflected this concern. In this context, one of the most influential papers has been that of Clarke and Gaydos (Chapter 16), which has not only been cited frequently, but the model reported is in widespread use.

Much spatial analysis has been criticised for being concerned with the geometric space as opposed to the humanly constructed objects and phenomena that occupy that space (Pickles, 1995). As Frank comments in his commentary in this volume (Chapter 14), we know relatively well how to analyse the space, but less well what occupies the space. This topic is partly the concern of semantics, and this has been a developing topic in the GIScience community. It was first introduced to the pages of *IJGIS* as an issue related to the interoperability of information (Bishr, Chapter 17). Smith and Mark (Chapter 19), among others, develop the topic into more fundamental issues of understanding the meaning in the geography.

Where to end this compilation of papers was a concern when the book was first proposed. I decided that the work of Llobera (Chapter 20) was an appropriate paper because he is taking a well-known GIS function, reevaluating it, and suggesting ways in which it can be related to the human cognitive experience. The paper represents an interesting blend of characteristics, and stretches the traditional view of analysis from geographical information systems. Widespread interest in this paper is illustrated by it being the most downloaded paper from the electronic version of the journal when the papers were being chosen.

The final chapter of this book is an original contribution that uses the database of citation indices to analyse the authors and the author relationships revealed in the *International Journal of Geographical Information Science*. In an interesting analysis, Arciniegas and Wood (Chapter 21) look at the networking and the connectivity of authors and nations, as well as the influence of the neighbouring (sister) field of remote sensing.

The commentaries

Each article (except one) is accompanied by a commentary written by at least one of the original authors. When commissioning these commentaries, I asked the authors to consider covering the following topics:

- What motivated the original work but is not documented in the original paper?
- Have there been further developments since the work was done, or has there not?
- If things have happened subsequently, comment on why they have happened.
- If there have been no further developments, comment on why that might be so.
- Has the subject remained an exclusive concern of GIScientists or has it always been broader?
- Is the topic still relevant to the GIScience agenda? Is it a worthwhile area to continue working in?
- How might the future look (within GIScience) if your work is widely adopted?
- I was not prescriptive, however, and I also invited authors to discuss any other topics that may be of interest to them or to readers.

Perhaps it is not surprising that the commentaries presented here represent a cross-section of what might have been written. The research of a number of authors has moved away from the work reported in the papers reprinted here (Skidmore, Chapter 5; Kiiveri, Chapter 15), although the commentaries nonetheless provide interesting reviews of some of the related work that has developed on their theme. Other commentaries present very personal perspectives on the research that led up to the paper and its circumstance (Charlton, Chapter 2; Raper and Livingstone, Chapter 11; Frank, Chapter 14). Some authors have done extensive further work in the area of the original paper and the methods have been adopted by other researchers (Weibel and Brassel, Chapter 3; Fotheringham, Charlton, and Brunsdon, Chapter 13; Clarke, Gazulis, Dietzel, and Goldstein, Chapter 16; Andrienko and Andrienko, Chapter 18). Still other commentaries point out what the authors still consider open and important questions (Fisher, Chapter 10).

References

FISHER, P., 1997, Editorial. *International Journal of Geographical Information Science*, **11**, 1–3.

FISHER, P., 2001, Citations to the *International Journal of Geographical Information Systems and Science*: The first 10 years. *International Journal of Geographical Information Science*, **15**, 1–6.

PICKLES J., 1995, Representations in an electronic age: Geography, GIS, and democracy. In *Ground Truth*, Pickles, J., Ed. Guilford, New York, 1–30.

International Journal of Geographical Information Systems,
1987, Vol. 1, No. 4, 335–358

2 A Mark 1 Geographical Analysis Machine for the Automated Analysis of Point Data Sets

Stan Openshaw, Martin E. Charlton, Colin Wymer, and Alan Craft

Abstract. This paper presents the first of a new generation of spatial analytical technology based on a fusion of statistical, GIS and computational thinking. It describes how to build what is termed a Geographical Analysis Machine (GAM), with high descriptive power. A GAM offers an imaginative new approach to the analysis of point pattern data based on a fully automated process whereby a point data set is explored for evidence of pattern without being unduly affected by pre-defined areal units or data error. No prior information or specification of particular location-specific hypotheses is required. If geographical data contain strong evidence of pattern in geographical space, then the GAM will find it. This technology is demonstrated by an analysis of data on cancer for northern England.

1 Introduction

The computerization of society and the development of Geographical Information Systems (GIS) are greatly increasing the supply of point-referenced data sets. The commodification of information is also emphasizing the importance of geographical analysis as value-adding technology (Openshaw and Goddard 1987). Yet there are few signs of much renewed methodological interest in the development of better spatial analytical techniques despite growing recognition of their importance. The Chorley Committee of Enquiry into opportunities for Geographic Information Handling in the United Kingdom comments that 'Existing statistical methodology which can be applied to spatial data is little developed: hence there is a need to develop this field'. (Department of the Environment 1987, p. 108). In particular, there is need for spatial statistical techniques that can both be utilized within a GIS framework and make best use of the new opportunities for exploratory analysis being created by GIS developments.

There are seemingly several reasons for this neglect of basic methodological research (Openshaw 1986). There is also a fundamental misunderstanding as to what spatial analytical techniques can and cannot reasonably be expected to achieve. It needs to be recognized that the task of detecting substantive processes from spatial patterns is always going to be difficult or uncertain and may often be impossible. The justification for spatial analysis returns to the more traditional objective analysis of map patterns itself. Indeed, it is precisely here where there is a demand for both better techniques and new opportunities. There is, of course, a well-established set of spatial statistical techniques that would claim to meet these needs (see, for example, Ripley 1981, Diggle 1983, Upton and Fingleton 1985). The problem is that, with the development of GIS technology and the vast growth in spatially-referenced data sets, the available tools are often no longer adequate for the tasks of data processing that they now face. There is also a tendency to limit the spatial analytical tool-kit to established statistical methods and thereby, implicitly, adopt a blinkered view of what is both possible and scientific. It should not be assumed that only statistical techniques operated in the traditional manual fashion are of interest; new methods, based on the linking of statistical spatial analysis to a GIS in an automated way, might well be possible and potentially far more useful as the basis for the exploratory analysis of point data sets.

Openshaw (1987) urged the development of fully-automated geographical analysis systems in order (1) to make best use of the information the data contain without being restricted by the dictates of state-of-the-art theories and methods, (2) to handle efficiently the growing number of important data sets of interest to many different subjects, (3) to allow the development of forms of analysis which can explicitly handle the special characteristics of spatial data without making untenable assumptions for purposes of statistical or mathematical tractability, (4) to better meet the needs of applied users who often want trustworthy answers to quite basic questions concerning spatial patterns and (5) to develop effective and unbiased search techniques capable of both generating and testing hypotheses related to spatial phenomena in an automated manner in an exploratory context under conditions of little and uncertain prior theoretical knowledge. There was also a desire to investigate new forms of spatial analysis which can exploit both the growing richness and size of spatial data sets and the great computational power of modern computers.

The objective then is clear enough. The aim is to develop a new automated approach to the analysis of point pattern data as a means of meeting current and future needs of geographical analysis whilst overcoming many of the statistical and scientific problems that currently exist. The immediate context for this work is the analysis of data on cancer but the general results are widely applicable to other types of point pattern data. In §2 the thinking that lay behind the idea for a conceptual Geographical Analysis Machine is outlined. Its implementation on computer hardware is described in §3 and its application to a point data set is discussed in §4. Finally §5 provides speculations on future developments.

2 Some outstanding technical problems

Point patterns analysis is complicated because it is not purely a statistical problem but also involves fundamental questions about the manner of conducting analyses and making inferences based on non-experimental research in geographical domain. Additionally, there is some need to take into explicit account geography as an endogenous variable rather than one which is neutral and exogenous to the chosen method of analysis. These issues are discussed here with particular reference to epidemiological studies of cancer patterns because this is a good example of point data sets which are of considerable topical interest. However, the same set of problems also applies to other point data sets.

The basic question being addressed in many cancer studies is whether or not a point pattern shows signs of clustering in space and, perhaps, in time also. Typically, these studies have used various measures of incidence to identify areas with excess cancer rates and a few localized areas of excess incidence or 'clusters' have been identified (see, for example, Craft *et al.* 1985). There are problems with conventional studies because a number of technical issues undermine the validity of the results. Indeed, purely statistical methods are increasingly being viewed as unable to provide either accurate or unbiased scientific answers to the problem of detecting clusters (see, for example, Croasdale and White 1987). It is helpful, therefore, to enumerate the problems briefly because they provide design objectives that any new form of analysis has to meet.

The list of technical problems includes the following: (i) the results may well have been biased by the conscious or accidental selection of time periods, categories of disease, age-groups, study-regions and (if circular statistics are used) the radial distance; (ii) the hypothesis being tested on spatial data is often invalidated by prior knowledge of the data; (iii) there are often problems associated with determining the significance of results; (iv) the presence and potential impact of errors of both measurement and spatial representation in point data sets are ignored and (v) in formulating point hypotheses, there is a heavy reliance on the prior identification of point locations assumed to be related or causally linked to the point pattern in some way which renders the process of hypothesis formulation biased against whatever point sources are used whilst ignoring all those which are not used.

In addition, there are several more general criticisms which can be applied. In particular, there are major problems in deciding what type of analysis is required. In the real world where the distribution of people, towns and points is not uniform but highly discontinuous, it hardly makes much sense to use statistics which try to summarize a complete map pattern for some arbitrary study region because of the high degree of abstraction from the geographical domain. Boundary effects, due to the inevitably subjective selection of a study region, may also influence the results. What value is there in averaging out all the information contained in a map pattern for a whole study region when it is the 'geography' of the map pattern within the area of interest and the locations of deviations from some overall measure of map pattern that are of most interest? So there needs to be some means of identifying both the existence and the location within the map where some general hypothesis or summary statistics breaks down rather than try to devise better descriptors of whole map patterns.

Finally, it is noted that the problem of *post hoc* hypothesis testing is particularly serious. Leamer (1978) discusses the difficulties caused when hypotheses are created or modified after examining the data on which they are to be tested. It seems that the spatial data should be viewed only as a means of generating, rather than testing, the very hypotheses they inspire. This is a major unresolved problem in spatial analysis where data and exploratory analysis usually precede more detailed subsequent studies, because it renders the results of inferential procedures for testing hypotheses invalid. What is needed is some means of extracting confessions from spatial data sets so as to be confident that the inferences are valid.

3 Building a Geographical Analysis Machine

Many of these problems result from selectivity in the hypotheses being tested and from fears of bias. Traditionally, only a small number of hypotheses could actually be formulated and tested, partly because of the nature of the scientific method widely employed as the basis for inference and partly because of the intellectual effort involved and the utter impracticality of listing all possible hypotheses before any data analysis. The increasing availability of databases in advance of any hypotheses suggests that the construction and modification of post-data models is emerging as an important problem as GIS start to provide vastly improved spatial data sets independent of the formulation of hypotheses and the specification of models. In short, a new era of *ad hoc* inference is dawning.

Openshaw (1987) suggests that the most general and least biased solution to these and other problems of creating knowledge by hypothetico-deductive means is simply to generate and test all possible geographical hypotheses relevant to a particular problem. Bias is excluded and prior knowledge rendered irrelevant because no selectivity is required. The universe of all possible hypotheses also contains the totality of all deductive knowledge about a given problem of spatial analysis, both known and unknown. It is now simply a matter of enumerating this universe and, armed with a procedure for assessing significance, tabulating the distribution of all meaningful or significant results that exist. This approach comes with guarantees that it is totally unbiased with respect to all knowledge both known and as yet undiscovered; it treats all locations equally; it is totally objective and capable of replication and it cannot fail to find any meaningful results if any exist because, within the design limits of the method, all relevant hypotheses have been examined.

This approach is termed a Geographical Analysis Machine (GAM). It was imagined initially that it might take 5–10 years to develop (Openshaw 1987). Six months later a prototype machine for point pattern analysis existed and was working (Openshaw *et al.* 1988). In principle, it is just another variant of an automated modeling system (Openshaw 1988) but this time applied to automating the process of hypothetico-deductive inference in the context of point data analysis. Another important difference is that there is no need for an elaborate search technique, provided the universe of all possible hypotheses is sufficiently small to be fully enumerated (as is usually the case).

The concept of a GAM is based on a very simple but computationally intensive type of statistical spatial analysis, yet it constitutes what is the most advanced hypothetico-deductively inferencing engine yet developed. There are four basic components to a GAM: (1) a spatial hypothesis generator, (2) a procedure for assessing significance, (3) a GIS to handle retrieval of spatial data and (4) a geographical display and map processing system.

3.1 Generating a universe of all possible hypotheses

The general hypothesis of interest here is whether there is an excess of observed points within x km of a specific location. A test statistic would be computed for this circular search area and some measure of significance obtained. The concept underlying the GAM is to take this generic hypothesis and generalize the locational aspects by examining circles of all sensible radii for all possible point locations in a given study region. In this way, the universe of spatial hypotheses of this general type can be defined and enumerated. This philosophy can be readily modified to handle different geometries of search area; for example, a wedge shape may be used to represent diffusion of atmospheric material. Here, attention is focused on a circular search region, which is the traditional geometry used in cancer studies. Squares could also be used but offer no significant computational benefits.

The following algorithm is used to generate a universe of all possible circle-based hypotheses.

Step 1

Define an initial two-dimensional grid lattice with the grid size (g). Define also a minimum circle radius, a maximum value and a radial size increment. Assume that a circle is to be located on each grid intersection, thereby completely covering the study region with a regular and even coverage of circles of some initial size. The lattice is sufficiently close-grained so that the circles overlap to a large degree. This requires that the spacing of the grid mesh (g) is some fraction of circle radius, namely, $g = z^*r$ where z is the circle overlap parameter and r the current circle radius.

Step 2

For each lattice point within the study region, retrieve the data needed to compute a spatial pattern measuring test statistic for circles of radius r. Store the results for all locations that pass a significance test set at a given threshold (see § 3.2).

Step 3

Increase the radius of the circle by a specified amount and change the grid mesh to reflect the increase in radius.

Step 4

Repeat steps *2* and *3* until all radii considered relevant have been examined.

In steps *1* and *3*, it is necessary to make the grid mesh a function of the radius of the circle so that the circles overlap. This is required so that the resulting sequence of circles provides a good discrete approximation to testing all possible point locations in two-dimensional space. The overlap between successive circles also serves another very important purpose. The overall effect in moving from one point on the grid lattice to another nearby is to perturb the circle centroids by an amount which is proportional to the radius of the circle. This is most useful because it provides a means of auto-sensitivity analysis which will identify the effects of change sin the data resulting from a slight but scaled shift in the location of circles and thus allows the GAM to take into account the effects of both possible data error and edge effects in the process of spatial data retrieval. This is necessary because few point data sets are both completely accurate and exist as zero-dimensional points in two-dimensional space. It is particularly relevant to data on the incidence of cancer which are typically locationally-referenced in the best practicable but still imprecise manner (for example, using 100 m point references for postcodes). This is a potential problem because uncertainties in the location of even a small number of cancers can produce spurious results.

Step 1 can also be modified to handle edge effects. Ideally, the study region needs to be surrounded either by an internal or by an external corridor equal in width to the radius of the largest circle. Circles located within the study region but overlapping the external boundary corridor will then be allocated the correct data values. This requires either that additional data are available for the corridor region outside the study region or that the analysis stops short of the boundary of the study region. The land-sea boundary offers no problem and can be ignored.

3.2 A procedure for assessing significance

For each of the circular search regions it is necessary to compute a test statistic and then decide whether it contains statistically significant excess of observed points. The choice of test statistic is problematical. Previous epidemiological studies have used both incidence rates and Poisson probabilities (see, for example, Craft *et al.* 1985). Both statistics suffer from problems in assessing the populations at risk when the cancer data relate to a 10- to 20-year period and the census-based counts of the child population refer to one particular year. This problem can be solved only if better population data for small areas allow the adoption of a demographically-sound concept of population at risk (Rees and Wilson 1977). However, this is unlikely to be possible in the short term and the analysis has to proceed in the knowledge that some of the key data are almost certainly in error for perhaps the major part of whatever period is used. This is a problem that cannot be overcome at present although estimates of the possible magnitude of the effects can be obtained by sensitivity analysis for instance, by exploring the effects of different denominator populations. The aim would be to identify the most robust results given the likely levels of uncertainty in estimates of the population at risk.

An additional difficulty is determining what might be a suitable threshold for measuring significance. One solution is to use Poisson probabilities although this assumes a particular null hypothesis, namely, that the point pattern is generated by a Poisson process. A more general approach is to use a Monte Carlo procedure for significance testing which shifts the decisions back in the direction of the specification of a particular null hypothesis regarding an underlying generating process. The preference is to use a Monte Carlo significance test based on a simple count of points within a circle. The number of observed points in a circle of any specific size based on a particular point location can be directly compared with the number that would be expected under a specific null hypothesis. This count statistic has the benefit of computational simplicity and it is easy to explain.

The standard Monte Carlo procedure for significance testing, used here to identify anomalously high counts of points, was developed by Hope (1968) and is widely used. Given a null hypothesis, it involves ranking the value of an observed test statistic, u_1, amongst a corresponding set of $n - 1$ values generated by random sampling from the null distribution of u. When the test statistic is a real number, then the rank of the observed value u_1 amongst the complete set of values (u_1, i = 1, 2, ... n) determines an exact significance level for the test since, under the null hypothesis, each of the n possible rankings of u_1 is equally likely. When, as here, the distribution of u is discrete (i.e. integer counts) then tied ranks can occur. Following Besag and Diggle (1977), an estimate of the upper bound for the significance level can be obtained by choosing the most conservative ranking. Since it is not necessary to obtain a precise estimate of the null distribution function of the test statistic, the number of simulated samples (n) can be quite small, typically $n = 100$. Indeed it is usually considered that a small value is sufficient (Besag and Diggle 1977), but whether this is so depends on the desired significance level. A value of $n = 100$ offers a significance level only down to the 0·01 level. However, lower significance levels can always be obtained by increasing the value of n to 500 or more. This Monte Carlo procedure is most useful because it allows the use of any test statistic and any null hypothesis without being constrained by known distribution theory.

The choice of null hypothesis is very important. The purpose of the GAM is to identify locations where a particular hypothesis breaks down within the study region. The utility of the procedure depends on the extent to which it can handle a wide range of different null hypotheses. However, for purposes of exploratory study, significance in a spatial context has traditionally been based only on an assessment of the probability of obtaining a given level of pattern if the points were randomly distributed over those parts of the study region where data can exist. The suggestion is, therefore, that an initially useful general purpose null hypothesis for benchmark purposes is to compare the observed pattern of points with what might be expected if they were generated by a Poisson process, and then to test for departures from this state. With this particular null hypothesis, all the Monte Carlo test of significance is doing is to provide an upper bound on the estimate of the Poisson probability of an observed number of points appearing in a particular circle. It would be far easier to calculate this statistic analytically rather than to estimate it indirectly by numerical methods, however, this would preclude the possibility of considering other types of null hypotheses. The limitation is best avoided as further research into the results obtained for different assumptions about types of process is likely to be important in the future.

A final issue is how to generate realistic random samples of point pattern data sets that reflect the chosen null hypothesis. It is not simply a matter of picking random sets of points anywhere within a study region, because the observed point patterns are selections from spatially discontinuous point populations at risk. It is necessary to take into account geographical variations in the distribution and density of whatever is considered to be the population at risk of becoming a member of the point data set under study. For the data on cancer in §4, these random point distributions are produced by setting up a virtual vector of k elements, where k is the total number of children in the study region. For any fixed number of cancers (m) and under the null hypothesis used here, uniformly distributed random numbers between 1 and k can be used to select m children from this list and then assign them the locational reference of, in this case, the census enumeration district containing their houses. This procedure is repeated to provide n sets of m cancers generated under the null hypothesis of a Poisson generating process. It should be noted that the generation and storage of these simulated point data sets increase the size of the database by a factor of n.

This procedure for testing statistical significance can be summarized as follows. For a circle of specified radius drawn around a specific point, the number of observed points lying within it is obtained. This observed count is then compared with a reference set of 499 different and separately generated point data sets reflecting the null hypothesis. The rank of the observed test statistic gives some measure of the significance of the observed number of points. Each circle that is evaluated requires the retrieval of data from 500 different data sets. It is here that a good GIS is needed.

3.3 Linking in a KDB tree GIS data structure

Creating a conceptual GAM is straightforward. Implementing it on a computer is more difficult because of the amount of machine time that is required. Most of the computer time is spent in retrieving data for often millions of circular search regions from a large database containing both the observed data and perhaps several hundred sets of simulated data. For any reasonably large study region, this task of spatial data retrieval would be virtually impossible without a highly sophisticated geographical data structure that allows for very fast retrieval of spatial data.

There are several possibilities. The preferred choice will largely reflect the available hardware. The design objective used here was that the GAM should ultimately be capable of being run on a 32-bit microcomputer as well as on general purpose mainframes. It was therefore assumed that there were restrictions on the amount of memory that would be available to the GAM and that, as a consequence, the data would have to be held on disk. If the GAM was to be ported onto a supercomputer (see §5.3) then a rather different approach could be adopted.

There are several hierarchical data structures which can be used for the efficient storage and retrieval of point data sets (Samet 1984), although not all are suitable for GAM applications. Probably the most efficient, and also the most neglected, is a multidimensional, multi-way search tree known as a KDB tree (Robinson 1981). This allows all the data (observed cancers and randomly generated cancers) to be stored on a single tree and retrieved simultaneously in a highly efficient manner. The KDB tree is attractive because it combines the search efficiency of the KD tree (Bentley 1975) with the I/O efficiency of the B tree (Bayer and McCreight 1972).

The GAM data are characterized by easting and northing coordinates, a simulation number and other basic information that is not currently used but which will be exploited in the future. The data are indexed by, at present, three keys rather than the more usual two. The KDB tree is used here because (i) it can deal with records with many keys and allows all the information of interest to be stored in a single integrated database of keys without affecting performance; (ii) it can perform efficient range searches based on those keys; (iii) it can cope efficiently with very large data sets which must reside on disk; (iv) the tree itself is optimally balanced in the sense that the length of path to any record is the same for all records and usually considerably shorter than in a KD tree and (v) prior knowledge of search requirements may be used to determine the 'shape' of a tree (i.e. depth and spread) when it is built and thus optimize retrieval times.

A KDB tree consists of a set of fixed pages, each with an identifier, and initial access to a tree is via its root page. Each page corresponds to a node of the tree and represents a (hyper-) rectangular region of the key space, with the root page representing the total domain of the key space. Pages corresponding to "leaf" nodes of the tree are called point pages and contain index records, together with pointers to locations in the database. The remaining pages are called region pages. Each entry in a region page (1) defines a rectangular subregion of the region represented by that page and (2) has an associated pointer giving the identifier of the page which represents that subregion. The subregions within a region page are disjoint and their union is the region represented by that page.

Effectively then, a KDB tree produces a hierarchy of partitions of the domain of the key space. Each level in the tree completely covers the domain and the deeper and level, the finer the partition. The precise structure of a KDB tree is determined by the sequence of insertion of index records into the tree. Details of algorithms for insertion, splitting, deletion and range search are to be found in Robinson (1981) and a full description of the implementation of the KDB tree can be found in Wymer *et al.* (1988).

In the current application, there are three key variables, namely, the x and y coordinates of the point and a 'type' variable which distinguishes between observed cancers, child population and simulated cancers. The range search, to which the KDB tree is ideally suited, consists of finding all index records whose key values lie within specified ranges, i.e., finding all points lying within a given hyper-rectangle. Thus to perform a search of a circular region, it is necessary first to perform a range search using the bounding square of the given circle and then to filter out those points found which do not lie within the required circle. Without use of the KDB tree or something equally efficient, the GAM would suffer from severe computational problems in terms of processing times and be restricted to small data sets. In principle, the KDB-tree-based GAM also opens up the prospect of dedicated micro-based systems some time in the future.

3.4 A geographical display, evaluation and map processing system

The GAM approach therefore, offers an examination of all geographical hypotheses of a particular type (in this case circular test statistics) with a built-in automated analyzer of data sensitivity and a means of testing for statistical significance. The final step is to put the results back into an explicitly geographical context. This could hardly be simpler and is achieved merely by identifying all the locations with significant circles and presenting them as a map. The resulting map has the advantage of being easily communicated to third parties who do not possess detailed technical knowledge.

The distribution of significant circles provides a visual representation of all locations where the null hypothesis breaks down. This is the principal function offered by the GAM. Meaningful locations where such breakdowns occur are expected to be characterized by a large number of circles drawn around closely-spaced centroids covering a range of different radii. When the null hypothesis is spatial randomness, then a cluster can be visualized as a geographically located and dense concentration of significant circles. By contrast, it is thought that any spurious circles that survive the significance test will occur in the form of a more scattered distribution of lower intensity.

The maps offer a means of identifying heavy concentrations of significant circles in localized areas which are unusual in their intensity, robust against uncertainly in the data and strong enough to stand out against a background of random noise. These qualities are desirable because the number of hypotheses being evaluated is sufficiently large for a possibly large number of significant circles to exist as type I errors. It is useful, therefore, to probe the patterns further by, for example, focusing only on certain sizes of radii and by raising the minimum number of points in those circles that are plotted. The changes in the resulting map patterns may provide indications of both the persistence and the strength of the resulting patterns of circles as well as some indication of possible scale effects and other clues about causal processes. This reliance on a system of eyeball information may appear unscientific but the GAM is basically a descriptive technique designed for an exploratory purpose, that is, to identify areas of interest where further work will be necessary to either validate the findings or to test more specific hypotheses. The pattern of significant circles can be further processed to offer a means of validating the results.

One potential problem with the statistical analysis used in the GAM is that it involves testing multiple hypotheses based on overlapping circles with at least some of the data in common. This will influence the estimated significance levels. It is also very difficult to ascertain the total significance of the complete distribution of significant circles as shown on a map. For example, what is the probability of a given pattern of three or four dense concentrations of significant circles occurring by chance? The problem is that little is known as yet about what patterns may be discovered by the GAM in purely random data and this question also involves problems of map pattern detection. Other difficulties may be caused by the discreteness of the test statistic. The risks are thought to be small but further experimentation is needed before they can be quantified and the results of a GAM run considered as having been validated in a purely statistical sense.

A final aspect concerns various prospects of further processing to enhance the visual presentation. One useful technique would be to use a GIS to manipulate the pattern of significant circles, for example, to display only non-overlapping circles. Other forms of map presentation could also be investigated. Cluster analysis could be used to classify the two-dimensional patterns to further simplify the picture. Additionally, it is possible to ask the question "Where are the most significant circles?" or "How does this cluster of circles compare with that cluster of circles?" Some of these questions can be answered by summarizing the map results for different sets of areal units which have connotations with place names. This might help to answer questions as to whether the patterns are influenced by rural-urban differences in the distribution of children. A major cluster in a rural area may survive over a wider range of circle radii than an equivalent cluster in an urban area where its effects may be quickly diluted by a greater density of population at risk. It would also be possible to alter the null hypothesis to take rural-urban differences into account. Various other solutions can be formulated to assess the relative strengths of the circles and thus obtain a quantified measure of local excess. One approach would be to computer a Poisson probability for the significant circles and display the top 10, 20, etc. per cent most significant ones on a map. Another would be to add a third dimension, in the form of a Poisson probability, to the distribution of circles and view the results as a surface or as some other type of three-dimensional display.

4 An application to cancer data for northern England

4.1 Data

The development of the GAM concept as a practical tool was largely a reaction to problems with more traditional statistical approaches to point pattern analysis when applied to data on cancer. With these methods, the basic questions as to whether there are any 'real' as distinct from 'illusory' clusters of cancer and their geographical locations still remain largely unanswered. It seems particularly appropriate therefore, to demonstrate the utility of a GAM on some data on cancer for northern England. The study region consists of the Newcastle and Manchester cancer registries and the observed data on cancer relates to the location of residence at the time of diagnosis for all 0–15 year olds in the period 1968–1985. The child population in this study region was 1 544963 in 1981 and this is used at the level of the census enumeration district (see Rhind 1983) as the basis for generating the random data sets needed by the procedure for significance testing. Both data sets are regarded as two-dimensional point data, the cancers being given postcodes and then converted to 100 m grid references whilst the census enumeration districts are already coded by 100 m point references for the centroids.

The null hypothesis of interest here is that of spatial randomness. Two types of cancer are examined; acute lymphoblastic leukaemia, which is thought to cluster, and Wilms' tumour, which has not previously been reported as showing strong (if any) tendencies for clustering.

4.2 Search parameters

The GAM algorithm outlined in §3 has a number of parameters that have to be set before it can be run. Optimal settings have yet to be determined and it is hoped to do so soon based on simulation experiments with a supercomputer version of GAM. Meanwhile, best estimates are made on the basis of experience so far. It is not thought that these settings are too critical but they might conceivably be data-dependent and further experimentation is needed to understand their effects on the GAM's power of detection.

The minimum and maximum radii of circles need to be specified and these could lie anywhere between 0·1 km (determined by the spatial resolution of the particular data used) and 75 km (determined by size of study region). In fact it seemed reasonable to run GAM using only radii between 1 and 20 km. The lower limit of 1 km and the 1 km increment in radial size used here are due to the large amounts of computer time that are required to run GAM should smaller values be used. It should also be pointed out that circles of large radii are of relatively little interest and that the increment in radial size is probably not too critical because a smaller value would duplicate the effect of circle overlap. This parameter was set to 0·2, again mainly to reduce computer time, although this figure is thought to be adequate for a prototype system. Another parameter is the minimum count of observed points in a circle; this is set to 1 but can be changed during the mapping of the results. Finally, it is necessary to set the number of random simulations to be used for the Monte Carlo significance test. Here, n is set to 500; a large value was considered unnecessary.

All these decisions affect the number of hypotheses that are evaluated and the spatial resolution of the results. Table 2.1 indicates the number of hypotheses that would need to be evaluated for different settings for two of the key GAM search parameters; increment in size of radius and degree of overlap. Currently, it is not known whether the optimal settings for the search parameters will tend towards the smaller values in Table 2.1, or whether the results will prove to be fairly insensitive across a broad range of values. It should also be noted that the total numbers of circles generated are about three times larger than shown in Table 2.1, but that only those circles with centroids located within the study region are currently evaluated. Moreover, only those circles with non-zero counts of children and one or more observed cancers are subjected to the statistical significance test. At present, no use is made of boundary corridors (see §3.1).

4.3 Results

Table 2.2 reports some statistics relating to the building of the KDB tree used by the GAM with the cancer data for northern England. These times could have been reduced by sorting the data prior to input. The other noteworthy aspect is the very shallow nature of the tree; this is one of the reasons why it is so efficient at spatial data retrieval.

TABLE 2.1
Estimates of the number of circles to be evaluated for different search parameters.

Number of hypotheses	Radial size increment (km)	Circle overlap parameter
12271889	0·2	0·1
4502960	0·5	0·1
3271971	1·0	0·1
2476478	2·0	0·1
2830474	0·2	0·2
1263889	0·5	0·2
812993[a]	1·0	0·2
615995	2·0	0·2
1163490	0·2	0·3
527496	0·5	0·3
358497	1·0	0·3
272498	2·0	0·3
641994	0·2	0·4
306498	0·5	0·4
198999	1·0	0·4
152499	2·0	0·4

Note: The total number of hypotheses given here is greater than the number that the GAM would actually evaluate since zero population and zero observed point circles need not be examined.

[a] Results for these values reported later.

TABLE 2.2
KDB tree build statistics.

Cancer	Total points	Page size	Total pages	Tree depth	I/O activity	Processing time (seconds)
Leukaemia	412925	400	1595	3	543000	835
Wilms' tumour	93853	87	1714	3	37000	87

Note: Time on Amdahl 5860.

Table 2.3 gives details of the numbers of hypotheses tested and the run times for the three data sets used here. The computer times are not excessive but this is largely a reflection of the GAM search parameters that were used. One run with an increment of circle size of 0·2 km and a parameter of circle overlap of 0·1 required 26 hours of CPU time with the same leukaemia data.

TABLE 2.3
GAM run statistics.

Cancer	Total cancers	Number of hypotheses	Processing time (seconds)
Leukaemia	853	812993	22758
Wilms' tumour	163	812993	5595

Note: Time on Amdahl 5860.

TABLE 2.4
Some aspects of KDB tree performance in a typical GAM run.

Circle search radius (km)	Number of search circles	Total number of disk I/Os	Number of disk I/Os per circle
1	465801	19968	0·0429
5	54099	53290	0·985
10	8571	258682	30·2
15	4705	349023	74·0
20	2485	466185	188·0
25	1097	281055	256·0

Note: A disk I/O is a measure of system disk activity.

One problem with the KDB tree version of GAM is that disk I/O activity rapidly increases once there is a large disparity between sizes of point page region and circle radius. Table 2.4 shows this effect for a typical GAM run. It begins as highly efficient and then deteriorates as the size of circle increases. Fortunately, there is a simple solution. It is possible to design an adaptive KDB tree which rebuilds itself with a size of point page region appropriate to the size of search areas being used once it predicts that the improvement in efficiency of future retrieval is greater than the cost of reorganization. A simple procedure for monitoring should be able to determine when to rebuild because the GAM's spatial retrieval activity is predictable and the decision rule is that of a classic discounted cost-benefit analysis.

Table 2.5 summarizes the number of circles that were found to be statistically significant departures from the expected Poisson pattern for radii in the range 1–20 km. There are clear differences between the cluster-prone leukaemia and the other data in terms of the persistence of the distribution of significant circles. However, it should be noted that these results will vary slightly every time the GAM is run because of variability in sampling in the Monte Carlo significance test.

Figure 2.1 and Figure 2.2 show plots of those circles significant at the $p = 0{\cdot}002$ level for each of the three data sets. This lower-than-usual significance level is the smallest that can be achieved by 499 simulations (namely, $0{\cdot}002 = 1/500$). It is also thought useful in order to reduce the presence of type I errors in the maps and to

TABLE 2.5
Distribution of significant circles.

Circle radius (km)	Total circles generated	Acute lymphoblastic leukaemia		Wilms' tumour	
		$p = 0{\cdot}01$	$p = 0{\cdot}002$	$p = 0{\cdot}01$	$p = 0{\cdot}002$
1	510367	549	164	430	49
2	127572	338	142	298	24
3	56719	311	153	220	17
4	31910	302	139	221	16
5	20428	298	116	171	7
6	14195	273	81	110	2
7	10423	238	53	68	1
8	7983	202	45	36	
9	6304	165	33	29	
10	5112	142	30	20	
11	4207	120	32	12	
12	3557	97	32	10	
13	3028	92	30	10	
14	2602	90	27	12	
15	2269	88	27	9	
16	1993	94	26	11	
17	1766	94	25	7	
18	1580	83	26	4	
19	1408	81	29	5	
20	1280	74	31	2	

focus attention on the most significant circles. The results are remarkable for their clarity. The maps for leukaemia show evidence of intense clustering in a small number of locations whilst, as expected, the Wilms' tumour map patterns are considerably weaker. There are only very slight indications of a few localized clusters. The qualitative differences between these maps are very large indeed, confirming the power of the GAM as a means of detecting departures from a Poisson distribution.

These maps are unique in that this is the first time that the complete set of (nearly) all possible geographical hypotheses of a particular type have been tested and the significant locations displayed. The pattern of significant circles of acute lymphoblastic leukaemia in Figure 2.1 shows the well-known cluster at Seascale but also identifies one other cluster that appears even stronger (Gateshead) but which was not previously identified, probably because of boundary effects in the small-area studies previously used and because attention was focused mainly on the Sellafield area (see Craft *et al.* 1985, Craft and Openshaw 1987). The weaker, less dense clusters at Sedbergh, Whittingham, Bishop Auckland and Macclesfield are probably spurious while the speculatively-identified Springfield cluster is missing. It is thought that the weaker clusters that survive the significance testing may well prove to be sensitive to the different assumptions used to generate the null hypothesis data sets so that, without further research into GAM power levels, it might be unwise

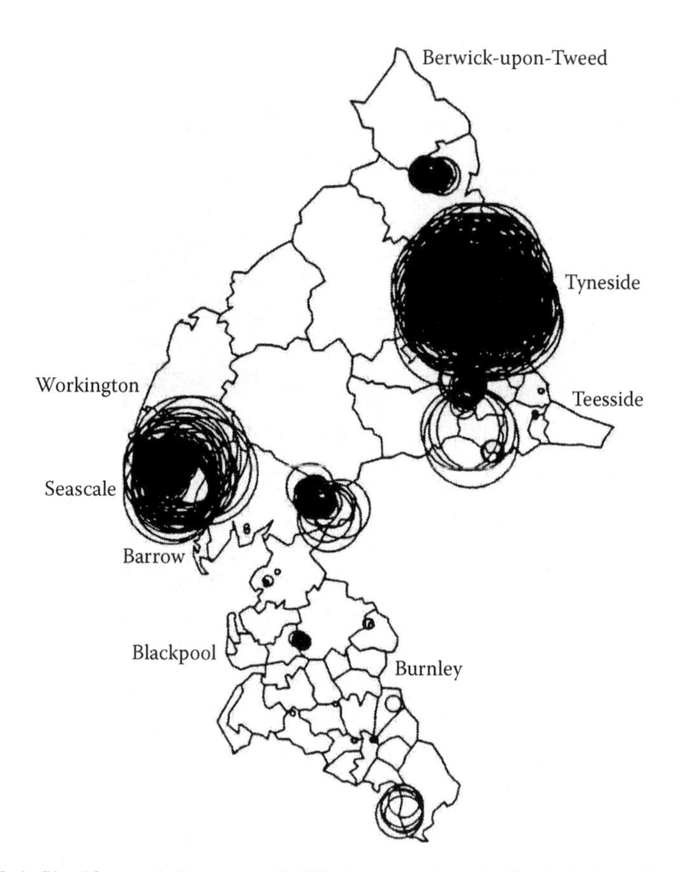

FIGURE 2.1 Significant circles at $p = 0{\cdot}002$ for acute lymphoblastic leukaemia.

to put too much emphasis on these particular areas. However, what is far more remarkable here is the lack of significant circles in both Cleveland and Manchester. This would indicate that there might be some value in running the GAM in reverse so as to identify areas with the greatest significant deficiency of cancer.

The results are surprising, not because they confirm the existence of the cluster at Seascale, but because of the much larger cluster that seems to exist in Tyneside focused on Gateshead. There is no known local major source of low-level radiation in this area so that, for the first time, there would appear to be a possible link with some other form of environmental pollution. Indeed, it might even be that a common non-radiation link might be responsible for both Seascale and Gateshead. This is a matter for further research. However, the significance of finding a major new leukaemia cluster in an area where there are no known local discharges of radiation, at a time when seemingly low-level discharges of radiation from nuclear installations are being blamed for all leukaemia clusters, is considerable.

One of the advantages of a map-based display system is that various filters can be imposed on the data to reveal different aspects of the patterns. Table 2.6 shows the effect of deleting all circles with an observed cancer count of less than two. This choice can be justified on the grounds that clusters of cancer need to have at least

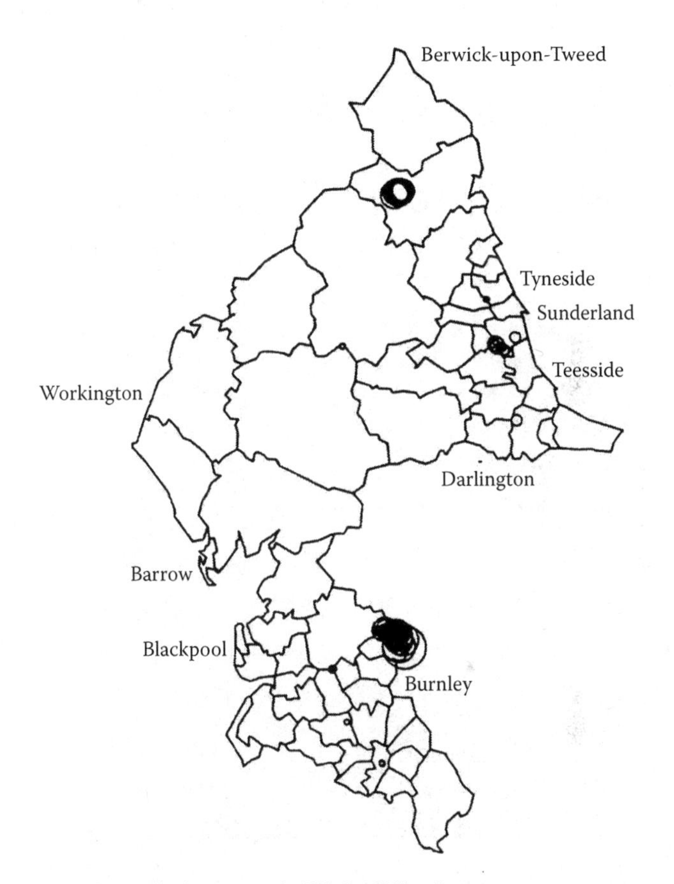

FIGURE 2.2 Significant circles at $p = 0{\cdot}002$ for Wilms' tumour.

two observed cases. The counter argument is that any such constraints tend to discriminate against rural areas. The effects are small for the $p = 0{\cdot}002$ level of significance.

Figure 2.3 shows the effects on the patterns of leukaemia of disaggregation by size of radii; namely, (*a*) 1–5 km, (*b*) 6–10 km, (*c*) 11–15 km and (*d*) 16–20 km. The idea here is that a 'real' cluster will be sufficiently robust to persist across a range of different radii, whereas spurious ones that survive a significance test will be much less able to do so. This is certainly the case since the Wilms' tumour circles are not persistent across many changes of circle radii. The more localized nature of the Seascale cluster, compared with that on Tyneside may well reflect the different scale of possible causes and also differences in the distribution of children.

A final analysis involves trying to measure the relative strength of the various circles. In the light of the discussion in §3.4, Figures 2.4a,b,c,d show the top 50, 100, 500 and 750 circles from figure 1 respectively when ranked by the Poisson probabilities of the observed number of cancers occurring purely by chance. These maps tend to emphasize the existence of a dense localized core at both Seascale and Gateshead and that parts of both areas have similar levels of elevated risk.

TABLE 2.6
Distribution of significant circles with two or more observed cancers.

Circle radius (km)	Total circles generated	Acute lymphoblastic leukaemia		Wilms' tumour	
		$p = 0{\cdot}01$	$p = 0{\cdot}002$	$p = 0{\cdot}01$	$p = 0{\cdot}002$
1	510367	452	153	238	47
2	127572	255	141	123	22
3	56719	281	153	62	17
4	31910	296	137	44	10
5	20428	294	116	25	4
6	14195	270	81	12	2
7	10423	238	53	10	1
8	7983	202	45	3	
9	6304	165	33	6	
10	5112	142	30	4	
11	4207	120	32	1	
12	3557	97	32	3	
13	3028	92	30	5	
14	2602	90	27	9	
15	2269	88	27	8	
16	1993	94	26	10	
17	1766	94	25	7	
18	1580	83	26	4	
19	1408	81	29	5	
20	1280	74	31	2	

5 Further developments of GAM technology

This case study has demonstrated only one aspect of GAM. It concentrated on a data set with no disaggregation and this example does not fully illustrate the versatility and power of this method of point pattern analysis. The GAM can be extended to handle most other aspects of a research design and to include a search for clustering over time. There are no obvious restrictions on the magnitude of the problems or data sets it can handle other than those enforced by computer hardware, disk space and the user's tolerance of long run times. Uniquely, GAM has properties that make it suitable for use on both a supercomputer for processing national data sets, where the explicit parallelism in the grid search and possibilities for fully-vectorized data structures can be exploited, and with continuously-running micro-computers as database scavengers, forever on the alert for the appearance of new disease clusters in local frequently updated databases.

5.1 Handling permutations of data categorization

When GAM is run on all the data without disaggregation, it is in effect concentrating on overall geographical effects. It is, of course, possible to run it on various disaggregations of the data; for example by age group and time period when the interest

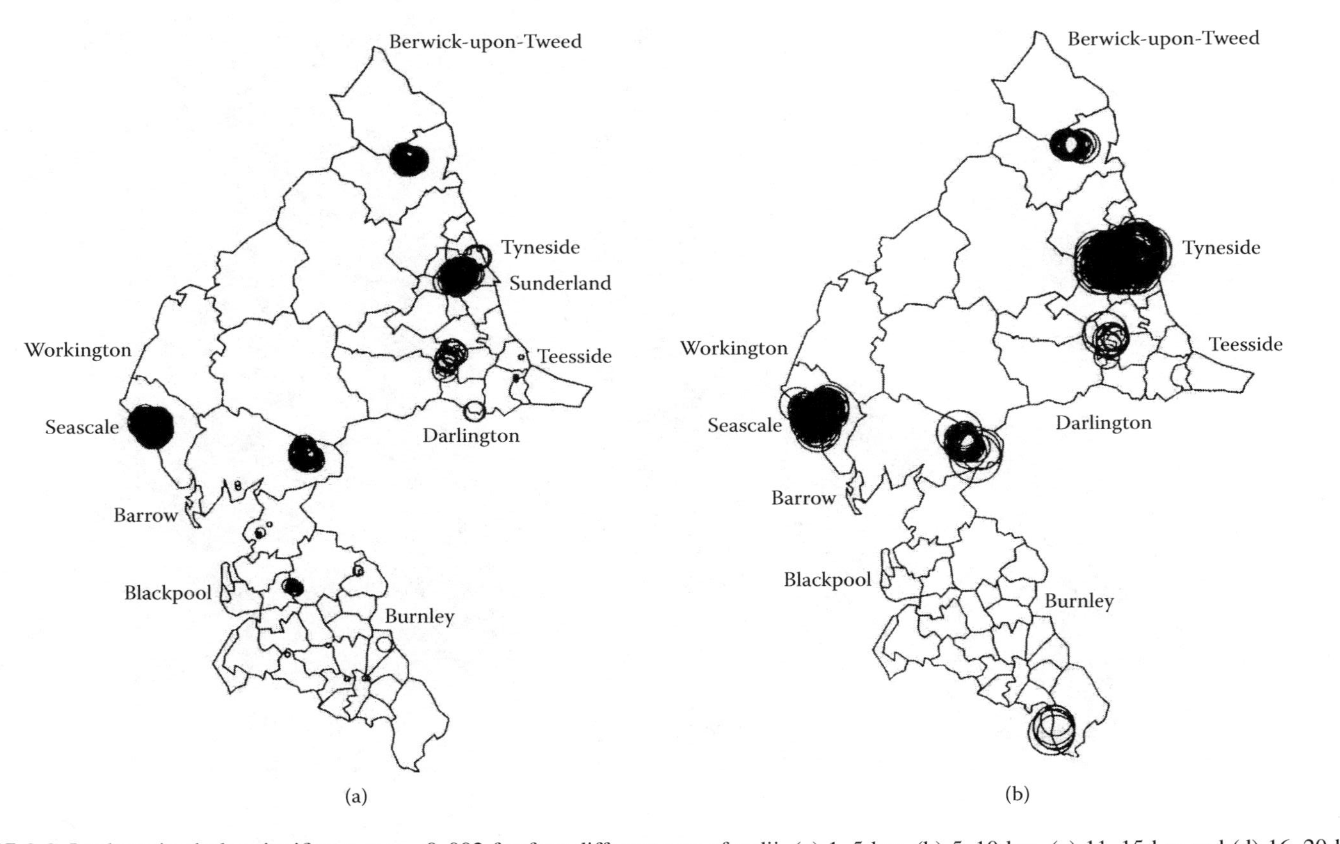

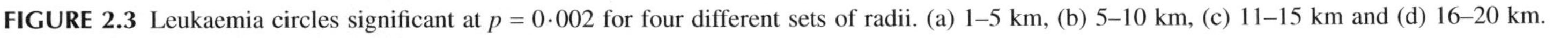
FIGURE 2.3 Leukaemia circles significant at $p = 0{\cdot}002$ for four different sets of radii. (a) 1–5 km, (b) 5–10 km, (c) 11–15 km and (d) 16–20 km.

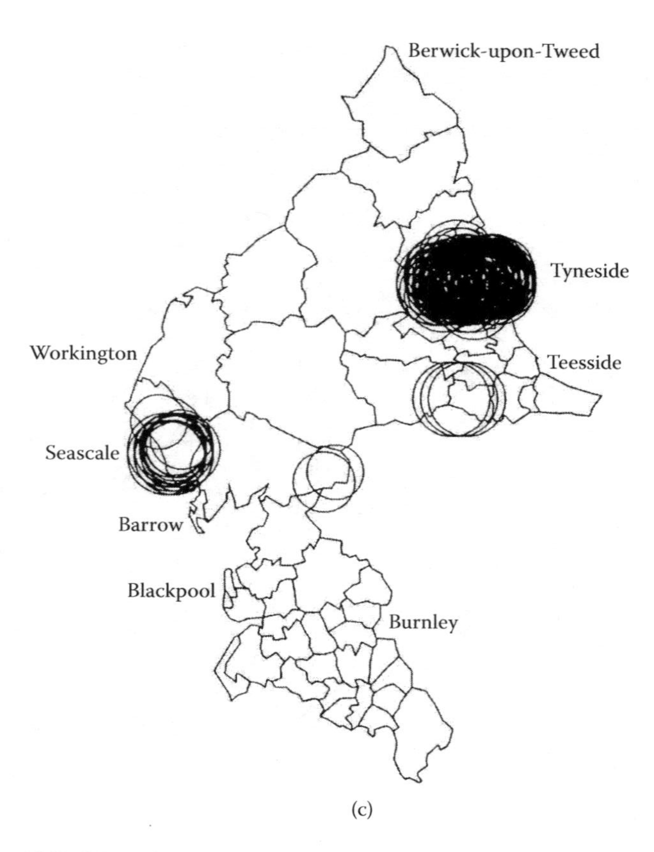

FIGURE 2.3 (continued)

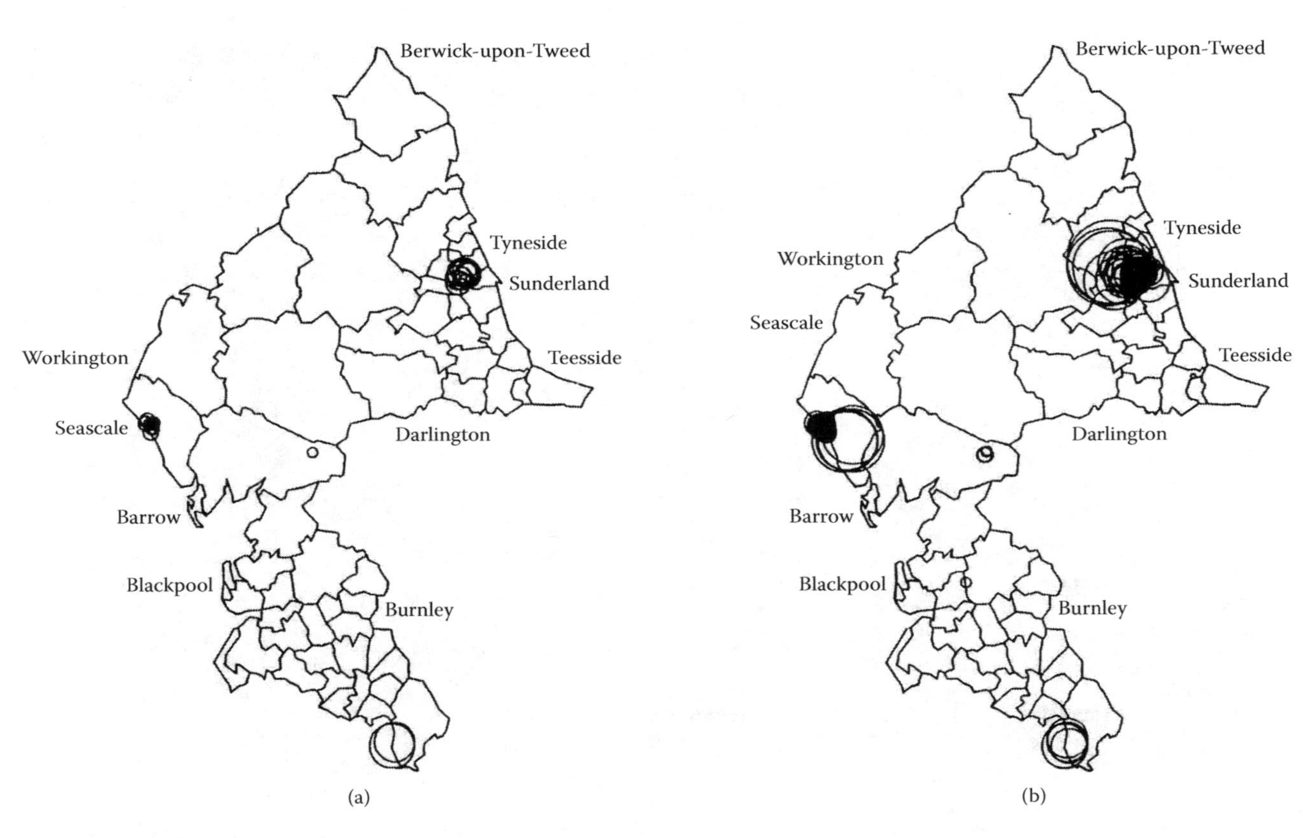

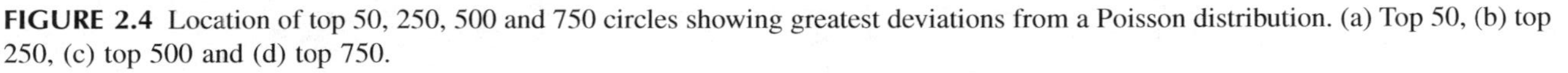

FIGURE 2.4 Location of top 50, 250, 500 and 750 circles showing greatest deviations from a Poisson distribution. (a) Top 50, (b) top 250, (c) top 500 and (d) top 750.

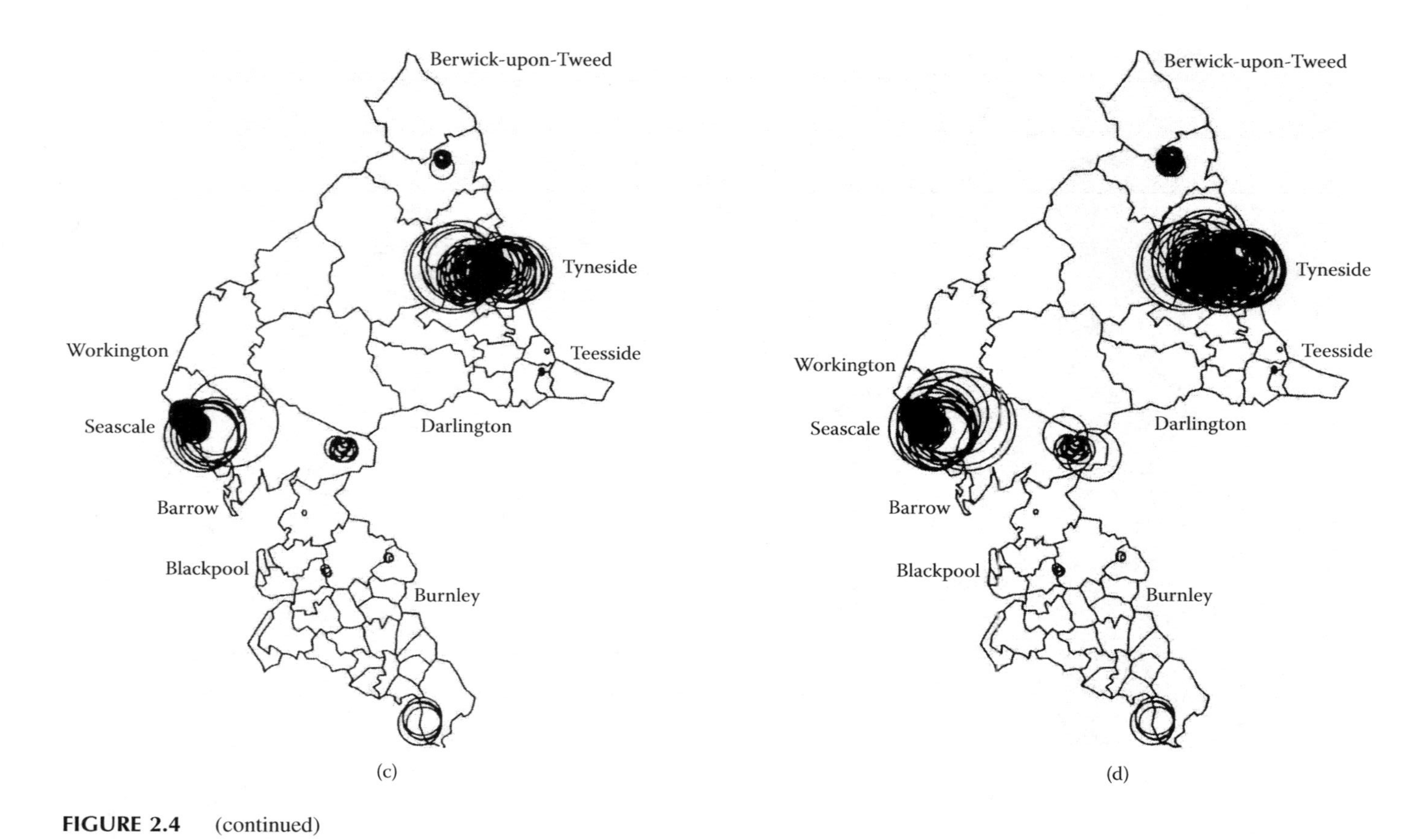

FIGURE 2.4 (continued)

is in retrospective explorations of historic data. A more disaggregated approach might also greatly reduce the effects of errors in estimates of the populations at risk used in the simulations and greatly improve both the sensitivity of GAM and its ability to suggest further testable hypotheses.

In the case study, the data refers to 0–15 year olds and to an 18-year period. Obviously, the least biased way of handling disagreegations would be to examine all possible permutations of age group and period. Unfortunately there are too many, although it would be possible to examine samples. However, if it is assumed that it is necessary only to split both the period and age variables into sub-groups of contiguous years, then there would be a more manageable number of permutations. In fact, there would be a maximum of $(m/2)*(m + 1)*(t/2)*(t+1)$ possible subsets of contiguous age groups and years, where m is the number of year age groups ($m = 15$) and t the number of periods ($t = 18$). If run separately, there would be a requirement for 20 520 GAM runs. This would not take 20 520 times as long to run because the data sets are smaller that those used here, but total computer times would be considerably greater than at present. Fortunately, there is a more effective solution. Nearly all the computer time is associated with data retrieval. There is no reason why all the subset hypotheses cannot be evaluated simultaneously, requiring only a single pass through the data. However, there would now be about 16 682 million hypotheses to evaluate for the standard search parameters used here. Clearly this would be an exercise for a supercomputer but it would seem to be within the ability of current technology to handle this scale of problem.

5.2 Identifying space-time clustering

The search for space-time clustering requires only a slightly different approach. There are two different definitions of a space-time cluster. One definition is to look for a significant excess of cancers or points in the same circle over two or more continuous periods. This implies only a very limited degree of contagion, one that does not extend outside a fixed circular boundary, for example, connected with a point source. If the effect of spatial contagion due, for instance, to a virus is considered, then the radii of circles may well increase for each subsequent period and, of course, in keeping with the GAM philosophy, all permutations of radius would need to be examined. These search models could be programmed for simultaneous evaluation with or without the other forms of search.

5.3 Running GAM on a supercomputer

The GAM could also be run on a supercomputer, primarily as a means of reducing run times in order to analyse a given data set more quickly, to allow for disaggregated searches by age group and period, to reduce the development times associated with the fine tuning and extension of the technology, to investigate its power in handling synthetic situations and allow large parts or all of the United Kingdom to be processed in a single run. Initially, a key requirement is for further experimentation. One topic of particular relevance is a study of the effects of different sizes of Monte Carlo samples and other GAM search parameters in order to build up more experience about the GAM's powers of detection and discrimination quickly. Another area of investigation is the power of the test statistic being used. It is most important to know what level or strength of deviation from a Poisson (or some other distribution) can be reliably detected. Only if this is known can greater confidence be placed in the results of epidemiological research. One of the advantages of GAM is that it is possible to measure these detection levels by experimentation and this is an in-built feature of this approach. These experiments can most readily be performed on a supercomputer, leaving more leisurely production runs with optimized search parameter settings for mainframe and micro environments.

It is useful to note briefly those aspects of the GAM that make it intrinsically suitable for use on a supercomputer. In particular, the search process on a grid is suitable for multi-processor machines (such as the Cray XMP) and (even more so) array processors. On a Cray XMP the database would be held in memory rather than on disk and parts of the KDB tree retrieval process are thought to be vectorizable. Research is also under way to develop an alternative, intelligent, totally vectorizable GIS data structure; indeed, initial experiments with a nine-level nested two-dimensional structure are quite promising. The idea is that the data structure monitors itself and adaptively switches the retrieval strategy whenever the current method becomes suboptimal.

5.4 Running GAM on a microcomputer

At the same time, the availability of fast, cheap, powerful 32-bit microcomputers with a hundred megabytes or so of hard-disk storage opens up the prospect of developing dedicated GAMs that concentrate on data analysis for specific areas, e.g. for regional health authorities in the United Kingdom. This is feasible with the KDB tree version of GAM because (1) the needs for memory and disk storage are not excessive; (2) the GIS data structure is portable; (3) the data structure can be tuned for optimal performance on any given hardware; (4) computer times are a function of size of data set and it is far quicker to handle data for a single health region that for larger areas and (5) the supercomputer-dependent disaggregated searches are far less relevant. GAM was also meant to be an automatic system. It was originally developed as a 'run-and-forget' system and its principal applied uses in the future are thought to be as database scavengers. For example, it might be programmed to explore a cancer database for a study area, taking into account all relevant aggregations or disaggregations of the data. It would run for 24 hours a day, 7 days a week. Once a month, its findings would be checked and, when it had finished, it would be loaded with more recent data and re-run. The run times are a function of size of study region, speed of microcomputer and the search parameters used. They could always be reduced by the simple expedient of sharing the total search over two or more machines or by setting the GAM search parameters to slightly cruder values.

This sort of 'database trawler', programmed to be on the lookout for interesting and significant results, offers many advantages. If there are, for example, localized and persistent environmental causes for certain diseases, then the patterns might be spotted quickly enough for them to be of some practical prescriptive use rather than, as is often the case at present, being of only retrospective and historical value.

5.5 *Identifying relationships and testing other hypotheses*

A further application is to use the GAM output as a basis for formulating and testing more specific and detailed hypotheses relating to possible explanatory causes. The main purpose of the GAM is to detect areas of deviation from a null hypothesis. It cannot generate and test new hypotheses simultaneously. However, it can indicate where to concentrate subsequent research effort and it can provide clues for the formulation of new, location-specific hypotheses, for example, by using GIS to overlay the centroids of significant circles on top of other geographical information and to look for interesting and recurrent relationships with, for example, different types of land use. The results, of course, constitute only circumstantial evidence but they do offer a very focused geographical basis for further research. This ability to point others in the right direction is a useful exploratory function.

6 Conclusions

This paper has outlined the design of a prototype geographical analysis machine, designed specifically for point pattern analysis. The implications of a working GAM may be quite profound in that it is necessary to consider whether all previous analyses of point pattern data in epidemiology, geography and other disciplines should now be re-examined using a GAM, so as to be certain that the conclusions that were drawn were not in fact spurious or biased and that important patterns did not remain undiscovered.

The GAM offers a radically new approach to spatial analysis that combines geostatistical thinking with GIS and a computational philosophy. Its great strength lies in its sophistication of spatial analytical technology. It is seemingly able to detect clusters worthy of further investigations and it is unlikely that any major clusters will be missed. As such, it is the first application of a largely post-statistical technique in a geographical context. However, it is also emphasized that the GAM is offered as a Mark I system and, no doubt, other variants of the basic technology will be developed which are able to offer even more confident results. It would seem to suggest that the alternative to simple science is a far more complex, machine-based science with computational procedures reducing the importance of human imagination, statistical theory and knowledge as thc basis for inference. The objective that the GAM addresses is how to improve dramatically the power and usefulness of spatial analytical technology in both exploratory and confirmatory modes of operation. It answers most of the problems raised in §2 and offers some prospect of a radically new approach to spatial analysis which is very relevant to a GIS environment.

Acknowledgments

The authors wish to acknowledge the helpful comments made by R. Flowerdew, J. M. Birch, R. Wakeford, D. Wilkie, M. Croasdale and K. Binks. Terry Coppock and an anonymous referee also made a number of very useful suggestions. The data came from two cancer registries supported by the North of England Children's Cancer Research Fund and the Cancer Research Campaign. Finally, the research into building GAM was sponsored partly by the Department of Health and Social Security and partly by the Economic and Social Research Council's Northern Region Research Laboratory at Newcastle University.

References

Bayer, R., and McCreight, E., 1972, Organisation and maintenance of large ordered indexes. *Acta Informatica*, **1**, 173.

Bentley, J. L., 1975, Multidimensional binary search trees used for associative searching, *Communications of the Association for Computing Machinery*, **18**, 509.

Besag, J., and Diggle, P. J., 1977, Simple Monte Carlo tests for spatial pattern, *Applied Statistics*, **26**, 327.

Craft, A. W., Openshaw, S., and Birch, J. M., 1985, Childhood cancer in the Northern Region, 1968–82: incidence in small geographical areas. *Journal of Epidemiology and Community Health*, **39**, 53.

Craft. A. W., and Openshaw, S., 1987, Children, radiation, cancer and the Sellafield nuclear reprocessing plant. In *Nuclear Power in Crisis*, edited by A. Blowers and D. Pepper (London: Croom Helm).

Croasdale, M. R., and White, A. A. L., 1987, A critical review of statistical evaluations of the clustering of rare diseases with particular reference to the frequency of cancers around nuclear sites in Great Britain. Presented at the British Nuclear Energy Society Conferences held in London in May 1987 (mimeo).

Department of the Environment, 1987, *Handling Geographic Information*. Report of the Committee of Enquiry chaired by Lord Chorley (London: Her Majesty's Stationery Office).

Diggle, P. J., 1983, *Statistical Analysis of Spatial Point Patterns* (New York: Academic Press).

Hope, A. C. A., 1968, A simplified Monte Carlo significance test procedure. *Journal of the Royal Statistical Society*, B, **30**, 582.

Leamer, E. E., 1978, *Specification Searchers:* ad hoc *Inference with Nonexperimental Data* New York: Wiley).

Openshaw, S., 1986, Modelling relevancy. *Environment and Planning*, A, **18**, 143.

Openshaw, S., 1987, An automated geographical analysis system. *Environment and Planning* A, **19**, 431.

Openshaw, S., 1988, Building an automated modeling system to explore a universe of spatial interaction models. *Geographical Analysis,* **20**, 31–46.

Openshaw, S., Charlton, M. E., and Craft, A. W., 1988, Searching for cancer clusters using a Geographical Analysis Machine. *Papers and Proceedings of the Regional Science Association,* **64**, 95–106.

Openshaw, S., and Goddard, J. B., 1987, Some implications of the commodification of information and the emerging information economy for applied geographical analysis in the United Kingdom. *Environment and Planning* A, **19**, 1423.

Rees, P. H., and Wilson, A. G., 1977, *Spatial Population Analysis* (London: Arnold).

Rhind, D., 1983, *A Census Users Handbook* (London: Methuen).

Ripley, B., 1981, *Spatial Statistics* (Chichester, Sussex: Wiley).

Robinson, J. T., 1981, The KDB Tree: A search structure for large multidimensional indexes. Research report CMU-CS-81-106, Carnegie-Mellon University, Pittsburgh, U.S.A.

Samet, H., 1984, The quadtree and related hierarchical data structures. *Association for Computing Machinery Computing Surveys*, **16**, 187.

Upton, G., and Fingleton, B., 1985, *Spatial Data Analysis by Example* (London: Wiley).
Wymer, C., Charlton, M. E., and Openshaw, S., 1988, An implementation of the KDB tree and its application to the GAM. Northern Regional Research Laboratory research report, Centre for Urban and Regional Development Studies, University of Newcastle upon Tyne, England (in preparation).

A Mark 1 Geographical Analysis Machine for the Automated Analysis of Point Data Sets: Twenty Years On

Martin E. Charlton

By rights, the commentator on Openshaw et al. (1987) should be Stan Openshaw himself. Unfortunately he had a severe stroke in May 1999, so I feel privileged to be writing in his place. Although it has been 20 years since we collaborated on the Geographic Analysis Machine (GAM), writing this commentary has brought on a pleasant nostalgia for those days at Newcastle University.

Dr. Alan Craft of the Department of Child Health had been researching the epidemiology of childhood cancer in northern England in the early 1980s. He had sought Stan's assistance several times in attempting to identify whether there were spatial patterns observable in the distribution of various types of childhood cancer (Craft et al., 1984; 1985). When drawn on a map, there were observable agglomerations of children with cancer. These largely followed the distribution of the at-risk population in the study area. There were more children with acute lymphoblastic leukaemia than any other type. No one knew why they had cancer, but it had been observed that the cases did appear to form clusters. No one quite knew what a cluster was, but pins placed on a map did appear to be in small, distinct groups.

Around that time there was also media interest in the incidence of childhood cancer, with a television documentary in 1983 (Yorkshire Television [YTV], 1983) alleging a link between cancer excess in southwest Cumbria and the nearby nuclear waste plant at Sellafield. A paper in Lancet (Urquhart et al., 1984) had postulated a link. Sir Douglas Black's report (Black, 1984) suggested that the link was not proven and that more research was needed to decide the case.

The problem with the Sellafield hypothesis is that it was only one possibility in the region. Doctors in Gateshead had observed local spatial variations in the residence of their young leukaemia patients. There had already been several attempts to demonstrate raised incidence around particular sites elsewhere in the U.K. How did we know whether there was a link between cancer clusters and nuclear waste plants or waste incinerators, when we perhaps ought to be looking at fish and chip shops or telephone boxes? Was there, enquired Alan, some way of identifying clusters in the data without having to suggest a location around which we could do a focused experiment?

I was appointed to a research associate post to work on local cancer registry data in late 1984. During the ensuing months, Stan's ideas about identifying patterns in the cancer data started to come together. At the time we had no geographic information (GIS) software, and everything was written in-house. The data existed in flat files. The computing system at Newcastle University was based around an operating system called MTS (the Michigan Terminal System, which was actually part of the University of Michigan Multiprogramming System [UMMPS]), but which included IBM's FORTRAN compilers. Interactive graphics were virtually nonexistent — we did have access to large-format pen plotters and some rudimentary graphics screens. Stan wrote the code for GAM, Colin Wymer assisted with the coding of the multidimensional search-tree algorithms, and I wrote the graphics software. The data from the cancer registries covered the counties of Cumbria, Northumberland, Tyne and Wear, Durham, and Cleveland. From this was extracted for each child, the type of cancer from which they were suffering, and the postcode of their residence at the time of diagnosis. A match against the POSTZON file provided grid references, and at-risk population data was extracted from the 1981 Census of Population data.

The essence of GAM was that a local excess of cancer would lead to the identification of dozens, if not hundreds, of significant circles located on top of one another. The trouble with this is that lots of circles drawn one after another would cause local saturation of the Calcomp 1039 plotter paper, leading to tearing of the sodden paper by the pen. Such plots were known as *operator's nightmare plots*. We plotted a couple of these before deciding to draw the circles in random order to minimise the chances of this happening — it took longer, but at least we had something potentially useable. Significance was determined using a Monte Carlo test. GAM was hugely computationally intensive. The director of the computing service at Newcastle was sufficiently benevolent to allow Stan something akin to unlimited computer time on the Amdahl 5860 mainframe. The lengthy run times were achieved by using a technique known as *checkpointing*. The job would be submitted to a batch queue to run for just under 1000 seconds. At the end of this period, GAM wrote its current state to a file, halted itself, and the job was automatically resubmitted, to start from where it left off. In this way, CPU cycles during the quiet periods on the system, usually in the early hours of the morning, could be used without inconveniencing students and staff using the system during the day. There were occasional accidents: a premature program failure one night led to the job queue being filled with rogue GAM jobs within the space of a few seconds. This meant that no else could submit any further jobs until the queue had been cleared — the luckless operators who had to issue the 2000 or so commands from their console must have wondered about our sanity. Our appearance in the machine room at Christmas bearing bottles of wine did much to restore our reputation.

One of the drawbacks of GAM was its computational load. During a run there were hundreds of thousands of spatial queries on a dataset that was too large to fit in the 8MB address space of the Amdahl. I had found Robinson's (1981) paper on KDB trees in the computing laboratory's library a few months earlier, and Colin Wymer came up with some elegant code that solved the spatial retrieval problem for GAM. It is perhaps a timely reminder how quickly the computing environment changes, and how easily we take technological changes for granted. On installation in 1985, Newcastle University's Amdahl 5860 with 40MB of main memory and some 10GByte of disk space was a hefty machine. Compared with the laptop on which I am typing this, it seems a puny machine in all but physical size and cost: the comparison is not entirely fair because the Amdahl would also support over 100 simultaneous users at interactive text-based terminals. At least I can carry my laptop onto one of Ryanair's Boeing 737-200s and run GAM, probably.

There are some photographs of the Amdahl console, CPU, and disk drives at http://www.cs.ncl.ac.uk/events/anniversaries/40th/images/amdahl2/index.html.

The paper appeared in the last issue of the first volume of *IJGIS* not long after Stan and I had travelled to Athens to present some initial results at the meeting of the European Regional Science Association in summer 1987. In those days there were no GISRUK, no GeoComputation or GIScience conferences, and the Spatial Data Handling Symposia had only just started. It was, however, not long before GAM came to the attention of the wider academic and epidemiological community. One of the benefits of this was that Julian Besag brought his considerable expertise to examine the clustering problem. James Newell was appointed to a research position in the Department of Child Health to assist.

Besag and Newell (1991) have commented that "GAM has attracted much criticism," although they point out that it was "the first major attempt to identify clusters of a rare disease" given the particular characteristics of the data available. They observed that there was no control for multiple testing, that it is not always easy to define the at-risk population in the circular zone, that there was no allowance for age-sex variation, and that it required an "enormous computational workload." However, they point out that it avoided the problem of preselection bias, used at-risk data at a fine spatial resolution, and avoided either fixed-size or fixed population subregions. They then presented an alternative test for those unable or unwilling to use GAM.

In 1992, Stan took up the chair of human geography at the University of Leeds, and founded the Centre for Computational Geography (CCG) within the Department of Geography. Gathering a group of enthusiastic and like-minded researchers together, the development of GAM continued; this included a web interface and developments of GAM to deal with space-time data, and the use of genetic algorithms in pattern search. However, these are but a part of the corpus of activity that has taken place in the CCG. The CCG continues an active programme of research into spatial analysis and modelling, "and is concerned with the development and application of tools for analysing, visualising and modelling geographical systems " (CCG, 2005).

What has been the influence of this paper? Papers from the first five years of *IJGIS* are not in the Web of Science, so an easily gathered citation count is not available. A Google search for Geographical Analysis Machine returns 508 hits; adding the authors' surnames reduces this to 185 hits. Of these, many refer to epidemiology, public health, the detection of clusters in spatial data, crime analysis, as well as spatial data mining. Not long after the *IJGIS* paper was published, we published a brief item on GAM in the Lancet (Openshaw et al., 1988). While we felt that the *IJGIS* paper might reach a nonmedical audience, it was important to reach medical researchers. The Web of Science reveals that there are 66 papers that cite the Lancet item, an annual average of between 3 and 4 citations; most of these papers deal with disease clustering or spatial statistics. Internet hits reveal descriptions of GAM in a number of pedagogic PowerPoint presentations, often with enthusiastic advocates.

There is a rapidly expanding literature on spatial statistical methods in epidemiology and health studies. The literature on cluster detection makes the distinction between general and focused tests. A focused test deals with the situation of detection in one or more locations, while general tests deal with the pattern of disease over a large area; Besag and Newell (1991) appear to be the first to make this distinction. If the literature on cluster detection in the mid-1980s was sparse, this is no longer the case in the mid-2000s. I shall not attempt an exhaustive review because there are many practitioners adopting a wide variety of approaches; Bailey (2001) provides a good overview. However, while there have been a set of techniques aimed initially at the detection of disease clusters, we may note that GAM provides the first example of a scan statistic. Martin Kulldorff's scan statistic is perhaps the most well known (Kulldorff, 1997; Kulldorff and Nargwalla, 1995; Kulldorff et al., 2003). Kulldorff is also notable for the creation of SaTScan (Software for the Spatial and Space-Time Scan Statistics). Kulldorff (2005) cites GAM and Turnbull et al. (1990) as inspiration for his scan statistic. Another researcher working with scan statistics is Tango, who has been working with noncircular scan statistics to look for irregularly shaped clusters (Tango and Takahashi, 2005) following on from similar work by Patil and Taillie (2004) and Duczmal and Assuncão (2004).

A welcome recent development has been the availability of software for cluster detection. SaTScan (Kulldorff, 2005) is available for download from www.satscan.org. The bibliography for SaTScan reveals applications across a wide range of fields, including epidemiology and studies of infectious diseases, cancer, paediatrics, sclerosis, lupus, alcohol, domestic and wild veterinary medicine, forestry, toxicology, psychology, criminology, and tomography.

Another cluster detection package, ClusterSeer, is marketed by TerraSeer Inc. of Crystal Lake, IL. (TerraSeer, 2005) The website for ClusterSeer lists 27 different methods for the detection of spatial anomalies, covers the areas of spatial cluster detection, temporal cluster detection, surveillance, and space-time cluster detection. Might ClusterSeer be the SPSS of cluster detection?

A third package is somewhat different in approach. The DCluster package for the R statistical software implements a range of cluster detection methods (Gomez-Rubio et al., 2005). The user needs to be familiar with data analysis and manipulation in R. Among the many commonly used methods implemented in this package is GAM. R is an extremely powerful system, running on a range of platforms, which include UNIX, Linux, MacOS, and Windows. The other two packages run on the Windows platform. While R offers a great deal of flexibility, it is perhaps more difficult to use. Like SaTScan, R is free (R, 2005).

Because of its importance in epidemiology, the art of cluster detection has not stayed strongly in GIScience. If we are to consider ideas that begin with geography and in geography, then the influence of the paper has been far and wide, and in many cases indirect. If there are users of ClusterSeer employing Kulldorff's scan statistic, then they may be unaware of its inspiration. The cultural turn in human geography has led many geographers to reject analyses such as that in GAM and many similar papers. Methods for the detection of disease clustering have yet to find analogues in physical geography. The explosion of activity in the area of spatial statistics in the last 20 years has perhaps left GIScience lagging somewhat. Although there were a number of prominent geographers contributing papers to the GeoMed 2005 conference at the University of Cambridge, there were many more contributions from researchers outside GIScience. This may not be a bad thing: good ideas should diffuse widely.

There are a number of other influences. GAM was concerned with the *local*. A whole-map clustering statistic would not suffice to deal with the problem. I wonder if the first use of the term *whole map* is to be found in section 2 of the GAM paper. Twenty years on, there are a plethora of techniques that deal with local analysis. A fair part of the paper deals with the problems of implementing the machine. Stan had been pushing computer technology to its limits ever since I arrived as a student at Newcastle in the mid-1970s. The first GeoComputation conference was at the University of Leeds in 1996 — the definition of GeoComputation as the "art and science of solving complex spatial problems with computers" (www.geocomputation.org) seems to be at the heart of GAM. The GeoComputation series has matured with the 2005 conference held in Ann Arbor at the University of Michigan. Coincidentally, the operating system on the Amdahl, which made GAM possible in the mid-1980s, MTS, had been developed at the University of Michigan Computer Center over a quarter of a century from 1967 onward. I am writing this in my office at the National Centre for Geocomputation, funded by Science Foundation Ireland. Who knows what 20 years on will bring?

If Alan Craft had not walked across to Stan's office in the early 1980s …

Dramatis personae: I have already mentioned that Stan suffered a severe stroke in 1999; this caused him speech and motor problems, and he has since retired from Leeds University. There's a thoughtful summary of his work and influence in Longley et al. (2001, 140). Professor Sir Alan Craft is now Sir James Spence Professor of Child Health at the University of Newcastle upon Tyne, and Consultant Paediatrician at the Royal Victoria Infirmary in Newcastle. Colin Wymer has retired from the School of Architecture Planning and Landscape at Newcastle University. Julian Besag is Professor in the Department of Statistics at the University of Washington, and James Newell is Senior Lecturer in Epidemiology and Public Health at the University of Leeds.

References

BAILEY, T.C., 2001, Spatial statistical analysis in health. *Cadernos de Saude Publica*, **17**, 1083–1098.

BESAG, J. AND NEWELL, J., 1991, The detection of clusters in rare diseases. *Journal of the Royal Statistical Society, Series A*, **154**, 143–155.

Black, D., 1984, *Investigation of the Possible Increased Incidence of Cancer in West Cumbria: Report of the Independent Advisory Group*, HMSO, London.

Centre for Computational Geography (CCG), 2005, University of Leeds. http://www.ccg.leeds.ac.uk/, accessed 16 December 2005.

Craft, A.W., Openshaw, S., and Birch, J.M., 1984, Apparent clusters in childhood lymphoid malignancy in Northern England. *Lancet* 2, **8394**, 96–97.

Craft, A.W., Openshaw, S., and Birch, J.M., 1985, Childhood cancer in the Northern Region 1968–82 — incidence in small geographical areas. *Journal of Epidemiology and Community Health*, **39**, 53–57.

Duczmal, L. and Assuncão, R., 2004, A simulated annealing strategy for the detection of arbitrarily shaped spatial clusters. *Computational Statistics & Data Analysis*, **45**, 269–286.

Gómez-Rubio, V., Ferrándiz-Ferragud, J., and López-Quílez, A., 2005, Detecting clusters of disease with R. *Journal of Geographical Systems,* **7**, 189–206.

Kulldorff, M., 1997, A spatial scan statistic. *Communications in Statistics*, **26**, 1481–1496.

Kulldorff, M., Fang, Z., and Walsh, S., 2003, A tree-based scan statistic for database disease surveillance. *Biometrics*, **59**, 323–331.

Kulldorff, M., 2005, *SaTScan User Manual v6.0*, http://www.satscan.org, accessed 16 December 2005.

Kulldorff, M. and Nagarwalla, N., 1995, Spatial disease clusters: detection and inference. *Statistics in Medicine*, **14**, 799–810.

Longley, P.A., Goochild, M.F., Maguire, D.J., and Rhind, D.W., 2001. *Geographic Information Systems and Science* (Chichester: John Wiley & Sons).

Openshaw, S., Charlton, M., Wymer, C., and Craft, A., 1987, A mark 1 geographical analysis machine for the automated analysis of large point data sets. *International Journal of Geographical Information Systems*, **1**, 335–358.

Openshaw, S., Charlton, M.E., Craft, A.W., and Birch, J.M., 1988, Investigation of cancer clusters by use of a geographical analysis machine. *Lancet* 1, **8580**, 272–273.

Patil, G.P. and Taillie, C., 2004, Upper level set scan statistics for detecting arbitrarily shaped hotspots. *Environmental and Ecological Statistics*, **11**, 183–197.

R, 2005, The R Project for statistical computing. http://www.r-project.org/, accessed 16 December 2005.

Robinson, J.T., 1981, *The KDB tree: a search structure for large multidimensional indexes.* Research report CMU-CS-81-106, Carnegie-Mellon University, Pittsburgh.

Simon, J., 2002, Michigan Terminal System, http://www.clock.org/~jss/work/mts/index.html, accessed 10 April 2006.

Tango, T. and Takahashi, K., 2005, A flexibly shaped spatial scan statistic for detecting clusters. *International Journal of Health Geographics*, 4 , 11, available at http://www.ij-healthgeographics.com/content/4/1/11, accessed 16 December 2005.

TerraSeer, 2005, ClusterSeer product description. http://www.terraseer.com/products/clusterseer.html, accessed 16 December 2005.

Turnbull, B.W., Iwano, E.J., Burnett, W.S., Howe, H.L., and Clark, L.C., 1990, Monitoring for clusters of disease: Application to leukemia in upstate New York. *American Journal of Epidemiology*, **132**, 136–143.

Urquhart, J., Palmer, M., and Cutler, J., 1984, Cancer in Cumbria: the Windscale connection. *Lancet* 1, 217–218.

Yorkshire Television (YTV), 1983, *Windscale — the Nuclear Laundry.* Leeds: Yorkshire Television, broadcast on 1 November 1983.

International Journal of Geographical Information Systems,
1988, Vol. 2, No. 3, 229–244.

3 A Review and Conceptual Framework of Automated Map Generalization

Kurt E. Brassel and Robert Weibel

Abstract. This paper reviews the prospects of computer-assisted generalization of spatial data. Generalization as a general human activity is first considered in a broad context and map generalization is defined as a special variant of spatial modelling. It is then argued that in computer-assisted generalization, the spatial modelling process can be simulated only by strategies based on understanding and not by a mere sequence of operational processing steps. A conceptual framework for knowledge-based generalization is then presented which can be broken down into five steps: structure recognition, process recognition, process modelling, process execution and display. With reference to the goals of map generalization we identified tasks of statistical and cartographic generalization. The use of these types of tasks is discussed in relation to the concepts of digital landscape models (DLM) and digital cartographic models (DCM). A literature review is then presented in the context of this conceptual framework. It considers theoretical aspects of generalization and technical procedures on attributes and geometrical generalization. Specific sections of this review include statistical generalization, structure recognition and processes of cartographic generalization (point, line, area features, surfaces) and efforts for system integration. The paper concludes with an evaluation of the state of the art and an outlook on the future. Major efforts have to be devoted to developing an understanding of structures and processes involved in generalization (structure recognition, process recognition) and to modelling these processes. New data models will be a prerequisite for success in this field.

1 Introduction

Automation is one of the most important and revolutionary technical developments of the past two decades and cartography is heavily involved in this process. One of the operations which is not easily automated is map generalization which is a complex process executed traditionally by artisan cartographers. It involves a great deal of intuitive spatial modelling for the communication of spatial concepts and is, therefore, not easily replicated by a sequence of analytical steps. Generalization is a fuzzy concept and is not well defined. However, the 1980s have brought forth a broad interest in the analysis and instigation of map generalization. In this paper we review the prospects of computer-assisted generalization of spatial data. We first consider the general character of generalization (§ 2) and then present a conceptual framework for generalization of map data (§ 3). We follow this with a review of recent research activities (§ 4) and draw some conclusions on future developments (§ 5).

2 Generalization as a human activity

Map generalization is one of the central concepts in map design. A map is a generalized, simplified abstraction of reality. A large body of literature and a number of practical rules for cartographic generalization exist, even though a clear understanding and a theory of the map generalization process are still lacking. Generalization is not only used in cartography; it represents a basic human activity and is intrinsically related to the term 'abstraction'. Abstraction (latin, *abstrahere*) emphasizes the removal of the 'specific', 'random' and 'unimportant' in order to concentrate on the 'important' and on the general aspects of reality. The term generalization focuses on the extraction of the general, crucial elements of reality. Both terms describe a process of separating concrete reality into two parts, i.e., general aspects and specific aspects; so

$$\text{concrete reality} \approx \text{general aspects} + \text{specific aspects}$$

The goal of this separation is to select important and disregard unimportant features. The notion of importance, however, is always related to purpose. Dependent upon the purpose of the generalization, general aspects may be of importance in some cases, and specific aspects in others. The separation of the important (be it specific or general information) from the unimportant is a complex mental process involving perception, cognition and other intellectual functions. Generalization is the mental processing of information and it involves such functions as ordering, distinction, comparison, combination, recognition of relations, drawing conclusions and abstraction. It represents a task which requires the highest intellectual and creative capacity as well as an understanding of reality. Generalization as an abstraction process is a central concept of inductive scientific reasoning and underlies all modelling in science, art and technology. It is one of the most widely used activities in science and everyday life. Journalists and artists generalize, extracting and representing 'important' information, be it specific or general. Caricaturists are generalizers, as are historians, systematic botanists, politicians and, of course, Earth scientists. Modelling in the Earth sciences has, as a common feature, the spatial domain. In that sense, map generalization can be considered a special variant of spatial modelling.

3 General issues in computer-assisted map generalization

To underline its general character, we have considered generalization in a broad context and now wish to emphasize the modelling aspects of map generalization.

Dependent on purpose (and scale) the cartographer visualizes 'important' and ignores 'unimportant' information. In mapping, the aim of modelling is also supplemented by the communication aspect. Model construction (i.e. generalization) is heavily influenced by the fact that the major goal is not mere analysis of space but communication of spatial concepts. Map generalization theory, therefore, has to consider not only spatial modelling but also visual communication theory.

In practice, procedures for generalization are perceived as pragmatic processing rules based on experiences and traditions and the theoretical background is quite often neglected. The literature also emphasizes the operational aspects. Definitions such as the one by the International Cartographic Association (ICA) are representative: 'Selection and simplified representation of detail appropriate to the scale and/or the purpose of the map' (ICA 1973, p. 137). In this tradition the process of generalization can be analytically separated into the thematic, spatial and temporal domains (Hake 1975). Other authors (e.g. Robinson *et al.* 1984) emphasize the operational functions (selection, classification, simplification, symbolization, induction) and the controls of the generalization process (objectives, scale, graphic limits, data quality). While this analytical separation into domains, functions and controls is a useful and necessary strategy for the implementation of procedures for automated generalization, we should be aware that these parameters are interrelated and not entirely separable.

Generalization, as an intellectual process, structures experienced reality into a number of individual entities, then selects important entities and represents them in a new form. If we want to simulate this process automatically we have first to understand the process of recognizing essential features to model this process and to implement component extraction and representation. Process definition by a number of analytical steps is an essential requirement for automated generalization, but a mere succession of mechanical steps can suffice only in exceptional cases. What is needed is *processing based on understanding*. To 'understand' generalization means to extract the essential structures of the spatial information available (in the thematic, spatial and temporal domains), to identify the essential processes for modifying these structures and to formalize these processes of modification adequately as a number of operational steps.

The following *conceptual framework* for generalization is proposed (Figure 3.1). The original database is first subjected to a process of *structure recognition* (*a*). This process aims at the identification of objects or aggregates, their spatial relations and the establishment of measures of relative importance. Structure recognition is controlled by the objectives of generalization, the quality of the original database, the target map scale and the communication rules (graphic and perceptual limits). It represents a process of intellectual evaluation which is traditionally performed by visual inspection of a map. Some processes may be simulated by specific computer vision procedures, but an adequate automation of this step is non-trivial. Once the major structure of the original database is known, the generalization process has to be defined. Based on the control parameters, the types of data modification and the parameters of the target structures have to be established (what is to be done with the original database? which types of conflicts have to be identified and resolved? which types of objects and structures are to be carried in the target database?). We

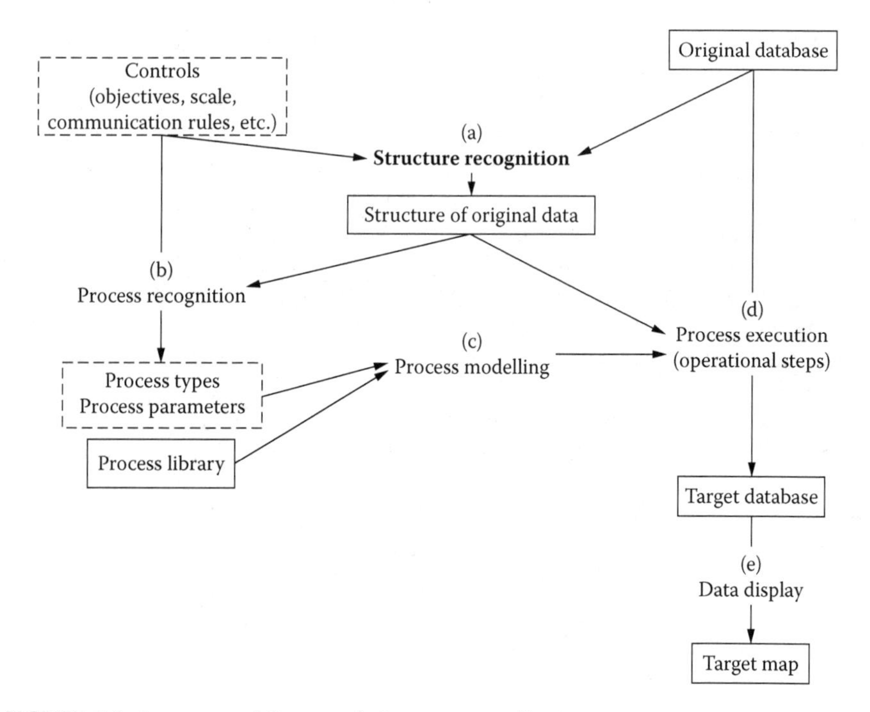

FIGURE 3.1 A conceptual framework for map generalization.

call this step *process recognition* (Figure 3.1b). Again, it represents a multifaceted recognition and decision process which, in several instances, is not easily automated. Once the desirable generalization process has been defined, it is modeled as a sequence of operational steps. This *process modelling* (Figure 3.1c) can be considered as a compilation of rules and procedures from a process library and the pre-setting of process parameters that were established in process recognition. This original database and information structures are then subjected to this process (Figure 3.1d) and converted into the target or generalized database. This step is called *process execution*. It consists of a sequence of operational steps as compiled from process functions stored in the process library. Examples of generalization processes are selection/elimination, simplification, symbolization, feature displacement and feature combination. Some of these functions may be simple and straightforward (selection, line simplification), others are more complex and involve synoptic processing (e.g., feature displacement). *Data display* then converts the target data into the target map (Figure 3.1e).

Generalization in this framework thus consists of five steps; structure recognition, process recognition, process modelling, process execution and data display (the last may well be considered external to actual generalization). The implementation of such a framework requires a good understanding of structure and process recognition, a comprehensive process library for various generalization actions and a process modelling mechanism. Existing approaches usually provide software for process execution driven by a number of user-defined parameters (see §4). The quality of these procedures cannot be judged in absolute terms, only in relation to their actual use. This means that the various generalization algorithms perform adequately if, and only if, they are used in the appropriate context (e.g., it makes a difference whether a line simplification algorithm is applied to a coastline or to a road). What is required is a controlled environment for the procedure which is based on an understanding of what should be achieved. At this time, not enough evidence is available to decide if, and under what circumstances the 'understanding process' (structure recognition, process recognition, process modelling) is to be run only once for an entire generalization task (global use) or to be repeated for sub-areas or specific features (local use).

In this framework, controls (e.g. objectives, communication rules) have been considered important ingredients for generalization. In practice it may be useful to distinguish between two major groups of *objectives* for spatial modelling; (1) spatial modelling for the purposes of data compaction, spatial analysis and the like (we call this process statistical generalization) and (2) spatial modelling for visual communication (cartographic generalization). *Statistical generalization* is mainly a filtering process. It represents processes of data reduction under statistical control. Concepts of reliability, tolerance or error may be applied. It aims at minimum average displacement. It is never used for display, but strictly for data reduction (e.g., to obtain a subset of an original database for data analysis). *Cartographic generalization*, in contrast, aims to modify local structure and is non-statistical. It operates on bias and elimination, emphasis and shape distortion. Its purpose is not minimum average displacement but representativeness in a holistic rather than an analytical sense. Cartographic generalization is used only for graphic display and therefore has to aim at visual effectiveness.

Both cartographic and statistical generalization, as subgroups of map generalization, have their proper place in spatial modelling. On the basis of studies by a group in Hannover, Germany (Brüggemann 1985, Grünreich 1985, Meyer 1986), we may illustrate these relationships in Figure 3.2. We call a spatial database a digital landscape model (DLM). A DLM is a generic term for a comprehensive description of a portion of 'reality' (landscape in the case of topographic mapping or some other spatially-distributed phenomenon in thematic mapping). It is characterized by a specific thematic content and a related accuracy parameter (A_i). This parameter indicates the level of detail or a measure of maximum error of the database and it is desirable that the measure of accuracy be approximately equal for various themes of the same database. It is clearly related to the data source. A DLM may now be generalized in two ways: either by reducing it into a DLM of lesser information content (e.g., $DLM_{A1} \rightarrow DLM_{A2}$) or else by converting it into a digital cartographic model (DCM) which may be displayed at a certain scale (e.g., S_1 = 1:25 000). A DCM is the digital equivalent of a specific map amenable for visual communication. On the basis of this discussion we call DLM–DLM conversion statistical generalization and DLM–DCM conversion cartographic generalization. To produce a generalized map at a certain scale (e.g., S_3 = 1: 100 000) we may use a DLM of an acceptable accuracy (e.g., A_1 = 10 m maximum displacement). We may subject this

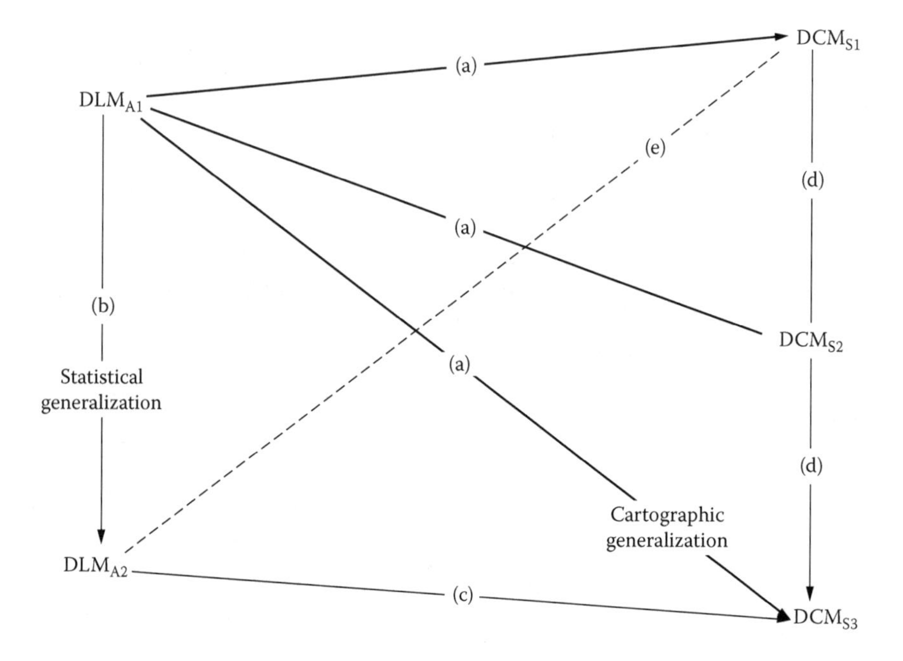

FIGURE 3.2 Digital landscape models and digital cartographic models.

DLM to a (direct) cartographic generalization process (Figure 3.2a) or else first produce statistically a DLM of 50 m resolution (Figure 3.2b) and then generalize it cartographically (Figure 3.2c). A third type of process is the DCM–DCM conversion. This, again, is a cartographic generalization process (Figure 3.2d). Since cartographic models do not vary too much for DCMs of similar scales (e.g., DCM 1:25 000 and DCM 1:50 000), statistical procedures may also be adequate in special instances. Quite often, spatial databases are DCMs rather than DLMs, since they were digitized from existing maps. DCM–DLM conversion (Figure 3.2e) is possible only if a respective reduction of accuracy is accepted and this is not legitimate in certain instances.

We conclude this section with an overview of the various generalization processes (Figure 3.3). Map (data) generalization as a spatial modelling operation can be broken down into cartographic and statistical generalization tasks. They both involve the thematic, temporal and spatial domains. Attribute generalization relates to the thematic and temporal domains, geometrical generalization to the spatial domain. The former operates on descriptor data, the latter on image data describing point, line, area and/or volume features. Generalization involving both attribute and geometrical modification may be called topological.

After presenting a conceptual framework for generalization based on understanding (Figure 3.1) and an overview of the various generalization processes (Figure 3.3) we proceed to present the state of the art of computer-assisted generalization. The literature will be discussed in the light of the concepts presented.

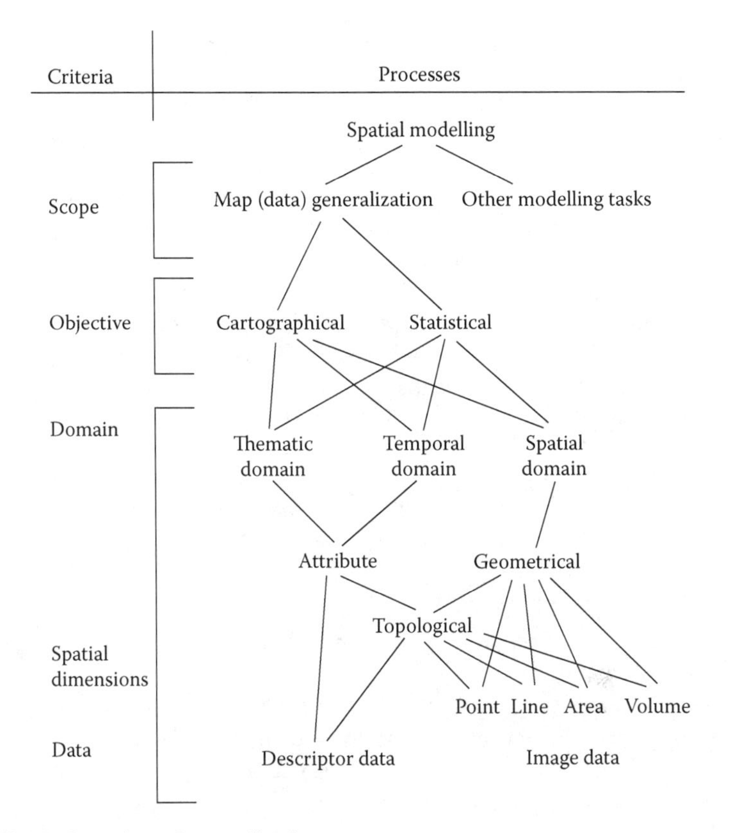

FIGURE 3.3 Overview of generalization processes.

4 Research activities in computer-assisted generalization

In this section we focus on the present situation of automated generalization. For this purpose we proceed from the presentation of the literature covering general principles and theoretical aspects to the description of technical procedures. Our compilation does not claim to be complete; in some respects it must remain superficial. Other survey papers will be more detailed on specific aspects and also include extensive bibliographies (Weber 1982, Zoraster *et al.* 1984, Weibel 1986).

Textbooks are still a major source of knowledge about the principles and operations of cartography and generalization. Imhof (1982) is mainly a translation of his 1965 German-language original and therefore neglects aspects of automation almost entirely; but it describes the principles of manual generalization in topographic mapping in a relatively detailed way. Another classic textbook is by Robinson *et al.* (1984). It has been updated several times and covers some aspects of automation. It presents the American view of generalization which differs somewhat from the European tradition as presented in Imhof (1982) in that it emphasizes simplification and smoothing processes within generalization. Another viewpoint is given by Töpfer

(1974). His theory of 'objective' generalization is based on empirically deduced parameters. He presents the application of these principles of quantitative generalization (e.g., the 'radical law') to the individual processes of geometric and attribute generalization. Another textbook deals exclusively with computer-assisted cartography (Monmonier 1982) but concentrates on technical aspects rather than on theoretical foundations or practical rules. Finally, the monograph by Steward (1974) explains the general concept of generalization (cartographic model definitions, processes of generalization, perception, automation aspects).

Other studies relevant to a discussion of generalization are contributions to the theoretical background of *map design*. They include communication, theoretical approaches (Board 1981, Knöpfli 1983), linguistics (Taketa 1979, Eastman 1987) and semiotics (Bertin 1967). At the moment, linguistic methods are gaining increased attention because of their potential application to knowledge representation for expert systems.

Other empirical studies focus on the *perception* aspect of various schemes of generalization. Such studies — which are still rare and in an early stage of development — serve to enhance existing procedures or to develop new ones. Examples are those by Phillips and Noyes (1982), who investigated visual clutter in order to optimize the topographic base of geological maps, and by Eastman (1981), who examined perception of scale changes as a function of symbol size, symbol density and line sinuosity.

In the discussion of algorithms and technical implementations we focus on the problems of geometrical rather than attribute generalization. Attribute generalization is a relatively simple process with a stronger formal foundation and can be automated in a straightforward fashion; appropriate solutions have already been developed. Examples in the *thematic domain* are classification (see, for example, Gibson 1987, Tobler 1973, Peterson 1979) or seriation (Bertin 1967, Muller and Honsaker 1983). Generalization in the *temporal domain* has received only minor attention in cartography (Hake 1982). It will, however, be a more pertinent issue as the mapping of high-resolution time series data (e.g., automatically collected weather variables) comes of age. Procedures integrating attribute and geometrical modelling (e.g., topological generalization) are not well advanced. Some approaches will be discussed implicitly with geometrical generalization and system integration.

We now concentrate on technical efforts for *geometrical generalization*. The first attempts at automating the generalization of individual cartographic elements were made at an early stage in the development of digital cartography (e.g., line smoothing by moving averages, Tobler 1964; line smoothing by epsilon filtering, Perkal 1966). In the 1970s, this early work was extended and more appropriate schemes of generalization were developed which tried to bring out the salient character of map elements. These algorithms mainly focused on the generalization of point or line features and used generalization processes of lower complexity (e.g., selection, simplification). Only a few approaches tried to incorporate complex processes such as combination and feature displacement or the handling of generalization of more than one class of feature. A technical review of several approaches of this period for point, line, area and surface generalization is presented by Weber (1982) who distinguishes between algorithms for information-oriented, filter-oriented and heuristic generalization. Another survey which is restricted to the reduction and displacement of linear features is reported by Zoraster *et al.* (1984).

Statistical generalization is a process of reducing the volume of data aimed at efficient processing. It contrasts with data compaction where data are compressed (e.g., for transmission) and reconstructed and where loss of information is critical. Statistical generalization is guided by quantitative criteria. Therefore, most procedures for selection and simplification based on a statistical model can be applied for this purpose (e.g., quantitative investigation of the performance of line reduction algorithms, McMaster 1986; filtering of triangulated surfaces, Weibel *et al.* 1987). Processes of statistical generalization must not be used for cartographic generalization (which also involves feature displacement, etc.). On the other hand, one cannot expect a database which was generalized using procedures for cartographic generalization to be a reliable source of predictable quality for analysis, because manipulation by cartographic generalization will introduce unpredictable errors through processes such as feature displacement. Again, the purpose of an intended map or database largely controls the selection of procedures to be used for generalization. Related to statistical generalization is the problem of errors in cartographic databases and their propagation through manipulation processes (e.g., map overlay or generalization). This issue has been of growing concern over the past few years (e.g., Blakemore 1984, Chrisman 1982).

We have pointed out that structure recognition is important for the automation of generalization. For the description of planar elements (i.e., points, lines, areas) quite extensive work has been done in the field of computer vision (e.g., Ballard and Brown 1982, Marr 1982, Bunke and Sanfeliu 1986; these sources also list further relevant literature). Research into recognition of planar features in digital cartography has been conducted by Saalfeld (1986 — shape description of polygons), Buttenfield (1985, review of several geometric measures for cartographic lines; 1986, quantitative description of scale-dependent line structure based on strip-tree decomposition), O'Neill and Mark (1987, line description by means of the Ψ-s curve), and others (e.g., Gaile and Willmott 1984). An approach to object recognition has been presented by De Simone (1986). Structure recognition of three-dimensional surfaces (usually topographic surfaces) is reviewed by Weibel *et al.* (1987). They distinguish between the recognition of global relief character (terrain roughness, major directions of landforms, etc., Evans 1979) and the extraction of local surface features (structure lines, drainage basins, etc., for formal definitions see Frank *et al.* 1986, for procedures see, e.g., Weibel *et al.* 1987, Douglas 1986). Structure recognition aims not only at identifying the individual elements but also at extracting knowledge about their context. For this purpose, strategies of syntactic pattern recognition from computer vision may be used (e.g., Bunke and Sanfeliu 1986). However, as cartographic structures are usually more complex than the scenes commonly modelled in computer vision, it may not be so easy to devise adequate solutions.

To make recognition of more complex structures feasible for generalization, more appropriate *cartographic data models* have to be used. Grelot (1986) argues that current data models in cartography are not capable of handling complex processes but rather represent a digital version of instructions for map drafting. New concepts considering spatial proximity, fuzziness (probabilist depiction), addressing composite entities and the like should be developed instead. The more modern data models holding promise for covering spatial relationships are based on hierarchical concepts (Samet 1984, Matsuyama *et al.* 1984, Peuquet 1986) or on hybrid structures (Peuquet 1983). Whereas hierarchical models have already shown feasibility in generalization related processes (e.g., name placement, Freeman and Ahn 1984), hybrid structures first have to be proven to be practicable. On the whole, current data models have to be improved further to include such concepts as fuzziness. Also, more descriptive information relating to the nature and structure of the captured features ought to be included in the cartographic databases (e.g., geomorphological information for digital terrain models). This would greatly simplify the task of structure recognition.

There is as yet no explicit literature on *process recognition* and *process modelling*. Implicitly, such concepts are contained in existing procedures. We, therefore, discuss here all aspects of generalization processes jointly. We group this discussion by feature type (point, line, area, surfaces).

Point features are the map elements most easily handled in generalization because they offer the greatest degree of freedom with respect to their manipulation. Several algorithms for selection and displacement of point features were developed in the 1970s (Weber 1982 describes some of these approaches). Mackaness and Fisher (1987) recently reported on a knowledge-based approach to displacement of point symbols *vis-à-vis* other points, lines and areas.

Line generalization has received a considerable amount of attention. Much recent research still concentrates on approaches to line simplification. This might suggest that cartographic generalization is merely a simplification process (or low-pass filtering). The classic line smoothing algorithms are by Perkal (1966) and Douglas and Peucker (1973). A number of authors have tried to improve these algorithms (e.g., Van Horn 1985, Deveau 1985, enhancement of Douglas-Peucker algorithm, Chrisman 1983, enhancement of epsilon-filtering). Some supporting work has been done in the quantitative assessment of various algorithms (McMaster 1986, White 1985). Other research has focused on the inverse process of generalization — fractal enhancement for display (Dutton 1981). The generalization of linear networks (e.g., rivers) has been addressed by, among others, Catlow and Du (1984). Recent research suggests operations for major scale reduction factors. Beard (1987) has proposed a model for the derivation of a series of small scale databases from a single detailed database. Included are steps for selection, re-classification, simplification, coarsening and collapsing (but no explicit displacement), but it is not clear whether this strategy is aiming at statistical or cartographic generalization. Another approach presents a design for a scale-independent database (Jones and Abraham 1986). The database consists of various scale-dependent levels containing subsets of points selected by the Douglas-Peucker algorithm at increasing tolerances and stored in a tree structure.

Examples of research in *generalization of areal features* are publications by Monmonier (1983) addressing generalization of land use and land cover maps and Lay and Weber (1983) addressing generalization of woodland areas. Both strategies are raster-based. Monmonier applies a series of geometrical filters for gap bridging and gap deletion; Weber and Lay use a series of expansion and contraction operations. Most other procedures applied to area generalization are in fact algorithms for line generalization (i.e., Douglas-Peucker-like procedures). This approach reduces the shape of an areal feature to its outline, which we strongly doubt to be feasible. Even advanced approaches to map generalization (e.g., Beard 1987, Jones and Abraham 1986, Nickerson and Freeman 1986) treat areas in this way. Contours are inappropriate data models for areal units and, similarly, line generalization algorithms are not adequate to process closed shapes. They operate relatively locally and do not consider the area's overall shape. A better concept for shape description is provided by the medial axis of polygons. Ho and Dyer (1986) present a shape smoothing approach based on the medial axis transform. This strategy used in computer vision might also be adapted to cartography.

As is the case with areal features, the *generalization of surfaces* (usually topographical surfaces) is also handled by inappropriate strategies. Existing approaches (i.e., smoothing of contours, digital elevation model (DEM) filtering, filtering of triangulated surfaces) do not address individual landforms for elimination and simplification and are therefore restricted to minor scale reduction tasks. In recent years, concepts of relief generalization have begun to use structure lines as the basis for the generalization model. Wu (1981) presented a strategy of enhanced contour generalization based on elimination of insignificant (short) valley lines. Wolf (1984) proposed a model based on a graph-theoretical approach based on the concept of the surface network (Pfaltz 1976). The total surface network (consisting of all structure lines) is reduced through network contraction; this generalization is performed in a purely lexical manner. In order to overcome the shortcomings of existing approaches Weibel (1987) developed a strategy based on heuristic concepts related to a suggestion by Yoeli (1988). It selects adaptively an appropriate generalization procedure for a given relief type and map purpose. Methods used include several filtering techniques (position-variant and position-invariant filters) and heuristic generalization of the structure line network, including processes for elimination, simplification, combination and displacement.

Trends in current research for automated map generalization point towards the treatment of more complex map features (i.e., areas, surfaces) as well as towards the development of integrated approaches to generalization. This last section of our research review is accordingly devoted to efforts to build *integrated generalization systems*.

Integration includes the incorporation of procedures for the handling of multi-layered map databases. Common to all integrative approaches is the application of knowledge-based concepts. There are several introductory texts for knowledge-based systems (or expert systems) and reviews of applications in cartography (e.g., Robinson and Frank 1987). One of the first integrated generalization systems is described by Leberl *et al.* (1985). This commercial system (called ASTRA) is based on algorithms developed in the 1970s (Lichtner 1979). It handles the generalization of topographic maps from 1:24 000 to 1:50 000, including the tasks of contour smoothing, building generalization, line smoothing, feature symbolization and feature displacement. An enhanced version of this system for generalizing large-scale maps (i.e., 1:5 000 to 1:25 000) was presented by Meyer (1986). Nickerson and Freeman (1986) have incorporated existing generalization procedures into a strategy for generalizing digital line graph (DLG) files using concepts of processing order and displacement hierarchy. Processes include elimination, simplification, combination and displacement. Other systems are still in a state of design or development; one paper describes an effort to develop a general expert system for cartographic design where generalization is a subset of all design processes (Mackaness *et al.* 1986) and the authors try to base their system on theoretical considerations of map design. Another paper (Monmonier 1986) identifies the requirements for an operational system for line generalization. He stresses the necessity of using appropriate data models and recommends a hybrid structure similar to the 'vaster' structure as suggested by Peuquet (1983). He further proposes the use of supervised classifiers similar to those used for the selection of thresholds in remote sensing.

Integration of various algorithms and concepts into comprehensive systems for map generalization is still at an early stage. Concepts are still relatively limited and process recognition and modelling are sometimes arbitrary. This is mainly caused by the lack of theory in cartographic communication and map generalization. This fact leads Fisher and Mackaness (1987) to the view that existing so-called expert systems do not deserve this title because their schemes of reasoning are not based on a clear understanding of the processes to be modelled.

5 Evaluation and outlook

Generalization is a basic human activity involving intellectual functions. In cartography the task is to extract relevant structured components from experienced reality and process them for visual communication. We have identified map generalization as a variant of spatial modelling. Automation of this process is not a trivial matter and short-term solutions should not be expected. To promote computer-assisted generalization we have presented a conceptual framework which structures map generalization into five steps: structure recognition, process recognition, process modelling, process execution and display. We have identified cartographic and statistical generalization as two distinct groups for spatial modelling and discussed their respective application range. In the light of these concepts we have then given a review of literature and empirical studies up to this time. In the remainder of this section we try to evaluate the present state of the art and give an outlook for future needs for development.

The approaches to automated map generalization in the 1970s and early 1980s had some substantial limitations.

(1) Research focused on problems of lower complexity, such as feature selection and line generalization.
(2) Algorithms almost exclusively concentrated on one single class of feature, an exception being a system for generalization at large scales (Lichtner 1979).
(3) Data models used were not well suited for generalization because they optimized plotting and graphical display and did not support concepts amenable to structure recognition (e.g., spatial proximity, addressing of composite entities, fuzziness, etc.)

(4) In contrast to manual practice, procedures worked sequentially rather than in a synoptic (i.e., parallel) manner. This was partly due to the inadequate data models.

(5) There was a general lack of theoretical foundation for the algorithms. Exceptions were the theory of the cartographic line (Peucker 1975; this theory was developed *a posteriori* to describe the model underlying the Douglas-Peucker line reduction algorithm, Douglas and Peucker 1973), Töpfer's (1974) empirically-deduced quantitative rules and the general framework of Rhind (1973). This lack of theory restricted generalization processes to the use of simple measures such as minimal distances and thresholds.

In recent years these weaknesses were acknowledged by a growing number of authors (e.g., Brassel 1984, Weibel 1986, Nickerson and Freeman 1986, Monmonier 1987, Eastman 1987, Grelot 1987). In current research we can observe two major trends to enhance existing models and algorithms.

A first group of authors strives for enhancement and modification of existing algorithms for specific applications; this is mainly the case with pragmatic production-oriented systems (see Zoraster *et al.* 1984) for handling specific tasks of limited complexity (e.g., line reduction after digitizing) and map features of limited graphical complexity (e.g., line features). Examples are the adaptations of line smoothing algorithms (e.g., Van Horn 1985, Deveau 1985) or the incorporation of several existing generalization procedures into a comprehensive system (e.g., the expert system for the DLG file generalization of Nickerson and Freeman 1986).

A second group argues that the concepts of generalization need rethinking and must be set into an environment of formal rules supported by a theoretical foundation. They further postulate that for handling more complex generalization models not only do algorithms have to be redesigned, but also the underlying data models have to be adjusted (e.g., to allow synoptic processing). This view arises in research projects envisioning the generalization of map features with high graphical complexity (i.e., areas, surfaces) and the generalization of maps with an overlay of multiple layers of feature classes or for extensive scale reduction factors.

Both approaches are legitimate. Generalization systems based on straightforward processing strategies and simple data structures can cover several tasks such as statistical generalization, selection and simplification, and displacement of point and line features. For some specific applications this may be adequate. However, for complex maps with multiple layers, cartographic generalization as a mere reduction and simplification process is too simplistic. For these cases, we have to devise schemes that can model the complexity of the task and that are based on theoretical understanding of cartographic communication and generalization. They involve the development of processes for structure recognition, process recognition and process modelling and are to be calibrated by empirical tests of map perception and generalization effects. Only such a strategy can lead to the development of real knowledge-based systems (and not just rule-based systems) using adaptive, intelligent algorithms and flexible data models.

In conclusion, we see future research efforts in advanced cartographic generalization along two major lines. First we need to develop a theoretical basis for the understanding of the function and structure of maps (i.e., structure recognition) and of the individual generalization processes (i.e., process recognition and process modelling). This can be achieved by theoretical reasoning using new concepts to model cartographic communication (e.g., linguistics), and at the same time, empirical studies with workbench systems for the testing of generalization concepts and algorithms (Brassel 1984). Secondly, we have to focus on the aspect of the data models underlying generalization. Today's data models are oriented towards graphical display. Complex generalization operations will involve synoptic and parallel spatial processing (e.g., for displacement). To make such processes possible, we have to develop new data models which emulate concepts such as spatial proximity and composite entity addressing and which lend themselves to structure recognition (Grelot 1986, Monmonier 1987). We also have to improve the databases we use for cartography by including more structural information at the beginning.

Acknowledgments

Professor H. Kishimoto has kindly revised this article, Ms C. Hofstetter, Ms C. Karrer, Mr M. Braendli and Mrs D. Koller have helped with the technical aspects of figures and text editing. These contributions are gratefully acknowledged.

References

Ballard, D. H., and Brown, C. M., 1982, *Computer Vision* (Englewood Cliffs, New Jersey: Prentice-Hall).

Beard, M. K., 1987, How to survive on a single detailed data base. *Proceedings of Auto-Carto 8 held in Baltimore, Maryland, on 29 March–3 April 1987* (Falls Church, Virginia: American Society of Photogrammetry and Remote Sensing, American Congress on Surveying and Mapping).

Bertin, J., 1967, *Semiologie Graphique* (Paris: Gauthier-Villars).

Blakemore, M., 1984, Generalization and error in cartographic databases. *Cartographica*, **21**, 131.

Board, C., 1981, Cartographic communication. In *Maps in Modern Cartography*, edited by L. Guelke, *Cartographica*, **18**, 42.

Brassel, K. E., 1984, Strategies and data models for computer-aided generalization. *International Yearbook of Cartography*, **25**, 11.

Brüggemann, H., 1985, Die topographisch-kartographische Datenbank in der Standardisierung. *Nachrichten aus dem Karten-und Vermessungswesen, Series I*, **95**, 43.

Bunke, H., and Sanfeliu, A., (editors), 1986, Advances in syntactic pattern recognition. *Pattern Recognition*, **19**, 249.

Buttenfield, B. P., 1985, Treatment of the cartographic line. *Cartographica*, **22**, 1.

Buttenfield, B. P., 1986, Digital definitions of scale-dependent line structure. *Proceedings of Auto-Carto London held in London on 14–19 September 1986*, Vol. 1 (London: Auto-Carto London), p. 497.

Catlow, D. R., and Du, D., 1984, The structuring and generalization of digital river data. *Technical Papers of the 44th Annual Meeting of the ACSM held in Washington, D.C., on 11–16 March 1984* (Falls Church, Virginia: American Congress on Surveying and Mapping), p. 511.

Chrisman, N. R., 1982, Methods of spatial analysis based on errors in categorical maps. Ph.D. Thesis, University of Bristol, England.

Chrisman, N. R., 1983, Epsilon filtering: A technique for automated scale changing. *Technical Papers of the 43rd Annual Meeting of the ACSM held in Washington, D.C., on 13–18 March 1983* (Falls Church, Virginia: American Congress on Surveying and Mapping), p. 322.

De Simone, M., 1986, Automatic structuring and feature recognition for large scale digital mapping. *Proceedings of Auto-Carto London held in London on 14–19 September 1986*, Vol. 1, (London: Auto-Carlo London), p. 86.

Deveau, T. J., 1985, Reducing the number of points in a plane curve representation. *Proceedings of Auto-Carto 7 held in Washington, D. C., on 11–14 March 1985* (Falls Church, Virginia: Society of Photogrammetry, American Congress on Surveying and Mapping), p. 152.

Douglas, D. H., and Peucker, T. K., 1973, Algorithms for the reduction of the number of points required to represent a digitized line or its caricature. *The Canadian Cartographer*, **10**, 112.

Douglas, D. H., 1986, Experiments to locate ridges and channels to create a new type of digital elevation model. *Cartographica*, **23**, 29.

Dutton, G. H., 1981, Fractal enhancement of cartographic line detail. *The American Cartographer*, **8**, 23.

Eastman, J. R., 1981, The perception of scale change in small-scale map series. *The American Cartographer*, **8**, 5.

Eastman, J. R., 1987, Graphic syntax and expert systems for map design. *Technical Papers of the 1987 ASPRS-ACSM Annual Convention held in Washington, D.C., on 29 March–3 April 1987*, Vol. 4 (Falls Church, Virginia: American Society of Photogrammetry and Remote Sensing, American Congress on Surveying and Mapping), p. 87.

Evans, I. S., 1979, An integrated system of terrain analysis and slope mapping. Final Report on DA-ERO-591-73-G0040: Statistical Characterization of Altitude Matrices by Computer, Department of Geography, University of Durham, England.

Fisher, P. F., and Mackanes, W. A., 1987, Are cartographic expert systems possible? *Proceedings of Auto-Carto 8 held in Baltimore, Maryland, on 29 March–3 April 1987* (Falls Church, Virginia: American Society of Photogrammetry and Remote Sensing, American Congress on Surveying and Mapping), p. 530.

Frank, A., Palmer, B., and Robinson, V., 1986, Formal methods for the accurate definition of some fundamental terms in physical geography. *Proceedings of the Second International Symposium on Spatial Data Handling held in Seattle, Washington, on 5–10 July 1986* (Williamsville, New York: International Geographical Union Commission on Geographical Data Sensing and Processing), p. 583.

Freeman, H., and Ahn, J., 1984, AUTONAP — An expert system for automatic name placement, *Proceedings of the First International Symposium on Spatial Data Handling held in Zurich, Switzerland on 20–24 August 1984*, Vol. 2 (Zurich: Department of Geography, University of Zurich), p. 544.

Gaile, G. L., and Willmott, C. J., 1984, *Spatial Statistics and Models* (Dordrecht: D. Reidel Publishing Company).

Gibson, A. E., 1987, A model describing options for parallel color/data structuring. *Technical Papers of the 1987 ASPRS-ACSM Annual Convention held in Washington, D.C., on 29 March–3 April 1987* Vol. 4 (Falls Church, Virginia: American Society of Photogrammetry and Remote Sensing, American Congress on Surveying and Mapping), p. 97.

Grelot, J.-P., 1986, Archaic data models or hardware as a concept killer. *Proceedings of Auto-Carto London held in London on 14–19 September 1986* Vol. 1 (London: Auto-Carto London), p. 572.

Grunreich, D., 1985, Ein Vorschlag zum Aufbau einer grossmassstäbigen topographisch-kartographischen Datenbank unter besonderer Berücksichtigung der Grundrissdatei des ALK-Systems. *Nachrichten aus dem Karten- und Vermessungswesen, Series I*, **95**, 55.

Hake, G., 1975, Zum Begriffssystem der Generalisierung. *Nachrichten aus dem Karten- und Vermessungswesen, Series I*, **85**, 53–62.

Hake, G., 1982, *Kartographie,* Vol. 1, Sixth edition (Berlin, New York: W. de Gruyter).

Ho, S.-B., and Dyer, C. R., 1986, Shape smoothing using medial axis properties. *I.E.E.E. Transactions on Pattern Analysis and Machine Intelligence*, **8**, 512.

Imhof, E., 1982, *Cartographic Relief Presentation* (Berlin: W. de Gruyter). And 1965 *Kartographische Geländedarstellung* (original German language edition) (Berlin: W. de Gruyter).

International Cartographic Association (ICA) 1973, *Multilingual Dictionary of Technical Terms in Cartography* (Wiesbaden, F. R. Germany: Franz Steiner Verlag).

Jones, C. B., and Abraham, I. M., 1986, Design considerations for a scale-independent cartographic database. *Proceedings of the Second International Symposium on Spatial Data Handling held in Seattle, Washington, on 5–10 July 1986* (Williamsville, New York: International Geographical Union Commission on Geographical Data Sensing and Processing), p. 384.

Knöpfli, R., 1983, Communication theory and map generalization. In *Communication and Design in Contemporary Cartography*, edited by D. R. F. Taylor (New York: John Wiley), p. 177.

Lay, H.-G., and Weber, W., 1983, Waldgeneralisierung durch digitale Rasterverarbeitung. *Nachrichten aus dem Karten- und Vermessungswesen, Series I*, **92**, 61.

Leberl, F. L., Olson, D., and Lichtner, W., 1985, ASTRA — A system for automated scale transition. *Technical Papers of the 51st Annual Meeting of the ASP held in Washington, D.C., on 10–15 March 1985,* Vol. 1 (Falls Church, Virginia: American Society of Photogrammetry), p. 1.

Lichtner, W., 1979, Computer-assisted processes of cartographic generalization in topographic maps. *Geo-Processing*, **1**, 183.

Mackaness, W. A., Fisher, P. F., and Wilkinson, G. G., 1986, Towards a cartographic expert system. *Proceedings of Auto-Carto London held in London on 14–19 September 1986*, Vol. 1 (London: Auto-Carto London), p. 578.

Mackaness, W. A., and Fisher, P. F., 1987, Automatic recognition and resolution of spatial conflicts in cartographic symbolization. *Proceedings of Auto-Carto 8 held in Baltimore, Maryland, on 29 March–3 April 1987* (Falls Church, Virginia: American Society of Photogrammetry and Remote Sensing, American Congress on Surveying and Mapping), p. 42

Marr, D., 1982, *Vision* (New York: Freeman).

Matsuyama, T., Hao, L. V., and Nagao, M., 1984, A file organization for geographic information systems based on spatial proximity. *Computer Vision, Graphics and Image Processing,* **26**, 303.

McMaster, R B., 1986, A statistical analysis of mathematical measures for linear simplification. *The American Cartographer*, **13**, 103.

Meyer, U., 1986, Software developments for computer-assisted generalization. *Proceedings of Auto-Carto London held in London held in London on 14–19 September 1986*, Vol. 2 (London: Auto-Carto London), p. 247.

Monmonier, M. S., 1982, *Computer-Assisted Cartography: Principles and Prospects* (Englewood Cliffs, New Jersey: Prentice-Hall).

Monmonier, M. S., 1983, Raster-mode area generalization of land use and land cover maps. *Cartographica*, **20,** 65.

Monmonier, M., 1986, Toward a practicable model of cartographic generalization. *Proceedings of Auto-Carto London held in London on 14–19 September 1986*, Vol. 2 (London: Auto-Carto London), p. 257.

Muller, J.-C., and Honsaker, J. L., 1983, Visual versus computerized seriation: the implications for automated map generalization. *Proceedings of Auto-Carto 6 held in Ottawa, Canada, on 16–21 October 1983*, Vol. 2 (Ottawa: The Steering Committee for the Sixth International Symposium on Automation in Cartography), p. 277.

Nickerson, B. G., and Freeman, H., 1986, Development of a rule-based system for automatic map generalization. *Proceedings of the Second International Symposium on Spatial Data Handling held in Seattle, Washington, on 5–10 July 1986* (Williamsville, New York: International Geographical Union Commission on Geographical Data Sensing and Processing), p. 537.

O'Neill, M. P., and Mark, D. M., 1987, The psi-s plot: A useful representation for digital cartographic lines. *Proceedings of Auto-Carto 8 held in Baltimore, Maryland, on 29 March–3 April 1987* (Falls Church, Virginia: American Society of Photogrammetry and Remote Sensing, American Congress on Surveying and Mapping), p. 231.

Perkal, J., 1966, An attempt at objective generalization. Discussion Paper, No. 10, Inter-University Community of Mathematical Geographers, Ann Arbor, Michigan.

Peterson, M. P., 1979, An evaluation of unclassed crossed-line choropleth mapping. *The American Cartographer*, **6**, 21.

Peucker, T. K., 1975, A theory of the cartographic line. *Proceedings of Auto-Carto II held in Reston, Virginia, on 21–25 September 1975* (Falls Church, Virginia: American Congress on Surveying and Mapping), p. 508.

Peuquet, D. J., 1983, A hybrid structure for the storage and manipulation of very large spatial data sets. *Computer Vision, Graphics and Image Processing*, **24**, 14.

Peuquet, D. J., 1986, The use of spatial relationships to aid spatial database retrieval. *Proceedings of the Second International Symposium on Spatial Data Handling held in Seattle, Washington, on 5–10 July 1986* (Williamsville, New York: International Geographical Union Commission on Geographical Data Sensing and Processing), p. 459.

Pfaltz, J. L., 1976, Surface networks. *Geographical Analysis*, **8**, 77.

Phillips, R. J., and Noyes, L., 1982, An investigation of visual clutter in the topographic base of geological maps. *The Cartographic Journal*, **19,** 122.

Rhind, D., 1973, Generalisation and realism within automated cartographic systems. *The Canadian Cartographer*, **10**, 51.

Robinson, A. H., Sale, R. D., Morrison, J. L., and Muehrcke, P. C., 1984, *Elements of Cartography*, Fifth edition (New York: John Wiley).

Robinson, V. B., and Frank, A. U., 1987, Expert systems applied to problems in geographic information systems: Introduction, review and prospects. *Proceedings of Auto-Carto 8 held in Baltimore, Maryland, on 29 March–3 April 1987* (Falls Church, Virginia: American Society of Photogrammetry and Remote Sensing, American Congress on Surveying and Mapping), p. 510.

Saalfeld, A,. 1986, Shape representation for linear features in automated cartography. *Technical Papers of the 1986 ASPRS-ACSM Annual Convention held in Washington, D.C., on 16–21 March 1986* (Falls Church, Virginia: American Society of Photogrammetry and Remote Sensing, American Congress on Surveying and Mapping), p. 143.

Samet, H., 1984, The quadtree and related hierarchical data structures. *ACM computing Surveys*, **16,** 187.

Steward, H. J., 1974, Cartographic generalization: some concepts and explanation. *Cartographica Monograph*, No. 10.

Taketa, R. A., 1979, *Structure and Meaning in Map Generalization.* PhD Dissertation (Ann Arbor, Michigan: University Microfilms International).

Tobler, W. R., 1964, An experiment in the computer generalization of maps. Technical Report, No. 1, Office of Naval Research, Task No. 389–137.

Tobler, W. R., 1973, Choropleth maps with class intervals. *Geographical Analysis*, **3**, 262.

Topfer, R., 1974, *Kartographische Generalisierung* (Gotha, Leipzig, German Democratic Republic: VEB Hermann Haack).

Van Horn, E. K., 1985, Generalizing cartographic data bases. *Proceedings of Auto-Carto 7 held in Washington, D.C., on 11–14 March 1985* (Falls Church, Virginia: American Society of Photogrammetry, American Congress on Surveying and Mapping), p. 532.

Weber, W., 1982, Automationsgestützte Generalisierung. *Nachrichten aus dem Karten- und Vermessungswesen, Series I*, **88**, 77.

Weibel, R., 1986, Automated cartographic generalization. In *A Selected Bibliography on Spatial Data Handling: Data Structures, Generalization, and Three-Dimensional Mapping*, Vol. 6, *Geoprocessing Series,* edited by R. Sieber and K. E. Brassel (Zurich: Department of Geography, University of Zurich), p. 20.

Weibel, R., 1987, An adaptive methodology for automated relief generalization. *Proceedings of Auto-Carto 8 held in Baltimore, Maryland, on 29 March–3 April 1987* (Falls Church, Virginia: American Society of Photogrammetry and Remote Sensing, American Congress on Surveying and Mapping), p. 42.

Weibel, R., Heller, M., Herzog, A., and Brassel, K., 1987, Approaches to digital surface modelling. *Proceedings of the First Latin American Conference on Computers in Geography held in San José, Costa Rica on 5–9 October 1987* (Heredia, Costa Rica: International Geographical Union Commission on Geographical Data Sensing and Processing, Escuela de Geografia Universidad Nacionale), p. 143.

White, E. R., 1985, Assessment of line-generalization algorithms using characteristic points. *The American Cartographer*, **12**, 17.

Wolf, G. W., 1984, A mathematical model of map generalization. *Geo-Processing*, **2**, 271.

Wu, H.-H., 1981, Prinzip und Methode der automatischen Generalisierung der Reliefformen. *Nachrichten aus dem Karten- und Vermessungswesen, Series I*, **85,** 163.

Yoeli, P., 1988, Entwurf einer Methodologie für computergestütztes kartographisches Generalisieren topographischer Reliefs. In *Kartographisches Generalisieren,* Vol. 9 (Zurich: Swiss Society of Cartography), (in the press).

Zoraster, S., Davis, D., and Hugus, M., 1984, Manual and automated line generalization and feature displacement. Report, ETL-0359 (+ETL-0359-1) U.S. Army Engineer Topographic Laboratories, Fort Belvoir, Virginia, U.S.A.

Map Generalization: What a Difference Two Decades Make

Robert Weibel and Kurt E. Brassel

1 Introduction

Almost twenty years ago, we presented a review of research in automated map generalization that spanned the period of the 1960s until the mid-1980s (Brassel and Weibel, 1988). In our review we identified progress, yet also saw clear limitations of the approaches prevalent at the time. The early methods of automated map generalization still fell a long way short of what manual cartography could deliver. In fact, automated procedures of the time took a mechanistic approach and neglected many of the fundamental principles of cartography. In order to help remedy the situation, we developed a conceptual framework of the map generalization process that departed from an analysis of (manual) cartographic practice and attempted to transpose those elements into the context of automation. It is mainly that conceptual framework that made the original paper popular in the research community and eventually brought it to the list of *Classics from IJGIS*.

In the remainder of the commentary, we will first take a look at how the context in which generalization research is embedded has changed over the years (Section 2). We will then discuss how our conceptual framework has inspired other models of the generalization process (Section 3). And finally, we will very briefly review present-day generalization research (Section 4) and attempt a projection into the future (Section 5). Note, however, that this commentary can by no means serve as a comprehensive review of the literature. That purpose is served far better by review articles that document the continuous progress over the years (Muller, 1991; Weibel, 1997; Weibel and Dutton, 1999), the special journal issues devoted to the topic (Weibel, 1995a; Weibel and Jones, 1998; Richardson and Mackaness, 1999; Jones and Ware, 2005a), as well as the various books that have been published to date (Buttenfield and McMaster, 1991; McMaster and Shea, 1992; Muller et al., 1995; Ruas, 2002; Mackaness et al., in press).

2 The context: Map generalization matters

Map generalization is one of the most fundamental processes of cartography. Likewise, scale issues are generally of fundamental importance to GIScience (UCGIS, 1996; UCGIS, 2002). Not surprisingly, map generalization was among the first topics to be addressed by research in computer cartography and GIS in the 1960s (cf. Brassel and Weibel, 1988), yet for many it remained still very much a niche topic. It was only shortly after our 1988 paper had been published that the situation started changing — mainly for two reasons. First, since the late 1980s a continuous stream of technological innovations in IT paved the way for innovative and more powerful generalization tools. These IT innovations included personal computing, graphical user interfaces, object-oriented software techniques, methods from computational intelligence (for example, genetic algorithms, neural nets, multiagent systems), or constraint-based numerical techniques, along with the general growth of computing power and storage capacity, making it possible to solve large computational problems efficiently. The second reason for an improved context of map generalization research was of a more political and economical nature. By around 1990 a growing number of large organizations involved in mapping and GIS had completed their Mark I spatial databases and were now ready to utilise those resources. In order to be able to flexibly derive multiple products from their databases, however, generalization functionality was needed, among other things. Flexible product generation was indispensable to amortise the immense investments made previously into data collection and database building, but the required generalization functionality did not exist at the time. Hence, large organizations, in particular national mapping agencies (NMAs), started to develop a keen and growing demand for better generalization tools.

The research community reacted to this trend by putting map generalization more prominently on the research agenda. In the years since we published our paper, sessions devoted specifically to the topic were present at most conferences related to methodological issues of GIScience. Research initiatives on generalization were created by a number of research organizations and working groups established on the international level. Perhaps the longest lasting and most continuous effect to date was achieved by the Commission on Generalization and Multiple Representation of the International Cartographic Association (ICA), established in 1991. Since then, this commission has held regular research workshops that generated and disseminated a lot of the main innovations of the specialised research community. The papers and presentations of these workshops, along with other resources, can be downloaded from the commission's Web sites (http://ica.ign.fr and http://www.geo.unizh.ch/ICA). The most recent output of the ICA commission's membership is a book that documents the state of the art and future issues of generalization and multiple representation (Mackaness et al., in press).

On the commercial end, GIS software companies, particularly those related to the mapping market, responded by developing specialised generalization software. MGE MapGeneralizer, an interactive generalization system by Intergraph, became available in the 1990s, along with CHANGE, a batch-oriented suite of programs for large-scale topographic map generalization developed by the University of Hanover. In the meantime, these original products have been superseded by more powerful systems that have, in large part, benefited from the academic research sector. Probably the most comprehensive generalisation system to date is Clarity by Laser-Scan. Intergraph offers a solution called DynaGEN, while ESRI (Environmental Systems Research Institute, Inc.) has extended its toolbox in ArcGIS by several generalization tools. These are but a few of the better known products. In short, customers now have several options on the market to choose from, and most NMAs in industrialised countries now routinely use software to automate the generalization process in their map production workflows at least partially. As NMAs are continuously extending their lines of products, including spatial database products and online services, and since new markets are developing (e.g., location-based services and mobile mapping), the demand for further generalization technology is continuing or even increasing, creating a favourable context for the specialised research community.

3 Conceptual frameworks for map generalization: The holistic view

As mentioned above, one of the key reasons why our IJGIS article has been widely cited in the literature was the conceptual framework that was proposed. Back in the days when our paper was published, this was one of very few models that tried to describe the generalization process on the conceptual level. We would further argue that our conceptual framework was perhaps the most comprehensive one of the time.

Over the course of the years, some of the elements of our framework have been renamed, changed, or even superseded by subsequent models published in the literature. For instance, what we called a generalization process (to denote individual transformations such as selection, simplification, aggregation, displacement) is now routinely termed a generalization operator (McMaster and Shea, 1992) to avoid confusion with the overall process. Likewise, statistical generalization to denote the controlled reduction of detail and accuracy of spatial databases not related to map graphics is now commonly called model generalization (Grünreich, 1992; Weibel, 1995b) or sometimes database generalization (Molenaar 1996). Also, the final step of data display in our conceptual framework, which we had viewed as a straightforward symbolization of a generalised target database to a target map, is certainly more complex, owing to the subtleties of map graphics as well as the need for feature displacement induced by enlarged map symbols. Nevertheless, the main elements of our model are still valid, including:

- The distinction between graphics-oriented and nongraphical generalization (*cartographic generalization* vs. *model generalization*, see above).
- The key role of *generalization controls* (map purpose, map scales, map symbology, communication rules, etc.) in forming the overall constraints governing the cartographic generalization process.
- The importance of *structure recognition*, that is, the recognition of specific patterns, object priorities, etc., that help to inform and guide the generalization process, implicitly contained in the original map data.
- The decomposition of the overall process into individual *generalization operators* (see above).
- And, perhaps most importantly, the need of performing generalization as "processing based on understanding" (Brassel and Weibel, 1988, 231).

Our conceptual framework has inspired others to propose further conceptual models of generalization. McMaster and Shea (1992) have presented a more detailed model responding to the three questions why, when, and how to generalise. In particular, their list of twelve generalization operators (that is, spatial and attribute transformations) is relatively well accepted in the literature. In more recent years attention has shifted more toward conceptually modelling the flow of control and guidance in generalization. Based on previous work by Beard (1991), a model that viewed generalization as a search for a solution subject to constraints was proposed by Ruas and Plazanet (1996). Their paper initiated a general move of the research community toward constraint-based modelling and processing in generalization (for a review, see Harrie and Weibel, in press). Perhaps the most sophisticated model of the generalization process is that of Ruas (1999), which laid the groundwork for the agent-based generalization system developed in the AGENT project (Barrault et al., 2001). The sum of all these models has helped the research community to collectively conceptualise and better understand the various aspects of the generalization process, that is, to understand what it takes to generalize a map or database. This improved understanding has in turn lead to continuous enhancements of the generalization techniques available.

4 Existing solutions: Where are we now?

In the editorial to their special issue of *IJGIS* on map generalization, Jones and Ware (2005b) briefly review the recent literature and the current state of the art in what they perceive as the main elements of map generalization.

Constraint-based techniques: The constraint-based view of map generalization, which has become prevalent in the last couple of years, has enabled significant progress in various respects. It has allowed succinct formulation of design criteria that define what a well-generalised map should look like. It enabled formulating map generalization as an optimization problem. It fostered the use of spatial data structures, such as triangulations and Voronoi diagrams, which could efficiently cope with spatial constraints. It gave rise to a stream of research on measures and methods to assess potential constraint violations, including methods for evaluating generalization quality.

Control and optimization methods: Owing to the interdependence of individual generalization operators, means of control of their application are necessary. While early research systems used simple, batchlike sequencing of operations (and many commercial systems still do today), modern generalization systems use more sophisticated methods to find a suitable generalization solution without the need of predefined operator sequences. The interplay of different generalization operators and constraints can be formulated as an optimization problem. A variety of optimization methods have been applied to map generalization, including discrete iterative improvement techniques (for example, simulated annealing, genetic algorithms), continuous optimization methods (least squares adjustment, snakes, and elastic beams), and agent-based approaches (for example, multiagent systems). A review of optimization techniques is presented in Harrie and Weibel (in press).

Online map generalization and multiscale databases: To date, most research has concentrated on generalization for paper map or cartographic database production. While these conventional applications are not time-critical, more modern uses of generalization in Web or mobile mapping definitely are. Unfortunately, few generalization methods exist today that are efficient enough to be performed in real-time. One of the ways to cope with online (that is, real-time) generalization is to use a combination of pregeneralization (different levels of details of the map data stored in a multiscale database) and online generalization. Apart from real-time applications, multiscale (or multiresolution) databases are also of interest to facilitate spatial database updating. A variety of NMAs in Europe have recently started integrating their previously separate databases into multiresolution databases by object matching between different levels of scale.

Beyond the brief discussion of Jones and Ware (2005b), more detailed information on the above topics can be found in Mackaness et al. (in press) and Ruas (2002). The latter two books also contain articles on issues not specifically addressed in the condensed review by Jones and Ware, issues that are nevertheless important. In Mackaness et al. (in press) this includes, for instance, a discussion of specific generalization algorithms for different object classes, methods for evaluation in the map generalization process, database integration in generalization, or experiences of embedding generalization software in map and spatial database production workflows in NMAs. Ruas (2002) includes chapters on database design methods, techniques for object matching in multiresolution databases, methods for cartographic structure and pattern recognition, real-time generalization methods, and machine learning procedures for cartographic knowledge formalization.

5 New challenges and opportunities: A look into the future

As discussed above, present-day generalization research has advanced quite a way from the state described in Brassel and Weibel (1988). Results from academic research have been transferred into commercial software, putting vastly improved generalization functionality into the hands of users of GIS and cartography systems, enabling significant productivity gains. Despite the encouraging progress made, however, continuous improvements are still required in all areas discussed in the previous section.

Apart from extensions of the existing research streams, new challenges are brought by new applications (Mackaness et al., in press). One of them is the fact that methods for 3-D data collection (e.g., laser scanning) have matured, which has lead to the development of a growing number of 3-D databases in a increasing number of organizations. Similar to the situation in 2-D databases in the late 1980s, it is foreseeable that there will be a dramatically increased demand for 3-D generalization functionality to efficiently and flexibly cope with product derivation from 3-D databases. Required functionality encompasses techniques for surface generalization (e.g., terrain) as well as for 3-D object generalization (e.g., buildings).

Another relatively new application that holds potential for significant future development is mobile cartography, needed for location-based services (LBS). Mobile mapping applications are characterised by several requirements that go beyond what is standard in conventional mapping. Not only are they time-critical, they also feature small screen sizes and limited bandwidth. Most importantly, however, mobile services require that the selection and display of information are tailored and adapted to the location and profile of the user. Such personalised information displays require flexible generalization strategies and techniques that allow delivery of highly abstracted information (i.e., simple map displays) efficiently. Initial solutions have been developed (Mackaness et al., in press) but research still has a long way to go.

The final example of a new application domain is the generalization of way-finding instructions (Mackaness et al., in press). Here, research challenges include the definition and extraction of landmarks as well as the generation of route descriptions. Both tasks clearly entail a great deal of abstraction and generalization because route descriptions (and their visualizations) have to be simple enough to be understandable by a human. Generalization procedures required will be of a nonstandard cartographic nature as they involve processes of spatial cognition.

The new applications above are but three that are presently entering the GIScience scene. New applications such as these are challenging our creativity as researchers, which naturally turns them into opportunities.

References

Barrault, M., Regnauld, N., Duchêne, C., Haire, K., Baeijs, C., Demazeau, Y., Hardy, P., Mackaness, W., Ruas, A., and Weibel, R., 2001, Integrating multi-agent, object-oriented, and algorithmic techniques for improved automated map generalization. In *Proceedings 20th International Cartographic Conference*, Beijing, 2210–2216.

Beard, M.K., 1991, Constraints on rule formation. In *Map Generalization: Making Rules for Knowledge Representation*, B.P. Buttenfield and R.B. McMaster, Eds., Longman, London, 121–135.

Brassel, K. and Weibel, R., 1988, A review and framework of automated map generalization. *International Journal of Geographical Information Systems*, **2**(3), 229–244.

Buttenfield, B.P. and McMaster, R.B., Eds., 1991, *Map Generalization: Making Rules for Knowledge Representation*, Longman, London.

Grünreich, D., 1992, ATKIS — A topographic information system as a basis for GIS and digital cartography in Germany. In *From Digital Map Series in Geosciences to Geo-information Systems*, R. Vinken, Ed., Geologisches Jahrbuch, Series A, **122**, 207–216, Federal Institute of Geosciences and Resources, Hanover.

Harrie, L. and Weibel, R., in press, Modelling the overall process of generalisation. In *Generalisation of Geographic Information: Models and Applications*, W.A. Mackaness, A. Ruas and L.T. Sarjakoski, Eds. Elsevier Science.

Jones, C.B. and Ware, J.M., Eds., 2005a, Special Issue: Generalisation. *International Journal of Geographical Information Science*, **19**(8/9).

Jones, C.B. and Ware, J.M., 2005b, Map generalization in the Web age. *International Journal of Geographical Information Science*, **19**(8/9), 859–870.

MACKANESS, W.A., RUAS, A., AND SARJAKOSKI, L.T., Eds., in press, *Generalisation of Geographic Information: Models and Applications*, Elsevier Science.

MCMASTER, R.B. AND SHEA, K.S., 1992, *Generalization in Digital Cartography*, Association of American Geographers, Washington, DC.

MOLENAAR, M., 1996, The role of topologic and hierarchical spatial object models in database generalization. In *Methods for the Generalization of Geo-databases*, M. Molenaar, Ed., Publications on Geodesy, **43**, 13–36, Netherlands Geodetic Commission, Delft.

MULLER, J.-C., 1991, Generalization of spatial databases. In *Geographical Information Systems: Principles and Applications*, D.J. Maguire, M.F. Goodchild, and D. W. Rhind, Eds., 457–475, Longman, London.

MÜLLER, J.-C., LAGRANGE, J.-P., AND WEIBEL, R., Eds., 1995, *GIS and Generalization: Methodological and Practical Issues*, Taylor & Francis, London.

RICHARDSON, D.E. AND MACKANESS, W., Eds., 1999, Computational Methods for Map Generalization. *Cartography and Geographic Generalization* (special issue on map generalization), **26**(1).

RUAS, A., 1999, *Modèle de généralisation de données géographiques à base de contraintes et d'autonomie*, Doctoral thesis, Université de Marne-la-Vallée.

RUAS, A., Ed., 2002, *Généralisation et représentation multiple*, Hermès Science Publications, Paris.

RUAS, A. AND PLAZANET, C., 1996, Strategies for automated generalization. In *Advances in GIS Research II* (Proceedings 7th International Symposium on Spatial Data Handling), M.J. Kraak and M. Molenaar, Eds., Taylor & Francis, London, 6.1–6.18.

UCGIS (UNIVERSITY CONSORTIUM FOR GEOGRAPHIC INFORMATION SCIENCE), 1996, Special issue: Research priorities for geographic information science. *Cartography and Geographic Information Systems*, **23**(3).

UCGIS (UNIVERSITY CONSORTIUM FOR GEOGRAPHIC INFORMATION SCIENCE), 2002, 2002 research agenda. Available online at: http://www.ucgis.org/priorities/research/2002research-agenda.htm (accessed 20 November 2005).

WEIBEL, R., Ed., 1995a, Special Issue: Automated Map Generalization, *Cartography and Geographic Information Systems,* **22**(4).

WEIBEL, R., 1995b, Three essential building blocks for automated generalization. In *GIS and Generalization: Methodological and Practical Issues*, J.-C. Müller, J.-P. Lagrange and R. Weibel, Eds., Taylor & Francis, London, 56–69.

WEIBEL, R., 1997, Generalization of spatial data. In *Algorithmic Foundations of Geographical Information Systems*, P. Widmayer, J. Nievergelt, T. Roos and M. van Kreveld, Eds., Lecture Notes in Computer Science, **1340**, Springer-Verlag, Berlin, 99–152.

WEIBEL, R. AND JONES, C.B., Eds., 1998, Special Issue on Map Generalization, *GeoInformatica,* **2**(4).

WEIBEL, R. AND DUTTON, G., 1999, Generalizing spatial data and dealing with multiple representations. In *Geographical Information Systems: Principles, Techniques, Management and Applications*, P. Longley, M.F. Goodchild, D.J. Maguire and D.W. Rhind, Eds., Second edition, John Wiley, Chichester, 125–155.

International Journal of Geographical Information Systems,
1989, Vol. 3, No. 4, 303–322.

4 Propagation of Errors in Spatial Modelling with GIS

Gerard B.M. Heuvelink, Peter A. Burrough, and Alfred Stein

Abstract. Methods are needed for monitoring the propagation of errors when spatial models are driven by quantitative data stored in raster geographical information systems. This paper demonstrates how the standard stochastic theory of error propagation can be extended and applied to continuously differentiable arithmetic operations (quantitative models) for manipulating gridded map data. The statistical methods have been programmed using the Taylor series expansion to approximate the models. Model inputs are (*a*) model coefficients and their standard errors and (*b*) maps of continuous variables and the associated prediction errors, which can be obtained by optimal interpolation from point data. The model output is a map that is accompanied by a map of prediction errors. The relative contributions of the errors in the inputs (model coefficients, maps of individual variables) can be determined and mapped separately allowing judgments to be made about subsequent survey optimization. The methods are illustrated by two case studies.

1 Introduction

Many users of geographical information systems (GIS) not only need to retrieve spatial entities and their associated attributes but also wish to link GIS databases to environmental models in order to predict the possible outcomes of particular processes. As they gain experience, users often become less content to judge the quality of the results of GIS analyses solely in terms of the quality of the graphical output. Users are beginning to realize that information about the intrinsic quality of results in terms of (*a*) the reliability of input data and (*b*) the confidence limits that can be associated with the end results is critical if the results of the GIS modelling are to carry any weight in decision making. Yet it is only relatively recently (e.g., Burrough 1986, Chrisman 1984, Goodchild and Dubuc 1987, Goodchild and Min-hua 1988, Mead 1982, NCGIA 1988) that much attention has been paid to the problems of errors and error propagation in GIS.

Mapped data are often held in a GIS as a series of raster overlays, as for example in the Map Analysis Package (Tomlin 1983a,b), a commonly used system for handling spatial data for quantitative analysis. Raster overlays are particularly useful for modelling quantitative attributes and the data they contain are the starting point for deriving many kinds of secondary data needed for decision making. These secondary data include attributes that could be measured directly but which are expensive to determine, such as the amounts of heavy metal in soil or the hydrological properties of the subsoil; or that are difficult to measure, such as potential crop yields under given conditions of climate and management or 'land qualities' (complex attributes of land) such as 'trafficability' or 'erodibility' used in land evaluation methodology (FAO 1976). In many cases it is sensible to use models to calculate the values of these attributes from cheaper or more readily-available data, providing the models yield reliable results.

Quantitative models can be used to derive the values of secondary attributes in two main ways in a raster GIS. In the first case the model is applied to each cell independently. Each cell is treated as a spatially independent location, x, where the model inputs are taken from the vertical vector of attribute values at x through the gridded overlays to yield an output P for the cell x. The model can be expressed generally as

$$P = f(A_1, A_2, A_3, \ldots, A_n) \tag{4.1}$$

where $A_1, A_2, A_3, \ldots, A_n$ are the values at x of the original properties used to calculate P at x. Figure 4.1a illustrates this type of operation. We call these type 1 operations. The maps of P over the whole area are obtained by carrying out the operation (4.1) independently for all cells. Note that any spatial contiguity in the values of P over all x positions arises only from the spatial contiguity of the input attributes A_i.

In the second case, spatial contiguity of the output results from deriving new data as a function of attribute values within a given range or window (Figure 4.1b; i.e., the value of the new attribute P' at the point x depends on spatial associations within an area W surrounding x,

$$P' = f(A_{1\text{W}}, A_{2\text{W}}, \ldots, A_{n\text{W}}) \tag{4.2}$$

where $\text{A}_{1\text{W}}, \text{A}_{2\text{W}}, \ldots, \text{A}_{n\text{W}}$ are the values of the original properties in the whole area W. We call these type 2 operations. In this case spatial contiguity of the map of P' is a result of the specific kind of the function (4.2) and the size of the window, W, in which it is carried out.

2 Model implementation and GIS

Both equations (4.1) and (4.2) can be considered as general statements of models for computing new attributes on a cell-by-cell basis. Transformation functions can take many forms and they can be implemented in a GIS in several ways.

Simple functions can be invoked directly by means of appropriate command language statements, whereas complex transformations can be achieved by combining

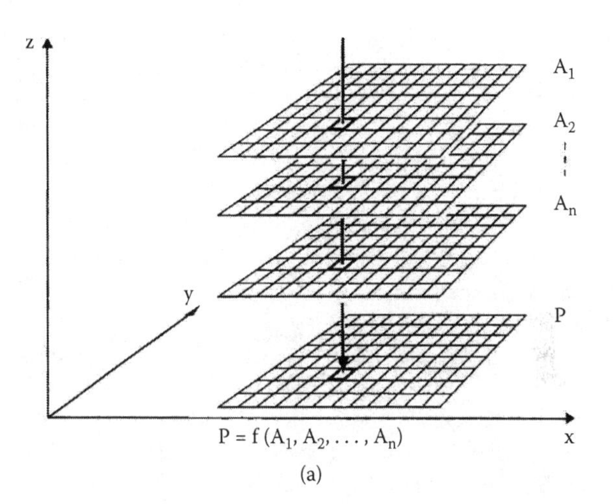

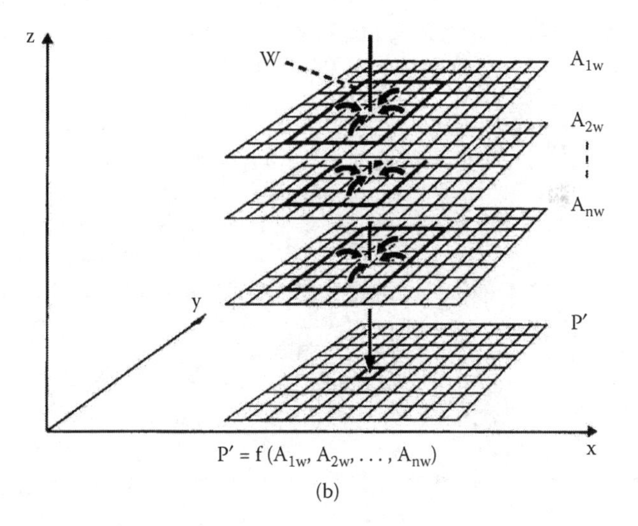

FIGURE 4.1 GIS operations: (a) type 1 — no spatial interaction (one vertical vector per cell); (b) type 2 — spatial interactions within window W.

simple functions serially. These methods are well known through the widespread use of the Map Analysis Package and the associated ideas of 'cartographic algebra' (e.g., Berry 1987, Burrough 1986, Tomlin 1983 a,b). Examples of GIS operations of the first type are, ADD, MULTIPLY, CROSS, MAXIMIZE and AVERAGE; examples of operations of the second are DIFFERENTIATE, FILTER, SPREAD and CLUMP.

The transformation function may also be defined by an empirical model that has been developed within a particular application discipline. Examples are the well-known Universal Soil Loss Equation and the many simple and multiple regression equations relating the values of expensive or difficult-to-measure attributes to data that are cheaper or simpler to obtain (Bouma and van Lanen 1987, Burrough 1986, 1989). For reasons of computational efficiency empirical models are often run outside the GIS, being linked by file transfer, though some specialized GIS allow frequently-used empirical models to be invoked directly from the GIS command language (e.g., Valenzuela 1988, Schultink 1987). Alternatively, some GIS allow users to key in mathematical formulae which are then compiled and run directly (TYDAC 1987). Most commonly-used empirical models are type 1 operations (see below) that treat each spatial unit, be it cell or polygon, independently from its surroundings.

Models that are too complex to be run directly from present state-of-the-art GIS, including models of crop growth, surface runoff, groundwater movement and atmospheric circulation (Beasley and Huggins 1982, Dumanski and Onofrei 1989, Jousma and van der Heyde 1985, McDonald and Harbaugh 1984), are usually run outside a GIS. These models attempt to describe a particular physical process and are often computationally too demanding to be run directly on the same machine as a standard GIS. In these cases the GIS is used to supply the input data at an appropriate level of resolution and to display the results graphically in combination with other relevant spatial data.

Complex models can be subdivided into those in which spatial interactions play no role (type 1), and those in which spatial interactions are an integral part of the modelling (type 2). Spatial interactions usually play no part in simulation of crop growth; these models require inputs of soil-water data, soil-nutrient status, climatic conditions and plane physiognomy for a given location, and so are usually type 1 operations. Spatial variations in crop growth can be estimated by measuring or estimating the values of site-specific parameters at several locations for which the model results are obtained. The spatial variation of calculated crop growth can then be depicted by interpolating from the sample sites over the area of interest (van Diepen *et al.* 1989). The quality of the maps of the model results is then dependent on the quality of the data about the site-specific parameters, the quality of the crop growth model itself, the kind of spatial variation over the area and the methods used for interpolating the results.

Models of groundwater, pesticide leaching, surface runoff and atmospheric circulation often require data about the spatial variation of the properties of the volume or surface through or over which they work. Because sampling is expensive and data points are sparse, the attribute values for each spatial element (a grid cell or cube) must be interpolated and are consequently never known exactly. These models often contain elements of both types of GIS operations.

3 Models and errors

This paper concentrates on the problem of error propagation in raster GIS when quantitative data from various overlay are combined by a continuous differentiable mathematical function of type 1 to yield an overlay (or map) of an output variable. In this paper the discussion is limited to models that only use simple arithmetical operations (+, –, /, *, exp, power, etc.), or combinations of these, such as occur in multiple regression equations, to compute model outputs on a cell-by-cell basis. The problem that is addressed here is: given values and standard deviations of input attributes, find the level of error δP associated with the model outputs P. If this can be done on a cell-by-cell basis then maps of both P and δP can be made.

Note that this paper does not cover error propagation through the type 2 operation as defined above nor through qualitative models and logical models. Neither does it extend to complex models such as those used for simulating crop growth, ground-water movement or erosion.

3.1 *The sources of error in* P

By definition a model is an approximation of reality. Some models describe reality better than others, and thus may be considered to be more appropriate. But more complex models often require more and better data. The model bias will be large when an inappropriate or incompletely defined model is chosen. The choice of the model thus plays an important role in the error δP. We assume here than an appropriate model has been selected (i.e., that bias is negligible) and that the model error can be described solely in terms of the errors of the model parameters. As an example, consider a multiple regression model of the form given as equation (4.3), in which the A_i values are grid maps of input variables and the α_i values are coefficients.

$$P = \alpha_0 + \alpha_1 A_1 + \alpha_2 A_2 + \ldots + \alpha_n A_n \tag{4.3}$$

There are two sources of error in the map of P. These are

(1) the errors associated with the model coefficients α_i and
(2) the errors associated with the spatial variation of the input attributes A_i.

3.1.1 The errors associated with the model parameters

For any given model the values of its parameters need to be chosen, estimated or adjusted to fit the area in which it is to be used. The values of model parameters can rarely be known exactly; usually their values must be estimated and there is always an associated error of estimation. It is not always easy to determine the errors associated with the model parameters without having a thorough knowledge of the model and the real situation where it is applied. However, when the model is a multiple regression equation like equation (4.3), the model error is given by the standard errors of the coefficients α_i and their correlations.

3.1.2 The errors associated with the spatial variation of the input data

Because the values of each input attribute A_i can vary spatially, the model will encounter a different set of input values at each location x. The spatial variation of the A_i and the associated errors can be described in two ways.

The first and most commonly encountered method divides the area of interest into polygons on the basis of other information such as land use or soil type. If observations of the A_i are available for points within each delineated area then average values and standard deviations can be calculated for each A_i for each polygon or mapping unit (Marsman and de Gruijter 1984). The polygons are converted to raster format yielding two overlays per attribute, the first overlay contains the average value of A_i and the second overlay contains the variance, both calculated for each polygon separately.

The second method involves interpolating the A_i from point observations directly to the grid. This could yield better predictions than simple polygon averages providing sufficient point observations are available. If optimal interpolation techniques based on regionalized variable theory (kriging) are used then each prediction of the value of A_i at a point or cell x is accompanied by an estimate of the variance of that prediction (known as the kriging variance), so the problem of determining the errors associated with each value of A_i at each cell x is solved directly.

Because some readers may not be familiar with regionalized variable theory a short review follows. Further details have been given by Burgess and Webster (1980a,b), Burrough (1986), Davis (1986), Journal and Huijbregts (1978), and Webster (1985). Note that we are using kriging only as one way to obtain estimates of the values of the A_i and their kriging variances for every cell x; for this paper it is not necessary to consider the reliability of these estimates and how they may vary with the size of the data set or the exact method used.

Regionalized variable theory assumes that the spatial variation of any variable can be expressed as the sum of three major components. These are

(*a*) a structural component, associated with a constant mean value or a polynomial trend,
(*b*) a spatially-correlated random component, and
(*c*) a white noise or residual error term that is spatially uncorrelated.

Let x be a position in one, two or three dimensions. Then the spatial variable A_i at x is given by

$$A_i(x) = m(x) + \in' (x) + \in'' \tag{4.4}$$

where $m(x)$ is a deterministic function describing the structural component of A_i at x, $\in'(x)$ is the term denoting the stochastic, locally-varying spatially-dependent residuals from $m(x)$, and $\in''$ is a residual, spatially-independent white noise term having a mean of zero. Optimal interpolation methods concentrate on the stochastic, spatially-correlated variation represented by $\in'(x)$.

The variation of the noise terms over space is summarized by a function known as the variogram. This function, which describes how one-half the variance of differences $\frac{1}{2}\mathrm{Var}[A_i(x) - A_i(x - h)]$ (known as the semivariance) varies with sample spacing (or lag), h, provides estimates of the weights used to predict the value of the A_i at unsampled points x_0 as a linear weighted sum of measurements made at sites x_i nearby. The weights are chosen to yield predictions of A_i at x_0 with the lowest possible kriging variance. The kriging variances can also be derived.

The kriging method yields predictions for sample supports (areas) that are the same size as the original field samples. For many purposes it is preferable to obtain predictions of average values of the attribute being mapped for blocks of a given size, for example, when combining data from optimal interpolation with data that have been transformed to a raster overlay at a given level of resolution, or that have already been collected for pixels of a given size (e.g., satellite imagery). In this case the kriging equations can be modified to predict an average or smoothed value of A_i over a block or cell. This procedure is known as block kriging (Burgess and Webster 1980 b). In many cases the kriging variances obtained by either point or block kriging are lower than those obtained by computing general averages for each polygon (Burrough 1986, Stein *et al.* 1988).

3.2 The balance of errors

The errors δP in the output maps of P accrue both from the errors in the model coefficients and from spatial variation in the input data. Knowledge of the relative balance of errors that accrue from the model and the separate inputs allows rational decisions to be made about whether extra sampling effort is needed (*a*) to determine the model parameters more precisely, or (*b*) to map one or more of the inputs more exactly. Comparing the error analyses for several different models (e.g., a first-order regression versus a second-order regression) would allow users of the GIS to decide which analysis route yielded the most reliable results. If the relative contributions to the overall error are known, then the allocation of survey resources between the tasks of model calibration and mapping can be optimized.

4 The theory of error propagation and its application to models and map overlays

Consider the following equation:

$$y = g(z_1, z_2, \ldots, z_n), \tag{4.5}$$

where g is a continuously differentiable function from R^n into R. Equation (4.5) is a generalized form of equations (4.1) and (4.3), more suited for the mathematical analysis that will follow. The arguments z_i of g may consist of both the input attributes A_i at some grid cell and the model coefficient α_i, but input attributes at neighbouring cells are excluded here. Therefore models represented by equation (4.2) are not covered here and will be the subject of a subsequent paper.

In order to study the propagation of errors we assume that the inputs, z_i, are not exactly known, but are contaminated by error (model coefficient errors and spatial variation of the inputs). The problem is to determine the magnitude of the error in y caused by the errors in the z_i. Therefore we interpret y as a realization of a random variable, Y. This random variable is fully characterized by its probability density function (p.d.f.) p_y. Important parameters of Y are its mean $\mu \equiv E[Y]$ and variance $\sigma^2 \equiv R[(Y - E[Y])^2]$. The standard deviation (σ) of Y, defined as the square root of the variance, may be interpreted as a measure for the absolute error within y. The relative error, also termed the coefficient of variation, is represented by the quotient of σ and μ.

The problem of error propagation involves determining the variance σ^2 of Y, given the means μ_i and variances σ_i^2 of the Z_i values, their correlation coefficients τ_{ij} and the type of operation, g. We note that we limit ourselves to the situation that the Z_i are *continuously* distributed random variables. In the case of operations on classified data the corresponding random variables would be *discretely* distributed. In that case a full probabilistic description would require the specification of all class-membership probabilities (Goodchild and Min-hua 1988, Drummond 1987). Unfortunately the desirable properties of the continuous case that we describe below are not directly transferable to the discrete case.

4.1 The derivation of error propagation equations using Taylor series

Consider equation (4.5) again, but since we now assume that the arguments are random variables, we write

$$Y = g(Z_1, Z_2, \ldots, Z_n). \tag{4.6}$$

For notational convenience we define vectors $\boldsymbol{\mu} = [\mu_1, \mu_2, \ldots, \mu_n]^T$ and $\mathbf{Z} = [Z_1, Z_2, \ldots, Z_n]^T$.

If the joint p.d.f. of $\mathbf{Z}$ were known, then it is possible in principle to calculate the p.d.f. of Y, and its mean and variance. However, this quickly leads to problems with numerical complexity because numerical integration is required (unless g is linear in its arguments). Even for the relatively simple case of the product of two normally-distributed random variables much computing time is required (Meeker *et al.* 1980). An alternative approach to estimating error propagation is to run a model many times using Monte Carlo simulations of all inputs and parameter values; but apart from the considerable computing problems, these methods do not satisfy an analytical form and the results cannot be transferred to new situations (Dettinger and Wilson 1981).

Because we are interested only in the mean and variance of Y, rather than its p.d.f., we propose that the best alternative to the above approaches uses a Taylor series expansion of g around $\boldsymbol{\mu}$. Fortunately, in the neighbourhood $\boldsymbol{\mu}$, the higher-order terms of the Taylor polynomial are small in comparison with lower-order terms. Therefore, neglecting these higher order terms leads to the approximation of g by a Taylor series of finite order, which is a computable polynomial.

Using a first order Taylor series expansion of g around $\boldsymbol{\mu}$ makes the calculation of the mean and variance of Y straightforward (the second-order expansion, which assumes the Z_i to be normally distributed, is given in the Appendix).

$$Y = g(\mathbf{Z}) = g(\boldsymbol{\mu}) + \sum_{i=1}^{n} \left\{ (Z_i - \mu_i) \left(\frac{\partial g}{\partial z_i}(\boldsymbol{\mu}) \right) \right\} + \text{rest term} \tag{4.7}$$

Assuming that the rest term of equation (4.7) may be neglected, the mean μ and variance σ^2 of Y follow by

$$\mu = E[Y] = E\left[g(\boldsymbol{\mu}) + \sum_{i=1}^{n} \left\{ (Z_i - \mu_i) \left(\frac{\partial g}{\partial z_i}(\boldsymbol{\mu}) \right) \right\} \right] = g(\boldsymbol{\mu}) \tag{4.8}$$

$$\begin{aligned}
\sigma^2 = E\left[(Y - E[Y])^2 \right] &= E\left[\left(g(\boldsymbol{\mu}) + \sum_{i=1}^{n} \left\{ (Z_i - \mu_i) \left(\frac{\partial g}{\partial z_i}(\boldsymbol{\mu}) \right) \right\} - g(\boldsymbol{\mu}) \right)^2 \right] \\
&= E\left[\left(\sum_{i=1}^{n} \left\{ (Z_i - \mu_i) \left(\frac{\partial g}{\partial z_i}(\boldsymbol{\mu}) \right) \right\} \right) \left(\sum_{j=1}^{n} \left\{ (Z_j - \mu_j) \left(\frac{\partial g}{\partial z_j}(\boldsymbol{\mu}) \right) \right\} \right) \right] \\
&= \sum_{i=1}^{n} \left\{ \sum_{j=1}^{n} \left\{ \tau_{ij} \sigma_i \sigma_j \frac{\partial g}{\partial z_i}(\boldsymbol{\mu}) \frac{\partial g}{\partial z_j}(\boldsymbol{\mu}) \right\} \right\}
\end{aligned} \tag{4.9}$$

It can be easily verified that equations (4.8) and (4.9) are equivalent to the results presented by Burrough (1986, p. 128–131) or other standard statistical texts (e.g., Parratt 1961).

We note an interesting feature resulting from equation (4.9). If the correlation between the arguments Z_i is zero, then the variance of Y is simply,

$$\sigma^2 = \sum_{i=1}^{n} \left\{ \sigma_i^2 \left(\frac{\partial g}{\partial z_i}(\boldsymbol{\mu}) \right)^2 \right\} \tag{4.10}$$

This shows that the variance of Y is the summation of parts, each to be attributed to one of the inputs Z_i. This partitioning property allows one to analyse how much each input contributes to the final error. Being able to do this is of great importance when deciding which of the inputs' errors should be diminished to reduce the final error. In practice, it is not always known if the various inputs are correlated or are truly independent.

5 Applications of the theory

5.1 A prototype software package for error propagation

In order to obtain a map showing how both the results of the model and the errors are distributed over space it is necessary to carry out the above procedure for each cell in the raster overlay. The ideas expressed above have been incorporated into a prototype software module called MAPCALC2, a module of the geostatistical software package PC-Geostat (*Rijksuniversiteit Utrecht* 1987). The MAPCALC2 program is written in TURBO Pascal for IBM-AT and IBM-PS2 computers and equivalents.

With MAPCALC2 it is possible to carry out any arithmetical operation on a maximum number of ten input maps. A map of predicted cell means and a map of the associated variances is required for each input attribute. In addition, it is possible to add a maximum of ten model coefficients to the operation. Once all inputs and their correlations have been specified, the calculation can be carried out, using either a first- or second-order Taylor series expansion. The output is a map of cell means and a map of variances.

5.2 Examples

Two examples of the method of error propagation analysis are given. The first is a simple example in which an external model is used (often uncritically) to calculate the value of a derived variable, as happens frequently in GIS analysis. The second example explores the situation where the model coefficients need to be estimated and it examines the relative balance of errors that accrue from the model and from the spatial variation of the inputs.

5.2.1 Case study 1

The floodplain of the Geul Valley, in the south of The Netherlands, is strongly polluted by heavy metals deposited with the stream sediments (Leenaers 1989). The degree of the threat to public health from the heavy metal pollution is determined both by the actual concentration of single elements and the combination of elements present. Figure 4.2 shows the location of the area. The soil was sampled at depths of 0–20 cm, 50–60 cm and 100-110 cm at 74 sites and was analysed for concentrations of heavy metals. Figures 4.3a,b,c,d show maps of the mean and relative error of the 0–20 cm layer (topsoil) for concentrations of lead (Pb) and cadmium (Cd) obtained by point kriging (Leenaers 1989).

Locally, there is a demand for land for growing vegetables. Epidemiological research suggests that large concentrations of heavy metals in the soil can yield crops that, if consumed, might endanger public health, especially that of children between one and five years old. Public health authorities in the region use simple models to assess the overall level of danger to public health caused by the cocktail of heavy metals in the soil. The model used assumes that the cadmium is 13 times more harmful than lead. Obviously there is an error (bias) associated with the value 13, but we are not able to explore this here. Areas with the highest risk can be found from a map of a new attribute of soil risk, R, which is defined as

$$R = \mathrm{Pb} + 13 \times \mathrm{Cd} \tag{4.11}$$

In this area the correlation coefficient between Pb and Cd was estimated as 0.84. Maps of the mean and relative error or R (Figure 4.3e and Figure 4.3f) were derived from the maps of lead and cadmium. The relative error of R is of the same order of

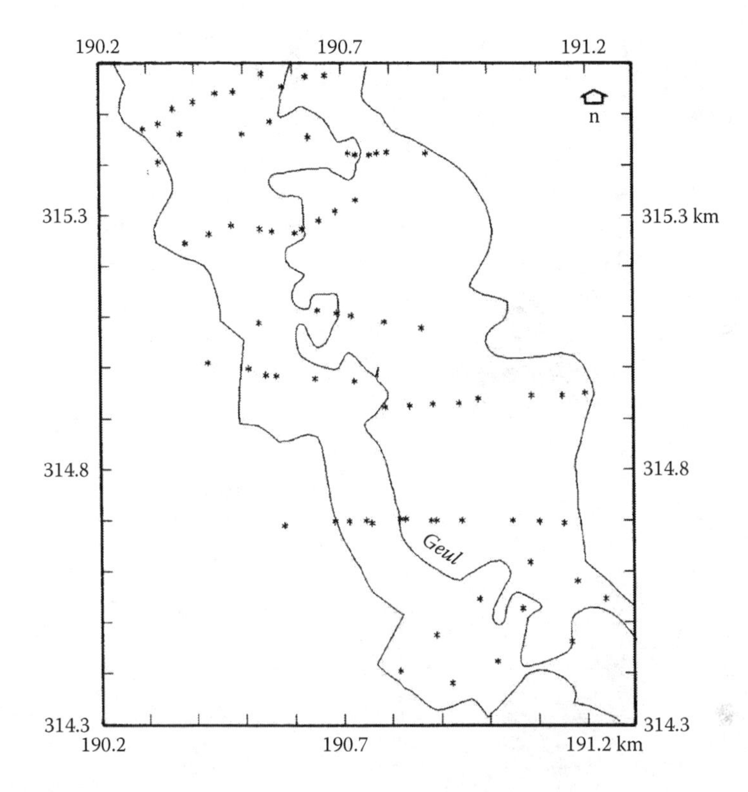

FIGURE 4.2 The Geul study area showing sampling points.

magnitude as that of Pb and Cd. This is because of the large positive correlation between the inputs. If the correlation between Pb and Cd had been lower or negative, the relative error in R would have been smaller because the errors in Pb and Cd would neutralize each other. The large positive correlation found here occurs because both heavy metals are deposited by the same fluvial process. The maps of the mean levels and relative error of R allow the probability that the true summed concentration exceeds a prescribed maximum level to be assessed, which provides a rational basis for deciding where land for vegetable growing can best be situated.

5.2.2 Case study 2 — analysing the errors associated with estimating the values of soil-water inputs for the WOFOST crop yield model at a site near Allier in central France

As part of a research study in quantitative land evaluation, the WOFOST crop simulation model (van Diepen *et al.* 1987, 1989) was used to calculate potential crop yields for river floodplain soils in the Limagne rift valley, near Allier in France. The WOFOST model was run at a number of sample sites from which maps of the variation of potential yield were obtained by optimal interpolation (Weterings 1988, Stein *et al.* 1989). The moisture content at wilting point (Θ_{wp}) is an important input parameter for the WOFOST model. Because Θ_{wp} varies considerably over the area

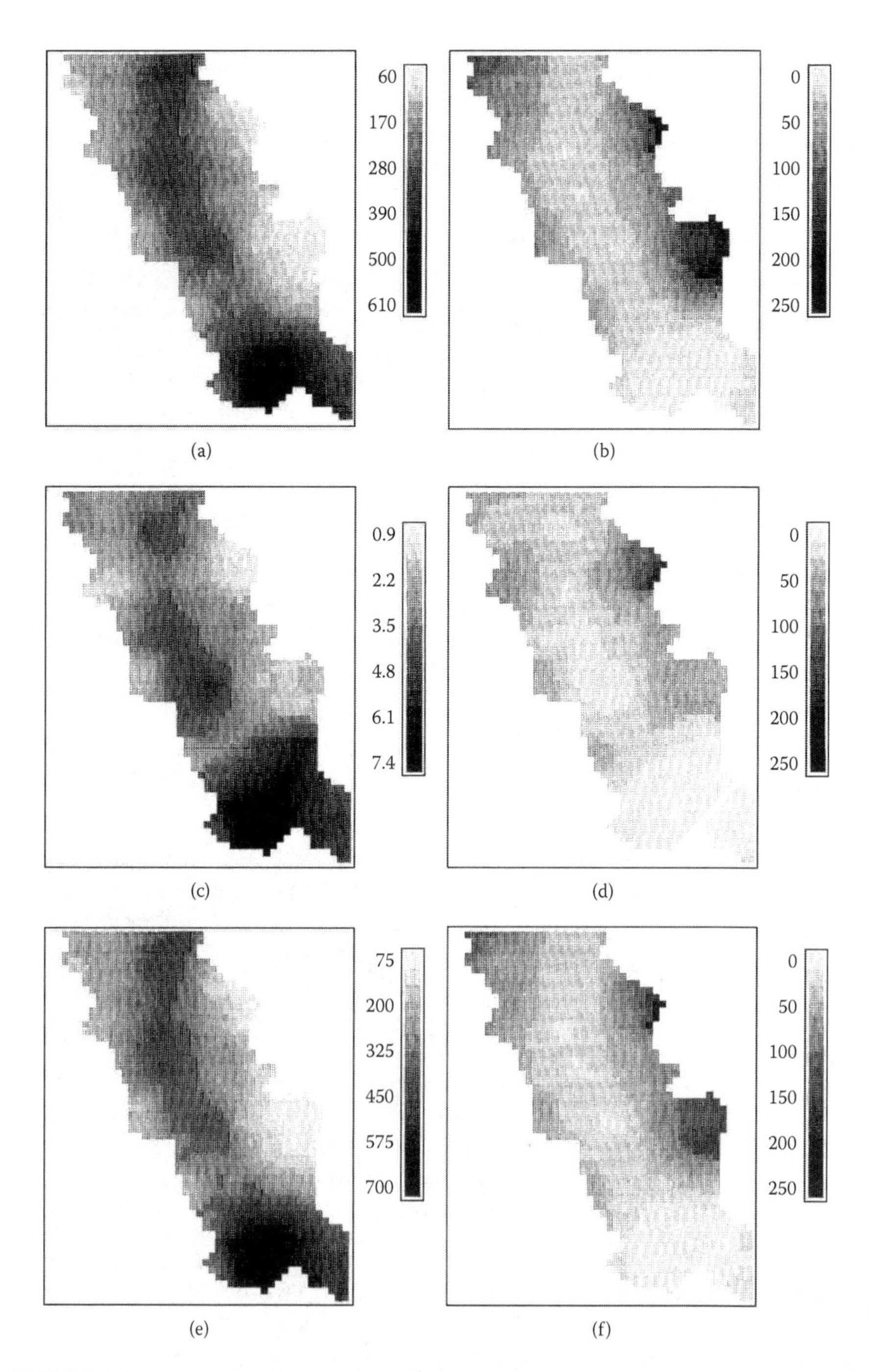

FIGURE 4.3 Heavy metal concentration in the topsoil (0–20 cm) and risk factor, *R*, of the Geul study area: (a) kriging map of lead (ppm), (b) relative error map of lead predictions (× 100 per cent), (c) kriging map of cadmium (ppm), (d) relative error map of cadmium predictions (× 100 per cent), (e) map of calculated risk factor, *R*, and (f) relative error map of *R* (× 100 per cent).

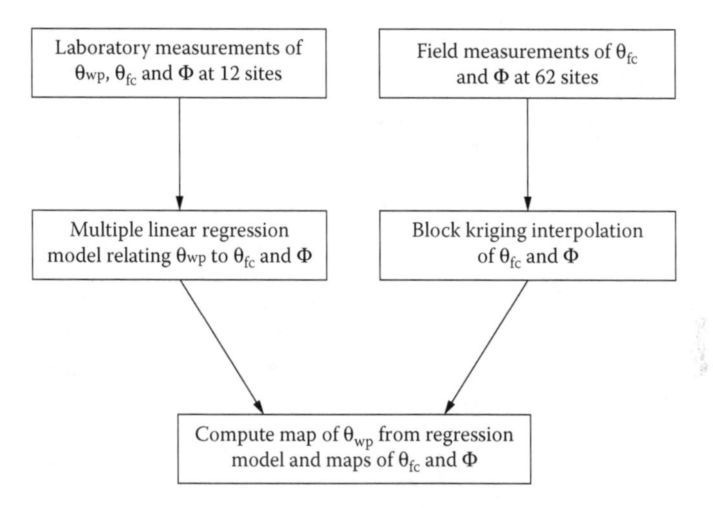

FIGURE 4.4 Flowchart of mapping procedure for Θ_{wp}.

in a way that is not linked directly with soil type, it was necessary to map its variation separately to see how moisture limitations affect calculated crop yield.

Unfortunately, because Θ_{wp} must be measured on samples in the laboratory, it is expensive and time-consuming to determine it for a sufficiently large number of data points for mapping. An alternative and cheaper strategy is to calculate Θ_{wp} from other attributes which are cheaper to measure. Because the moisture content at wilting point is often strongly correlated with the moisture content at field capacity (Θ_{fc}) and the bulk density (BD) of the soil — both of which can be measured more easily with acceptable precision — it was decided to investigate how errors in measuring and mapping these data would work through to a map of calculated Θ_{wp}. In this study we report the results in terms of the porosity (Φ) which was determined from laboratory measurements of the bulk densities of field samples as $\Phi = (1 - BD/2.65)$. This conversion has no effect on the results of the study of error propagation reported here and was made as part of the study of land evaluation (Weterings 1988).

The following procedure, illustrated in Figure 4.4, was used to obtain a map of calculated Θ_{wp} and the associated error surface.

(1) The properties Θ_{wp}, Θ_{fc} and Φ were determined in the laboratory for samples taken from the topsoil (0–20 cm) at 12 selected sites shown as the circled points in Figure 4.5; the results are given in Table 4.1.

(2) These results were used to set up an empirical model relating Θ_{wp} to Θ_{fc} and Φ, which took the form of a multiple linear regression:

$$\Theta_{wp} = \beta_1 + \beta_2 \Theta_{fc} + \beta_3 \Phi + \epsilon \tag{4.12}$$

The coefficients β_1, β_2 and β_3 of this regression equation were estimated using the Statgraphics package (Statgraph 1987). The values for the regression coefficients

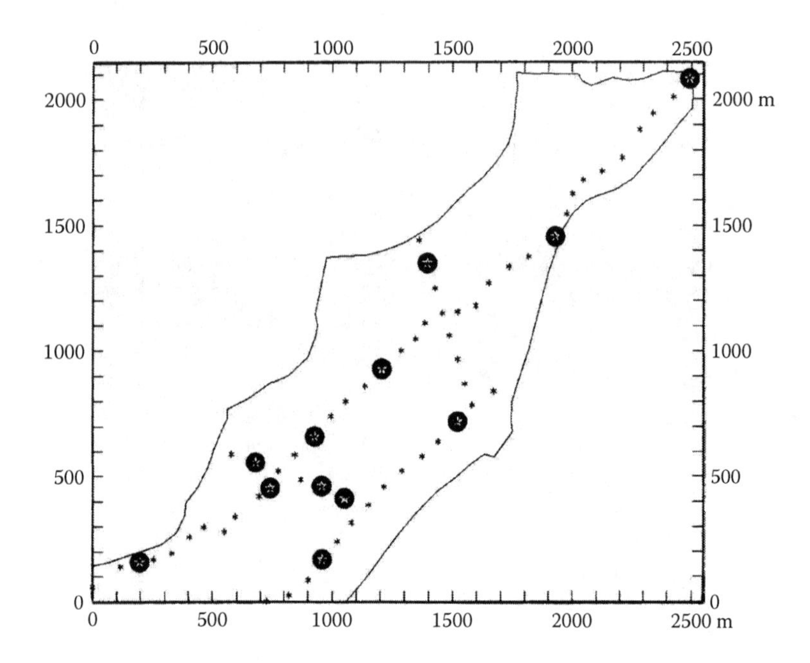

FIGURE 4.5 The Allier area showing sampling points. Circled sites are those used to estimate the regression model coefficients.

TABLE 4.1
Laboratory results of Θ_{wp}, Θ_{fc} and Φ at 12 selected sites (0–20 cm). All units are cm³/cm³.

Point	Θ_{wp}	Θ_{fc}	Φ
1	0·072	0·272	0·419
2	0·129	0·369	0·491
3	0·189	0·392	0·566
4	0·103	0·334	0·464
5	0·086	0·304	0·453
6	0·114	0·328	0·532
7	0·205	0·363	0·634
8	0·199	0·451	0·566
9	0·108	0·299	0·509
10	0·103	0·337	0·491
11	0·112	0·318	0·509
12	0·103	0·337	0·479

and their respective standard errors were found to be $\beta_1 = -0.263 \pm 0.031$, $\beta_2 = 0.408 \pm 0.096$, $\beta_3 = 0.491 \pm 0.078$: the standard deviation of the residual ϵ was 0.0114. As can be seen from these figures, the relative error of the model parameters

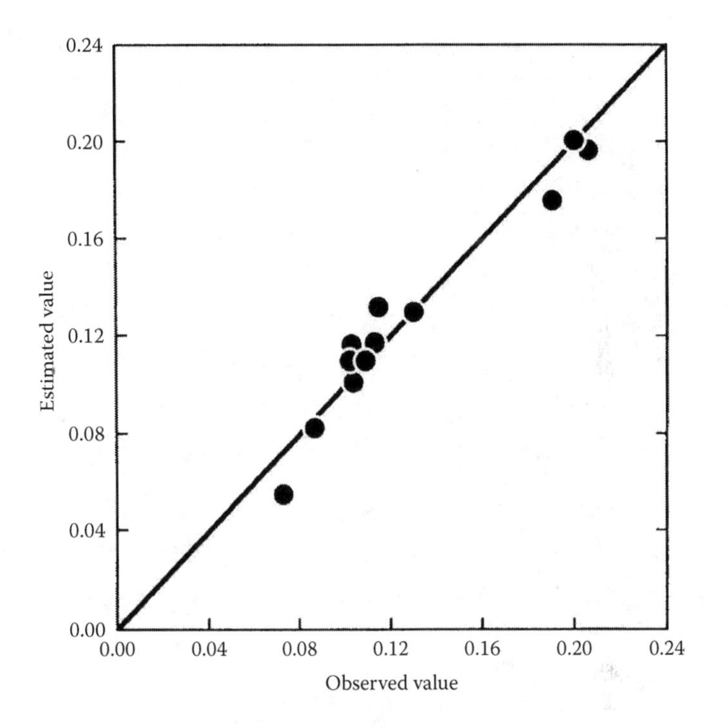

FIGURE 4.6 Observed values versus the estimated values of Θ_{wp} using the multiple regression model $\Theta_{wp} = \beta_1 + \beta_2\,\Theta_{fc} + \beta_3\,\Phi$ (cm^3/cm^3).

vary from 12 to 24 per cent. The correlation coefficients between these coefficients are $\tau_{12} = -0.22$, $\tau_{13} = -0.59$, $\tau_{23} = -0.66$. The goodness-of-fit of this regression (R^2 is 93.6 per cent) indicates that the model is satisfactory (see Figure 4.6). Note that these estimates of the standard errors of the model parameters are probably a slight underestimate because spatial covariance was not taken into account.

(3) Sixty-two measurements of Θ_{fc} and Φ were made in the field at the sites indicated in Figure 4.5. For the purposes of this study the input data for the regression model were mapped to a regular 50 m × 50 m grid using block kriging with a block size of 50 m × 50 m. The block kriging yielded raster maps of means and variances for both Θ_{fc} and Φ. Figure 4.7 displays the maps of means, together with maps of the relative error. For Θ_{fc} the relative error varies from 0 to 21 per cent, for Φ from 0 to 11 per cent. In the neighbourhood of the sampled points the error is comparatively small.

(4) The maps of the calculated values and relative errors of Θ_{wp} were obtained by means of the MAPCALC2 program using the regression model given above. The standard errors of, and the correlations between, the regression coefficients β_1, β_2 and β_3 were taken into account. Because the inputs to the error propagation are averages for blocks that are much larger than the sample support, under the assumption of spatial independence, the residual errors ϵ cancel out, and therefore were

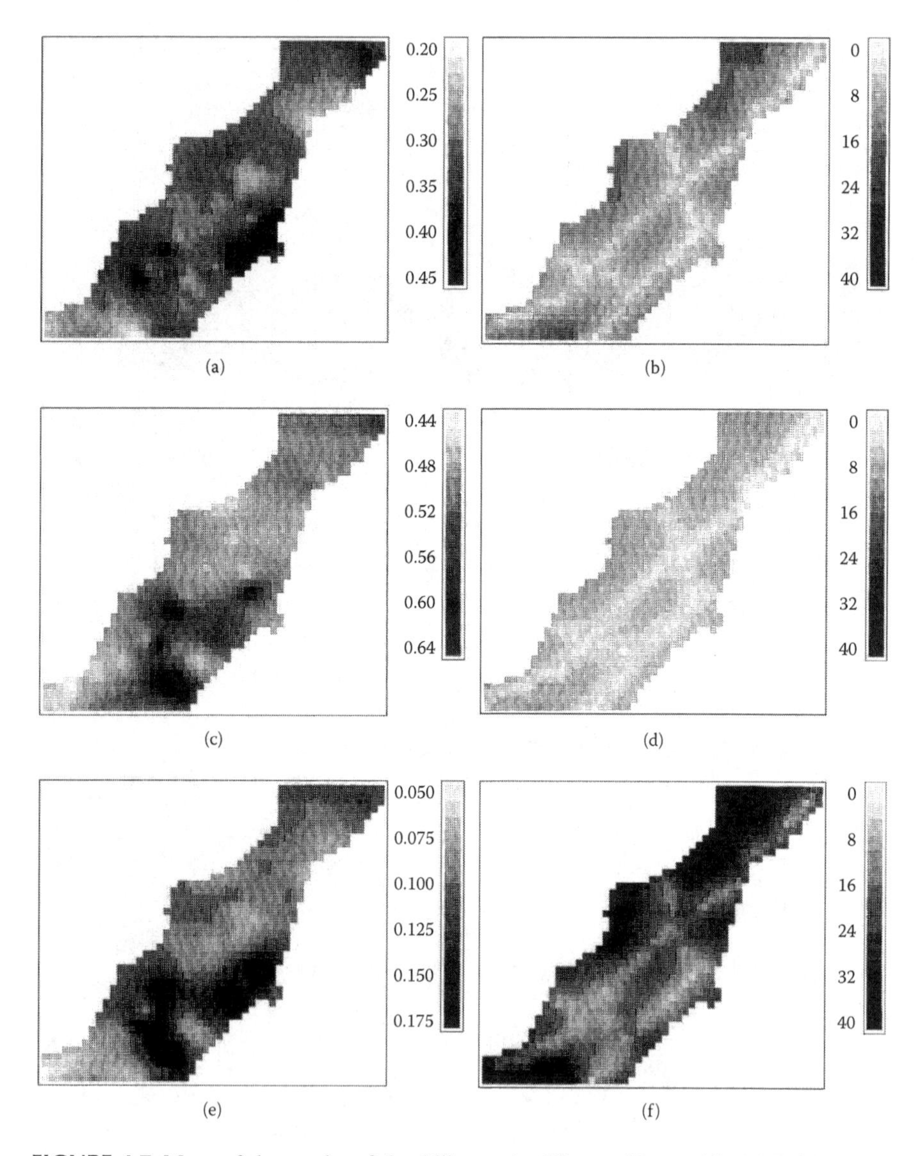

FIGURE 4.7 Maps of the results of the Allier study (50 m × 50 m grid). (a) Kriging map (cm^3/cm^3) and (b) relative error map (× 100 per cent) of soil moisture content at field capacity (c) Kriging map (cm^3/cm^3) and (d) relative error map (× 100 per cent) of soil porosity (e) Map of calculated moisture content at wilting point (cm^3/cm^3) obtained using the regression model. (f) Relative error map of the calculated moisture content surface (× 100 percent).

not further included in the error propagation analysis. There was no need to use a Taylor series expansion of the model, since equation (4.12) is a second-order polynomial. Using the field measurements, the correlation between Θ_{fc} and Φ was estimated as 0.09. Because the model parameters and the field measurements were determined independently the correlation between them is by definition zero. From the resulting maps (maps of means and variances of Θ_{wp}) a map of the relative error was calculated. The map of means and of the relative error are given in Figure 4.7e and Figure 4.7f.

(5) The relative contributions to the error in Θ_{wp} from (*a*) the regression model and (*b*) the spatial variation of the two input variables were determined, making use of the partitioning property of equation (4.10). When there is no correlation between the parts, the variance of Θ_{wp} is the sum of separate parts, each of which due to the model or the moisture content at field capacity, or to the porosity. Here we assumed that the correlation between the input maps was zero, instead of 0.09. Figure 4.8 presents results which show that both Θ_{fc} and Φ form the main source of error. Only in the immediate vicinity of the data points was the model the largest source of uncertainty, as would be expected because there the kriging variances of Θ_{fc} and Φ are the smallest.

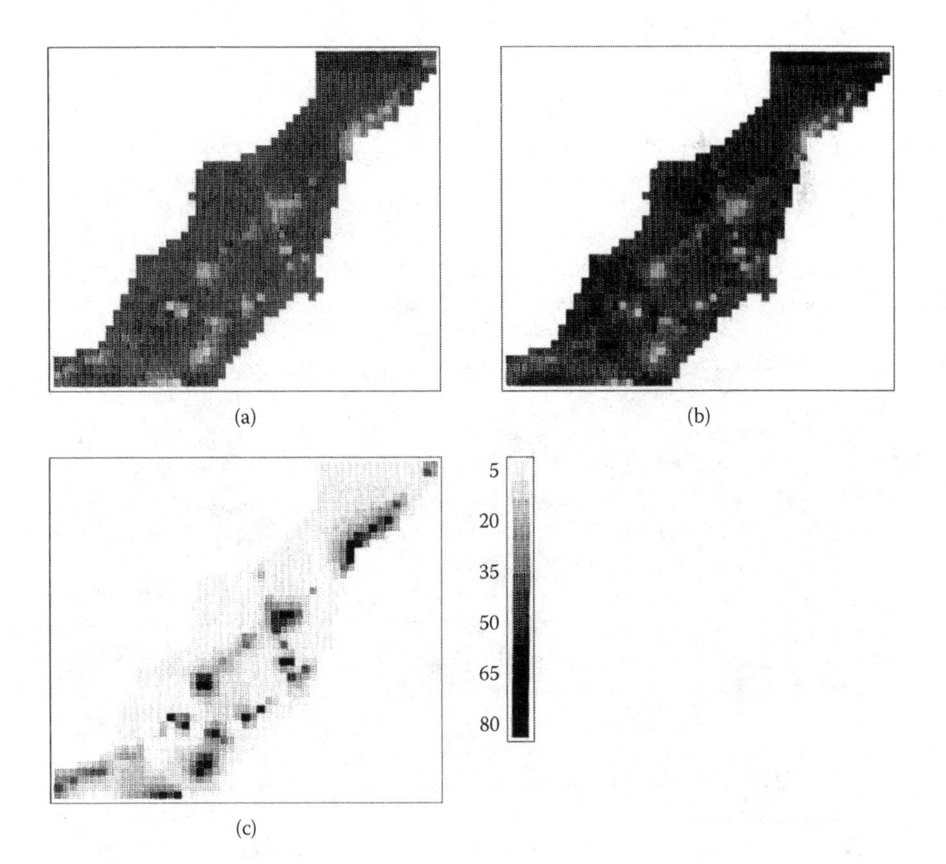

FIGURE 4.8 Maps showing the relative contributions (per cent) of the different inputs to the variance of calculated moisture content at wilting point Θ_{wp}: (a) due to Θ_{fc}, (b) due to Φ and (c) due to model.

5.2.2.1 Conclusions from this case study

The reliability of the map of Θ_{wp} is reasonable: the relative error never exceeds 40 per cent. The errors associated with this map could be used as the basis of a sensitivity analysis of the WOFOST model with respect to errors in Θ_{wp}. If the sensitivity analysis showed that the errors of Θ_{wp} cause large deviations in the output of WOFOST, then ways to improve the accuracy of the estimates for Θ_{wp} should be sought.

The main source of the error in Θ_{wp} is that associated with the spatial variation of Φ and Θ_{fc}. If the reliability of the map of Θ_{wp} needs to be improved, then it can best be done by improving the maps of Φ and Θ_{fc}, by taking more measurements over the study area. The semivariograms of Φ and Θ_{fc} could be used to assist in optimizing sampling (McBratney and Webster 1981).

6 Discussion and Conclusions

This paper has demonstrated the application of the theory of error propagation for continuously distributed random variables to the quantitative analysis of gridded data in raster geographical information systems. The method is limited to models that consist of sets of continuous and differentiable functions or regression-type equations. In the case studies presented here, the input data were maps of quantitative properties in which mean attribute values had been obtained for each grid cell by kriging from point measurements. In other situations where fewer spatial data are available it might only be possible to use single estimates of means and standard deviations for all cells belonging to a given polygon, but the principle is the same. It is essential, however, that the input maps contain measurable quantitative attributes and not classified qualitative attributes. The procedures illustrated here show how the theory and practice of methods of geostatistical interpolation can be used to advantage in a GIS.

For computational reasons it is necessary to approximate the model by a Taylor expansion, which is of course also a potential source of error. Our experience to date suggests that a second-order expansion may be sufficient for most purposes, given that the order of magnitude of the relative errors of the model is often of more interest than exact estimates of those errors. Even a first-order expansion can give a good idea of the general levels and distribution of the sources of error; this conclusion will be clarified by further research. The thinking behind the approach we have presented here is that users of GIS and models should be able to obtain an indication of the intrinsic quality of model results that transcends those currently available, but the present approach makes no attempt to be exhaustive.

In the second case study presented here, we deliberately chose a data set in which the data for formulating and calibrating the model and the data from which the variables were mapped had been collected independently by the same staff. This was done to illustrate how the uncertainties associated with the model coefficients and the inputs can both be handled. Unfortunately, in many practical situations the user has no information about the errors associated with model coefficients. Empirical models are often taken over from work in other geographical areas using data that have at best an uncertain pedigree. In many practical applications of empirical models the user will not be able to formulate and calibrate area-specific values of the coefficients but will have to make do with relationships that have been established elsewhere. If the uncertainties associated with the model coefficients are unknown, then the relative contribution of the model to the total error in the output map is also unknown. Our procedure allows a user to use 'eyeball' estimates of the standard errors of the model coefficients should these not be available from actual measurements, but informed guesses are no substitute for real values. Accordingly, we strongly recommend that in the future all developers of empirical models publish not only the values of the coefficients, but also the associated standard errors, together with information about the data sets on which the models are based so that the contributions of model error can be correctly assessed.

This paper has not addressed several important aspects of the error problem. First we have assumed that the errors estimated for the attribute values at each cell are themselves free of uncertainties. This is clearly not so and there is much work to be done to assess the integrity of the geostatistical techniques used to estimate the error surfaces. Studies of the errors associated with estimating and fitting the variogram are particularly relevant here. Secondly, no attention has been paid to errors associated with type 2 models and both topics will be objects of future research.

The ability to examine how various sources of errors contribute to the errors in maps of model outputs is potentially of great value for resource managers who need to get reliable information for realistic budgets. Error analysis shows which inputs need to be measured more accurately and how the quality of the output will improve. Similarly, and by no means less importantly, it is also possible to determine which of the inputs is known with too high precision; decreasing the number of measurements of such an attribute may substantially reduce costs without causing a deterioration of the intrinsic quality of the output. Applying error analysis to different models can reveal which approach is most suitable for any given situation. It could demonstrate clearly whether or not benefits are likely to accrue from using more complex models that require large quantities of data in situations where the spatial variation is large or where the data themselves are highly unreliable.

Acknowledgments

The authors are indebted to Ms M. H. W. Weterings of the Department of Soil Science, Agricultural University, Wageningen for supplying the Allier data set, to Dr H. Leenaers (University of Utrecht) for supplying the Geul data and to Dr J. J. de Gruijter, of Het Onderzoeksinstituut voor het Landelijk Gebied, Wageningen, for constructive criticism. This research was supported by the Netherlands Technology Foundation (STW).

Appendix

Error propagation using a second-order Taylor expansion

In the case of a second-order Taylor expansion we have instead of equation (4.7),

$$
\begin{aligned}
Y = g(Z) = g(\boldsymbol{\mu}) &+ \sum_{i=1}^{n}\left\{(Z_i - \mu_i)\left(\frac{\partial g}{\partial z_i}(\boldsymbol{\mu})\right)\right\} \\
&+ \frac{1}{2}\sum_{i=1}^{n}\sum_{j=1}^{n}\left\{(Z_i - \mu_i)(Z_j - \mu_j)\left(\frac{\partial^2 g}{\partial z_i \partial z_j}(\boldsymbol{\mu})\right)\right\} + \text{rest term}
\end{aligned}
\tag{4.A1}
$$

If the rest term of (4.A1) is neglected, then the mean and variance of Y can be determined for the case in which the Z_i are *jointly normally distributed* thus:

$$
\begin{aligned}
\mu = E[Y] = E[g(\boldsymbol{\mu})] &+ E\left[\sum_{i=1}^{n}\left\{(Z_i - \mu_i)\left(\frac{\partial g}{\partial z_i}(\boldsymbol{\mu})\right)\right\}\right] \\
&+ \frac{1}{2}E\left[\sum_{i=1}^{n}\sum_{j=1}^{n}\left\{(Z_i - \mu_i)(Z_j - \mu_j)\left(\frac{\partial^2 g}{\partial z_i \partial z_j}(\boldsymbol{\mu})\right)\right\}\right] \\
&= g(\boldsymbol{\mu}) + \frac{1}{2}\sum_{i=1}^{n}\sum_{j=1}^{n}\left\{\tau_{ij}\sigma_i\sigma_j\left(\frac{\partial^2 g}{\partial z_i \partial z_j}(\boldsymbol{\mu})\right)\right\}
\end{aligned}
\tag{4.A2}
$$

$$
\begin{aligned}
\sigma^2 = E[(Y - E[Y])^2] = E&\left[\left(\sum_{k=1}^{n}\left\{(Z_k - \mu_k)\left(\frac{\partial g}{\partial z_k}(\boldsymbol{\mu})\right)\right\}\right.\right. \\
&\left.\left. + \frac{1}{2}\sum_{i=1}^{n}\sum_{j=1}^{n}\left\{((Z_i - \mu_i)(Z_j - \mu_j) - \tau_{ij}\sigma_i\sigma_j)\left(\frac{\partial^2 g}{\partial z_i \partial z_j}(\boldsymbol{\mu})\right)\right\}\right)^2\right] \\
= &\sum_{k=1}^{n}\sum_{l=1}^{n}\left\{\tau_{kl}\sigma_k\sigma_1\left(\frac{\partial g}{\partial z_k}(\boldsymbol{\mu})\right)\left(\frac{\partial g}{\partial z_l}(\boldsymbol{\mu})\right)\right\} \\
&+ \sum_{k=1}^{n}\sum_{i=1}^{n}\sum_{j=1}^{n}\left\{E[(Z_k - \mu_k)((Z_i - \mu_i)(Z_j - \mu_j) - \tau_{jk}\sigma_i\sigma_j)]\left(\frac{\partial g}{\partial z_k}(\boldsymbol{\mu})\right)\right.
\end{aligned}
$$

$$\times\left(\frac{\partial^2 g}{\partial z_i \partial z_j}(\boldsymbol{\mu})\right)\Bigg\}+\frac{1}{4}\sum_{i=1}^{n}\sum_{j=1}^{n}\sum_{k=1}^{n}\sum_{l=1}^{n}\Big\{(E[(Z_i-\mu_i)(Z_j-\mu_j)$$

$$\times(Z_k-\mu_k)(Z_l-\mu_l)-\tau_{ij}\sigma_i\sigma_j\sigma_{kl}\sigma_k\sigma_l)]\left(\frac{\partial^2 g}{\partial z_i \partial z_j}(\boldsymbol{\mu})\right)\left(\frac{\partial^2 g}{\partial z_k \partial z_l}(\boldsymbol{\mu})\right)\Bigg\}$$

which, using properties of the normal distribution (Parzen 1962, pp. 93–95), reduces to

$$\sigma^2=\sum_{k=1}^{n}\sum_{l=1}^{n}\left\{\tau_{kl}\sigma_k\sigma_l\left(\frac{\partial g}{\partial z_k}(\boldsymbol{\mu})\right)\left(\frac{\partial g}{\partial z_l}(\boldsymbol{\mu})\right)\right\} \tag{4.A3}$$

$$+\frac{1}{4}\sum_{i=1}^{n}\sum_{j=1}^{n}\sum_{k=1}^{n}\sum_{l=1}^{n}\left\{(\tau_{ik}\sigma_i\sigma_k\tau_{jl}\sigma_j\sigma_l+\tau_{il}\sigma_i\sigma_l\tau_{jk}\sigma_j\sigma_k)\left(\frac{\partial^2 g}{\partial z_i \partial z_j}(\boldsymbol{\mu})\right)\left(\frac{\partial^2 g}{\partial z_k \partial z_l}(\boldsymbol{\mu})\right)\right\}$$

References

Beasley, D. B., and Huggins, L. F., 1982, *ANSWERS — User's Manual* (West Lafayette: Department of Agricultural Engineering, Purdue University).

Berry, J. K., 1987, Fundamental operations in computer-assisted map analysis. *International Journal of Geographical Information Systems*, **1**, 119.

Bouma, J., and van Lanen, H. A. J., 1987, Transfer functions and threshold values: from soil characteristics to land qualities. In *Quantified Land Evaluation Procedures*, edited by K. J. Berry, P. A. Burrough and D. E. McCormack, ITC Publication No. 6 (Enschede: ITC), pp. 106–110.

Burgess, T. M., and Webster, R., 1980 a, Optimal interpolation and isarithmic mapping I. The semivariogram and punctual kriging. *Journal of Soil Science,* **31**, 315.

Burgess, T. M., and Webster, R., 1980 b, Optimal interpolation and isarithmic mapping II. Block kriging, *Journal of Soil Science*, **31**, 333.

Burrough, P. A., 1986, *Principles of Geographical Information Systems for Land Resources Assessment* (Oxford: Clarendon Press).

Burrough, P. A., 1989, Matching spatial databases and quantitative models in land resource assessment. *Soil Use and Management*, **5**, 3.

Chrisman, N. R., 1984, The role of quality information in the long-term functioning of a geographical information system. *Cartographica*, **21**, 79.

Davis, J. C., 1986, *Statistics and Data Analysis in Geology*, second edition (New York: John Wiley & Sons).

Dettinger, M. D., and Wilson, J. L., 1981, First order analysis of uncertainty in numerical models of groundwater flow. Part I. Mathematical development. *Water Resources Research*, **17**, 149.

Diepen, C. A. van, Rappoldt, C., Wolf, J., and Keulen, H. van, 1987, CWFS crop growth simulation model WOFOST version 4.1. Staff working paper SOW-87-0, Centre for World Food Studies, Wageningen, The Netherlands.

Diepen, C. A. van, Berkhout, J., and Keulen, H. van, 1989, WOFOST: a simulation model of crop production. *Soil Use and Management*, **5**, 16.

Drummond, J., 1987, A framework for handling error in geographic data manipulation. *ITC Journal,* 1987–**1**, 73.

Dumanski, J., and Onofrei, C., 1988, Crop yield models for agricultural land evaluation. *Soil Use and Management*, **5**, 9.

Food and Agriculture Organization (FAO), 1976, *A Framework for Land Evaluation,* Soils Bulletin No. 32 (Rome: United Nations FAO).

Goodchild, M. F., and Dubuc, O., 1987, A model of error for choropleth maps, with applications to geographic information systems. *Proceedings of Auto-Carto 8, held in Baltimore, Maryland, on 29 March–3 April 1987* (Falls Church, Virginia: American Society Photogrammetry and Remote Sensing, American Congress of Surveying and Mapping), pp. 162–172.

Goodchild, M. F., and Min-hua, W., 1988, Modelling error in raster-based spatial data. *Proceedings of the Third International Symposium on Spatial Data Handling, held in Sydney, Australia, on 17–19 August 1988*, pp. 97–106.

Journel, A. G., and Huijbregts, Ch. J., 1978, *Mining Geostatistics* (New York: Academic Press).

Jousma, G., and Heyde, P. van der, 1985, A review of numerical modelling of groundwater flow and pollution. International Groundwater Modelling Center, Delft, The Netherlands and Indianapolis, U.S.A.

Leenaers, H., 1989, Deposition and storage of heavy metals in river flood-plains. *Environmental Monitoring and Assessment* (in the press).

Marsman, B., and de Gruijter, J. J., 1984, Dutch soil survey goes into quality control. In *Soil Information Systems Technology*, edited by P. A. Burrough and S. W. Bie (Wageningen: PUDOC), pp. 127–134.

McBratney, A. B., and Webster, R., 1981, The design of optimal sampling schemes for local estimation and mapping of regionalized variables. II Program and Examples. *Computers and Geosciences*, **7**, 335.

McDonald, M. G., and Harbaugh, A. W., 1984, *A Modular Three-Dimensional Finite-Difference Ground-Water Flow Model* (Washington, D.C.:U.S. Department of the Interior, U.S. Geological Survey, and Scientific Publications).

Mead, D. A., 1982, Assessing data quality in geographic information systems. In *Remote Sensing for Resource Management*, edited by C. J. Johannsen and J. L. Sanders (Ankeny, Iowa: Soil Conservation Society of America), pp. 51–62.

Meeker, W. Q., Cornwell, L. W., and Aroian, L. A., 1980, The product of two normally distributed random variables. In *Selected Tables in Mathematical Statistics*, Vol. 7.

National Center for Geographic Information and Analysis (NCGIA), 1988, National Center for Geographical Information and Analysis research proposals. NCGIA, University of Santa Barbara, California, U.S.A.

Parratt, L. G., 1961, *Probability and Experimental Errors* (New York: John Wiley & Sons).

Parzen, E., 1962, *Stochastic Processes* (San Francisco: Holden-Day, Inc.).

Rijksuniversiteit Utrecht, 1987, PC-Geostat manual. Department of Physical Geography, University of Utrecht internal publication, Utrecht, The Netherlands.

Schultink, G., 1987, The CRIES resource information system: computer-aided spatial analysis of resource development potential and development policy alternatives. In *Quantified Land Evaluation Procedures*, edited by K. J. Beek, P. A. Burrough and D. E. McCormack, ITC Publication No. 6 (Enschede: ITC), pp. 95–99.

Sokolnikoff, I. S., and Sokolnikoff, E. S., 1941, *Higher Mathematics for Engineers and Physicists* (New York: McGraw-Hill).

Statgraph, 1987, Statgraphics Manual (Statistical Graphics Corporation).

Stein, A., Hoogerwerf, M., and Bouma, J., 1988, Use of soil map delineations to improve (co)kriging of point data on moisture deficits. *Geoderma,* **43**, 163.

Stein, A., Bouma, J., Mulders, M. A., and Weterings, M. H. W., 1989, Using cokriging in variability studies to predict physical land qualities of a level river terrace. *Soil Technology* (submitted).

Tomlin, C. D., 1983a, Digital cartographic modelling techniques in environmental planning. Unpublished Ph.D. dissertation, Yale University, Connecticut, U.S.A.

Tomlin, C. D., 1983b. A map algebra. *Proceedings Harvard Computer Conference 1983, held in Cambridge, Massachusetts, on 31 July–4 August,* Laboratory for Computer Graphics and Spatial Analysis, Harvard.

TYDAC, 1987, *Spatial Analysis System Reference Guide, Version 3.6* (Ottawa: TYDAC Technologies Inc.).

Valenzuela, C., 1988, The integrated land watershed management information system (ILWIS). In *Status report, Geo Information System for Land Use Zoning and Watershed Management,* edited by A. M. J. Meijerink, C. R. Valenzuela and A. Stewart, ITC Publication No. 7 (Enschede: ITC), pp. 3–14.

Webster, R., 1985, Quantitative spatial analysis of soil in the field. *Advances in Soil Science,* Vol. 3 (New York: Springer-Verlag).

Weterings, M. H. W., 1988, Variabiliteit van een landkarakteristiek en van gesimuleerde gewasopbrengsten binnen een rivierterras van de Allier in de Limagneslenk, Frankrijk. Students' Report of the Department of Soil Science and Geology, Agricultural University Wageningen, The Netherlands.

Developments in Analysis of Spatial Uncertainty Since 1989

Gerard B.M. Heuvelink, Peter A. Burrough, and Alfred Stein

When our paper on spatial error propagation was published in 1989, it proved to be well timed. It was then that the geographical information system (GIS) was transformed from a tool for specialists to a routine instrument for all users of geographical information. GIS offered the possibility of analysing and manipulating spatial data in a way that had not been possible before. MAP algebra and coupling of GIS with environmental models allowed almost unlimited creation of new maps from existing and previously created maps. However, digital databases and presentation on the computer screen created a sense of perfection and absoluteness about the maps produced. We warned against such mistaken beliefs and argued that there are virtually no error-free spatial data, and that errors propagate in spatial modelling. We claimed that it is of crucial importance that error propagation in spatial modelling is traced to assess the quality of modelling results. The point was well taken by the readers of the journal. This, together with the fact that we presented a practical error propagation method and demonstrated its potential with realistic case studies, explains the success of the paper.

In the more than 15 years that have passed since the paper was published, spatial accuracy assessment has, in fact, never disappeared from the research agenda. New audiences also recognised the importance of the subject and statements that accuracy assessment and error propagation analysis need attention are made regularly in the geographical information (GI) literature. The interest is clearly there, but spatial error propagation analysis is not a routine exercise within a GIS environment. Much progress has been made since 1989, both in theory and application, as we will review below, but the degree with which error propagation analysis has been integrated and employed in GI systems and science is disappointing. For this we also attempt to give an explanation.

Theoretical developments in spatial error propagation analysis

The most striking theoretical development is the advancement of numerical simulation or Monte Carlo techniques. In our paper we only paid attention to analytical approaches to error propagation (that is, the Taylor series method), but nowadays Monte Carlo methods have almost completely taken over (for an exception, see Arbia et al., 2003). This is not only because computers have become more powerful, which is to the advantage of the computationally demanding numerical approaches, but also because spatial models executed within a GIS have become more complex. This renders analytical methods much less attractive because they become cumbersome and need to make use of simplifying approximations. In 1989 there were no proper and efficient techniques for generating realisations of spatially correlated random fields, but now we have various kinds of sequential simulation algorithms (Deutsch and Journel, 1992). More recent simulation techniques, summarised under the name Markov Chain Monte Carlo, offer great flexibility in simulating from complex probability distribution functions, even if the probability distribution is not explicitly defined. Our objection against simulation methods — that "these methods do not satisfy an analytical form and the results cannot be transferred to new situations" (Heuvelink, Burrough, and Stein, this volume, Section 4.4.1) — remains valid, but with sufficient computer resources one can simply redo the simulation for each new situation. Monte Carlo simulation is perfectly suited for running on parallel computers and can benefit much from GRID computing methodology (see http://www.gridcomputing.org).

In our paper we briefly reviewed the most basic geostatistical model of spatial dependence. Much progress has been made in this area, which has led to a much richer set of geostatistical models. For example, models have been developed that make use of information derived from auxiliary variables (i.e., regression kriging), which can simulate spatial objects rather than fields, or can deal with categorical variables (that is, indicator geostatistics, stochastic cellular automata). In our paper we anxiously stayed away from including uncertain categorical data because "the desirable properties of the continuous case are not directly transferable to the discrete case" (Heuvelink, Burrough, and Stein, this volume, Section 4.4). Indeed, it is true that for discretely distributed categorical variables, many more parameters are needed to characterise the probability distribution, but once this is done, error propagation analysis follows much the same procedure as for continuously distributed variables (Kros et al., 1999).

An interesting aspect of the original paper is that it includes model error as a source of uncertainty, and that it examines how various sources of errors contribute to the errors in the output. Quantification of model error remains a difficult subject, although progress has been made here as well (Beven, 2000). Specific algorithms to efficiently estimate the contribution of different error sources in a Monte Carlo framework have also been developed (Jansen, 1999; Tarantola and Saltelli, 2003).

These theoretical developments mean that we now have the methods to analyse how error propagates in spatial modelling for a wide variety of situations — much wider than we studied at the time.

Practical developments in spatial error propagation analysis

Our paper was one of the first to present a real-world case study of spatial error propagation. The focus was on a soil study, where an important soil physical property was related to basic soil properties. The model was a simple linear regression model, and error propagation could easily reveal interesting information on how uncertainty in input variables propagates. Since then numerous case studies have been reported. For example, a recent paper related error propagation to utility values, including cost models linked in a GIS (Van Oort et al., 2005). The variety within the collection of case studies demonstrates the flexibility of spatial error propagation methodologies. Also, the many case studies published confirm the need for applied spatial error propagation analysis. An important stimulus to developments and applications has been the International Symposium on Spatial Data Quality and the Spatial Accuracy Assessment in Natural Resources and Environmental Sciences symposia series. These symposia have produced many books (i.e., Lowell and Jaton, 1999; Mowrer and Congalton, 2000; Heuvelink and Burrough, 2002; Shi et al., 2002) and continue as formally organised international events (i.e., see www.spatial-accuracy.org).

One obvious practical issue concerns the development of software for spatial error propagation analyses. Software tools typically have a much shorter life than statistical methodologies, and indeed this is confirmed here. In our paper we described the software tool MAPCALC2, which was "written in TURBO Pascal for IBM-AT and IBM-PS2 computers and equivalents" (Heuvelink, Burrough, and Stein, this volume, Section 4.5.1). This software was soon replaced by ADAM (Wesseling and Heuvelink, 1991), and in turn has been replaced by Data Uncertainty Engine (DUE) (Heuvelink et al., in press) and modules in PCRaster (Karssenberg and De Jong, 2005). Progress has been made with respect to user friendliness and flexibility of spatial error propagation software, although we are still waiting for large-scale incorporation of error propagation functions in commercial GIS. Note that IDRISI (http://www.clarklabs.org/) does have some rudimentary error propagation functionality.

Challenges

The theoretical and practical developments reviewed above make clear that we now have a much more powerful toolbox to carry out a spatial uncertainty analysis. It is indeed remarkable how often in our paper we noted the limited applicability of the methodology, by statements such as "we limit ourselves to the situation that the uncertain inputs are continuously distributed random variables" (Heuvelink, Burrough, and Stein, this volume, Section 4.4) and "[t]he method is limited to models that consist of sets of continuous and differentiable functions or regression-type equations" (this volume, Section 4.6). Nowadays we have methods and tools that apply to a much wider set of problems, including complex spatially distributed models, uncertain categorical input data, and positional error; this given, why do we see so few of the methods implemented in mainstream GIS, and why are spatial error propagation analyses still the exception rather than the rule? We think there are four main reasons for this that cannot easily be resolved.

First, in order to carry out an error propagation analysis, one must know the errors associated with the inputs to the analysis and, if a model is used, to the model itself. Most spatial data have no or only rudimentary information about the associated accuracy. Also, most models used in the earth and environmental sciences have no uncertainty information attached to them. Our plea that "in future all developers of empirical models publish not only the values of the coefficients, but also the associated standard errors, together with information about the data sets on which the models are based so that the contributions of model error can be correctly assessed" (this volume, Section 4.6) remains as valid now as then, because it is only rarely being followed. Gathering and storing information about accuracy of data and models, however, is a difficult and cumbersome activity.

Second, an error propagation analysis takes a lot of work. For example, for Monte Carlo studies the computational time and data storage requirements are increased by a factor 100 at least. Also, the manual labour involved is more than double that required for a conventional analysis, which makes an error propagation analysis very costly. Budget controllers will ask themselves whether it is worth the extra costs. In spite of the many valuable insights that one gets from a spatial error propagation analysis, it would be naïve to claim that error propagation analysis always pays off.

Third, no matter how much we would like to, it is also naïve to think that we can develop a single "error propagation button." Responsible use of error propagation analysis requires that the users have sufficient background in statistics and know what they are doing. Many of the people who now do the spatial analyses do not satisfy these requirements, and so organisations will have to look for people with the right expertise. Courses will have to be organised to train them. Again, a costly affair that must be justified against the reward that comes from the error propagation analysis.

Fourth, and maybe most importantly, the users of geographical information are not always equally interested in the uncertainty and quality of information. It only troubles their minds. They may not see its benefit and contributions to decision making. Much of the lack of enthusiasm for quantified spatial uncertainty may be due to the real difficulty of visualising and communicating uncertainty; some of it may be due to the often rather more conceptual than analytical minds of many decision makers. And, let us face it, although the equations in the appendix of the paper are really not very difficult, many users of spatial information are frightened by their appearance. However, without a strong mathematical backbone, geographical information science may be on a losing trail.

The future

In terms of scientific developments, we have seen the advance of Bayesian methods (that is, methods based on prior belief and collected data). These methods are promising when dealing with error propagation and explicit assessment of uncertainties in environmental data and models (Brown 2004). In the near future, increased use may be expected as the availability of these methods increases. They may benefit from the presence of (historical) data, which are currently available within a GIS, and from approximate information that can be obtained with approximate modelling or from remote sensing data.

We believe that error propagation analyses should leave the domain of the research environment and move toward the production and decision environments. In fact, in an environment where the same or similar analyses are done repeatedly, there may be little need to include an error propagation analysis in each and every case. One for each type of analysis will do, and will point to the weak spots and thus help to decide how to improve the procedure. We foresee that it may still take some time before this becomes standard practice, but perhaps after a range of court trials, error propagation and spatial accuracy assessment are going to be much more on the front page of users and decision makers. It can only work if users and decision makers are convinced of the importance of quantifying and visualising the uncertainty in geographic information, because it requires strong commitment to meet the challenges listed above.

References

ARBIA, G., GRIFFITH, D.A., AND HAINING, R.P., 2003, Spatial error propagation when computing linear combinations of spectral bands: the case of vegetation indices, *Environmental and Ecological Statistics*, **10**, 375–396.

BEVEN, K.J., 2000, On model uncertainty, risk and decision making, *Hydrological Processes*, **14**, 2605–2606.

BROWN, J.D., 2004, Knowledge, uncertainty and physical geography: towards the development of methodologies for questioning belief, *Transactions of the Institute of British Geographers*, **29**, 367–381.

DEUTSCH, C.V. AND JOURNEL, A.G., 1992, *GSLIB: Geostatistical Software Library and User's Guide*, Oxford University Press, New York.

HEUVELINK, G.B.M. AND BROWN, J.D., 2005. Handling spatial uncertainty in GIS: development of the Data Uncertainty Engine. In: *Proceedings GIS Planet 2005*.

HEUVELINK, G.B.M., BURROUGH, P.A., AND VAN LOON, E.E., in press. A probabilistic framework for representing and simulating uncertain environmental variables, *International Journal of Geographical Information Science*.

JANSEN, M.J.W., 1999, Analysis of variance designs for model output. *Computer Physics Communications*, **117**, 35–43.

KARSSENBERG, D. AND DE JONG, K., 2005, Dynamic environmental modelling in GIS. 2: Modelling error propagation. *International Journal of Geographical Information Science*, **19**, 623–637.

KROS, J., PEBESMA, E.J., REINDS, G.J., AND FINKE, P.F., 1999, Uncertainty assessment in modelling soil acidification at the European scale: a case study. *Journal of Environmental Quality*, **28**, 366–377.

LOWELL, K. AND JATON, A., Eds., 1999, *Spatial Accuracy Assessment. Land Information Uncertainty in Natural Resources*, Ann Arbor Press, Chelsea, MI.

MOWRER, H.T. AND CONGALTON, R.G., Eds., 2000, *Quantifying Spatial Uncertainty in Natural Resources: Theory and Applications for GIS and Remote Sensing*, Ann Arbor Press, Chelsea, MI.

SHI, W., FISHER, P.F., AND GOODCHILD, M.F., Eds., 2002, *Spatial Data Quality*, Taylor & Francis, London.

TARANTOLA, S. AND SALTELLI, A., Eds., 2003, SAMO 2001: methodological advances and innovative applications of sensitivity analysis. *Reliability Engineering & System Safety* **79**(2).

VAN OORT, P., STEIN, A., BREGT, A.K., DE BRUIN, S., AND KUIPERS, J., 2005, A variance and covariance equation for area estimates with a Geographical Information System. *Forest Science*, **51**(4), 347–356.

WESSELING, C.G. AND HEUVELINK, G.B.M., 1991, Semi-automatic evaluation of error propagation in GIS operations. In: *EGIS '91 Proceedings*, J. Harts, H.F.L. Ottens and H.J. Scholten, Eds., EGIS Foundation, Utrecht, 1228–1237.

International Journal of Geographical Information Systems,
1989, Vol. 3, No. 4, 323–334.

5 A Comparison of Techniques for Calculating Gradient and Aspect from a Gridded Digital Elevation Model

Andrew K. Skidmore

Abstract. Digital elevation data spaced on regular 30 m grid were generated over a region of moderate topography in southeast Australia. Six algorithms for calculating gradient and aspect from these data were compared. General linear regression models and the third-order finite difference methods were the most accurate.

1 Introduction

A digital elevation model (DEM) is a digital representation of a land surface in a computer, that is, a sample of elevation points for which the x (northing), y (easting) and z (elevation) values are recorded. Elevation data are often sampled irregularly, but then interpolated to a regular grid. Burrough (1986) and Skidmore (1990) refer to the technology involved with generating a random sample of irregular elevation points, and also discuss the methods for interpolating the sample to a regular grid.

Slope is defined by a plane tangent to the surface, as modelled by the DEM at a point (Burrough 1986). Slope has tow components, viz., gradient, which is the maximum rate of change in altitude, and aspect, which is the compass direction of this maximum rate of change.

Gradient and aspect are important environmental variables in many ecological models as well as in the management of natural resources. In particular, these variables have been used extensively in combination with remotely-sensed data for improving accuracies of thematic image mapping (for example, Strahler *et al.* 1978, Fleming and Hoffer 1979, Tom and Miller 1980, Justice *et al.* 1981, Hutchinson 1982, Franklin *et al.* 1986, Cibula and Nyquist 1987), as well as in geographical information systems (GIS) for modelling purposes (Johnston 1987, Skidmore 1989).

A number of methods for calculating gradient and aspect have been proposed for a DEM based on a regular grid. The aim of this study is to compare six methods which are commonly used to generate aspect and gradient from gridded DEMs.

2 Methodology

Irregularly-spaced elevation data were digitized over a 100 km^2 rectangle in southeast Australia using a New South Wales Department of Lands 1:25 000 series map ('Mount Imlay' sheet (8823-IV-S)) which conforms to the Australian National Mapping Council's standards of accuracy. Digitizing was performed on a Tektronix digitizing table with the digital elevation data being input to a VAX computer using the DIG (ANU 1988) software system. All stream lines were digitized (a total of 2115 points), except for first-order streams that did not significantly affect topography (i.e., in situations where contour lines did not significantly deviate when crossed by a stream line). Spot heights on significant hilltops and saddles were also digitized. A selection of other irregularly-spaced elevation points was taken, predominantly along the heavier lined 100 m contour intervals which were easier to follow when hand digitizing. In all 3306 spot heights were recorded.

This irregular data set was input to a program which calculated elevation values to 1 m contours, on a regular grid spaced at 30 m (Hutchinson 1989). The 30 m interval used in the DEM was selected as it matches Landsat Thematic Mapper pixels. The DEM and Landsat Thematic Mapper data were geometrically rectified and input to a geographical information system database at a grid spacing of 30 m. This database was accessed by an expert system, and forest cover types mapped (Skidmore 1989). Hutchinson's program imposes a global drainage condition which automatically removes spurious sinks, the degree of such removal being dependent on three user-defined thresholds. Data points which block drainage lines and are less than the first threshold above the drainage line are removed; with the DEM used in this study the first threshold was set to 10 m. The second threshold is the maximum difference in height of non-data (i.e., interpolated) point saddles — which may be considered as possible exits from the sink — and the sink itself. In this study the second threshold was set to 20 m. The third threshold was set to 50 m and is the maximum (ridge or saddle) elevation that a sink can 'push' through to maintain a connected stream.

Six different methods were tested to ascertain how effectively they calculated aspect and gradient from the regularly-gridded DEM. All methods are based on a three-by-three moving window which traverses the DEM.

(1) The first method defines aspect as the direction of the maximum drop (i.e., the maximum gradient) from the center pixel to the eight nearest cells (Travis *et al.* 1975, EPPL7 1987), that is,

$$\text{Gradient} = \max((z_{i,j} - z_{i-1,j-1}), (z_{i,j} - z_{i-1,j}), \ldots, (z_{i,j} - z_{i+1,j+1}))$$

where $z_{i,j}$ is the center cell of the window located at the *i*th row and *j*th column. In other words, aspect is the direction (in 45° intervals) of the maximum gradient. For example, if the maximum gradient is in the direction of cell $z_{i-1,j+1}$ then aspect = 45°.

(2) second method of calculating gradient and aspect is similar to the first, with gradient being defined as either the maximum gradient of the steepest drop or the steepest rise. Aspect is then the direction of the maximum gradient (EPPL7 1987).

(3) The third method tested was a second-order finite difference method (also called numerical differentiation) described by Dozier and Strahler (1983) for a two-by-two moving window, and also used by Fleming and Hoffer (1979). The first step in the algorithm is to calculate

$$[\delta_z/\delta x]_{i,j} = [z_{i+1,j} - z_{i-1,j}] / 2\Delta X$$

and

$$[\delta z/\delta y]_{i,j} = [z_{i,j+1} - z_{i,j-1}] / 2\Delta Y$$

where ΔX is the spacing between points in the horizontal direction, ΔY is the distance in the vertical direction, and i and j are not the peripheral row or columns.

For points on the end of a row of column, calculate

$$\{[\delta z / \delta x]_{i,j}|(u = 1)\} = [-3z_{1,j} + 4z_{2,j} - z_{3,j}]/2\Delta X$$

and

$$\{[\delta z / \delta y]_{i,j}|(u = u_n)\} = [z_{n-2,j} - 4z_{n-j,j} + 3z_{n,j}] / 2\Delta Y$$

where u is the row and/or column number and n is the total number of rows and/or columns.

The gradient is then defined as

$$\tan G = \sqrt{[(\delta z / \delta x)^2 + (\delta z + \delta y)^2]} \tag{5.1}$$

while aspect is defined as

$$\tan A = \frac{(\delta z/\delta x)}{(\delta z/\delta y)} \tag{5.2}$$

(4) A third-order finite difference method for calculating gradient and aspect proposed by Horn (1981) was the fourth method used, where

$$[\delta z / \delta x]_{i,j} = [(z_{i+1,j+1}) + 2(z_{i+1,j}) + (z_{i+1,j-1})] - [(z_{i-1,j+1}) + 2(z_{i-1,j}) + (z_{i-1,j-1})]/8\Delta X \quad (5.3)$$

and

$$[\delta z / \delta y]_{i,j} = [(z_{i+1,j+1}) + 2(z_{i,j+1}) + (z_{i-1,j+1})] - [(z_{i+1,j-1}) + 2(z_{i,j-1}) + (z_{i-1,j-1})]/8\Delta Y \quad (5.4)$$

Aspect and gradient are calculated for each cell as in equations (5.1) and (5.2). Sharpnack and Akin (1969) had earlier proposed a third-order finite difference method for calculating gradient and aspect that did not have a weighting factor for the non-diagonally adjacent cells. For the Sharpnack and Akin (1969) model, equation (5.3) would be rewritten as

$$[\delta z / \delta x]_{i,j} = [(z_{i+1,j+1}) + (z_{i+1,j}) + (z_{i+1,j-1}] - [(z_{i-1,j+1}) + (z_{i-1,j}) + (z_{i-1,j-1})]/6\Delta X$$

and equation (5.4) as

$$[\delta z / \delta y]_{i,j} = [(z_{i+1,j+1}) + (z_{i,j+1}) + (z_{i-1,j+1})] - [(z_{i+1,j-1}) + (z_{i,j-1}) + (z_{i-1,j-1})] / 6\Delta Y$$

In fact, any weighting scheme could be introduced into the third-order finite difference model, though the Horn (1981) model was used for comparative purposes in this study.

(5.5) The fifth and sixth methods tested were multiple linear regression models proposed by Travis *et al.* (1975) and reported by Evans (1980), where a surface is fitted to the nine-grid cells in a three-by-three window, using least squares (or orthogonal polynomials to improve computational efficiency) to minimize the sum of distances from the surface to the cells. The regression surface for the fifth model is

$$Z = \beta_0 + \beta_1 X + \beta_2 Y + \varepsilon_i \quad (5.5)$$

Assuming $E(\varepsilon_i) = 0$, the regression function for the model in equation (5.5) is

$$E(Z) = \beta_0 + \beta_1 X + \beta_2 Y \quad (5.6)$$

(Neter and Wasserman 1974). Taking partial derivatives of equation (5.6) with respect to X and Y yields $(\partial E(Z)/\delta X) = \beta_1$ and $(\partial E(Z)/\delta Y) = \beta_2$. Gradient and aspect are then calculated using equations (5.1) and (5.2), substituting $(\delta E(Z)/\delta X)$ for $(\delta z/\delta x)$ and $(\delta E(Z)/\delta Y)$ for $(\delta z/\delta y)$.

(6) The surface modelled by the sixth method is

$$Z = \beta_0 + \beta_1 X + \beta_2 Y + \beta_3 X^2 + \beta_4 Y^2 + \beta_5 XY + \varepsilon_i \tag{5.7}$$

Again, assuming $E(\varepsilon_i) = 0$, the regression function for the model in equation (5.5) is

$$E(Z) = \beta_0 + \beta_1 X + \beta_2 Y + \beta_3 X^2 + \beta_4 Y^2 + \beta_5 XY \tag{5.8}$$

(Neter and Wasserman 1974). The partial derivatives of equation (5.8) with respect to X and Y are

$$(\delta E(Z)/\delta X) = \beta_1 + \beta_3 2X + \beta_5 Y \text{ and } (\delta E(Z)/\delta Y) = \beta_2 + \beta_4 2Y + \beta_5 X$$

As for method 5, gradient and aspect are then calculated using equations (5.1) and (5.2), substituting $(\delta E(Z)/\delta X)$ for $(\delta z/\delta x)$ and $(\delta E(Z)/\delta Y)$ for $(\delta z/\delta y)$.

The linear multiple regression coefficients for methods 5 and 6 (β_1, β_2, β_3, β_4 and β_5) were calculated from covariance matrices (Neter and Wasserman 1974), using FORTRAN subroutines provided by the IMSL (1987) suite of statistical routines.

These six methods were compared quantitatively by taking a subsample grid of three-by-three cells and recording the elevation of the cells. The grid was plotted on graph paper and the center cell was connected to the adjacent positions in the three-by-three matrix that had the same elevation. The gradient and aspect were then calculated manually by drawing a tangent to the contour passing through the centre cell and graphing the perpendicular bisector of the tangent (Myers and Shelton 1980). Aspect was calculated from the direction of the perpendicular bisector and the gradient calculated by dividing the difference in height along the perpendicular bisector by the length of the perpendicular bisector. The manually-calculated values of slope and aspect were taken as the true values.

In all, 50 samples were taken at random. One sample occurred on a saddle, and therefore had no gradient and an undefined aspect. This sample was rejected, and another sample was randomly selected to replace it. The difference between true aspect and the aspect calculated by methods 1 to 6 were calculated, and plotted as histograms (Figure 5.3). The mean and standard error of the mean for these histograms are also given. The differences between true gradient and calculated gradient were similarly found and plotted as histograms (Figure 5.4). The aim was to detect visually those methods which had the least deviation from the true values of aspect and gradient.

As the true gradient and aspect measurements were paired with the gradient and aspect values calculated by the six methods, Kendall's tau measure of association test (Noether 1976) was used to test a research hypothesis that the true gradient was positively correlated with the calculated gradient and the true aspect was positively correlated with the calculated aspect, viz.,

H_o:X, Y are independent versus H_A:X, Y are positively correlated,

FIGURE 5.1 Original contour map of the study area.

where X is the true aspect of the cell and Y is aspect calculated by method 1. Note that the same null hypothesis was applied to the aspect values derived from methods 2, 3, 4, 5 and 6, and also to the gradient calculations using methods 1 to 6. The test was one-sided with $\alpha' = 0{\cdot}05$ and the test statistic $z_{\alpha'} = 1{\cdot}645$.

Spearman's rank correlation coefficient (Noether 1976) was used to measure the monotone relationship between the true values of aspect and the measured values of aspect, and the true and calculated values of gradient, for the six methods tested.

FIGURE 5.2 Contour map interpolated from the DEM by Hutchinson's (1989) program.

Tests for significant positive association between the true and calculated aspect and gradient variables were calculated using the Spearman's rank correlation coefficients. The same null and alternative hypotheses were used as described above. In this case, the test was one-sided with $\alpha' = 0{\cdot}05$, and H_O was rejected if

$$\sqrt{(n-1)}(r_S) > z_{\alpha'}$$

where n is the number of samples and r_S is the Spearman's rank correlation coefficient.

TABLE 5.1
Comparison of the six methods used to calculate aspect and gradient with the true values of aspect and gradient using Kendall's tau test.

Topographic feature	Method number					
	1	2	3	4	5	6
Aspect	5·595*	7·034*	9·163*	9·998*	9·934*	9·956*
Gradient	7·439*	7·965*	8·758*	8·973*	9·025*	9·034*

* Denotes that the standard τ values were statistically significant at $\alpha' = 0{\cdot}05$. Consequently, reject the null hypothesis and conclude that the true value is positively correlated with the calculated value.

4 Results

The contour map used to prepare the DEM is presented in Figure 5.1. The contour map interpolated from the DEM (Figure 5.2) compares closely with the original contour map (Figure 5.1).

The six methods used to calculate gradient and aspect were compared quantitatively with the true values, using the Kendall tau test, and the resulting standardized tau (τ) values are presented in Table 5.1. Aspect values calculated by methods 1 to 6 were significantly correlated with the true aspect at $\alpha' = 0{\cdot}05$. Figure 5.3 confirms diagrammatically the result from Table 5.1; that there is no significant deviation between the true aspect and the calculated aspect for any of the six methods.

The gradient values calculated by methods 1 to 6 were all significantly correlated with the true gradient (Table 5.1). Table 5.1 shows that the gradient values calculated by methods 5 and 6 have the highest correlation with the true values, and this is confirmed by Figure 5.4, where the deviation of the gradient values from the true values was smaller compared with methods 1, 2, 3 and 4.

The Spearman's rank correlation coefficients are listed in Table 5.2 and confirm the observations made from Figures 5.3 and 5.4 and Table 5.1. The correlation between the gradient and calculated gradient and true aspect and calculated aspect was highest for methods 4, 5 and 6 and lowest for methods 1 and 2. The statistical significance of the r_S values is summarized in Table 5.3. The results are similar to those presented in Table 5.1.

5 Discussion

There was a statistically significant correlation between true gradient and calculated gradient for all six methods, when tested with the Kendall tau and Spearman rank correlation coefficients tests (Tables 5.1 and 5.3).

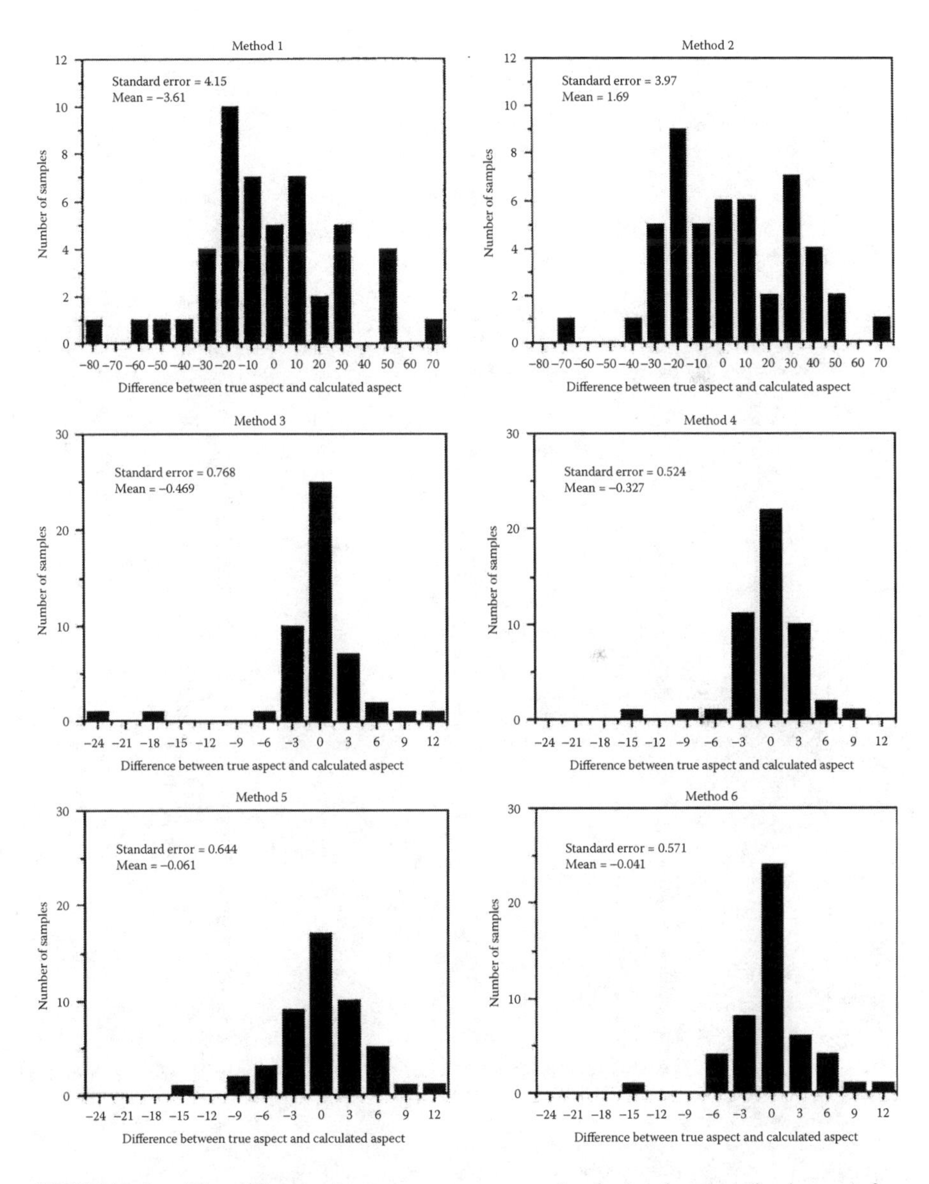

FIGURE 5.3 The difference between true aspect and calculated aspect (in degrees) for methods 1 to 6.

A statistically significant correlation was also apparent between true aspect and calculated aspect for all methods when using the Kendall tau and Spearman rank correlation coefficient tests (Tables 5.1 and 5.3).

Table 5.1 shows gradient calculated by method 6 had the highest tau value. Modelling the surface using linear regression models (methods 5 and 6) yielded slightly higher tau values than when using the difference methods (3 and 4). It is

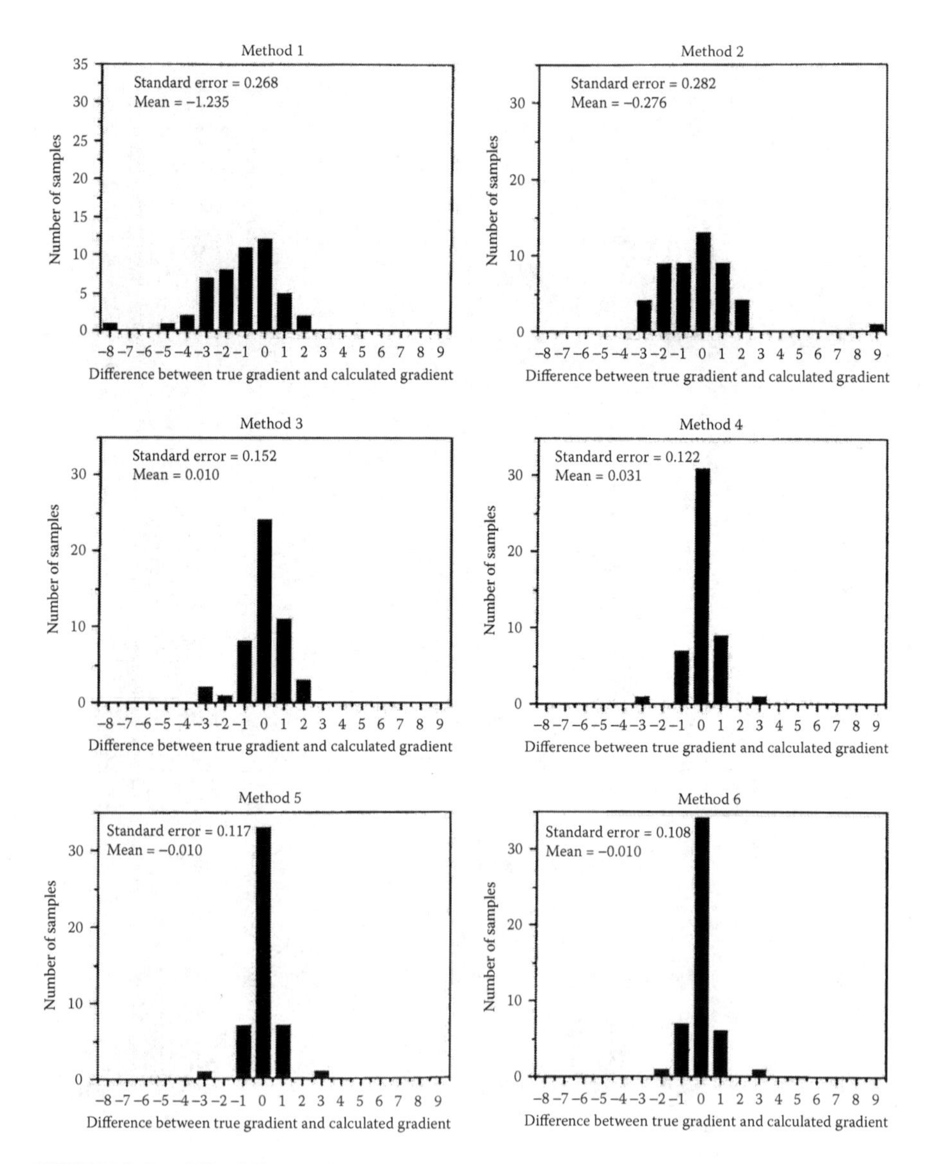

FIGURE 5.4 The difference between true gradient and calculated gradient (in degrees) for methods 1 to 6.

interesting to note that the r_S values were equal for methods 4, 5 and 6 (Table 5.2). As the differences in tau values for methods 4, 5 and 6 (Table 5.1) are small, it can be seen there is little difference between methods 4, 5 and 6.

There was no difference between true aspect and calculated aspect for methods 4, 5 and 6 when these were tested with the Spearman rank correlation coefficient test (all three had a correlation of 0·999). The correlation between the true aspect and calculated aspect was more accurately estimated by method 6 than method 5 according to the Kendall tau test. The subtle changes introduced by the quadratic

TABLE 5.2
Spearman's rank correlation coefficients for gradient and aspect.

Method	Gradient (r_S)	Aspect (r_S)
1	0·908	0·619
2	0·936	0·844
3	0·974	0·884
4	0·985	0·999
5	0·985	0·999
6	0·985	0·999

TABLE 5.3
Comparison of the six methods used to calculate aspect and gradient with the true values of aspect and gradient using Spearman's rank correlation coefficient.

Topographic feature	Method number					
	1	2	3	4	5	6
Aspect	4·333*	5·908*	6·188*	6·993*	6·993*	6·993*
Gradient	6·356*	6·552*	6·818*	6·895*	6·895*	6·895*

* Denotes that values from the Spearman's rank correlation coefficient were statistically significant at $\alpha = 0{\cdot}05$. Consequently, reject the null hypothesis and conclude that the true value is positively correlated with the calculated value.

terms in the liner regression model may have caused a slightly better estimate of changes in elevation in the *X* and Y directions. The correlation between true aspect and calculated aspect is greater for method 4 than methods 5 and 6. However, the difference in correlation values is small and it would be improper to conclude that method 4 is better.

Gradient and aspect were calculated more accurately by method 4 compared with method 3. This may be due to a larger area being sampled by the adjacent diagonals used by the third-order finite difference method (method 4). Gradient and aspect were more accurately estimated by method 2 than method 1.

Methods 1 and 2 gave poorer estimates of gradient and aspect. This is due to aspect and gradient being calculated in the direction of steepest rise or fall (for the eight cells adjacent to the test cell). The direction is therefore truncated to 45° intervals. In addition, in the event of two directions having an equal and maximum gradient, aspect calculated by methods 1 and 2 will be less accurate, as a choice about the 'correct' direction of maximum gradient will have to be made. The main reason for the third-order finite difference method (method 4) and the multiple linear regression methods (methods 5 and 6) more accurately calculating gradient and aspect compared with methods 1, 2 and 3, is that all cells adjacent to the centre cell contribute to the calculation.

Spurious values of gradient and aspect may be generated over some terrain features regardless of the method used to calculate gradient and aspect. Examples of terrain features for which aspect and gradient may be difficult to define include saddles and areas of flat terrain. (It was noted in § 2 that one sample occurred over a saddle and as a consequence was rejected.) In the forest mapping study (Skidmore 1989) for which gradient and aspect were included as data layers in a GIS, the effect of spurious gradient and aspect values was minimized by the use of prior probabilities in the expert system and the inclusion of ecotonal information. These issues are discussed in detail by Skidmore (1989).

Evans (1980) reported correlation coefficients between true aspect and calculated aspect, and true gradient and calculated gradient, for a six-parameter linear regression model (equivalent to method 6 reported here). The values of r were similar to those reported for this study (i.e., Evans reported an r value of 0·998 for aspect and 0·984 for gradient). Differences in reported r values could be expected because, as the spacing of the grid shortens, the gradient and aspect values will approach the true gradient and aspect values for any of the methods tested. It is not apparent from Evans (1980) which correlation coefficient was used.

The assumption of Hutchinson (1989) that all the streams drain was valid in the study area, as no natural sinks (i.e., ponds or lakes) occurred. The irregular data interpolated by Hutchinson's (1989) program were digitized from a topographic map over a forested region. The contours drawn on the map were assumed to be accurate (it was stated on the map that it was prepared to conform with Australian Mapping Council's standards of accuracy), though the map may have been another source of error in this study.

The assumption that the manually-measured aspect and gradient are the true aspect and gradient should be considered further. Fitting contour lines manually between elevation points permits the use of human intelligence in drawing smoothly-fitting contours that are sensible in the context of the grid points. The disadvantage is that different operators may produce different results when they fit slightly different curves between a given set of points. In other words, the curve fitted between points by two operators will never be the same, so an exact mathematical model cannot be developed even using higher-order linear regression models. Another problem with estimating gradient and aspect is that they will vary at different scales in the topography. Anyone who has measured gradient and aspect in the field will appreciate that gradient and aspect vary according to whether one measures them over a length of 1 m (rocks and boulders may influence the estimate) or over 100 m (a change in grade and aspect may occur some distance from the point being measured). The method used in this study follows the accepted field technique for recording gradient and aspect (Myers and Shelton 1980), by calculating the average gradient along the perpendicular bisector of the tangent to the contour. The results obtained here indicate that the linear regression models and the third-order finite difference method have the best potential for estimating gradient and aspect.

In conclusion, there is little difference between methods 4, 5 and 6 for calculating aspect and gradient. The general linear regression models or the third-order finite difference method (i.e., method 4) appear to be optimal for calculating gradient and aspect from a gridded DEM. If only methods 1 or 2 (i.e., direction of steepest rise and/or steepest fall from a central pixel) have been implemented in the GIS package you use, then be wary of the results. This work suggests that the use of regression or third-order finite difference methods will give accurate results in areas of unambiguous gradients.

Acknowledgments

Dr B. J. Turner of the Australian National University reviewed the manuscript. Dr M. Hutchinson, also of the Australian National University, made his interpolation program (SPLIN2H) available and provided comments on the manuscript.

References

ANU, 1988, Digitized program documentation. Computer Services Centre, Australian National University, Canberra, Australia.

Burrough, P. A., 1986, *Principles of Geographical Information Systems for Land Resources Assessment* (Oxford: Clarendon Press).

Cibula, W. G., and Nyquist, M. D., 1987, Use of topographic and climatological models in a geographical data base to improve Landsat MSS classification for Olympic National Park. *Photogrammetric Engineering and Remote Sensing*, **53**, 67.

Dozier, J., and Strahler, A. H., 1983, Ground investigations in support of remote sensing. In *Manual of Remote Sensing*, Vol. 1, edited by R. N. Colwell (Falls Church, Virginia: American Society of Photogrammetry).

EPPL, 1987, Environmental planning and programming language user's guide. Minnesota State Planning Agency, St Paul, Minnesota, U.S.A.

Evans, I. S., 1980, An integrated system of terrain analysis for slope mapping. *Zeitschrift für Geomorphologie*, **36**, 274.

Fleming, M. D., and Hoffer, R. M., 1979, Machine processing of Landsat MSS data and DMA topographic data for forest cover type mapping. LARS technical report 062879, Laboratory for Applications of Remote Sensing, Purdue University, West Lafayette, Indiana, U.S.A.

Franklin, J., Logan, T. L., Woodcock, C. E., and Strahler, A. H., 1986, Coniferous forest classification and inventory using Landsat and digital terrain data. *I.E.E.E. Transactions on Geoscience and Remote Sensing*, **24**, 139.

Horn, B. K. P., 1981, Hill shading and the reflectance map. *Proceedings of the I.E.E.E.*, **69**, 14.

Hutchinson, C. F., 1982, Techniques for combining Landsat and ancillary data for digital classification improvement. *Photogrammetric Engineering and Remote Sensing,* **48,** 123.

Hutchinson, M. F., 1989, A new procedure for gridding elevation and stream line data with automatic removal of spurious pits. *Journal of Hydrology,* **106**, 211.

IMSL, 1987, *STAT/LIBRARY User's Guide: Fortran subroutines for statistical analysis*, Vol. 1 (Houston, Texas: IMSL).

Johnston, K. M., 1987, Natural resource modelling in the geographic information system environment. *Photogrammetric Engineering and Remote Sensing*, **53**, 1411.

Justice, C. O., Wharton, S. W., and Holben, B. N., 1981, Application of digital terrain data to quantify and reduce the topographic effect on Landsat data. *International Journal of Remote Sensing*, **2**, 213.

Myers, W. L., and Shelton, R. L., 1980, *Survey Methods for Ecosystem Management* (New York: Wiley & Sons).

Neter, J., and Wasserman, W., 1974, *Applied Linear Statistical Models* (Illinois: Irwin).

Noether, G. E., 1976, *Introduction to Statistics: a Nonparametric Approach* (Boston: Houghton Mifflin Company).

Sharpnack, D. A., and Akin, G., 1969, An algorithm for computing slope and aspect from elevations. *Photogrammetric Engineering and Remote Sensing*, **35**, 247.

Skidmore, A. K., 1989, An expert system classifies eucalypt forest types using thematic mapper data and a digital terrain model. *Photogrammetric Engineering and Remote Sensing* (in the press).

Skidmore, A. K., 1990, Terrain position as mapped from a gridded digital elevation model. *International Journal of Geographical Information Systems*, **4** (to be published).

Strahler, A. H., Logan, T. L., and Bryant, N. A., 1978, Improving forest cover classification accuracy from Landsat by incorporating topographic information. *Proceedings of the 12th International Symposium on Remote Sensing of Environment held in Ann Arbor, Michigan* (Ann Arbor: Environmental Research Institute of Michigan), pp. 927–942.

Tom, C. H., and Miller, L. D., 1980, Forest site index mapping and modelling. *Photogrammetric Engineering and Remote Sensing*, **46**, 1585.

Travis, M. R., Elsner, G. H., Iverson, W. D., and Johnson, C. G., 1975, VIEWIT: computation of seen areas, slope, and aspect for land-use planning. US Department of Agriculture Forest Service General Technical Report PSW-11/1975, 70 p. Pacific Southwest Forest and Range Experimental Station, Berkeley, California, U.S.A.

Evolution of Methods for Estimating Slope Gradient and Aspect from Digital Elevation Models

Andrew K. Skidmore

Although by the late 1980s commercial and low-cost GIS packages were available, there had been limited critical review of the algorithms underlying the various GIS functions. As these functions were being used for environmental modeling, users had little or no idea of the errors (and bias) introduced by these functions, and the impact of the error on model output (Heuvelink et al., 1989, and this volume). Indeed, it was known that the quality of the input data was an important source of error, but the contribution of the algorithm to error was also to become an important research area (Goodchild and Jeansoulin, 1998). The motivation behind the original paper was twofold: first to understand the algorithms available to calculate slope and aspect from a digital elevation model and second, to evaluate the performance of the algorithms when tested against a true measure of slope and aspect.

In the following 15 years, most research around the topic of slope, gradient and aspect has been in the use of these terrain derivatives for a vast array of GIS applications and models. There have been some incremental improvements in the algorithms used, as well as research on methods to test the accuracy of digital elevation models (DEMs) and the terrain variables derived from DEMs, such as slope gradient and aspect. In order to continue with the gist of the original paper, I will discuss some of the technical developments of algorithms for calculating slope gradient and aspect, as well as improvements in the methods to estimate accuracy. Some interesting applications of slope gradient and aspect are cited. Even though the field of DEMs appears rather mature, I identify some areas for further work.

The names of the algorithms have become somewhat confused as different authors have attributed various authors and names to the algorithms. Table 5.4 shows the nomenclature used by some authors.

A breakthrough in the DEM field was the software written by Mike Hutchinson of the Australian National University (ANU) to generate gridded DEMs from contours and streamlines (using topographic maps or directly from the stereomodel of aerial photos) — the so-called contour-to-grid derived DEMs (Hutchinson, 1989). I was undertaking my Ph.D. at ANU, and Mike Hutchinson allowed me access to his code so I could generate elevation surfaces; the original paper was the first to

TABLE 5.4
Names of algorithms used by various authors.

Name of Method and Attribution in the Original Paper	Subsequent Name and Attribution
1 – Direction of Maximum Gradient (drop): Travis et al. 1975 EPPL7 (1987)	Jones (1998a) called this method Maximum downward gradient method (MDG) Zhou and Liu (2004b) called this method the Maximum Downhill Slope and attributed it to O'Callaghan and Mark (1984)
2 – Direction of Maximum Gradient (drop or rise): Travis et al. 1975 EPPL7 (1987)	No subsequent activity
3 – Second Order Finite Difference: Fleming and Hoffer (1979); Dozier and Strahler (1983)	Zhou and Liu (2004a) attributed this method to Zevenbergen and Thorne (1987); Ritter (1987) Jones (1998a) attributed this method to Fleming and Hoffer (1979)
4 – Third Order Finite Difference with weighting: Horn (1981) – it was noted in the original paper that this method is a weighted modification of that proposed by Sharpnack and Akin (1969)	Zhou and Liu (2004b) attributed this method to Unwin (1981) Jones (1998a) attributed this method to Sharpnack and Akin (1969)
5 – Linear Regression Model: Travis et al. 1975 – it was noted in the original paper that this method was reported by Evans (1980)	No subsequent activity
6 – Multiple Linear Regression Model: Travis et al. 1975 – it was noted in the original paper that this method was reported by Evans (1980)	Florinksy 1998 attribute this method to Evans (1980)

use contour-to-grid DEMs with analytical analysis of slope gradient and aspect. Mike Hutchinson's code has now been adopted by ESRI Arc-Info (for creating grid DEMs), has become the industry standard, and is used to derive DEMs by national mapping agencies such as the USGS (U.S. Geological Survey). Since then a number of techniques have been developed for generating DEMs, including airborne LIDAR (light detecting and ranging) and IFSAR (interferometric synthetic aperture radar), but recent work by Hodgson et al. (2003) indicates that the original contour-to-grid-derived DEM continues to show the highest overall absolute elevation accuracy, while LIDAR-DEMs are equally accurate, but are strongly affected by the land cover, particularly over forest and multistory vegetation cover.

The algorithms used to calculate slope and aspect described in the original paper have remained essentially unchanged. A few minor adaptations of the algorithms include:

- The Zevenbergen and Thorne (1987) paper describes an exact model based on a partial quartic expression to model a 3 by 3 local neighbourhood. In my original paper, a second-order polynomial (method 6 in the original paper) was tested. The Zevenbergen and Thorne (1987) algorithm has been evaluated by other authors.
- Wood (1996) constrained the quadratic surface method originally described by Horn (1981).
- Dunn and Hickey (1998) proposed using the maximum downhill slope from a 3 by 3 moving window (which is essentially the same as method 1 in the original paper), but they did not test the algorithm against a "true" estimate of slope.
- Jones (1998a) proposed a Diagonal Ritters method — a modification of the second-order finite difference method (method 3 in the original paper) — where the perpendicular partial gradients are taken at an angle of 45 degrees to the principle axis of the DEM grid.
- Jones (1998a) proposed a Simple method — a modification of the second-order finite difference method (method 3 in the original paper) — where the difference in elevation between a given elevation cell and its neighbour to the west calculates the "east-west" gradient, and with a similar technique the "north-south" gradient is calculated.
- Moore et al. (1993) modified the denominator weighting for the multiple linear regression model (method 6 in the original paper).

In fact, higher-order polynomials can be used to model more convoluted surfaces, but require 10 points in order to define the surface. In other words, a moving window of more than 3 by 3 cells is required to calculate slope gradient and aspect, but this causes the surface to become more generalized. Thus the conclusion in the original paper that the third-order finite difference method is optimal with a real-world DEM, or with a synthetic DEM surface with added error, has withstood scientific testing (Jones, 1998a; Zhou and Liu, 2004a). In the original paper I reported no difference between the third-order finite difference (Horn, 1981) and linear regression models (Evans, 1980), and that these methods appear to be optimal for calculating slope gradient and aspect from a gridded DEM, but the third-order finite difference method is simpler to implement. Jones (1998a) concurred, stating that Mike Goodchild (pers comm.) proved that the methods give identical results, though Florinsky (1998) concluded that the linear regression models are most precise. Subsequently, Horn's (1981) third-order, finite-difference algorithm has been implemented in ESRI Arc-Info/Arc-View products.

The algorithms tested in the original paper were repeated by Jones (1998a) and Zhou and Liu (2004a) with the difference that an artificial (mathematical) surface (usually the Morrison's Surface III) was used to test the algorithms instead of an actual DEM surface. Noise was added to the artificial surface, and error was measured as the difference between the slope estimates produced by the algorithms and the reference slope grids derived by analytic partial differentiation of the synthetic surface. With no noise, the second-order finite difference (method 4 in the original paper, but often referred to as the Fleming-Hoffer method of 1979 or the Zevenbergen-Thorne method from 1987) performed "best." With intermediate noise levels, the third-order finite difference method attributed to Horn (1981) was optimal, and with extreme noise added to the synthetic surface, the third-order finite difference method of Sharpnack and Akin (1969) performed "best" (as noted in table 1, the Sharpnack and Akin [1969] method is the same as Horn [1981] without a weighting factor for the non-diagonally-adjacent cells). Subsequently Jones (1998a) and Zhou and Liu (2004a) published similar papers confirming their results (Jones, 1998b; Zhou and Liu, 2004b).

In order to state which methodology is "best," researchers used, respectively, artificial surfaces (such as the Morrison's Surface III described above [Jones, 1998a]), or reference values derived manually from topographic maps (such as used in the original study), or field observation (Bolstad and Stowe, 1994; Giles and Franklin, 1996). It was debated in the original paper whether a true reference value may be calculated from topographic maps or from field observation yields, while Florinsky (1998) discusses in detail the pros and cons of various methods for comparing algorithms.

Even though Gao (1995, 1997) found that gentle terrain was more accurately represented than complex terrain when using a DEM of the same dimension, there has been little progress in calculating slope and aspect for areas of flat terrain — a problem identified in the original paper and reiterated by subsequent authors such as Florinsky (1998). With respect to sampling methods, purposive sampling, as implemented in Hutchinson's (1989) code, has been shown to be more accurate than systematic sampling (Gao, 1995). A number of researchers have found that errors in modeled slope angles were greatest in areas of higher slope (Chang and Tsai, 1991; Bolstad and Franklin, 1994; Gao, 1995; Gong et al., 2000; Hodgson et al., 2003). In other words, areas of higher slope need a denser sampling intensity (Gao, 1995). Elevation accuracy varies with land cover category as well as terrain: Hodgson et al. (2003) found that land cover with trees, or with higher variability such as scrub/shrub, have higher elevation error, which would also increase error in estimates of slope and aspect.

If you search for research papers using keywords such as slope, gradient, and aspect, hundreds of application papers appear across an array of disciplines including hydrology (Moore et al., 1991), geomorphology (Walsh et al., 1998; Alcantara-Ayala, 2004), natural resources (Cannon et al., 2001; Takken et al., 2001), geology and earth sciences (Onorati et al., 1992; Kuhni and Pfiffner, 2001), ecology (Skidmore, 1989; Skidmore et al., 2001), oceanography (McAdoo et al., 2000), environmental modeling (Skidmore, 2001) and so on. Clearly, digital terrain models have a broad acceptance in GIS analysis, and analysis using them is a keystone of GIS modeling.

In conclusion, it appears that there have been improvements in algorithms, but the development of the science has been one of evolution not revolution, with many authors retesting and confirming earlier results. A number of recent papers continue to compare the accuracy and efficiency of slope and aspect algorithms, confirming the main conclusions of the original paper.

Are there substantial research questions remaining to be answered? In the original paper I noted that a problem with estimating gradient and aspect is that they will vary at different scales in the topography. There has been great progress on the issue of scale in geography in general, especially with respect to the intensity of sampling, with limited work on the properties of terrain that may emerge at different scales (e.g., Wood, 1996). At what point does generalization in scale estimates become meaningless? And when do improvements in DEM resolution become irrelevant? Digital elevation models and their terrain derivatives are interesting tools for working on scale issues in the geographical sciences. A better understanding of scale in DEMs may prompt scientists to conceive new approaches for calculating terrain derivatives, such as slope and aspect, for example, using a scaleless approach similar to those developed by Hutchinson (1989).

References

Alcantara-Ayala, I., 2004, Hazard assessment of rainfall-induced landsliding in Mexico, *Geomorphology*, 61(1–2), 19–40.

Bolstad, P.V. and Stowe, T., 1994, An evaluation of DEM accuracy: elevation, slope, and aspect, *Photogrammetric Engineering and Remote Sensing* 60(11), 1327–1332.

Cannon, S.H., Kirkham, R.M., and Parise, M., 2001, Wildfire-related debris-flow initiation processes, Storm King Mountain, Colorado, *Geomorphology* 39(3–4), 171–188.

Chang, K. and Tsai, B., 1991, The effect of DEM resolution on slope and aspect mapping, *Cartography and Geographic Information Systems*, 9(4), 405–419.

Dozier, J. and Strahler, A.H., 1983, Ground investigations in support of remote sensing. In *The Manual of Remote Sensing*, Vol. 1, R. N. Colwell, Ed., American Society of Photogrammetry, Falls Church, VA, 959–986.

Dunn, M. and Hickey, R., 1998, The effect of slope algorithms on slope estimates within a GIS, *Cartography and Geographic Information Systems*, 27(1), 9–15.

Evans, I.S., 1980, An integrated system of terrain analysis for slope mapping, *Zeitshrift fur Geomorphologie*, 36, 274–295.

EPPL7 (Environmental Planning and Programming Language), 1987, *Environmental Planning and Programming Language Users Guide*, Minnesota State Planning Agency, St. Paul, MN.

Fleming, M.D. and Hoffer, R.M., 1979, Machine processing of Landsat MSS data and DMA topographic data for forest cover type mapping. Laboratory for Applications of Remote Sensing, Purdue University, West Lafayette, IN 47906.

Florinsky, I.V., 1998, Accuracy of local topographic variables derived from digital elevation models, *International Journal of Geographical Information Science*, 12(1), 47–61.

Gao, J., 1995, Comparison of sampling schemes in constructing DTMs from topographic maps, *ITC Journal* (1), 18–22.

Gao, J., 1997, Resolution and accuracy of terrain representation by grid DEMs at a micro-scale, *International Journal of Geographical Information Science*, 11(2), 199–212.

Giles, P.T. and Franklin, S.E., 1996, Comparison of derivative topographic surfaces of a DEM generated from stereoscopic SPOT images with field measurements, *Photogrammetric Engineering and Remote*, 62, 1165–1171.

GONG, J., LI, Z., ZHU, Q., SUI, H., AND ZHOU, Y., 2000, Effects of various factors on the accuracy of DEMs: an intensive experimental investigation, *Photogrammetric Engineering and Remote Sensing*, 66(9), 1113–1117.

GOODCHILD, M.F. AND JEANSOULIN, R., 1998, Editorial, *GeoInformatica*, 1(3), 211–214.

HEUVELINK, G.B.M., BURROUGH, P.A., AND STEIN, A., 1989, Propagation of errors in spatial modelling with GIS, *International Journal of Geographical Information Systems* 3(4), 303–322.

HODGSON, M.E., JENSEN, J.R., SCHMIDT, L., SCHILL, S., AND DAVIS, B., 2003, An evaluation of LIDAR- and IFSAR-derived digital elevation models in leaf-on conditions with USGS Level 1 and Level 2 DEMs, *Remote Sensing of Environment*, 84, 295–308.

HORN, B.K.P., 1981, Hill shading and the reflectance map, *Proceedings of the IEEE*, 69(1), 14–47.

HUTCHINSON, M.F., 1989, A new procedure for gridding elevation and stream line data with automatic removal of spurious pits, *Journal of Hydrology*, 106, 211–232.

JONES, K.H., 1998a, A comparison of algorithms used to compute hill slope as a property of the DEM, *Computers & Geosciences*, 24(4), 315–323.

JONES, K.H., 1998b, A comparison of two approaches to ranking algorithms used to compute hill slopes, *GeoInformatica*, 2(3), 235–256.

KUHNI, A. AND PFIFFNER, O.A., 2001, The relief of the Swiss Alps and adjacent areas and its relation to lithology and structure: topographic analysis from a 250-m DEM, *Geomorphology*, 41(4), 285–207.

MCADOO, B.G., PRATSON, L.F., AND ORANGE, D.L., 2000, Submarine landslide geomorphology, U.S. continental slope, *Marine Geology*, 169(1–2), 103–136.

MOORE, I.D., GRAYSON, R.B., AND LADSON, A.R., 1991, Digital terrain modelling: a review of hydrological, geomorphological, and biological applications, *Hydrological Processes*, 5(1), 3–30.

MOORE, I.D., GESSLER, P.E., NIELSEN, G.A., AND PETERSON, G.A., 1993. Soil attribute prediction using terrain analysis, *Soil Science Society of America Journal*, 57(2), 443–452.

O'CALLAGHAN, J.F. AND MARK, D.M., 1984, The extraction of drainage networks from digital elevation data, *Computer Vision, Graphics and Image Processing*, 28, 323–344.

ONORATI, G., VENTURA, R., CHIARINI, V., AND CRUCILLA, U., 1992, The digital elevation model of Italy for geomorphology and structural geology, *Catena*, 19(2), 147–178.

RITTER, P., 1987, A vector-based slope and aspect generation algorithm, *Photogrammetric Engineering and Remote Sensing*, 53(8), 1109–1111.

SHARPNACK, D.A. AND AKIN, G., 1969. An algorithm for computing slope and aspect from elevations, *Photogrammetric Engineering and Remote Sensing*, 35, 247–248.

SKIDMORE, A.K., 1989, An expert system classifies eucalypt forest types using Landsat Thematic Mapper data and a digital terrain model, *Photogrammetric Engineering and Remote Sensing*, 55(10), 1449–1464.

SKIDMORE, A.K., 2001, *Environmental modeling using GIS and remote sensing*, Taylor and Francis, London.

SKIDMORE, A.K., SCHMIDT, K.S., KLOOSTERMAN, H., KUMAR, L., AND VAN OOSTEN, H., 2001, *Hyperspectral imagery for coastal wetland vegetation mapping*. BCRS - Beleids Commissie Remote Sensing, Delft. PO Box 5023, 2600 GA, Delft.

TAKKEN, I., JETTEN, V., GOVERS, G., NACHTERGAELE, J., AND STEEGEN, A., 2001, The effect of tillage-induced roughness on runoff and erosion patterns, *Geomorphology*, 37(1–2), 1–14.

TRAVIS, M.R., ELSNER, G.H., IVERSON, W.D., AND JOHNSON, C.G., 1975, *VIEWIT: A computation of seen areas, slope and aspect for land use planning*. USDA Forest Service General Technical Report PSW-11/1975, Pacific Southwest Forest and Range Experimental Station, Berkeley, California.

Unwin, D., 1981, *Introductory Spatial Analysis*, Methuen, London and New York.

Walsh, S.J., Butler, D.R., and Malanson, G.P., 1998, An overview of scale, pattern, process relationships in geomorphology: a remote sensing and GIS perspective, *Geomorphology* 21(3–4), 183–205.

Wood, J.D., 1996, The geomorphological characterisation of digital elevation model, Ph.D. thesis, University of Leicester.

Zevenbergen, L.W. and Thorne, C.R., 1987, Quantitative analysis of land surface topography, *Earth Surface Processes and Landforms*, 12, 47–56.

Zhou, Q. and Liu, X., 2004a, Analysis of errors of derived slope and aspect related to DEM data properties, *Computers & Geosciences*, 30(4), 369–378.

Zhou, Q. and Liu, X., 2004b, Error analysis on grid-based slope and aspect algorithms, *Photogrammetric Engineering and Remote Sensing*, 70(8), 957–962.

International Journal of Geographical Information Systems,
1990, Vol. 4, No. 4, 369–383.

6 Object-Oriented Data Modelling for Spatial Databases

Michael F. Worboys, Hilary M. Hearnshaw, and David J. Maguire

Abstract. Data modelling is a critical stage of database design. Recent research has focused upon object-oriented data modes, which appear more appropriate for certain applications than either the traditional relational model or the entity-relationship approach. The object-oriented approach has proved to be especially fruitful in application areas, such as the design of geographical information systems which have a richly structured knowledge domain and are associated with multimedia databases. This article discusses the key concept in object-oriented modelling and demonstrates the applicability of an object-oriented design methodology to the design of geographical information systems. In order to show more clearly how this methodology may be applied, the paper considers the specific object-oriented data model IFO. Standard cartographic primitives are represented using IFO, which are then used in the modelling of some standard administrative units in the United Kingdom. The paper concludes by discussing current research issues and directions in this area.

1 Introduction

The choice of an appropriate representation for the structure of a problem is perhaps the most important component of its solution. For database design, the means of representation is provided by the data model. A data model provides a tool for specifying the structural and behavioural properties of a database and ideally should provide a language which allows the user and database designer to express their requirements in ways that they find appropriate, while being capable of transformation to structures suitable for implementation in a database management system. Data modelling is among the first stages of database design. The purpose of data modelling is to bring about the design of a database which performs efficiently; contains correct information (and which makes the entry of incorrect data as difficult as possible); whose logical structure is natural enough to be understood by users; and is as easy as possible to maintain and extend. Of course, different problems require different means of representation and a large number of data models is described in the database literature. Some are close to implementation structures, for example the relational model (Codd 1970). Others as yet have no directly corresponding implementation. This is the case for those which support a wide variety of modelling constructs as well as a high level of abstraction. Such models allow representations which are closer to the original problem as framed by the user. An example is the IFO (Is-a relationships, Functional relationships, complex Objects) model discussed later.

In this paper, emphasis is placed upon the so-called object-oriented data models, which are at the problem-oriented end of the scale. Object-oriented approaches originated in programming languages such as Simula and Smalltalk. The application of object-oriented ideas to databases was spurred on by the apparent limitations of traditional relational technology when applied to some of the newer applications. Typical examples are the applications of databases in computer-aided design (CAD), office information systems (OIS), software engineering and geographical information systems (GIS). A common difficulty in all of these application areas is the gulf between the richness of the knowledge structures in the application domains and the relative simplicity of the data model in which these structures can be expressed and manipulated. Object-oriented models have the facilities to express more readily the knowledge structure of the original application.

There is no clear definition of, or even general agreement in the computing community on what precisely is, an object-oriented data model. The area is still very new and individual ideas have not yet been synthesized into a general view. However, it is generally recognized (Peckham and Maryanski 1988) that there is a clear ascending chain from the relational model through the earlier semantic data models to object-oriented models and it is in this context that this paper considers object-oriented modelling methods. The major abstraction constructs are discussed and exemplified. One particular object-oriented formalism, IFO, developed by Abiteboul and Hull (1984) is considered, followed by a description of the application of the IFO formalism to two examples in the GIS area. These examples show how IFO can represent precisely the basic spatial elements (point, line and polygon) and, on a larger scale, represent and model relationships between administrative and postal area units in the United Kingdom. Such modelling is a key application for any GIS (see, for example, Egenhofer and Frank 1989).

To avoid confusion, it is necessary to differentiate object-oriented data modelling from object-oriented database management systems (OODBMSs). An OODBMS is a system upon which the database is implemented. It is possible (but not optimal) to model using an object-oriented methodology and implement in, for example, a relational DBMS. Of course, it is most desirable to use an OODBMS which can naturally implement all the constructs of the data model. However, owing to the newness of the technology, OODBMSs are only now emerging as viable systems. A recent description of some of the most innovative of such systems (e.g., Iris, ORION, OZ +, and GemStone) is given in Kim and Lochovsky (1989). Such systems will have an important impact on GIS technology. For example, most OODBMSs support version control, where the system can generate multiple different versions of an object, maybe corresponding to different time-slices. This would be a natural implementation of spatially referenced data (e.g., census data) where spatial boundaries can change. After this brief mention of the emergent OODBMS technology, the remainder of this paper will concentrate on the object-oriented data models.

2 Semantic data models

The relational model (Codd 1970) provides the database designer with a modelling tool which is independent of the details of physical implementation. However, the relational model is limited with respect to semantic content (i.e., expressive power) and there are many design problems which are not naturally expressible in terms of relations. Spatial systems are a case where the limitations become clear. To illustrate this point, consider the relational model of a polygon as originally given by van Roessel (1987), based upon the definitions proposed by the National Committee for Digital Cartographic Data Standards (Moellering 1986) and discussed in greater detail later in this paper:

POLYGON (Polygon, ID, Ring ID, Ring Seq)
RING (Ring ID, Chain ID, Chain Seq)
CHAIN1 (Chain ID, Point ID, Point Seq)
CHAIN2 (Chain ID, Start Node, End Node, Left Pol, Right Pol)
NODE (Node ID, Point ID)
POINT (Point ID, X Coord, Y Coord)

This model of a polygon as a set of relations, though complete, is low-level and some way from one which represents a user's normal view of such an object. Semantic data models aim to provide more facilities for the representation of the users' view of systems than the relational model, as well as to de-couple these representations from the physical implementation of the databases. Fundamental work in this area was undertaken by Chen (1976), who proposed a semantic data model and a diagrammatic technique known as the entity-relationship (E-R) model and diagram respectively. The concepts underlying the E-R model are described in some detail in the next section, since these concepts are required for an understanding of many later semantic data models.

3 Entity-relationship modelling

The entity-relationship model utilizes the concepts of entity, attribute and relationship. A distinction is made between a type and an occurrence of a type. An entity is an item about which the database is to record information. Such an item should be uniquely identifiable. For example, a particular point could be uniquely identified by its coordinates or a census tract by its census code. An entity type is an abstraction representing a class of entities of the same kind. For example, POINT and CITY are entity types. Occurrences of those types are particular points and cities, e.g., a point with coordinates (3,4) and a city named Oxford.

An attribute is an element of data associated with an entity. A city has a population, thus the entity type CITY has attribute type POPULATION. (In this

FIGURE 6.1 Entities and relationship.

paper, types will usually be printed in upper-case.) A particular city has a particular population. Such a population is an example of an attributed occurrence. To avoid a cumbersome presentation, we will omit the terms 'type' and 'occurrence' when no ambiguity is involved. The attribute(s) which identify an occurrence of an entity uniquely are termed identifiers or keys.

A relationship is an association between entities. For example, LIVES_IN is relationship between entities PERSON and CITY. Again, we may distinguish between types and occurrences of relationships. Relationships may have attributes, for example, the relationship LIVES_IN might have the attribute DURATION which gives the length of time that a person has lived in a city.

Chen (1976) proposed a diagrammatic means of representing this model. The diagrammatic form of the above example is shown in Figure 6.1. Rectangles depict entities and rhombi depict relationships. M and N indicate that the relationship is many-to-many, that is each person may live in more than one city and each city may have more than one person living in it.

Many systems may be modelled using entities, attributes and relationships, including systems with a dominating spatial component. Calkins and Marble (1987) apply the method to the design of a cartographic database. They describe the strengths of the method as being flexibility, control of database integrity and generality (i.e., not linked with particular implementations). An important feature of E-R modelling is the natural and well-understood method of the transformation from the E-R model to the rational model (Whittington 1988).

4 Extensions to the entity-relationship model

The entity-relationship approach is at present recognized as the prime tool for data modelling (see, for example, Whittington 1988). However, experience has shown that for many systems the initial set of modelling constructs (entity, attribute and relationship) is inadequate. For example, view integration (the process by which several local views are merged into a single integrated model of the database) is recognized by many workers (for example, Calkins and Marble 1987) as of great importance for GIS design. View integration is greatly facilitated by the introduction of abstraction concepts additional to the original E-R model. In the mid-1970s, Smith and Smith (1977) proposed the introduction of two abstraction constructs, generalization and aggregation, into the data modelling tool-kit. These constructs are provided by almost all contemporary semantic data models. We proceed to describe them in detail, (using some of the IFO notation which is explained in detail later in the paper), distinguishing between generalization and specialization, considering a further construct, association or grouping, and then discussing their relevance to spatial database design. A similar approach will be found in Egenhofer and Frank (1989).

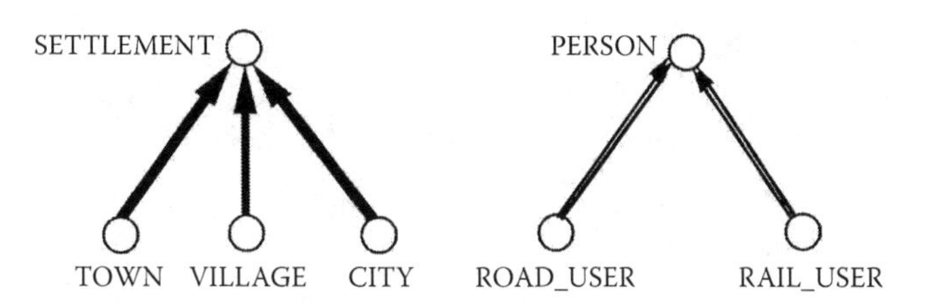

FIGURE 6.2 SETTLEMENT is a generalization of TOWN, VILLAGE and CITY. ROAD_ and RAIL_USER are specialization of PERSON.

4.1 Generalization

Generalization is the construct which enables groups of entities of similar types to be considered as a single higher-order type. For example, entities of types VILLAGE, TOWN, and CITY may be merged and considered as entities of the single type SETTLEMENT. SETTLEMENT is said to be a generalization of VILLAGE, TOWN, and CITY. The diagrammatic representation of this situation is shown in Figure 6.2. Formally, the generic higher-order type is the set-theoretic union of objects in the lower-order types. An object may be thought of, at this intermediate stage between classical and object-oriented data modelling, as an entity along with its attributes. This definition will be extended and made more formal later in the paper.

4.2 Specialization

Specialization is the construct which enables the modeller to define possible roles for members of a given type. For example, entities of type PERSON might be considered occurrences of type ROAD_USER or RAIL_USER, depending upon the context in which we see them. The diagrammatic representation of this situation is shown in Figure 6.2. Formally, the specialized type is made up of a subset of occurrences of the higher-order type.

It should be noted that, although generalization and specialization are in a sense inverse to one another, there are distinctions. A generic type inherits its structure from its lower-order types (and possibly adds some of its own). In the case of specialization, the specific types inherit structure from the higher-order type (and possibly add some of their own). In the case of both generalization and specialization, we say that the lower-order type is a subtype of the higher-order type.

4.3 Aggregation

Aggregation is the construct which enables types to be amalgamated into a higher-order type, the attributes of whose objects are a combination of the attributes of the objects of the constituent types. Formally, the objects which are occurrences of the aggregate type are tuples, the components of which are the objects of the constituent types. In short, aggregation corresponds to the mathematical operation of cartesian product. An example is the type POINT, which is the aggregate of type POINT_ID with two integer types named X_COORD and Y_COORD, thus representing a point

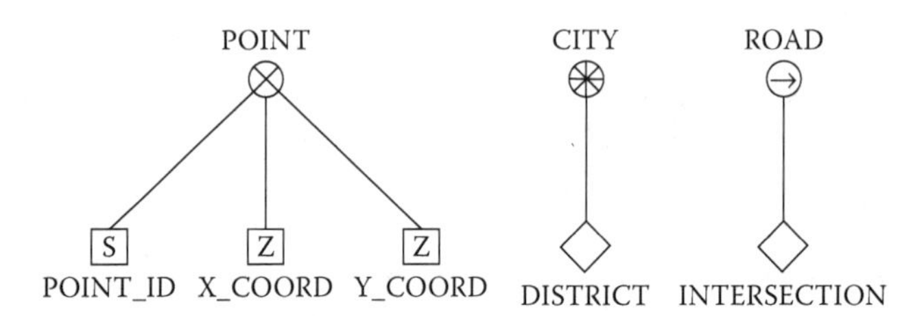

FIGURE 6.3 POINT is an aggregation of identifier and coordinates. CITY is an association of DISTRICT. ROAD is an ordered association of INTERSECTION.

as having two spatial coordinates. This relationship is represented diagrammatically in Figure 6.3.

4.4 Association

Association or grouping is the construct which enables a set of objects of the same type to form an object of higher level type. It is often stipulated that the sets are finite. The corresponding set-theoretic construct is the power-set operator. An example is the view of a city as, amongst other things, a collection of districts. CITY is an association of DISTRICT. This is shown diagrammatically in Figure 6.3.

4.5 Ordered association

Sometimes, it is important to take into account the ordering of a collection of objects. For example, the ordering of intersections making up a road may be critical. Ordered sets are called lists and we allow a higher-order type which is a collection of lists of the lower-order type. In our example we say that ROAD is an ordered association of INTERSECTION. This is shown diagrammatically in Figure 6.3.

5 Object-oriented data modelling

In describing the above abstraction constructs, we have gradually moved towards an object-oriented view. In the basic E-R model, the entities are conceived as having attributes, occurrences of which are drawn from atomic domains. That is, the underlying domains are of basic and indecomposable types such as INTEGER, REAL and STRING. As we bring in the abstractions above, we add a further dimension to the structure of the underlying domains, which no longer need be atomic.

In object-oriented data modelling, all conceptual entities are modelled as objects. An abstraction representing a collection of objects with properties in common is called an object type. Objects of the same type share common functions. The objects associated with an object type are called occurrences. INTEGER and STRING are object types, as is a complex assembly such as a CITY. Indecomposable object types are called primitive. Decomposable objects are called composite or complex objects. A composite object, therefore, is an object with a hierarchy of component objects.

We have seen how complex types may be formed from primitive types using generalization, specialization, aggregation and grouping. These are the primary object-type operations in object-oriented data modelling. Other operations have been introduced and can be found in the literature.

Object-oriented data models support the description of both the structural and the behavioural properties of a database. Structural properties concern the static organizational nature of the database. Behavioural properties are dynamic and concern the nature of possible allowable changes to the information in the database. This paper concentrates on the structural description.

The object-oriented approach to data modelling has proved to be especially fruitful in application areas which are not of the standard corporate database type. Complex molecular and engineering part-assembly databases are examples of systems which have been successfully modelled using these techniques. What such applications have in common is a richly-structured semantic domain, often with a hierarchical emphasis, and associated with multimedia database (e.g., text, numeric, graphical, audio). Since a GIS also shows these characteristics, it seems then that a GIS is an ideal application for object-oriented modelling.

In order to show more clearly how this methodology may be applied, we will concentrate on the specific recent data model IFO (Abiteboul and Hull 1987) which contains the above object-oriented constructs. It is concerned almost wholly with structural properties of a database.

6 IFO

The IFO model was introduced by Abiteboul and Hull (1984). A more condensed account of the model is given by Abiteboul and Hull (1987). IFO incorporates all the constructs so far introduced in this paper with the exception of 'relationship'.

6.1 *Object types*

IFO is truly object-oriented in that all its component types may be composite. Atomic types are of three kinds; printable, abstract and free. A printable type (shown in IFO by a square) corresponds to objects which may be represented directly as input and output. Examples of printable types are INTEGER, STRING, REAL and PIXEL. An abstract type (shown in IFO by a diamond) corresponds to physical or conceptual objects which are not printable. PERSON is an example of an abstract type. Free types (shown in IFO by circles) serve as links in generalization and specialization relationships. Representations of examples of atomic types are given in Figure 6.4. S and Z indicate STRING and INTEGER types respectively. Non-atomic types are constructed from atomic types using aggregation and grouping as already discussed. For diagrammatic clarity, it is sometimes convenient to treat complex types as atomic. For example, Figure 6.3 shows POINT is an aggregate of atomic types but later diagrams treat POINT as abstract atomic.

SETTLEMENT

FIGURE 6.4 Atomic types.

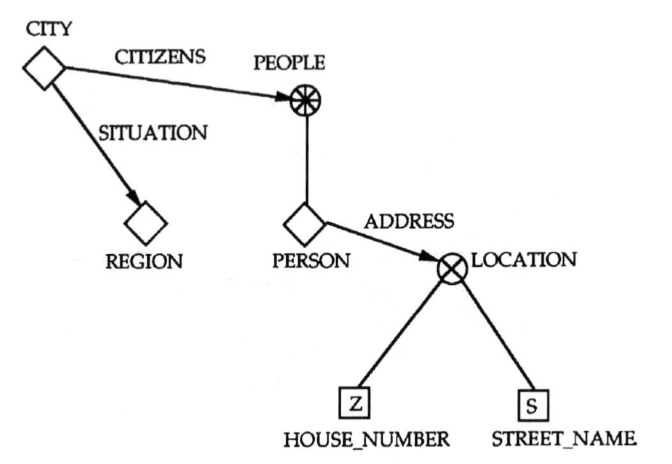

FIGURE 6.5 The CITY fragment if IFO.

6.2 *Functional relationships between objects*

So far the ways in which complex objects may be constructed from atoms have been described. We now discuss how types may be related. IFO provides a formalism for representing functional relationships between types. The means by which functional relationships are represented is the fragment. Informally, a fragment is a part of the IFO model, containing types and functions (but no generalization or specialization relationships), subject to certain constraints. We illustrate with an example, shown in Figure 6.5. This fragment shows functional relationships SITUATION and CITIZENS between object types CITY, REGION and PEOPLE. The structure of the knowledge being modelled here is that cities are situated in regions and are occupied by people, each of whom may have for an address a location which is an aggregation of a house number and street name. The function CITIZENS has the dependent function ADDRESS. Intuitively, this models the case where a person may live in more than one city and so have different addresses in different cities.

The E-R model allows the possibility of many-valued relationships between types and so appears to be more general than a functional model. However, the grouping operator can be used to provide the facility of representing many-valued functions. For example, the relationship shown in Figure 6.1, where a person may live in several cities and a city comprises many people, is represented functionally in IFO as shown in Figure 6.5, where the image of a city under the function CITIZENS is a set of persons, since it is an object of type PEOPLE, which is an association of PERSON. Formally, a fragment F is a rooted directed tree of types. The root of a fragment is called a primary vertex. Full details of fragments may be found in Abiteboul and Hull (1984).

6.3 *Schemas*

In IFO, fragments form the building blocks of schemas. A schema is the largest IFO unit and is a forest of fragments, possibly connected together at their primary vertices

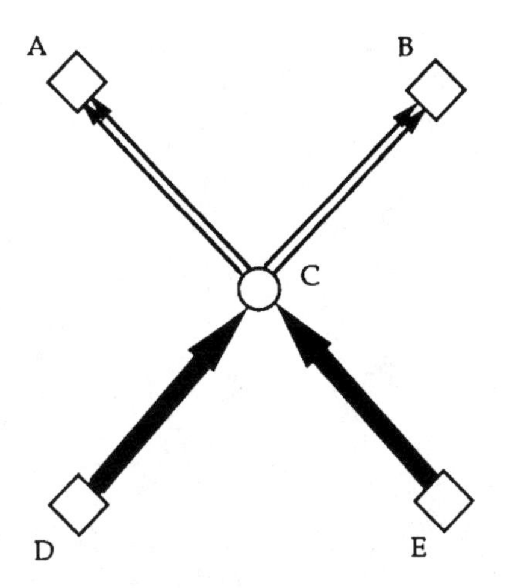

FIGURE 6.6 An inconsistent structuring of objects.

by generalization and specialization edges. Thus the schema allows the representation of all the components of the IFO model.

IFO places some constraints on the way that schemata may be constructed. These concern the directions in which objects are structured from other objects. For example, aggregated and grouped objects are constructed from their constituent elements. The non-commutativity of this structuring relationship leads to the distinction made between generalization and specialization. A generalized type is structured from its sub-types. A specialized type is structured from its super-types. The schema must have its object sources and sinks arranged in a consistent fashion. For example, the configuration shown in Figure 6.6 is not permissible as it results in inconsistent structuring of objects. Objects of type C result from specializing objects of types A and B and generalizing objects of types D and E. There is no guarantee that this can be done consistently and that clashes will not result. A further point here is that, even neglecting D and E, C is the result of specializing from two possibly quite distinct types A and B. Again, there is no guarantee that this can be done consistently and we would require that types A and B 'arise' from a common type. If all the arrows in Figure 6.6 were reversed, we would again have an impermissible schema. In this case, C is a source for objects of types A, B, D and E, but is not itself defined in terms of any other type. It could be considered a 'black hole' of the system. Abiteboul and Hull define the permissible configurations by stating five rules which they must satisfy. They also state a theorem showing that any schema satisfying their rules leads to a consistent structuring of objects at each vertex of the schema. The reader is referred to Abiteboul and Hull (1984) for details.

7 Fundamental spatial objects of IFO

The first application of IFO that we present is its use to represent the three fundamental spatial object types; point, line, and polygon. These representations are based upon the definitions proposed by the National Committee for Digital Cartographic

Data Standards (Moellering 1986), which are summarized in van Roessel (1987) as follows

> A point is a zero-dimensional spatial object with coordinates and a unique identifier within the map.
> A line is a sequence of ordered points, where the beginning of the line may have a special start node and the end a special end node.
> A chain is a line which is a part of one or more polygons and therefore also has a left and right polygon identifier in addition to the start and end node.
> A node is a junction or endpoint of one or more lines or chains.
> A ring consists of one or more chains.
> A polygon consists of one outer and zero or more inner rings.

Section 2 shows how it is possible to represent these spatial elements directly using the relational model. An IFO representation of POINT is given in Figure 6.3 and that for NODE, LINE and POLYGON in Figures 6.7, 6.8 and 6.9, respectively. Figure 6.7 shows a node as a special kind of point, with its own node identifier as well as its identifier as a point. In Figure 6.8, a line is modelled as an ordered association of points, with identifier and begin and end nodes. A polygon, in Figure 6.9, is an ordered association of rings, which in turn are ordered associations of chains. Polygons, rings and chains have identifiers. A chain is a special type of line with corresponding left and right polygons.

The aim is a presentation which accords with users' own views of an object and decouples the representation from the implementation.

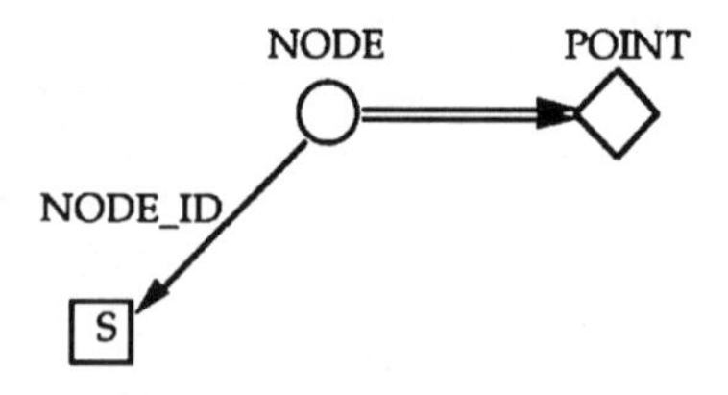

FIGURE 6.7 NODE modelled in IFO.

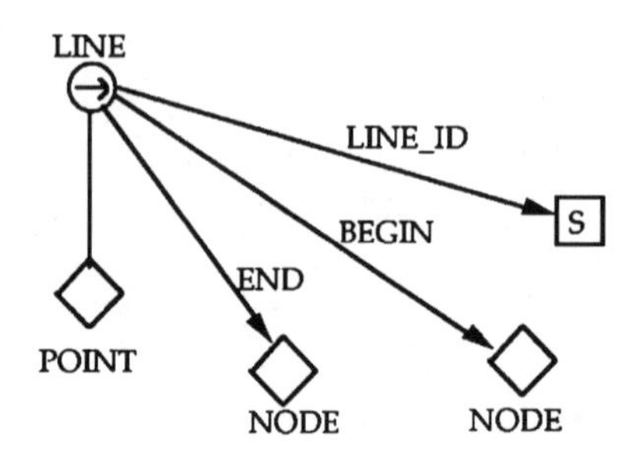

FIGURE 6.8 LINE modelled in IFO.

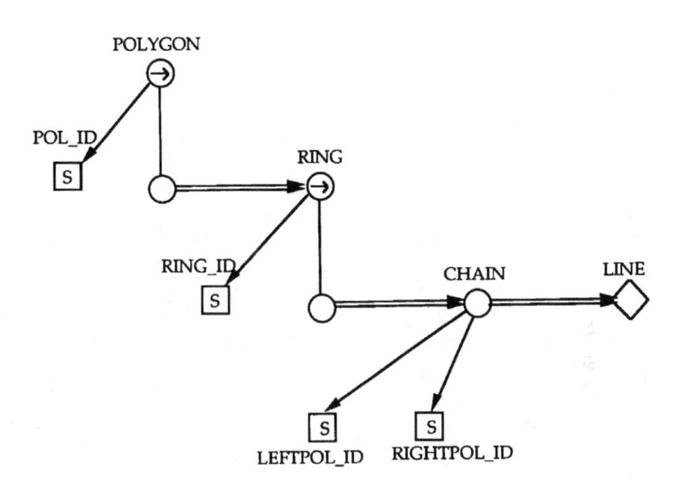

FIGURE 6.9 POLYGON modelled in IFO.

8 Spatial units for Leicestershire in IFO

The second and more substantial application is the modelling of spatial units and their relationships using our object-oriented formalism. Leicestershire is a county located in the Midlands, England. Its central city, Leicester, is the administrative and industrial heart of the county, with half a dozen county towns around the perimeter (Figure 6.10). At the Midlands Regional Research Laboratory (MRRL), in setting up a geographical database for Leicestershire which contains a wide range of data sets covering the county, one of the major problems is the diversity of spatial units on which the data are reported (Hearnshaw et al. 1989). The relationships between the spatial units bearing data and the lack of those relationships are of significance in the design of the database. This section gives examples of IFO models of the post-code-based and census/administrative units.

8.1 The post-code system

The United Kingdom is divided into 120 post-code areas, of which Leicester is one, designated LE. The post-code area, LE, is divided into 17 district post-codes (LEI to LE17). These in turn are divided into 86 sector post-codes (LE1 7, LE16 6, etc.). Sectors are divided into the smallest of the post-code areal units: the unit post-code (UPC) e.g., LE1 7RH, which is a unique identifier for all the points of delivery (addresses) on one postman's 'walk'. The post-code address file (PAF) provides the UPC for all addresses in the county. It also provides a coded form of the address: a 4-digit PREMCODE, which consists of the first four characters of information of an address, e.g., 3 Main Street has PREMCODE 3MAI. There are provisions in the PREMCODE for removing ambiguities, and so this, together with the UPC, can uniquely identify every address (Post Office 1985). It can be seen (Figure 6.11) that these units nest neatly into each other in a clearly defined, and clearly identifiable,

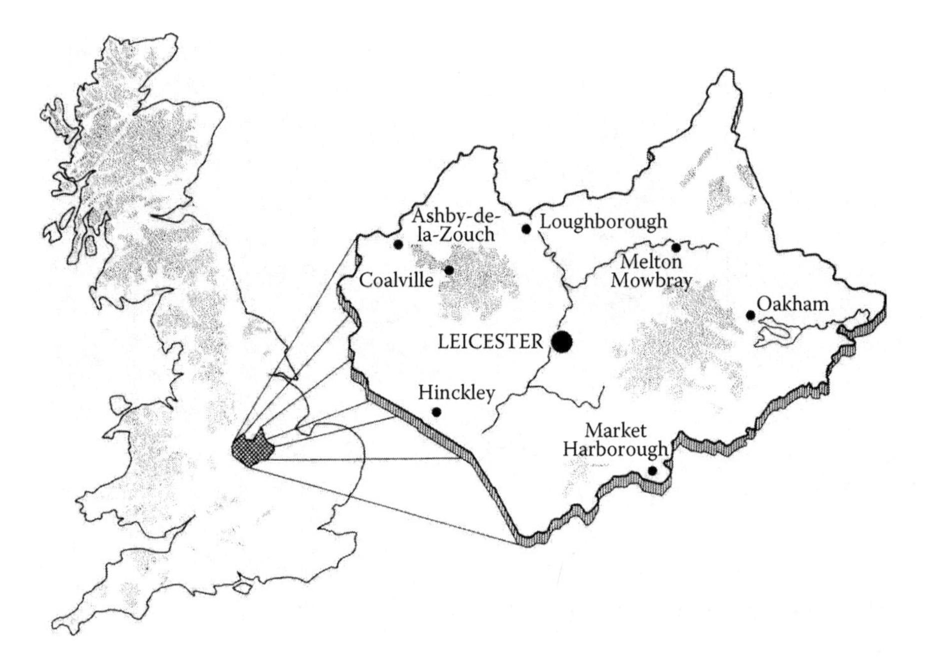

FIGURE 6.10 Map of Great Britain showing Leicestershire (Strachan 1985). Shaded areas show land above 200 m.

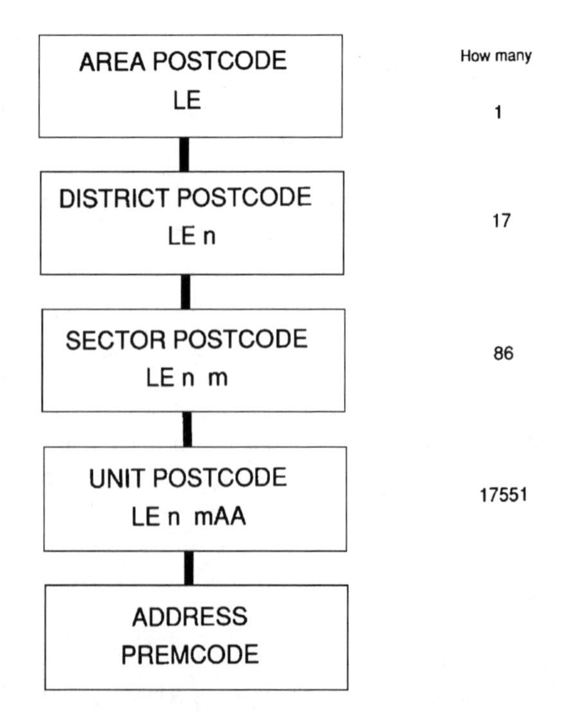

FIGURE 6.11 The post-code units for Leicestershire.

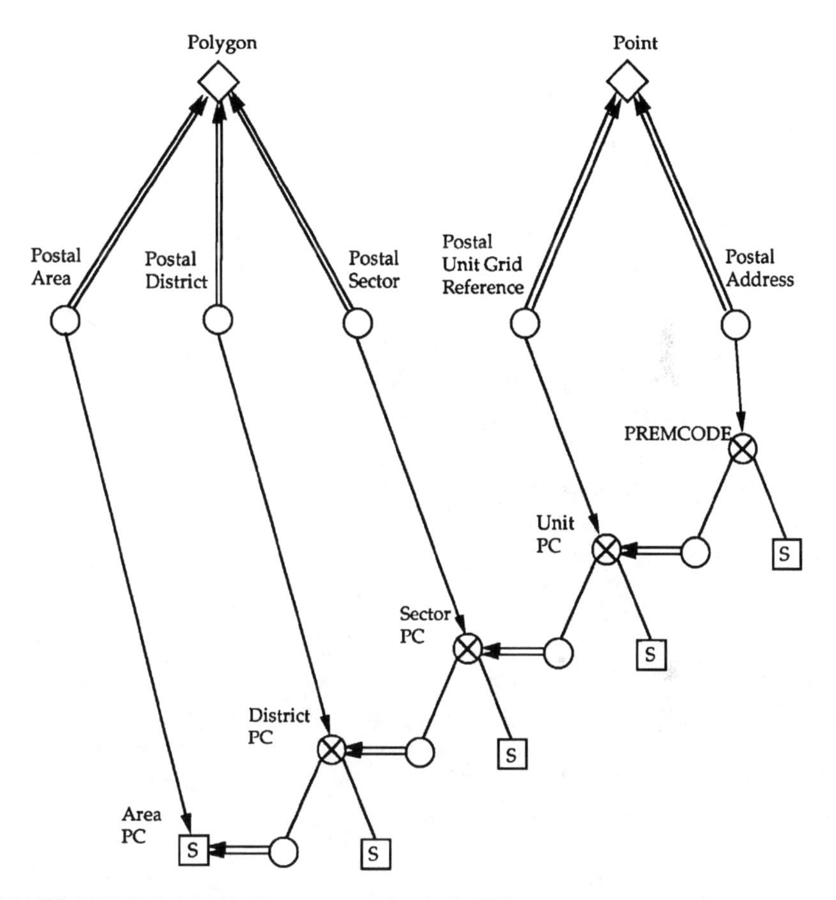

FIGURE 6.12 Relationship between postal units in IFO.

hierarchy. The Central Postcode Directory (CPD) provides the Ordnance Survey Grid reference for the SW corner of a 100m grid square for the first address in each UPC.

Figure 6.12 displays the postal units in IFO. POLYGON and POINT types, defined earlier, are taken as given abstract atomic types. It can be seen that each postal unit type (for, example, POSTAL DISTRICT) is a specialization of an abstract spatial type (for example, POLYGON), and is also the domain of a function whose co-domain is a composite string type which constitutes an identifier (for example, DISTRICT POSTCODE). This representation makes a clear and useful distinction between the spatial and non-spatial aspects of the postal units.

8.2 Census of population and local administrative units

The most recent, decennial Census of Population was taken in 1981. The basic spatial unit on which census data are collected and published in England and Wales is the Enumeration District (ED). The size of an ED varies from about 500 households, in densely populated urban areas, to about 150, in rural areas. The 1981 EDs were

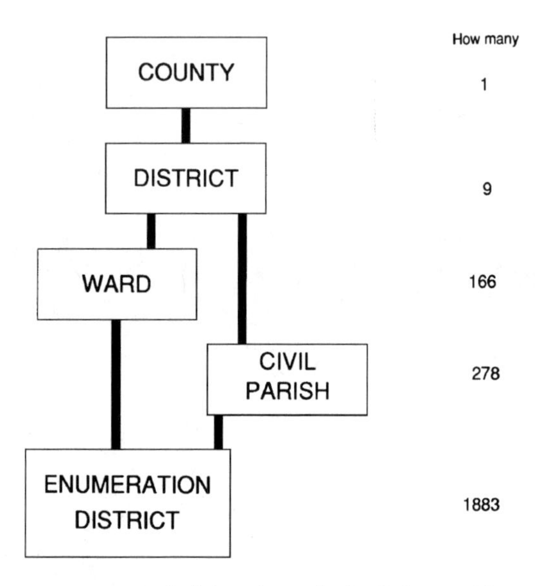

FIGURE 6.13 Some census and administrative units for Leicestershire.

designed to partition district electoral wards, that is, a collection of non-overlapping EDs exactly cover each ward. Wards also partition the districts. The units nest neatly into each other in a well-defined, and clearly identifiable, hierarchy, as shown in Figure 6.13. In the case of Leicestershire, those parts which used to be rural districts are divided into civil parishes. The 1981 EDs were also designed to partition civil parishes.

The Small Area Statistics (SAS) files of the 1981 Census of Population include a 12-digit grid reference, for the weighted centre population (centroid) for each ED. Thus the ED data can, in theory, be related to other spatial units, such as post-codes, via grid references (Gattrell 1988).

Each ED is uniquely identifiable by a six character code. The leftmost two characters define the district, the middle two characters define the ward, and the rightmost two define the ED. For example, ED code ABCD04 defines the fourth ED in ward CD of district AB. Each county also has a 2-digit code: that for Leicestershire is 32.

The administrative units corresponding to the 1981 census are represented in IFO in Figure 6.14. COUNTY, DISTRICT, WARD and PARISH are specializations of POLYGON. ENUMERATION DISTRICT has a CENTROID which is of type POINT. All the units have NAMEs, and there exist functions from all unit types, except PARISH, to identifiers which are nested as shown in the diagram. PARISHes may be associated with sets of enumeration district identifiers, (i.e., identifiers for those EDs which are contained in each parish).

An example of the relationships between postal and census units is given in Figure 6.15 where the functional relationship between postal units and wards provided by the Central Postcode Directory is given. With each postal unit is associated a unique ward.

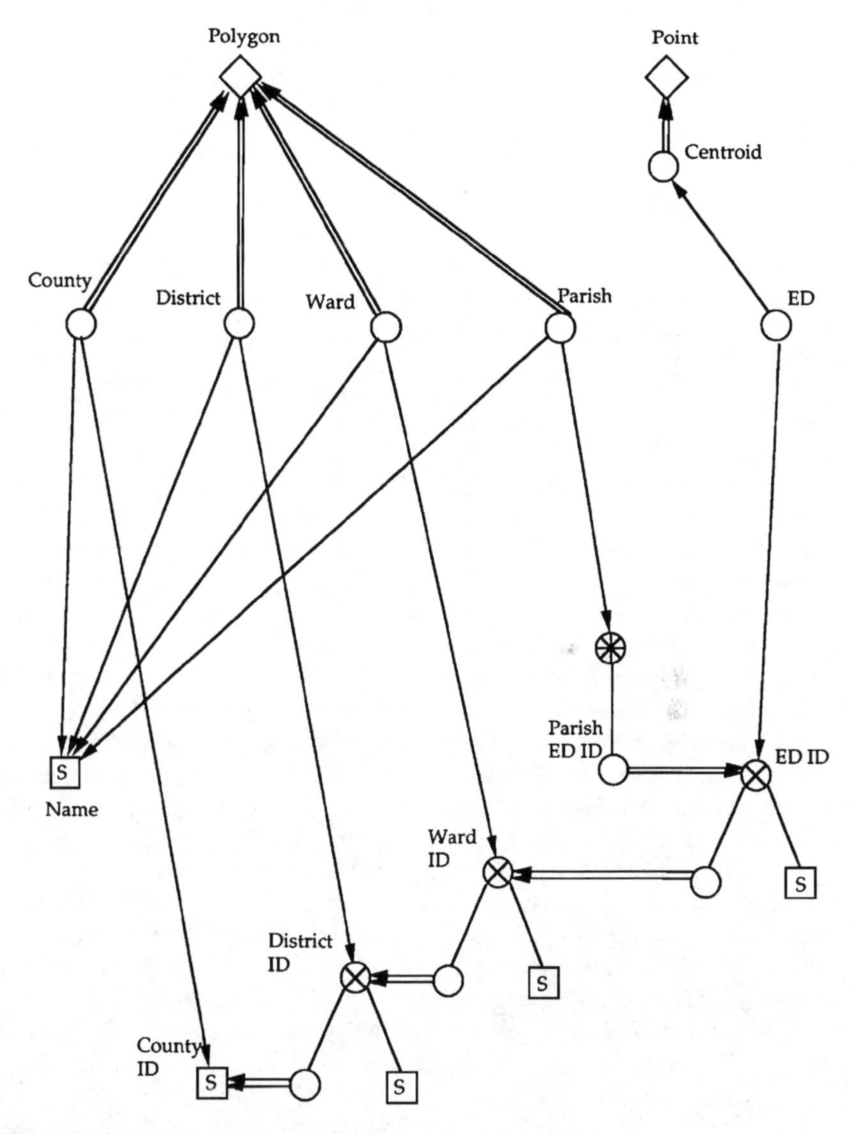

FIGURE 6.14 Relationship between 1981 census units in IFO.

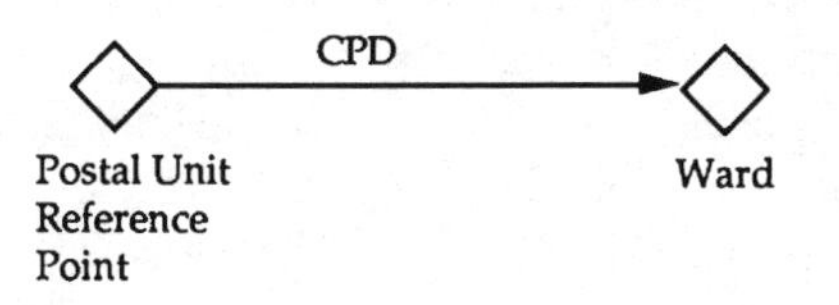

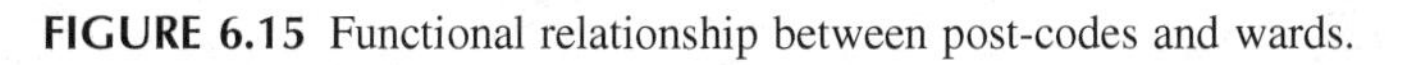

FIGURE 6.15 Functional relationship between post-codes and wards.

Relationships between different spatial units are considerably complicated by boundary redefinitions, for example, corresponding to different census years. These problems are handled in object-oriented systems by means of version control, where facilities exist for the definition of different versions of an object.

9 Implementation and research issues

The discussion up to this point has focused upon data modelling in an object-oriented setting. When the modelling is complete there follows the implementation stage. The system designer must then decide upon the DBMS which will implement the application. At present, the relational systems are by far the most popular for most applications, using standard commercial software such as IBM's DB2, ORACLE, INGRES and others. Under development, and shortly due on the market, are object-oriented systems. These provide a direct link with the modelling tools discussed here and, once proved efficient, will be ideal implementation tools.

If a relational DBMS is chosen for implementation, the object-oriented data model must be transformed into the relational model. It has been shown (Blaha et al. 1988) that many of the constructs discussed in this paper can be mapped to the relational model, although losing some of their natural meaning for the user and some efficiency in the process. However, a unified treatment of the relationship between IFO and the relational model has not yet been given.

At present, the user of a spatial database that is implemented on a general-purpose DBMS has available only the general purpose query language provided by the system. The industry standard query language SQL (Structured Query Language) is satisfactory for interrogating databases of the standard corporate type, but is less than ideal for databases with a complex object structure. It would be better to have a language which matched the data model in some way, with facilities for handling spatial and object-oriented constructs. This research area promises to be fruitful (Frank 1988). There are two main approaches to the construction of a spatial query language. Either one starts from fundamentals or one builds on top of an existing language such as SQL. At present, the latter is the favoured approach. However, it is well recognized within the database community (see, for example, Whittington 1988) that SQL itself has shortcomings. The authors are engaged in the construction of a spatial query language using the former approach.

A major concern of computer science at present is the problem of correctness of systems. Computers are now used to control many systems where safety is the highest priority. The correct functioning of software developed for such safety-critical systems should be guaranteed. The methodologies being developed to deliver such guarantees are based upon the formal specification of systems, including database systems. Research needs to be undertaken which leads to the formal specification of GIS. The IFO model, since it is itself formally defined, is an excellent vehicle for such work. Already, Abiteboul and Hull (1984) have described work which shows the effects of updates on the integrity of object-oriented systems. Research is in progress at the MRRL which considers these questions in the special light of spatial databases.

10 Conclusions

This paper has traced the development of data modelling from the relational model to a contemporary object-oriented method and suggested, with examples, some of the applications of object-oriented modelling to geographical information systems. It is argued that such methodologies offer clear advantages over traditional methods such as E-R modelling. In particular, object-oriented modelling allows database designers to incorporate more readily the complexities of spatial data.

Acknowledgment

Acknowledgment is made of the help provided by colleagues Alan Strachan and Bill Hickin of the MRRL, and Graham Winter of the Research and Information Group, Leicestershire County Council Department of Planning and Transportation.

References

Abiteboul, S., and Hull, R., 1984, IFO: A Formal Semantic Database Model, TR-84-304, Computer Science Department, University of Southern California.

Abiteboul, S., and Hull, R., 1987, IFO: A formal semantic database model. *Association for Computing Machinery Transactions on Database Systems*, **12,** 525.

Blaha, R. B., Premerlani, W. J., and Rumbaugh, J. E., 1988, Relational database design using an object-oriented methodology. *Communications of the Association for Computing Machinery*, **31,** 414.

Calkins, H. W., and Marble, D. F., 1987, The transition to automated production cartography: design of the master cartographic database. *American Cartographer*, **14**, 105.

Chen, P. P.-S., 1976, The entity-relationship model–toward a unified view of data. *Association for Computing Machinery Transactions on Database Systems*, **1**, 3.

Codd, E. F., 1970, A relational model of data for large shared data banks. *Communications of the Association for Computing Machinery*, **13,** 377.

Egenhofer, M. J., and Frank, A. U., 1989, Object-oriented modeling in GIS: inheritance and propagation. *Proceedings of Auto-Carto 9 held in Baltimore, Maryland, in April 1989*, edited by E. Anderson (Falls Church: American Congress on Surveying and Mapping American Society of Photogrammetry and Remote Sensing), pp. 588–598.

Frank, A. U., 1988, Requirements for a database management system for a GIS *Photogrammetric Engineering and Remote Sensing*, **54,** 1557.

Gattrell, A. C., 1989, On the spatial representation and accuracy of address-based data in the U.K., *International Journal of Geographical Information Systems*, **3**, 335.

Hearnshaw, H. M., Maguire, D. J., and Worboys, M. F., 1989, An introduction to area-based spatial units: a case study of Leicestershire. Midlands Regional Research Laboratory Research Report No. 1, University of Leicester.

Kim, W., and Lochovsky, F. H. (editors), 1989, *Object-Oriented Concepts, Databases, and Applications* (Berkshire: Addison-Wesley).

Moellering, H., 1986, A review and definition of 0-, 1- and 2-dimensional objects for digital cartography. *Proceedings of the 2nd International symposium on Spatial Data Handling* (Ohio: International Geographical Union Commission on Geographical Data Sensing and Processing).

Peckham, J., and Maryanski, F., 1988, Semantic data models. *Association for Computing Machinery Computer Surveys*, **20**, 153.
Post Office, 1985, *The Postcode Address File Digest* (London: Post Office).
Smith, J. M., and Smith, D. C. P., 1977, Database abstractions: aggregation and generalization. *Association for Computing Machinery Transactions on Database Systems*, **2**, 105.
Strachan, A. J., 1985, Atlas of Leicestershire, County and Employment Research Unit, University of Leicester.
van Roessel, J. W., 1987, Design of a spatial data structure using the relational normal form. *International Journal of Geographical Information Systems*, **1,** 33.
Whittington, R. P., 1988, *Database Systems Engineering* (Oxford: Oxford University Press).

Object-Oriented Data Modelling for Spatial Databases: After Fifteen Years

Michael F. Worboys and David J. Maguire

One of the key issues in geographic information system application development is modeling the real world in a digital computer system. Modeling takes place at several levels, from high-level conceptual domain modeling and user requirement analysis, through logical modeling for the database, down to low-level (implementation-specific) physical data structures. At the conceptual or domain level, the modeling phase is particularly critical because on it depends the final system's fitness for purpose. Therefore, the whole development process rests upon this first step. So how do we construct a conceptual model, and what approaches and tools are available to help us? It was these questions that prompted the three authors to undertake the research to write this paper (Worboys et al., 1990).

In the late 1980s the three authors found themselves involved in a relatively new enterprise, the Midlands Regional Research Laboratory (MRRL), part of the U.K.-wide funding of geographical information system (GIS) research initiated by the Economic and Social Research Council (ESRC). The MRRL was based at Leicester and Loughborough Universities, in the United Kingdom, and one of its projects was to consider the linkage of GIS with general-purpose database management systems, such as Oracle.

In the beginning we approached the development of geographic databases in much the same manner as we would develop any general relational system. In particular, we began to use the entity-relationship-attribute (ERA) model (Chen 1976) as the foundation for conceptual modeling. The ERA model is beautifully spare and simple, founded as it is just on the concepts of entity, relationship between entities, and attributes of entities, and it works very well for simple business applications. The problem came when we started to use it to model geographies. The geographic domain is characterized by complexity, both in terms of parthood, in which objects are components of other objects, and in terms of taxonomy, in which object types are subtypes of other types. ERA does not really have the machinery to fully deal with such complexity, and at the time it was found that several application domains could not be modeled with efficiency and completeness. These domains include CAD/CAM, engineering databases, text retrieval systems, and GIS.

So the question became, "What approaches are appropriate for modeling geographic phenomena?" and the answer proposed in the paper was based upon object-oriented approaches. The paper reviews the ERA model, discusses its shortcomings in geographic domains, and introduces the object-oriented approach as a solution. We chose to develop the presentation of the object-oriented approach, using a specific formalization due to Abiteboul and Hull (1987), called the IFO model, introduced originally to provide a sound theoretical basis for semantic data models. IFO is an acronym for IS-a relationships, Fragments, and Objects — the main constructs that are used to model complex systems. At that time, several kinds of semantic data models were being explored by the database community as a means of providing a richer array of modeling tools (see Peckham and Maryanski, 1988, for a review). Other modeling approaches being explored included the functional data model (Buneman and Nikhil, 1984) and extensions to the relational model (Codd, 1979). In the paper we demonstrated the power of the object-oriented approach by applying it to a model of administrative units in the United Kingdom. Although this was a relatively simple model, it served the purpose of illustrating the value of this approach.

So how has the object-oriented approach fared in the 15 years since the paper was published? From the point of view of domain conceptual modeling, there is no doubt that object-orientation, with its rich collection of semantic constructs that allow the bridging of the gap between user and machine, has been the preeminent modeling approach. Its impact is shown by the following quote from the *Open Geospatial Consortium Reference Model* document (Percivall, 2003, 48): "This document defines that set of vendor-neutral, object-oriented geometric and geographic object abstractions." We see also how modeling tools (e.g., the Unified Modeling Language UML [www.uml.org]), and programming languages (e.g., JAVA [java.sun.com] and C# [msdn.microsoft.com/vcsharp/]), are now well-accepted manifestations of the object-oriented approach.

From the database system perspective, pure object-oriented database management systems (ODBMS) have been less successful, not providing the level of performance required of a production database system for very large applications. Nevertheless, many object-oriented concepts can be found in general purpose DBMS, such as DB2, Informix, ORACLE, and SQL Server, and in specialized GIS software, such as ArcGIS. For example, the geodatabase of ArcGIS comprises a general object-oriented data model template that can be implemented in the users' choice of DBMS. The geodatabase embodies the key concepts outlined in our original article. The object-oriented constructs are implemented in a series of object-components (reusable software classes) that are bound together at run-time with data retrieved from a set of database tables. In this way the best elements of the relational model (simple, well understood, high performance data management) and the object model (rich semantic modeling) are combined. A further advantage of this approach is that complex GIS data models can be built as extensions of simpler relational models used in business systems, and in this way existing applications can be "spatially enabled."

The paper concluded by selecting some research issues that would be interesting avenues for further investigation. The topics identified were:

1. Object-oriented extensions to the relational model;
2. Object-oriented query interfaces to geographic databases;
3. The application of object-oriented approaches, and particularly the formal specification provided by IFO, to safety-critical systems.

The first of these has become an important issue. Object-oriented and spatial extensions to the relational model, discussed above, are now important components of many proprietary DBMS. The second has been indirectly realized as part of the first. For example, IBM DB2, Informix Spatial Extender, and Oracle Spatial have query language constructs that have been added to SQL. Indeed the SQL standard has been extended with spatial SQL constructs (ISO/IEC 2004).

The last issue regarding safety-critical GIS is interesting, and reflects something of the concerns in computer science in the late 20th century. While safety-critical features have not become a very big issue for GIS, we see something of the same kind of focus in today's discussions about information privacy for location-based systems (Petitcolas et al., 1999, Myles et al., 2003).

In spite of the considerable progress that has been made in data modeling in the past 15 years, much more remains to be done in this critical area of GIS. In particular, we note the following areas that require urgent attention:

1. Lack of good CASE (computer-aided software engineering) tools for modeling geographic processes. Rational Rose, Visio, and other commercial off-the-shelf products fall some way short of modeling rich geographic relationships. These tools still reflect the relational paradigm and cannot deal with geographic constructs such as topologies, linear referencing, and geometric networks.
2. Paucity of detailed knowledge about the performance problems of using complex models with large geographic databases. In general, a fully normalized data model with a large number of object classes (tables) and relationships will not provide good query performance for typical geographic queries because of the cost of relational joins. Denormalization of data models should be considered in order to meet query performance demands.
3. The relatively crude nature of spatial SQL for modeling sophisticated geographic patterns and processes. The restriction to simple features (point, line, and polygon) absence of support for 3-D and important constructs like topology and many key data types such as survey measurements, CAD, and addresses, are significant limitations (see Hoel et al., 2003 for a commentary on the topology case).
4. Only very recently have people started to extend generic data models into specific application domains (Maguire and Grise, 2001; Arctur and Zeiler, 2004) and much remains to be done here in the area of "best practices." Until a wide set of domain data models is available we will not be able to understand the efficiencies of specific approaches and each modeler will be forced to discover them for themselves.
5. More generally, work on conceptual data models can be seen within the broader area of ontology development (Fonseca et al., 2003). There is still plenty of scope for development of the role of ontologies in the context of geospatial data, particularly when taking account of the dynamic aspects of much geographic phenomena, as in the next item.

6. The dynamic nature of many geographic phenomena, from environmental features to transportation, are leading researchers to explore modeling techniques that are not only centered on static features, but also events and processes (Worboys and Hornsby, 2004; Worboys, 2005).

References

Abiteboul, S. and Hull, R., 1987, IFO: a formal semantic database model, *ACM Transactions on Database Systems*, 12 (4), 525–565.

Arctur, D. and Zeiler, M., 2004, *Designing geodatabases: case studies in GIS data modeling*, ESRI Press, Redlands, CA.

Buneman, O.P. and Nikhil, R., 1984, The functional data model and its uses for interaction with databases. In *On Conceptual Modelling, Perspectives from Artificial Intelligence, Databases, and Programming Languages*, M.L. Brodie, J. Mylopoulos, and J.W. Schmidt, Eds., Springer-Verlag, New York, 359–380.

Chen, P.P., 1976, The entity-relationship model — toward a unified view of data, *ACM Transactions on Database Systems*, **1**(1), 9–36.

Codd, E.F., 1979, Extending the database relational model to capture more meaning, *ACM Transactions on Database Systems*, 4 (4), 397–434.

Fonseca, F., Davis, C., and Camara, G., 2003, Bridging ontologies and conceptual schemas in geographic information integration, *Geoinformatica*, 7(4), 355–378.

Hoel, E., Menon, S., and Morehouse, S., 2003, Building a robust relational implementation of topology. In: *Advances in Spatial and Temporal Databases. Proceedings of 8th International Symposium, SSTD 2003 Lecture Notes in Computer Science*, Vol. 2750, T. Hadzilacos, Y. Manolopoulos, J.F. Roddick and Y. Theodoridis, Eds., Springer-Verlag, Berlin, 508–524.

ISO/IEC 2004, ISO/IEC 13249 Information technology — Database languages — SQL Multimedia and Application Packages: Part 3: Spatial.

Maguire, D.J. and Grise, S., 2001, Data models for object-component technology. In *Proceedings of GITA 2001*, http://www.gisdevelopment.net/proceedings/gita/2001/system/sa002.asp, accessed 26 April 2006.

Myles, G., Friday, A., and Davies, N., 2003, Preserving privacy in environments with location-based applications, *Pervasive Computing*, 2(1), 56–64.

Peckham, J. and Maryanski, F., 1988, Semantic data models, *ACM Computer Surveys*, 20(3), 153–189.

Percivall, G., 2003. *Open Geospatial Reference Model.* Version 0.1.3. Open Geospatial Consortium. http://www.opengeospatial.org, accessed 16 December 2005.

Petitcolas, F., Anderson, R., and Kuhn, M., 1999, Information hiding — a survey. *Proceedings of the IEEE*, 87(7), 1062–1078.

Worboys, M.F., 2005, Event-oriented approaches to geographic phenomena, *International Journal of Geographic Information Science*, 19(1), 1–28.

Worboys, M.F., Hearnshaw, H.M., and Maguire, D.J., 1990, Object-oriented data modelling for spatial databases, *International Journal of Geographical Information Systems*, 4(4), 369–383.

Worboys, M.F. and Hornsby, K., 2004, From objects to events: GEM, the geospatial event model. In *Proceedings of the Third International Conference on GIScience 2004,* Lecture Notes in Computer Science 3234, M. Egenhofer, C. Freksa, and H. Miller, Eds., Springer-Verlag, Berlin, 327–344.

International Journal of Geographical Information Systems, 1991, Vol. 5, No. 2, 161–174.

7 Point-Set Topological Spatial Relations

Max J. Egenhofer and Robert D. Franzosa

Abstract. Practical needs in geographic information systems (GIS) have led to the investigation of formal and sound methods of describing spatial relations. After an introduction to the basic ideas and notions of topology, a novel theory of topological spatial relations between sets is developed in which the relations are defined in terms of the intersections of the boundaries and interiors of two sets. By considering *empty* and *non-empty* as the values of the intersections, a total of sixteen topological spatial relations is described, each of which can be realized in R^2. This set is reduced to nine relations if the sets are restricted to spatial regions, a fairly broad class of subsets of a connected topological space with an application to GIS. It is shown that these relations correspond to some of the standard set theoretical and topological spatial relations between sets such as equality, disjointedness and containment in the interior.

1 Introduction

The work reported here has been motivated by the practical need for a formal understanding of spatial relations within the realm of GIS. To display, process or analyze spatial information, users select data from a GIS by asking queries. Almost any GIS query is based on spatial concepts. Many queries explicitly incorporate *spatial relations* to describe constraints about spatial objects to be analyzed or displayed. For example, a GIS user may ask the following query to obtain information about the potential risks of toxic waste dumps to school children in a specific area: "Retrieve all toxic waste dumps which are within 10 miles of an elementary school and located in Penobscot Country and its adjacent counties". The number of elementary schools known to the information system is restricted by using the formulation of constraints. Of particular interest are the spatial constraints expressed by *spatial relations* such as *within 10 miles*, *in*, and *adjacent*.

The lack of a comprehensive theory of spatial relations has been a major impediment to any GIS implementation. The problem is not only one of selecting the appropriate terminology for these spatial relations, but also one of determining their semantics. The development of a theory of spatial relations is expected to provide answers to the following questions (Abler 1987):

- What are the fundamental geometric properties of geographic objects needed to describe their relations?
- How can these relations be defined formally in terms of fundamental geometric properties?
- What is a minimal set of spatial relations?

In addition to the purely mathematical aspects, cognitive, linguistic and psychological considerations (Talmy 1983, Herskovits 1986) must also be included if a theory about spatial relations, applicable to real-world problems, is to be developed (NCGIA 1989). Within the scope of this paper, only the formal, mathematical concepts which have been partially provided from point-set topology will be considered.

The application of such a theory of spatial relations exceeds the domain of GIS. Any branch of science and engineering that deals with spatial data will benefit from a formal understanding of spatial relations. In particular, its contribution to spatial logic and spatial reasoning will also be helpful in areas such as surveying engineering, computer-aided design/computer-aided manufacturing (CAD/CAM), robotics and very large-scale integrated (VLSI) design.

The variety of spatial relations can be grouped into three different categories: (1) topological relations which are invariant under topological transformations of the reference objects (Egenhofer 1989, Egenhofer and Herring 1990); (2) metric relations in terms of distances and directions (Peuquet and Ci-Xiang 1987); and (3) relations concerning the partial and total order of spatial objects (Kainz 1990) as described by prepositions such as *in front of*, *behind*, *above* and *below* (Freeman 1975, Change *et al.* 1989, Hernández 1991). Within the scope of this paper, only topological spatial relations are discussed.

Formalisms for relations have so far been limited to simple data types in a one-dimensional space such as integers, reals, or their combinations, e.g., as intervals (Allen 1983). Spatial data, such as geographic objects or CAD/CAM models, extend in higher dimensions. It has been assumed that a set of primitive relations in such a space is richer, but so far no attempt has been made to explore this assumption systematically.

The goal of this paper is two-fold. First, to show that the description of topological spatial relations in terms of topologically invariant properties of point-sets is fairly simple. As a consequence, the topological spatial relation between two point-sets may be determined with little computational effort. Second, to show that there exists a framework within which any topological spatial relation falls. This does not state that the set of relations determined by this formalism is complete, i.e., humans may distinguish additional relations, but that the formalism provides a complete coverage, i.e., any such additional relation will be only a specialization of one of the relations described.

As the underlying data model, subsets of a topological space were selected. The point-set approach is the most general model for the representation of topological spatial regions. Other approaches to the definition of topological spatial relations using different models, such as intervals (Pullar and Egenhofer 1988), or simplicial complexes (Egenhofer 1989), are generalized by this point-set approach.

This paper is organized as follows. The next section reviews previous approaches to defining topological spatial relations. Section 3 summarizes the relevant concepts of point-set topology and introduces the notions used in the remainder of the paper. Section 4 introduces the definition of topological spatial relations and shows their realization in R^2. Section 5 investigates the existence of the relations between two spatial regions, subsets of a topological space with particular application to geographic data handling. In Section 6, the relations within $R^n (n \geq 2)$ and R^1 are compared.

2 Previous work

Various collections of terms for spatial relations can be found in the computer science and geography literature (Freeman 1975, Claire and Guptill 1982, Change *et al.* 1989, Molenaar 1989). In particular, designs of spatial query languages (Frank 1982, Ingram and Phillips 1987, Smith *et al.* 1987, Herring *et al.* 1988, Roussopoulos *et al.* 1988) are a reservoir for informal notations of spatial relations with verbal explanations in natural language. A major drawback of these terms is the lack of a formal underpinning, because their definitions are frequently based on other expressions which are not exactly defined, but are assumed to be generally understood.

Most formal definitions of spatial relations describe them as the results of binary point-set operations. The subsequent review of these approaches will show their advantages and deficiencies. It will be obvious that none of the previous studies has been performed systematically enough to be used as a means to prove that the relations defined provide a complete coverage for the topological spatial relation between two spatial objects. Some definitions consider only a limited subset of representations of 'spatial objects', whereas others apply insufficient concepts to define the whole range of topological spatial relations.

A formalism using the primitives *distance* and *direction* in combination with the logical connectors *AND*, *OR* and *NOT* (Peuquet 1986) will not be considered here. The assumption that every space has a metric is obviously too restrictive so that this formalism cannot be applied in a purely topological setting.

The definitions of relations in terms of set operations use pure set theory to describe topological relations. For example, the following definitions based on point-sets have been given for *equal*, *not equal*, *inside*, *outside* and *intersects* in terms of the set operations $=$, $\neq$ $\subseteq$ and $\cap$ (Güting 1988):

$$x = y: = \text{points}\ (x) = \text{points}\ (y)$$

$$x \neq y:= \text{points}\ (x) \neq \text{points}\ (y)$$

$$x \text{ inside } y:= \text{points}\ (x) \subseteq \text{points}\ (y)$$

$$x \text{ outside } y:= \text{points}\ (x) \cap \text{points}\ (y) = \emptyset$$

$$x \text{ intersects } y:= \text{points}\ (x) \cap \text{points}\ (y) \neq \emptyset$$

The drawback of these definitions is that this set of relations is neither orthogonal nor complete. For instance, *equal* and *inside* are both covered by the definition of *intersects*. In contrast, the model of point-sets *per se* does not allow the definition of those relations that are based on the distinction of particular parts of the point-sets such as the boundary and the interior. For example, the relation *intersects* is topologically different from that where common boundary points exist, but no common interior points are encountered.

The point-set approach has been augmented with the consideration of *boundary* and *interior* so that *overlap* and *neighbor* can be distinguished (Pullar 1988):

$$x \text{ overlay } y := \text{boundary }(x) \cap \text{boundary }(y) \neq \emptyset \text{ and}$$
$$\text{interior }(x) \cap \text{interior }(y) \neq \emptyset$$

$$x \text{ neighbor } y := \text{boundary }(x) \cap \text{boundary }(y) \neq \emptyset \text{ and}$$
$$\text{interior }(x) \cap \text{interior }(y) = \emptyset$$

In a more systematic approach, boundaries and interiors have been identified as the crucial descriptions of polygonal intersections (Wagner 1988). By comparing whether or not boundaries and interiors intersect, four relations have been identified: (1) *neighborhood* where boundaries intersect, but interiors do not; (2) *separation* where neither boundaries nor interiors intersect; (3) *strict inclusion* where the boundaries do not intersect, but the interiors do; and (4) *intersection* with both boundaries and interiors intersecting. This approach uses a single, coherent method for the description of topological spatial relations, but it is not carried out in all its consequences. For example, no distinction can be made between *intersection* and *equality*, because for both relations boundaries and interiors intersect.

3 Point-set topology

This model of topological spatial relations is based on the point-set topological notions of *interior* and *boundary*. In this section the appropriate definitions and results from point-set topology are presented. Some of the results are stated without proofs. Those proofs are all straightforward consequences of the definitions and can be found in most basic topology textbooks, e.g., Munkres (1966) and Spanier (1966).

Let X be a set. A *topology* on X is a collection $\mathcal{A}$ of subsets of X that satisfies the three conditions: (1) the empty set and X are in $\mathcal{A}$; (2) $\mathcal{A}$ is closed under arbitrary unions; and (3) $\mathcal{A}$ is closed under finite intersections. A *topological space* is a X with a topology $\mathcal{A}$ on X. The sets in a topology on X are called *open sets*, and their complements in X are called *closed sets*. The collection of closed sets: (1) contains the empty set and X; (2) is closed under arbitrary intersections; and (3) is closed under finite unions.

Via the open sets in a topology on a set X, a set-theoretic notion of closeness is established. If U is an open set and $x \in U$, then U is said to be a *neighborhood* of x.

This set-theoretic notion of closeness generalizes the metric notion of closeness. A metric d on a set X induces a topology on X, called the *metric topology defined by d*. This topology is such that $U \subset X$ is an open set if, for each $x \in U$, there is an $\varepsilon > 0$ such that the d-ball of radius ε around x is contained in U. A d-ball is the set of points whose distance from x in the metric d is less than ε, i.e., $\{y \in X | d(x, y) < \varepsilon\}$.

For the remainder of this paper let X be a set with a topology $\mathcal{A}$. If S is a subset of X then S inherits a topology from $\mathcal{A}$. This topology is called the *subspace topology* and is defined such that $U \subset S$ is open in the subspace topology if, and only if $U = S \cap V$ for some set $V \in \mathcal{A}$. Under such circumstances, S is called a *subspace* of X.

3.1 Interior

Given $Y \subset X$, the *interior* of Y, denoted by Y°, is defined to be the union of all open sets that are contained in Y, i.e., the interior of Y is the largest open set contained in Y. y is in the interior of Y if and only if there is a neighborhood of y contained in Y, i.e., $y \in Y^\circ$ if, and only if, there is an open set U such that $y \in U \subset Y$. The interior of a set could be empty, e.g. the interior of the empty set is empty. The interior of X is X itself. If U is open then $U^\circ = U$. If $Z \subset Y$ then $Z^\circ \subset Y^\circ$.

3.2 Closure

The *closure* of Y, denoted by $\overline{Y}$, is defined to be the intersection of all closed sets that contain Y, i.e. the closure of Y is the smallest closed set containing Y. It follows that y is in the closure of Y if and only if every neighborhood of y intersects Y, i.e., $y \in \overline{Y}$ if and only if $U \cap Y \neq \emptyset$ for every open set U containing y. The empty set is the only set with empty closure. The closure of X is X itself. If C is closed then $\overline{C} = C$. If $Z \subset Y$ then $\overline{Z} \subset \overline{Y}$.

3.3 Boundary

The *boundary* of Y, denoted by ∂Y, is the intersection of the closure of Y and the closure of the complement of Y, i.e. $\partial Y = \overline{Y} \cap \overline{X - Y}$. The boundary is a closed set. It follows that y is in the boundary of Y if and only if every neighborhood of y intersects both Y and its complement, i.e., $y \in \partial Y$ if and only if $U \cap Y \neq \emptyset$ and $U \cap (X - Y) \neq \emptyset$ for every open set U containing y. The boundary can be empty, e.g. the boundaries of both X and the empty set are empty.

3.4 Relationship between interior, closure, and boundary

The concepts of interior, closure and boundary are fundamental to the forthcoming discussions of topological spatial relations between sets. The relationships between interior, closure and boundary are described by the following propositions:

Proposition 3.1. $Y^\circ \cap \partial Y = \emptyset$.

Proof: If $x \in \partial Y$, then every neighborhood U of x intersects $X - Y$ so that U cannot be contained in Y. As no neighborhood U of x is contained in Y it follows that $x \notin Y^\circ$ and, therefore, $\partial Y \cap Y^\circ = \emptyset$. □

Proposition 3.2. $Y^\circ \cup \partial Y = \overline{Y}$

Proof: $Y^\circ \subset Y \subset \overline{Y}$ and, by definition, $\partial Y \subset \overline{Y}$. As Y° and ∂Y are both subsets of $\overline{Y}$ it follows that $(Y^\circ \cup \partial Y) \subset \overline{Y}$. To show that $\overline{Y} \subset (Y^\circ \cup \partial Y)$,let $x \in \overline{Y}$ and assume that $x \notin Y^\circ$. It is shown that $x \in \partial Y$ which, since $x \in \overline{Y}$, only requires showing that $x \in \overline{X - Y}$. $x \notin Y^\circ$ implies that every neighborhood of x is not contained in Y; therefore, every neighborhood of x intersects $X - Y$, implying that $x \in \overline{X - Y}$. So $x \in \partial Y$. Thus if $x \in \overline{Y}$ and $x \notin Y^\circ$, then $x \in \partial Y$ and it follows that $\overline{Y} \subset (Y^\circ \cup \partial Y)$. Thus $\overline{Y} = (Y^\circ \cup \partial Y)$. □

3.5 Separation

The concepts of separation and connectedness are crucial for establishing the forthcoming topological spatial relations between sets. Let $Y \subset X$. A *separation* of Y is a pair, A, B of subsets of X satisfying the following three conditions: (1) $A \neq \emptyset$ and $B \neq \emptyset$; (2) $A \cup B = Y$; and (3) $\overline{A} \cap B = \emptyset$ and $A \cap \overline{B} = \emptyset$. If there exists a separation of Y, then Y is said to be *disconnected*, otherwise Y is said to be *connected*. If Y is the union of two non-empty disjoint open subsets of X, then it follows that Y is disconnected. If C is connected and $C \subset D \subset \overline{C}$, then D is connected. In particular, if C is connected, then $\overline{C}$ is connected; however, ∂C and C° need not be connected.

Proposition 3.3. If A, B form a separation of Y and if Z is a connected subset of Y, then either $Z \subset A$ or $Z \subset B$.

Proof: By assumption, Z is a subset of the union of A and B, i.e. $Z \subset A \cup B$. It is shown that the intersection between Z and one of A or B is empty, i.e., either $Z \cap B = \emptyset$ or $Z \cap A = \emptyset$. Suppose not, i.e., assume that both intersections are non-empty. Let $C = Z \cap A$ and $D = Z \cap B$. Then C and D are both non-empty and $C \cup D = Z$. As $\overline{C} \subset \overline{A}$, $D \subset B$, and $\overline{A} \cap B = \emptyset$ (because A, B is a separation of Y), it follows that $\overline{C} \cap D = \emptyset$. Similarly, $C \cap \overline{D} = \emptyset$; therefore, C and D form a separation of Z, contracting the assumption that Z is connected. So either $Z \cap B = \emptyset$ or $Z \cap A = \emptyset$, implying that either $Z \subset A$ or $Z \subset B$. □

A subset Z of X is said to separate X if $X - Z$ is disconnected. The following separation result gives simple conditions under which the boundary of a subset of X separates X.

Proposition 3.4. Assume $Y \subset X$. If $Y^\circ \neq \emptyset$ and $\overline{Y} \neq X$, then Y° and $X - \overline{Y}$ form a separation of $X - \partial Y$, and thus ∂Y separates X.

Proof: By assumption, Y° and $X - \overline{Y}$ are non-empty. Clearly, they are disjoint open sets. Proposition 3.2 implies that $X - \partial Y = Y^\circ \cup (X - \overline{Y})$. It follows that Y° and $X - \overline{Y}$ form a separation of $X - \partial Y$. □

3.6 Topological equivalence

The study of topological equivalence is central to the theory of topology. Two topological spaces are *topologically equivalent* (*homeomorphic* or *of the same topological type*) if there is a bijective function between them that yields a bijective correspondence between the open sets in the respective topologies. Such a function, which is continuous with a continuous inverse, is called a *homeomorphism*. Examples of homeomorphisms are the Euclidean notions of translation, rotation, scale and skew. Properties of topological spaces that are preserved under homeomorphism are called *topological invariants* of the spaces. For example, the property of connectedness is a topological invariant.

4 A framework for the description of topological spatial relations

This model describing the topological spatial relations between two subsets, A and B, of a topological space X is based on a consideration of the four intersections of the boundaries and interiors of the two sets A and B, i.e. $\partial A \cap \partial B$, $A^\circ \cap B^\circ$, $\partial A \cap B^\circ$ and $A^\circ \cap \partial B$.

Definition 4.1. Let A, B be a pair of subsets of a topological space X. A topological spatial relation between A and B is described by a four-tuple of values of topological invariants associated to each of the four sets $\partial A \cap \partial B$, $A^\circ \cap B^\circ$, $\partial A \cap B^\circ$, and $A^\circ \cap \partial B$, respectively.

A topological spatial relation between two sets is preserved under homeomorphism of the underlying space X. Specifically, if $f: X \rightarrow Y$ is a homeomorphism and $A, B \subset X$, then $\partial A \cap \partial B$, $A^\circ \cap B^\circ$, $\partial A \cap B^\circ$, and $A^\circ \cap \partial B$ are mapped homeomorphically onto $\partial f(A) \cap \partial f(B)$, $f(A)^\circ \cap f(B)^\circ$, $\partial f(A) \cap f(B)^\circ$, and $f(A)^\circ \cap \partial f(B)$, respectively. Since the topological spatial relation is defined in terms of topological invariants of these intersections, it follows that the topological spatial relation between A and B in X is identical to the topological spatial relation between $f(A)$ and $f(B)$ in Y.

A topological spatial relation is denoted here by a four-tuple (_, _, _, _). The entries correspond in order to the values of topological invariants associated to the four set-intersections. The first intersection is called the *boundary–boundary* intersection, the second intersection the *interior–interior* intersection, the third intersection the *boundary–interior* intersection, and the fourth intersection the *interior–boundary* intersection.

4.1 *Topological spatial relations from empty/non-empty set-intersections*

As the entries in the four-tuple, properties of sets that are invariant under homeomorphisms are considered. For example, the properties *empty* and *non-empty* are set-theoretic, and therefore topologically invariant. Other invariants, not considered in this paper, are the dimension of a set and the number of connected components (Munkres 1966). Empty/non-empty is the simplest and most general invariant so that any other invariant may be considered a more restrictive classifier.

For the remainder of this paper, attention is restricted to the binary topological spatial relations defined by assigning the appropriate value of *empty* ($\emptyset$) and *non-empty* ($\neg\emptyset$) to the entries in the four-tuple. The 16 possibilities from these combinations are summarized in Table 7.1.

TABLE 7.1
The 16 specifications of binary topological relations based on the criteria of empty and non-empty intersections of boundaries and interiors.

	$\partial \cap \partial$	$^\circ \cap ^\circ$	$\partial \cap ^\circ$	$^\circ \cap \partial$
r_0	$\emptyset$	$\emptyset$	$\emptyset$	$\emptyset$
r_1	$\neg\emptyset$	$\emptyset$	$\emptyset$	$\emptyset$
r_2	$\emptyset$	$\neg\emptyset$	$\emptyset$	$\emptyset$
r_3	$\neg\emptyset$	$\neg\emptyset$	$\emptyset$	$\emptyset$
r_4	$\emptyset$	$\emptyset$	$\neg\emptyset$	$\emptyset$
r_5	$\neg\emptyset$	$\emptyset$	$\neg\emptyset$	$\emptyset$
r_6	$\emptyset$	$\neg\emptyset$	$\neg\emptyset$	$\emptyset$
r_7	$\neg\emptyset$	$\neg\emptyset$	$\neg\emptyset$	$\emptyset$
r_8	$\emptyset$	$\emptyset$	$\emptyset$	$\neg\emptyset$
r_9	$\neg\emptyset$	$\emptyset$	$\emptyset$	$\neg\emptyset$
r_{10}	$\emptyset$	$\neg\emptyset$	$\emptyset$	$\neg\emptyset$
r_{11}	$\neg\emptyset$	$\neg\emptyset$	$\emptyset$	$\neg\emptyset$
r_{12}	$\emptyset$	$\emptyset$	$\neg\emptyset$	$\neg\emptyset$
r_{13}	$\neg\emptyset$	$\emptyset$	$\neg\emptyset$	$\neg\emptyset$
r_{14}	$\emptyset$	$\neg\emptyset$	$\neg\emptyset$	$\neg\emptyset$
r_{15}	$\neg\emptyset$	$\neg\emptyset$	$\neg\emptyset$	$\neg\emptyset$

A set is either empty or non-empty; therefore, it is clear that these 16 topological spatial relations provide complete coverage, that is, given any pair of sets A and B in X, there is always a topological spatial relation associated with A and B. Furthermore, a set cannot simultaneously be empty and non-empty, from which follows that the 16 topological spatial relations are mutually exclusive, i.e. for any pair of sets A and B in X, exactly one of the 16 topological spatial relations holds true.

In general, each of the 16 spatial relations can occur between two sets. Depending on various restrictions on the sets and the underlying topological space, the actual set of existing topological spatial relations may be a subset of the 16 in Table 7.1. For general point-sets in the plane R^2, all 16 topological spatial relations can be realized (Figure 7.1).

4.2 Influence of the topological space on the relations

The setting, i.e. the topological space X in which A and B lie, plays an important role in the spatial relation between A and B. For example, in Figure 7.2 (left panel) the two sets A and B have the relation ($\emptyset$, $\neg\emptyset$, $\neg\emptyset$, $\neg\emptyset$) as subsets of the line. The same configuration shows a different relation between the two sets when they are embedded in the plane (Figure 7.2, right panel). As subsets of the plane, the boundaries of A and B are equal to A and B, respectively, and the interiors are empty, i.e., $\partial A = A$, $A^\circ = \emptyset$, $\partial B = B$, and $B^\circ = \emptyset$. It follows that in the plane the spatial relation between the two sets A and B is ($\neg\emptyset$, $\emptyset$, $\emptyset$, $\emptyset$).

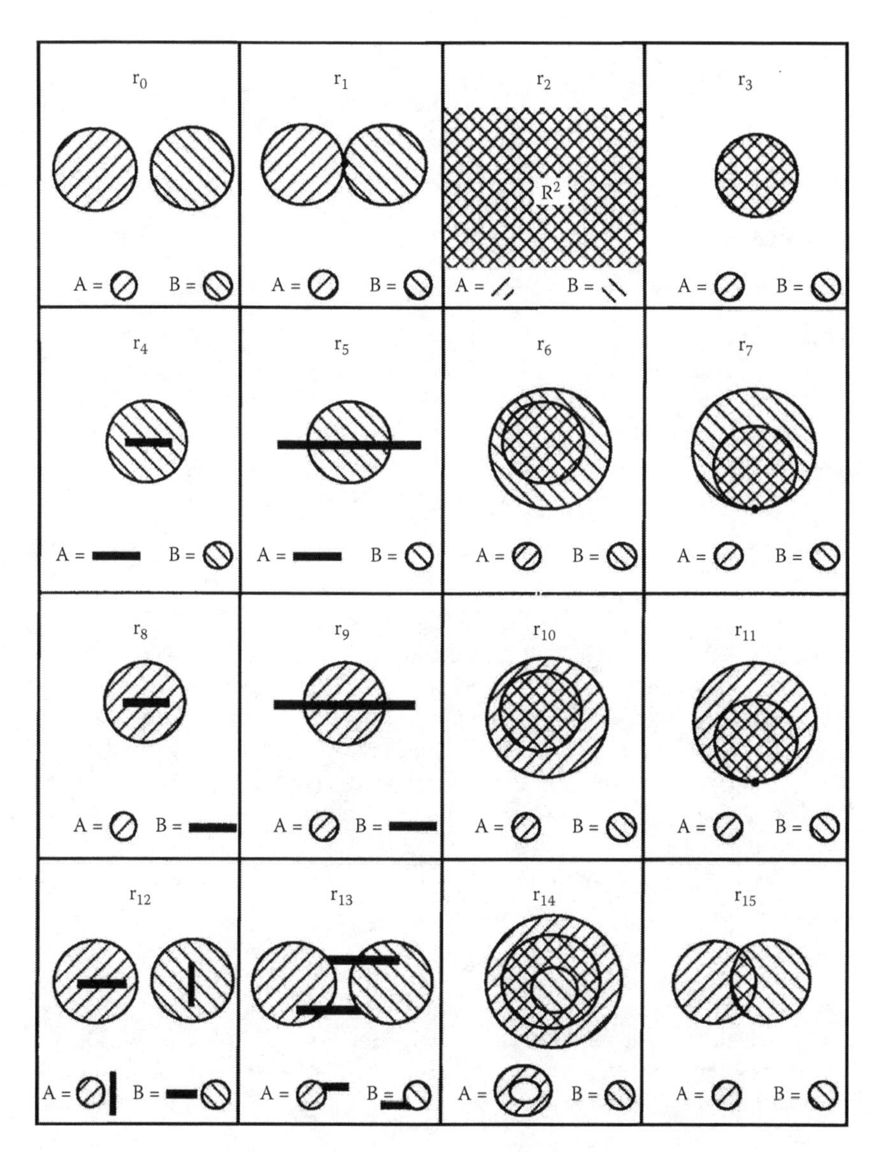

FIGURE 7.1 Examples of the 16 binary topological spatial relations based on the comparison of empty and non-empty set-intersections between boundaries and interiors.

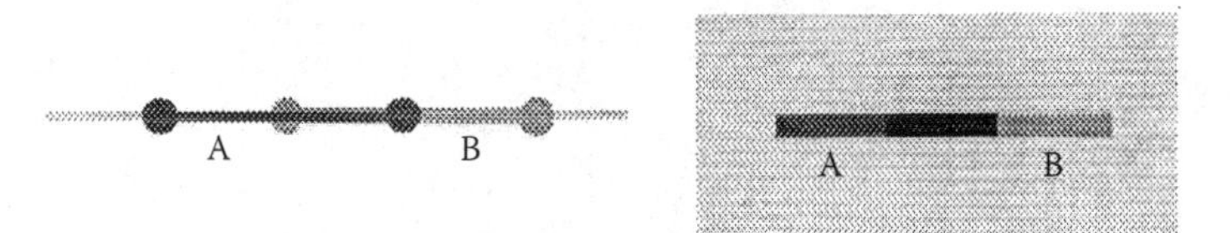

FIGURE 7.2 The same configuration of the two sets A and B with (left panel) the topological spatial relation ($\emptyset$, $\neg\emptyset$, $\neg\emptyset$, $\neg\emptyset$) when embedded in a line and (right panel) ($\neg\emptyset$, $\emptyset$, $\emptyset$, $\emptyset$) in a plane.

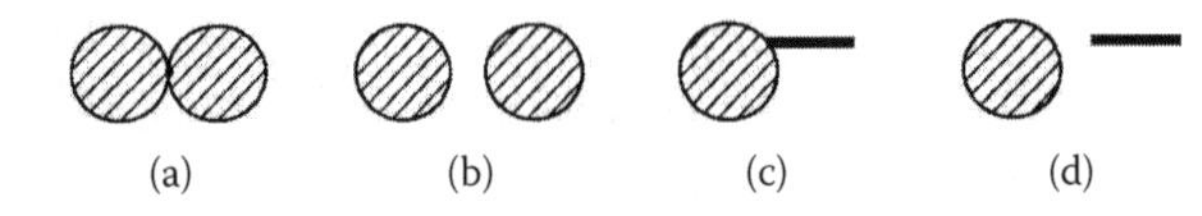

FIGURE 7.3 Sets in the plane that are not spatial regions.

5 Topological relations between spatial regions

It is the aim of this paper to model topological spatial relations that occur between polygonal areas in the plane; therefore, the topological space X and the sets under consideration in X are restricted. These restrictions are not too specific and the only assumption that is made about the topological space X is that it is connected. This guarantees that the boundary of each set of interest is not empty.

The sets of interest are the *spatial regions*, defined as follows:

Definition 5.1. Let X be a connected topological space. A spatial region in X is a non-empty proper subset A of X satisfying (1) A° is connected and (2) $A = \overline{A^\circ}$.

It follows from the definition that the interior of each spatial region is non-empty. Furthermore, a spatial region is closed and connected as it is the closure of a connected set. Figure 7.3 depicts sets in the plane which, by failing to satisfy either condition (1) or condition (2) in Definition 5.1, are not spatial regions. A and B are not spatial regions, because A° and B° are not connected, respectively. C and D are not spatial regions, because they fail to satisfy condition (2), i.e. $C \neq \overline{C^\circ}$ and $D \neq \overline{D^\circ}$. The latter sets are needed to realize the topological spatial relations r_2, r_5, r_8, r_9, r_{12} and r_{13} in the plane.

The following proposition implies that the boundary of each spatial region is non-empty.

Proposition 5.2. If A is a spatial region in X then $\partial A \neq \emptyset$.

Proof: $A^\circ \neq \emptyset$. $A = \bar{A}$ since A is closed, and $A \neq X$ by definition of a spatial region. From proposition 3.4 it follows that A° and $X - A$ form a separation of $X - \partial A$. If $\partial A = \emptyset$ then the two sets form a separation of X, which is impossible since X is connected; therefore, $\partial A \neq \emptyset$. □

5.1 Existence of region relations

The framework for the spatial relations between point-sets carries over to spatial regions, however, not all of the 16 relations between arbitrary point-sets exist between two spatial regions. From the examples in Figure 7.1 it is concluded that at least the relations r0, r_1, r_3, r_6, r_7, r_{10}, r_{11}, r_{14} and r_{15} exist between two spatial regions. The following proposition shows that these nine topological spatial relations are the only relations that can occur between spatial regions.

Proposition 5.3. For two spatial regions the spatial relations r_2, r_4, r_5, r_8, r_9, r_{12} and r_{13} cannot occur.

Proof: This begins by proving that if the boundary–interior or interior–boundary intersection is non-empty then the interior–interior intersection between the same two regions is also non-empty. This implies that the six topological spatial relations r_4, r_5, r_8, r_9, r_{12} and r_{13}, all with empty interior–interior and non-empty boundary–interior or interior–boundary intersections, cannot occur.

Let A and B be spatial regions for which $\partial A \cap B^\circ \neq \emptyset$. It is shown that $A^\circ \cap B^\circ \neq \emptyset$. Using proposition 3.2, $A^\circ \cup \partial A = \bar{A}$ and $A^\circ \cup \partial(A^\circ) = \overline{A^\circ}$. $\bar{A} = A = \overline{A^\circ}$, so $A^\circ \cup \partial A = A^\circ \cup \partial(A^\circ)$. Furthermore, by proposition 3.1, $A^\circ \cap \partial(A^\circ) = \emptyset$ and $A^\circ \cap \partial A = \emptyset$. It follows that $\partial(A^\circ) = \partial A$. Now let $x \in \partial A \cap B^\circ$, then $x \in \partial(A^\circ)$, and since B° is open and contains x, it follows that $A^\circ \cap B^\circ \neq \emptyset$. Thus if the boundary–interior intersection is non-empty, then the interior–interior intersection is also non-empty. It also follows that if the interior–boundary intersection is non-empty, then the interior–interior intersection is also non-empty.

Next it is proved that if the boundary–boundary intersection is empty and the interior–interior intersection is non-empty, then either the boundary–interior or the interior–boundary intersection is non-empty. This implies that the spatial relation r_2, with a non-empty interior–interior intersection and empty intersections for boundary–boundary, boundary–interior and interior–boundary, cannot occur. This will complete the proof of the proposition.

Let A and B be spatial regions such that $\partial A \cap \partial B = \emptyset$ and $A^\circ \cap B^\circ \neq \emptyset$. It is shown that if $\partial A \cap B^\circ = \emptyset$, then $A^\circ \cap \partial B \neq \emptyset$. Assume that $\partial A \cap B^\circ = \emptyset$. Since $B = B^\circ \cup \partial B$, it follows that $\partial A \cap B = \emptyset$ and, therefore, $B \subset X - \partial A$. Proposition 3.4 implies that A° and $X - A$ form a separation of $X - \partial A$, and since B is connected, proposition 3.3 implies that either $B \subset A^\circ$ or $B \subset X - A$. Since, by assumption, $A^\circ \cap B^\circ \neq \emptyset$, it follows that $B \subset A^\circ$ and, therefore, $\partial B \subset A^\circ$. Clearly, $\partial B \cap A^\circ \neq \emptyset$ and the result follows. □

5.2 Semantics of region relations

In Figure 7.1, examples were depicted for the topological spatial relations r_0, r_1, r_3, r_6, r_7, r_{10}, r_{11}, r_{14} and r_{15} between spatial regions. Each of these nine relations is considered in the definitions below and their semantics are investigated using the same notation as in Egenhofer (1989) and Egenhofer and Herring (1990).

Definition 5.4. The descriptive terms for the nine topological spatial relations between two regions are given in Table 7.2.

If the topological spatial relation between A and B is r_0 then, in the set-theoretic sense, A and B are disjoint and, therefore, the topological spatial relation *disjoint* coincides with the set-theoretic notion of disjoint. The following proposition and corollaries justify the other descriptive terms for the topological spatial relations defined in Table 7.2.

Proposition 5.5. Let A and B be spatial regions in X. If $A^\circ \cap B^\circ \neq \emptyset$ and $A^\circ \cap \partial B = \emptyset$, then $A^\circ \subset B^\circ$ and $A \subset B$.

Proof: A° is connected. Proposition 3.4 implies that B° and $X - B$ form a separation of X. Since $A^\circ \cap \partial B = \emptyset$, it follows by proposition 3.1 that $A^\circ \subset B^\circ \cup (X - B)$.

TABLE 7.2
Terminology used for the nine relations between two spatial regions.

	$\partial\cap\partial$	$^\circ\cap^\circ$	$\partial\cap^\circ$	$^\circ\cap\partial$	
r_0	$(\emptyset,$	$\emptyset,$	$\emptyset,$	$\emptyset)$	A and B are disjoint
r_1	$(\neg\emptyset,$	$\emptyset,$	$\emptyset,$	$\emptyset)$	A and B touch
r_3	$(\neg\emptyset,$	$\neg\emptyset,$	$\emptyset,$	$\emptyset)$	A equals B
r_6	$(\emptyset,$	$\neg\emptyset,$	$\neg\emptyset,$	$\emptyset)$	A is inside of B or B contains A
r_7	$(\neg\emptyset,$	$\neg\emptyset,$	$\neg\emptyset,$	$\emptyset)$	A is covered by B or B covers A
r_{10}	$(\emptyset,$	$\neg\emptyset,$	$\emptyset,$	$\neg\emptyset)$	A contains B or B is inside of A
r_{11}	$(\neg\emptyset,$	$\neg\emptyset,$	$\emptyset,$	$\neg\emptyset)$	A covers B or B is covered by A
r_{14}	$(\emptyset,$	$\neg\emptyset,$	$\neg\emptyset,$	$\neg\emptyset)$	A and B overlap with disjoint boundaries
r_{15}	$(\neg\emptyset,$	$\neg\emptyset,$	$\neg\emptyset,$	$\neg\emptyset)$	A and B overlap with intersecting boundaries

Proposition 3.3 implies that either $A^\circ \subset B^\circ$ or $A^\circ \subset (X - B)$. But $A^\circ \cap B^\circ \neq \emptyset$; therefore, $A^\circ \subset B^\circ$. Since $A^\circ \subset B^\circ$, it follows that $\overline{A^\circ} \subset \overline{B^\circ}$ which, by definition 5.1, implies that $A \subset B$. □

From proposition 5.5 it follows that if A is covered by B, then $A \subset B$; therefore, the spatial relation *is covered by* coincides with the set-theoretic notion of being a subset of.

The following corollary to proposition 5.5 shows that the spatial relation *equal* corresponds to the set-theoretic notion of equality.

Corollary 5.6. Let A and B be spatial regions. If the spatial relation between A and B is r_3, then $A = B$.

Proof: $A^\circ \cap B^\circ \neq \emptyset$ and $A^\circ \cap \partial B = \emptyset$; therefore, proposition 5.5 implies that $A \subset B$. Furthermore, $\partial A \cap B^\circ = \emptyset$. Again by proposition 5.5, $B \subset A$. Thus $A = B$. □

The following corollary to proposition 5.5 shows that if A is inside B, then $A \subset B^\circ$; therefore, the spatial relation *inside* coincides with the topological notion of being contained in the interior. Conversely, *contains* corresponds to contains in the interior.

Corollary 5.7. Let A and B be spatial regions. If the spatial relation between A and B is r_6, then $A \subset B^\circ$.

Proof: Proposition 5.5 implies that $A^\circ \subset B^\circ$ and $A \subset B$. By proposition 3.2, $A = A^\circ \cup \partial A$ and $B = B^\circ \cup \partial B$. So $\partial A \subset B$. Since $\partial A \cap \partial B = \emptyset$, it follows that $\partial A \subset B^\circ$. Together with $A^\circ \subset B^\circ$ this implies that $A \subset B^\circ$. □

6 Relations in *n*-dimensional spaces

It is natural to ask 'What further restrictions on the topological space X and the sets under consideration in X further reduce the topological spatial relations that can occur?' This section will explore this question by considering the case where X is a Euclidean space.

R^n denotes n-dimensional Euclidean space with the usual Euclidean metric. A subset of R^n is *bounded* if there is an upper bound to the distances between pairs of points in the set; otherwise, it is said to be *unbounded*.

The *unit disk* in R^n is the set of points in R^n whose distance from the origin is less than, or equal to, 1. The *unit sphere* in R^n is the set of points in R^n whose distance from the origin is equal to 1. For $n \geq 1$ the unit disk in R^n is connected. For $n \geq 2$ the unit sphere in R^n is connected. Let X be a topological space. An *n-disk* in X is a subspace of X that is homeomorphic to the unit disk in R^n. An *n-sphere* in X is a subspace of X that is homeomorphic to the unit sphere in R^{n+1}. n-disks in R^n are bounded and are spatial regions; the latter is a relatively straightforward consequence of the Brouwer theorem on the invariance of domain (Spanier 1966). Since n-disks in R^n are spatial regions, proposition 5.3 restricts the number of spatial relations that can occur between them.

In proposition 6.1 it is shown that if A and B are n-disks in R^n with $n \geq 2$, then the spatial relation *overlap with disjoint boundary* cannot occur. The proof of this proposition is based on the following two facts:

Fact 1. Let A be an n-disk in R^n with $n \geq 2$. Then ∂A is an $(n - 1)$-sphere in R^n and, therefore, connected.

This fact, also, is a consequence of the Brouwer theorem on the invariance of domain (Spanier 1966).

Fact 2. Let A be an n-disk in R^n with $n \geq 2$. Then $R^n - A^\circ$ is connected and unbounded.

This second fact is a (non-)separation theorem related to the Jordan–Brouwer separation theorem (Spanier 1966).

Proposition 6.1. The topological spatial relation r_{14}, overlap with disjoint boundaries, does not occur between n-disks in R^n with $n \geq 2$.

Proof: Let A and B be n-disks in R^n with $n \geq 2$. It is shown that if $\partial A \cap \partial B = \emptyset$, then A and B do not overlap and, therefore, the spatial relation *overlap with disjoint boundaries* cannot occur.

Assume $\partial A \cap \partial B = \emptyset$ and A and B overlap. A contradiction will be derived. B is a spatial region; therefore, proposition 3.4 implies that B° and $R^n - B$ form a separation of $R^n - \partial B$. As $\partial A \cap \partial B = \emptyset$ it follows that $\partial A \subset R^n - \partial B$. By fact 1, ∂A is connected, therefore, proposition 3.3 implies that either $\partial A \subset B^\circ$ or $\partial A \subset (R^n - B)$. Since A and B overlap, it follows that $\partial A \cap B^\circ \neq \emptyset$ and, therefore, $\partial A \subset B^\circ$.

$\partial A \subset B^\circ$ implies that $\partial A \cap (R^n - B^\circ) = \emptyset$. By fact 2, $R^n - B^\circ$ is connected. Using propositions 3.3 and 3.4 and arguing as above, it follows that either $(R^n - B^\circ) \subset A^\circ$ or $(R^n - B^\circ) \subset (R^n - A)$. The first case yields a contradiction because, by fact 2, $R^n - B^\circ$ is unbounded, but A° is not. The second case implies that $A \subset B^\circ$ and, therefore, $A^\circ \cap \partial B = \emptyset$, which contradicts the assumption that A and B overlap. Therefore, in either case a contradiction is obtained and it follows that the spatial relation r_{14} cannot occur between n-disks in R^n with $n \geq 2$. □

Note that for $n \geq 2$ the topological spatial relation r_{15}, *overlap with intersecting boundaries*, does occur between two n-disks (Figure 7.1).

The opposite situation occurs in R^1 where r_{14} can occur between 1-disks, while r_{15}, *overlap with intersecting boundaries*, cannot. It is clear that r_{14}, can occur between two 1-disks in R^1 (Figure 7.2). Proposition 6.2 shows that r_{15} cannot occur. Its proof requires the easily derived fact that a spatial region in R^1 is either a closed interval $[a, b]$ for some $a, b \in R^1$, or a closed ray $[a, \infty)$ or $(-\infty, a]$ for some $a \in R^1$.

Proposition 6.2. The topological spatial relation r_{15} does not occur between spatial regions in R^1.

Proof: Let A and B be spatial regions in R^1 and assume that A and B overlap. It is shown that $\partial A \cap \partial B = \emptyset$. Each of A and B is a closed interval or a closed ray; therefore, there are nine different cases to examine. One is selected; the others can be proven accordingly.

Assume $A = [a, \infty)$ and $B = (-\infty, b]$. Then $\partial A = \{a\}$ and $\partial B = \{b\}$. Since A and B overlap, it follows that $a < b$, which implies that $\partial A \cap \partial B = \emptyset$. □

7 Conclusion

A framework for the definition of topological spatial relations has been presented. It is based on purely topological properties and is thus independent of the existence of a distance function. The topological relations are described by the four intersections of the boundaries and interiors of two point-sets. Considering the binary values empty and non-empty for these intersections, a set of 16 mutually exclusive specifications has been identified. Fewer relations exist if particular restrictions on the point-sets and the topological space are made. It was proved that there are only nine topological spatial relations between point-sets which are homeomorphic to polygonal areas in the plane.

Although the nature of this work is rather theoretical, the framework has an immediate effect on the design and implementation of geographic information systems. Previously, for every topological spatial relation a separate procedure had to be programmed and no mechanism existed to assure completeness. Now, topological spatial relations can be derived from a single, consistent model and no programming for individual relations will be necessary. Prototype implementations of this framework have been designed and partially implemented (Egenhofer 1989), and various extensions to the framework have been investigated to provide more details about topological spatial relations, such as the consideration of the dimensions of the intersections and of the number of disconnected subsets in the intersections (Egenhofer and Herring 1990). Ongoing investigations focus on the application of this framework for formal reasoning about combinations of topological spatial relationships.

The framework presented is considered a start and further investigations are necessary to verify its suitability. Here, only topological spatial relations with co-dimension zero were considered, i.e. the difference between the dimension of the space and the dimension of the embedded spatial objects is zero, e.g. between regions in the plane and intervals on the one-dimensional line. Also of interest for GIS applications are the topological spatial relationships with co-dimension greater than zero, e.g. between two lines in the plane (Herring 1991). Likewise, the applicability of this framework to topological spatial relations between objects of different dimensions, such as a region and a line, must be tested.

Acknowledgments

The motivation for this work was given by Bruce Palmer. During many discussions with John Herring, these concepts have been clarified. Andrew Frank and Renato Barrera made valuable comments on an earlier version of this paper. This work was partially funded by grants from NSF under No. IST 86-09123, Digital Equipment Corporation under Sponsored Research Agreement No. 414 and TP-765536, Intergraph Corporation, and the Bureau of the Census under Joint Statistical Agreement No. 89-23. Additional support from NSF for the NCGIA under grant number SES 88-10917 is gratefully acknowledged.

References

Abler, R., 1987, The National Science Foundation National Center for Geographic Information and Analysis. *International Journal of Geographical Information Systems*, **1**, 303–326.

Allen, J. F., 1983, Maintaining knowledge about temporal intervals. *Communications of the ACM*, **26**, 832–843.

Change, S. K., Jungert, E., and Li, Y., 1989, The design of pictorial databases based upon the theory of symbolic projections. In: *Proceedings of Symposium on the Design and Implementation of Large Spatial Databases held in Santa Barbara, CA*, edited by A. Buchmann, O. Günther, T. Smith, and Y. Wang (New York: Springer-Verlag), *Lecture Notes in Computer Science*, Vol. 409, pp. 303–323.

Claire, R., and Guptill, S., 1982, Spatial operators for selected data structures. In: *Proceedings of Auto-Carto V, held in Crystal City, Virginia*, pp. 189–200.

Egenhofer, M., 1989, A formal definition of binary topological relationships. In: *Third International Conference on Foundations of Data Organization and Algorithms (FODO) held in Paris, France*, edited by W. Litwin and H.-J. Schek (New York: Springer-Verlag) *Lecture Notes in Computer Science*, Vol. 367, pp. 457–472.

Egenhofer, M., and Herring, J., 1990, A mathematical framework for the definition of topological relationships. In: *Proceedings of Fourth International Symposium on Spatial Data Handling held in Zurich, Switzerland*, edited by K. Brassel and H. Kishimoto pp. 803–813.

Frank, A., 1982, Map query — database query language for retrieval of geometric data and its graphical representation. *ACM Computer Graphics*, **16**, 199–207.

Freeman, J. 1975, The modelling of spatial relations. *Computer Graphics and Image Processing*, **4**, 156–171.

Güting, R., 1988, Geo-relational algebra: a model and query language for geometric database systems. In: *International Conference on Extending Database Technology held in Venice, Italy*, edited by J. Schmidt, S. Ceri, and M. Missikoff (New York: Springer–Verlag), *Lecture Notes in Computer Science*, Vol. 303, pp. 506–527.

Hernández, D., 1991, Relative representation of spatial knowledge: the 2-D case. In: *Cognitive and Linguistic Aspects of Geographic Space*, edited by D. Mark and A. Frank (Dordrecht Kluwer Academic) (in press).

Herring, J., 1991, The mathematical modeling of spatial and non-spatial information in geographic information systems. In: *Cognitive and Linguistic Aspects of Geographic Space*, edited by D. Mark and A. Frank (Dordrecht: Kluwer Academic) (in press).

Herring, J., Larsen, R., and Shivakumar, J., 1988, Extensions to the SQL language to support spatial analysis in a topological database. In: *Proceedings of GIS/LIS '88, held in San Antonio, Texas*, pp. 741–750.

Herskovits, A., 1986, *Language and Spatial Cognition — An Interdisciplinary Study of the Prepositions in English* (Cambridge: Cambridge University Press).

Ingram, K., and Phillips, W., 1987, Geographic information processing using a SQL-based query language. In: *Proceedings of AUTO-CARTO 8, Eighth International Symposium on Computer-Assisted Cartography held in Baltimore, Maryland*, edited by N. R. Chrisman pp. 326–335.

Kainz, W., 1990, Spatial relationships — topology versus order. In: *Proceedings of Fourth International Symposium on Spatial Data Handling held in Zurich, Switzerland*, edited by K. Brassel and H. Kishimoto, pp. 814–819.

Molenaar, M., 1989, Single valued vector maps — a concept in geographic information systems. *Geo-Informationssysteme*, **2**, 18–26.

Munkres, J., 1966, *Elementary Differential Topology* (Princeton, NJ: Princeton University Press).

National Center for Geographic Information and Analysis (NCGIA), 1989, The research plan of the National Center for Geographic Information and Analysis. *International Journal of Geographical Information Systems*, **3**, 117–136.

Peuquet, D., 1986, The use of spatial relationships to aid spatial database retrieval. In *Proceedings of Second International Symposium on Spatial Data Handling held in Seattle WA*, edited by D. Marble, pp. 459–471.

Peuquet, D., and Ci-Xiang, Z., 1987, An algorithm to determine the directional relationship between arbitrary-shaped polygons in the plane. *Pattern Recognition*, **20**, 65–74.

Pullar, D., 1988, Data definition and operators on a spatial data model. In: *Proceedings of ACMS-ASPRS Annual Convention, held in St. Louis, Missouri*, pp. 197–202.

Pullar, D., and Egenhofer, M, 1988, Towards formal definitions of topological relations among spatial objects. In: *Proceedings of Third International Symposium on Spatial Data Handling held in Sydney, Australia*, edited by D. Marble, pp. 225–242.

Roussopoulos, N., Faloutsos, C., and Sellis, T., 1988, An efficient pictorial database system for PSQL. *IEEE Transactions on Software Engineering*, **14**, 630–638.

Smith, T., Peuquet, D., Menon, S., and Agrawal, P., 1987, KBGIS-II: a knowledge-based geographical information system. *International Journal of Geographical Information Systems*, **1**, 149–172.

Spanier, E., 1966, *Algebraic Topology* (New York: McGraw-Hill).

Talmy, L., 1983, How language structures space. In: Spatial Orientation: Theory, Research, and Application, edited by H. Pick and L. Acredolo (New York: Plenum Press), pp. 225–282.

Wagner, D., 1988, A method of evaluating polygon overlay algorithms. In: *Proceedings of ACSM-ASPRS Annual Convention, held in St. Louis, Missouri*, pp. 173–183.

International Journal of Geographical Information Systems,
1991, Vol. 5, No. 3, 287–301.

8 Modelling Accessibility Using Space-Time Prism Concepts within Geographical Information Systems

Harvey J. Miller

Abstract. The space-time or time geographical framework is a powerful perspective from which to analyse human behaviour. One of the central concepts in this framework is the space-time prism, which models individual accessibility to an environment. In this paper, the derivation and manipulation of space-time prism concepts within a geographical information system (GIS) are discussed. The required system inputs and desired outputs are identified and a generic GIS based procedure is presented. Given these basic requirements, issues are discussed which can determine the feasibility of current GIS technology to handle the derivation of space-time prism concepts.

1 Introduction

The objective of this paper is to present an analysis of the requirements and feasibility of deriving and applying space-time prism constructs within a GIS. Space-time prisms model the ability of individuals to travel and participate in activities at different locations in an environment. GIS based derivations of these constructs could provide sophisticated and pragmatic representations of individual accessibility which may be of significant use in location analysis and transportation planning. The capability of the available GIS technology to accomplish the derivation of space-time prisms is also considered.

The space-time or time-geographical framework is a broad and powerful perspective from which to analyse human behavior. This framework, originally developed by Hägerstrand, focused on the behavioural possibilities of individuals. By recognizing that individuals must operate within very basic spatial and temporal constraints on their behaviour, the space-time framework can complement a wide variety of approaches of modeling human behaviour in addition to aiding the planning and location of activities and infrastructure (Hägerstrand 1970, Pred 1977).

At the core of the space-time framework is the notion that the events that comprise an individual's existence have both spatial and temporal attributes (Pred 1977, Golledge and Stimson 1987). An individual can directly experience or participate in events at a single location in space at one particular time. In turn, fixed or recurring activities in which an individual desires or is required to participate (such as working, shopping) occur only at discrete locations for limited durations. As time is finite, it is a resource which must be used by the individual to travel and participate in spatially dispersed and temporally limited activities (Hägerstrand 1970). At the broadest level, the space-time framework is concerned with the nature of the constraints that limit an individual's ability to participate in events in space and time, and can determine the necessary (but not sufficient) conditions for virtually all human interaction (Pred 1977).

One of the many ideas to emerge from this framework is the space-time prism. As a geometrical construct, the prism delimits what can be physically reached by an individual from specified locations during a given interval of time (Lenntorp 1976). The prism is determined by the locations(s) at which the individual must be at the beginning and end of the interval, any time required for participation in activities during that interval and the rates at which the individual can trade time for space in movement (i.e., travel velocities) through the environment. The prism models the accessibility of an individual within a particular spatial and temporal context and can provide a valuable measure of individual accessibility.

There have been several attempts to operationalize and apply the space-time prism in spatial analysis and planning. An especially significant contribution was made by Lenntorp (1976, 1978), who used these constructs to measure patterns of accessibility in an urban area. Lenntorp's simulation model does not calculate space-time prisms directly. Instead, the model accepts as the input variables the transportation characteristics of the study area, the locations and operating hours of activities, and a hypothetical activity schedule. The model then calculates all the feasible realizations of this schedules, which is regarded as a measure of accessibility. Lenntorp compared these patterns of accessibility for various combinations of transportation infrastructure, public transport schedules and locations of activities.

Burns (1979) measured the effects of different factors on space-time prisms. He compared space-time prisms that resulted from changes in travel velocities to those which resulted from changing the time constraints of an individual. This comparison was accomplished analytically by calculating the volume of the prism to measure 'space-time autonomy,' or the freedom which the prism provided to the individual to travel and participate in activities. Unlike Lenntorp's simulation model, Burns' analyses were conducted in the abstract and were not applied to any specific setting.

Several researchers have also attempted to use the space-time prism as an input to other models or to measure aspects of the prism empirically. Landau *et al.* (1981) developed a trip generation model which considers the spatial and temporal constraints when trips are made. Similarly, Landau *et al.* (1982) modeled choices of destinations under these constraints. Janelle and Goodchild (1983) analysed activity diaries to determine the space-time autonomy of different subpopulations in their study area.

In general, however, the full potential of the space-time prism concept has not been realized in spatial analysis and planning. There is as yet no general and widely applicable procedure for deriving space-time for empirical settings. As Lenntorp's (1976, 1978) simulation analysis illustrates, these constructs can have direct and valuable applications in locational analysis and transportation planning. However, the initial costs required to calculate space-time prisms for a specific empirical setting can be prohibitive. Whereas the prism does not necessarily require a large amount of data, it does require data at a detailed level of spatial resolution if it is to be effective in analysis. The basic data required are the time available for an activity, the distances between relevant locations and the velocities at which travel between locations can occur. The greater the spatial resolution of the latter two kinds of data, the more realistic and effective the prism becomes. Difficulty in manipulating and representing these data may be part of the reason why the space-time prism has not been used to its full potential in spatial analysis.

The rapidly increasing ability to store, display, manipulate and analyse spatial data through the use of GIS has the potential to alleviate the difficulties in operationalizing and applying space-time prism concepts. The ability of GIS to represent accurately the travel characteristics of an environment makes it useful for deriving these constructs for specific empirical settings. Especially valuable is the ability of GIS to represent the urban transportation network structure.

This paper explores the potential derivation and application of basic GIS based space-time prism constructs, focusing on the space-time prism itself rather than the broader aspects of the time-geographical framework. The broader framework would address issues such as the origin of the constraints which help to define the space-time prism. However, these broader concerns are not central to the derivation and application of the space-time prism in a GIS. The following section of this paper provides a more detailed discussion of space-time prism concepts, both in their traditional forms and operational definitions that are more amenable to implementation and application. Some of the potential applications of these constructs in a GIS are then identified. The requirements for deriving a space-time prism construct in a GIS are examined and a generic procedure for this derivation is presented. Given these requirements, implementation issues and the feasibility of current GIS technology to accommodate the space-time prism derivation are discussed.

2 Space-time prism concepts and operational definitions

2.1 Space-time prism concepts

The space-time prism determines the feasible set of locations for travel and activity participation in a bounded expanse of space and a limited interval of time. A bounded space-time region is represented by an orthogonal triad of axes, with a pair of axes labeled x and y defining two-dimensional planar space, and a z-axis representing time. A point object located at (x_i, y_i, z_i) shows the location coordinates of the object in two-dimensional space at time z_i. As time progresses (i.e. positive movement along the z-axis), the point object, usually representing an individual, can be traced with respect to its movement in space over time. An example of this space-time path can be seen in Figure 8.1. If the path is parallel to the z-axis, then the individual is not moving through space during that time interval. As the slope of the path becomes

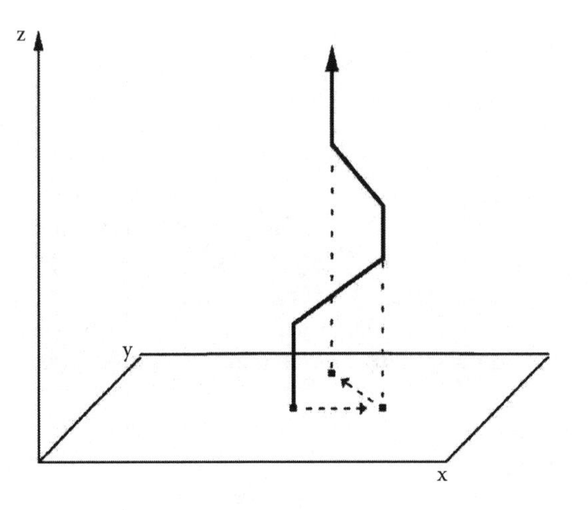

FIGURE 8.1 Schematic representation of the space-time path.

more horizontal, then the individual is moving through space at higher velocities, that is, trading less time for more space. Projecting the space-time path on the *x–y* plane (the broken arrows in Figure 8.1) provides the purely spatial movements of the individual over the time frame.

The space-time prism itself is an extension of the space-time path. However, the prism does not trace the observed movements through space of an individual over an interval of time; instead it shows what portions of space are possible for an individual to be in at specified times. The prism or potential path space (PPS) delimits locations for which the probability of being included in an individual's space-time path is greater than zero (Lenntorp 1976, p. 13). The PPS is a three-dimensional entity existing in the region of bounded space-time; an example is provided in Figure 8.2.

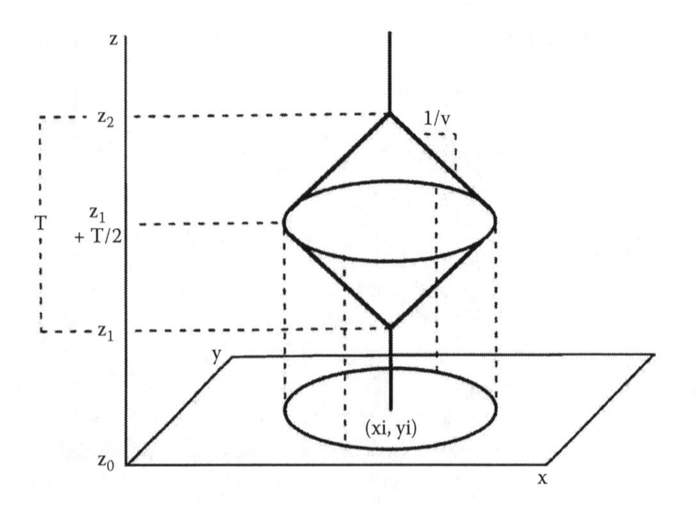

FIGURE 8.2 Schematic representation of the space-time prism.

The PPS is determined by an individual's time budget or the time available for travel and activity participation, the potential travel velocities in the environment, any stop time (time that must be spent at an unspecified location to participate in an activity), and spatial constraints such as the need to be at a certain location at the beginning of the available period (the travel origin) or at the end of the period (the travel destination). For example, in Figure 8.2, the individual must be at a fixed location (such as a workplace) located at (x_i, y_i) from z_0 to z_1. S/he can leave this travel origin at time z_1 but must return there at z_2, leaving $z_2 - z_1 = T$ time units for travel (which can be in any ratio measure). The specified stop time during the time interval is not considered in this example. Given the traditional PPS assumption of a constant and uniform velocity of travel (v), the PPS is determined by deviating the path from parallel to the z-axis in all spatial directions of a slope of $1/v$. This slope is away from the travel origin and positive along the z-axis starting at time z_1, and away from the travel destination and negative along the z-axis starting at time z_2, until both cones intersect at time $z_1 + T/2$ (Lenntorp 1976, Burns 1979). The prism encompasses all points in space-time that can be reached by the individual during the time interval z_1 to z_2. This simple example is the basic PPS with a defined origin and destination at the same location and no explicit stop time during the intervening travel time. The geometry of the PPS becomes more complex when the origin and destination are not coincident in space, or when the stop time during the interval is also considered.

Although the PPS is a revealing construct, of more direct concern in travel modeling, locational analysis, analysis, transportation planning or other applications is the potential path area (PPA). The PPA is the projection of the three-dimensional PPS onto two-dimensional planar space. It delimits the purely spatial extent or area within which the individual can travel. The circle in the x–y plane in Figure 8.2 is the PPA for that potential path space. The PPA can be calculated directly without reference to the PPS, with any stationary stop time for activity participation netted from the overall time budget to reflect the reduced amount of time available for spatial movement (Lenntorp 1976).

2.2 *Operational definitions*

The following operational definitions provide the basic entities of the space-time prism for derivation and use in a GIS. These basic entities are operational versions of the PPA construct. The time dimension of the space-time prism (and the bounded space-time region) is not considered beyond the time budget available for travel and activity participation. Although this simplification results in a loss of the individual's broader activity schedule as a context for the PPA, these simpler constructs are sufficient for practical applications. Extensions of the operational definitions to encompass the full temporal dimension of the space-time prism are warranted. However, given the current difficulties in representing time in a GIS and in spatial data structures (Langran 1989), these extensions will be less immediate than the basic constructs presented here.

Traditionally, the PPA is defined in continuous, two-dimensional space. When operationalizing and using this construct for empirical analysis, this planar space version of the PPA is too abstract and is in fact unnecessary. A large proportion of the space in a real environment is useless for travel or activity participation. Travel can occur only along the channels in the environment defined by the transportation network (e.g., streets, public transport routes), and activities can only take place at certain locations (e.g., shopping can only occur at stores). The planar form of the PPA can be discarded in favour of more meaningful and easily calculated discrete space forms.

Two discrete space versions of the PPA can be derived in an operational system such as a GIS. A PPA can be defined on the basis of the punctiform representations of locations at which activities can be undertaken. In this instance, the PPA is defined as a subset of the relevant activity locations in an environment, with this subset distinguished by being feasible for activity participation by the individual. No PPA boundary in the traditional sense can be extrapolated between the punctiform locations. Travel in the environment between any two defined locations (such as the travel origin and the location of an activity) is characterized as occurring at a constant rate of speed along a fixed line segment connecting the pair, although this line segment can summarize more complex combinations of travel velocities in the actual environment.

The punctiform based PPA can be extended to allow for more sophisticated consideration of the actual travel characteristics in an environment. These include variations in the possible travel velocity between different links in a transportation network that are due to factors such as variations in traffic congestion and differences in the legal speed limit. These considerations can be addressed explicitly and in detail by defining PPA structure within the format of the urban transportation network. A network representation can be formulated where arcs represent the individual streets in the transportation network and nodes represent intersections of these streets. A PPA defined in this format shows the streets (arcs) in the network that are feasible for travel and the intersections (nodes) which it is feasible to reach. This form of the PPA has the advantage of allowing the accessibility of an individual to be seen within the context of the structure that both facilitates and constrains movement in the environment.

There can be multiple paths between any two locations in a transportation network. In defining a network based PPA, the possibility of multiple paths can complicate matters as each path between two fixed locations can involve a different expenditure of time. For simplicity, the network based PPA can be based on the assumption of the shortest path through the network being used by the individual. Although there is evidence that individuals do not necessarily use the shortest paths in their movement (Garling and Garling 1988), the shortest path assumption is viable for present purposes. Modification of the network based PPA for multiple paths between locations is a worthwhile extension to the definition provided here.

The nature of the urban transportation network also means that the actual velocities attained when travelling will not be static. The main cause of fluctuation in travel velocities is network congestion: more traffic results in slower actual velocities. The amount of traffic in the urban network will vary according to the time of day (rush hour versus non-rush hours) and the day of the week (weekday versus non-weekday). Conceptually, travel velocities in the transportation network can be represented by velocity probability distributions corresponding to each arc (e.g., Sigal *et al.* 1980). Operationally it is more convenient to maintain a set of travel time for each path or network arc in the travel environment, with separate travel velocities corresponding to 'morning peak' and 'evening peak' (i.e., rush hour) travel, 'off peak' (non-rush hour, but weekday) travel, and 'weekend' travel. The distinction between morning and evening rush hours is necessary as the effects of congestion are directional with the flow of traffic through a network arc. This implies a separate set of travel velocities for each direction of travel through an arc.

Travel times through nodes in the network can also vary across the network and fluctuate with time. The nodes in the network representation correspond to intersections in the actual street network, and experience shows that the time required to more through these intersections varies depending on the direction of the movement. For example, driving straight through an intersection requires less time than making a left turn. To reflect this fact in the network model, turn times can be defined at each node. These turn times can also be conceptualized as stochastic functions, with the operational system containing separate turn times for each possible direction and corresponding to peak, non-peak and weekend travel. Turn times can also be set at very high values to reflect restrictions on turning (e.g., 'no left turns' at an intersection).

Stop times required for activity participation can also fluctuate as a result of factors such as the changing needs of individuals with each visit and congestion at the location (e.g. a crowded store). This can also be considered in deriving a PPA. Stop times can be represented by a set of stop times for each activity type reflecting (for example) 'short', 'medium' or 'long' participation times for each activity.

3 Potential applications

Using a GIS to derive space-time prism constructs such as the network based PPA can model individual accessibility within the format of a detailed and sophisticated representation of the travel environment. A GIS can allow accessibility patterns within the transportation network to be visualized. This alone provides a powerful perspective for the spatial analyst. The concept of 'accessibility' is central to many theories and models of spatial behaviour and location, and the ability to see this concept operating within its environmental context adds a valuable dimension to the intuition behind the construction of theories and models.

Beyond these beneficial insights, the derivation and manipulation of space-time prism constructs in a GIS have more pragmatic applications in areas such as travel modelling, transport planning and retail and/or facility location. A key aspect of several of these applications is that, whereas the space-time prism is a model of individual accessibility, a GIS can facilitate the combination of these individual constructs into aggregate measures of accessibility. Although the recognition that individual space-time constructs can serve as the basis for aggregate analysis is not new (Lenntorp 1976, 1978), the use of a GIS greatly expands this capability. In addition, with a GIS the space-time prism construct can be easily combined with other data in the analysis.

By deriving PPAs for an analysis of spatial choice, the analyst can use a GIS to address a primary factor in delimitation of the choice set: accessibility to alternatives within space and time constraints. Although this would not address all the factors that can define an individual's choice set (such as the lack of information about alternatives), it can nevertheless improve model specification and fit over the arbitrary delimitation or non-delimitation of choice sets. For example, Landau *et al.* (1982) found that restricting choice sets on the availability of time improved the average prediction of their model by 5 per cent, with possible improvements of 15–20 per cent also indicated. The idea that individuals are constrained in their abilities to participate in events at different locations has potential for addressing one of the problem areas in modelling spatial choice and travel behaviour; i.e. the identification of the choice set of individuals, or the set of alternatives that an individual considers when making a discretionary choice (such as the set of stores from which a person chooses a place to shop). The incorrect identification of the choice set can lead to serious mis-specification errors in the model as well as a general misrepresentation of the spatial choice process (Horowitz 1986).

The concept of the space-time prism recognizes that a major factor influencing an individual's participation in spatially dispersed activities is the ability to trade time for space in movement. The objective of transport planning can be seen as complementary to this perspective: to provide individuals with the ability to trade less time for more space when travelling. Space-time prism constructs in GIS can be used to assess progress towards this goal so as to analyse the effects on this relationship from changes in the transportation infrastructure.

The network PPA can be used to derive a network based measure of accessibility which can gauge the performance of the transportation infrastructure. This measure can also indicate variations in accessibility across the study area, thus directing planning efforts. To derive this measure, the analyst can derive a separate network PPA for each individual in a study area (or, more efficiently, an aggregate unit of analysis which represents a specified number of individuals) based on a specified time budget. These PPAs can be based on the most probable travel origins and/or destinations of an individual for the particular type of travel being analysed. These individual network PPAs can be combined in a single coverage, and the accessibility of different portions of the network can be measured based on the percentage of the PPAs that overlap at different network arcs. Characteristics of the transportation infrastructure can be manipulated and the resulting changes in patterns of accessibility can be assessed.

Similarly, the locations of retail stores or public facilities depend on accessibility to a spatially dispersed demand. Combining network based PPAs in a GIS can allow for this consideration in site selection. For example, assume that a retail analyst knows that individuals are most likely to come from their homes to the particular type of store that s/he wishes to locate, and that they are generally unwilling to travel more than 20 minutes to patronize that type of store. Network based PPAs with 20 minute time budgets centred on home locations can be combined, as in the previous example, and the network arc(s) where the greatest number of PPAs overlap are candidates for the store location. The PPAs can also be weighted on the basis of measures of demand such as population and income.

This analysis can be turned around by determining a PPA for each candidate location of a store or facility, but based on travel to (not from) those sites. The site that can 'capture' the largest amount of demand given the appropriate time budget is the best choice based on this measure. A series of PPAs based on specified increments of time can also be derived if it is believed that travel time has an effect on the amount of demand realized at a location.

4 GIS considerations

4.1 Required system inputs

Table 8.1 gives a summary of the required data for input to the GIS that will generate discrete space PPAs. For both types of discrete space PPAs three basic types of input data are needed. These are: (1) the locations of the origin of travel and destination (these can be the same); (2) the locations and characteristics of relevant activities; and (3) the characteristics of the travel environment.

In the punctiform example, both the origin of travel and destination and the locations of activities are treated as point objects in planar space; that is, each of these locations is considered as an entity with no spatial dimensions existing in some defined coordinate system. Each travel origin and destination pair is assigned labels which uniquely identify the pair among all sets of locations for individuals (or population aggregates) in the analysis. The activity locations are also identified by

TABLE 8.1
Data requirements for PPAs based on geographical information systems.

	Data characteristics	
Data requirement	**Punctiform version**	**Network version**
Travel origin/end location	Point objects in planar space	Nearest nodes in a defined transportation network structure of the actual locations
Activity locations	Point objects in planar space	Nearest nodes in a defined transportation network structure to the actual locations
Travel environment	Direct paths between travel origin/end locations and activity locations Each path characterized by a representative travel distance and a set of temporally dependent travel velocities	Arcs and nodes corresponding to linkages (streets, intersections) in the urban transportation network Each arc characterized by the length of the linkage and a set of temporally dependent velocities Each node characterized by a set of temporally dependent turn times

unique labels. Pairings of travel origins and destinations with activity locations define a network of direct paths. For example, a straight and direct path is represented between travel origins and activity locations, and then from activity locations to travel destinations. Each of these direct paths is labelled by the two locations that define it and is characterized by a travel distance and a set of temporally dependent (e.g., morning peak, evening peak, non-peak) travel velocities. Thus, in the punctiform case the locations of travel origins and destinations and activity locations define the underlying (and simplistic) travel network.

In the network based PPA, the transportation network acts as the structure within which the travel origin, travel destination and activity locations are represented. Each node in the representation of the network is identified with a unique label, and the arcs are identified by the nodes they connect in a FROM (node i) TO (node j) type format. These arcs are directed to allow for one-way streets as well as differences in travel velocity with direction of movement. A two-way street between nodes i and j is represented as an arc from node i to node j and an arc from node j to node i. Travel characteristics of this network are defined by the length of each directed arc and a set of temporally dependent travel velocities for the arc. The nodes are characterized by a set of temporally dependent turn time for each possible direction of travel out of the node.

The travel origins, destinations and activity locations are represented as coinciding with the nearest node (intersection) to their actual locations in the representation of the transportation network. The nodes in this network potentially maintain a great deal of information. A node record can contain its identification, a set of turn times for each possible direction and the identification of any travel origins, destinations and activity locations at the node.

4.2 Desired system outputs

The basic product that is desired from a GIS generating PPA structures is a coverage showing activity locations divided into 'feasible' and 'non-feasible' (in the punctiform example) or showing the arcs in the network that are feasible for travel (in the network example) for an individual or population aggregate. This basic product can then be used to build more complex PPA based structures for analysis and modelling applications.

An example of the desired system output for the network PPA can be seen in Figure 8.3. The top half of the figure provides the input data for a network based PPA: the calculated arc travel times (which are bi-directional), the node travel times and the location of the travel origin and destination (again, any stop time during the travel interval is not considered in this example). In the bottom half of Figure 8.3, a network PPA has been calculated using a time budget of 35 units. The solid lines indicate the subnetwork that is feasible for travel, assuming that each arc is completely traversed by the individual when travelling. The individual can travel from the origin to any node connected to a feasible arc (using the shortest path) and then to the travel destination (again using the shortest path) within 35 time units. Note that the budget only allows for a single trip to and from a specified node within the feasible subnetwork.

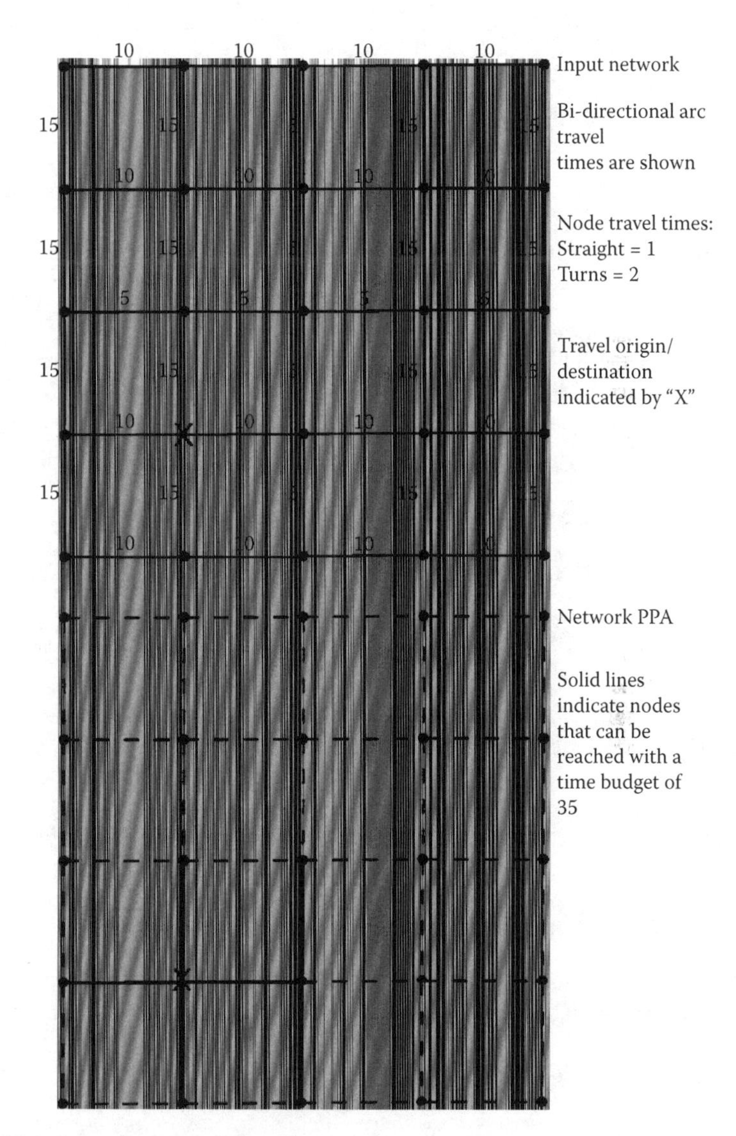

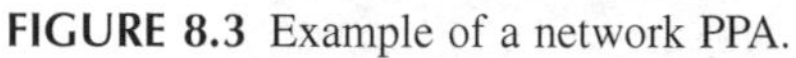
FIGURE 8.3 Example of a network PPA.

4.3 Generic procedure for building network based PPAs

The following generic procedure generates a basic network based PPA that does not consider stationary time at an activity location. It is assumed that the input data for the network PPA as identified in the preceding sections is present, and that each node in the network is uniquely identified by a non-zero number. It is also assumed that the arcs are completely traversed by individuals when travelling.

PHASE 0: INITIALIZATION

Step 0.1: define the following variables:

C = a pointer to a base node for considering arcs
N = a sequential counter labelling the current arc under consideration
$A(C,j)$ = a directed arc from node C to node j
NTIME(C) = the cumulative travel time along the shortest path from travel origin node to node C
ENTARC(C) = a pointer to the arc that enters node C in the shortest path from the travel origin node
CTIME(N) = the cumulative travel time along the shortest path from the travel origin node and through arc N
TRAVEL (N) = the travel time through arc N alone
TURN(ENTARC(C), N) = the turn time from ENTARC to arc N through node C
BUDGET = the time budget for the PPA
START = a set of nodes to be used as base nodes for considering arcs

Step 0.2: assign the appropriate travel times to the nodes and arcs in the network

Step 0.3: set $N = 1$; set C to the travel origin node; initialize START as an empty set

PHASE 1: ACCUMULATION OF ARCS FROM THE ORIGIN

Step 1.1: calculate the cumulative travel time along the shortest path from the travel origin node to node C, using the appropriate turn times and travel times. Set NTIME(C) to this value. Set ENTARC(C) to the arc that enters node C in the shortest path. If C is the travel origin, ENTARC(C) = (0, 0)

Step 1.2: choose an arc $A(C, j)$ that is incident to node C, is not currently in the tentative PPA, and has not been considered in conjunction with the current C. If no arcs can be chosen, then go to Step 1.7; otherwise, label this arc N (replacing any other arc previously labeled N)

Step 1.3: set TURN(ENTARC(C), N) to the appropriate turn time required to enter arc N from ENTARC through node C. If ENTARC(C) = (0, 0), TURN(ENTARC(C), N) = 0. Set TRAVEL(N) to the travel time for arc N calculated in Phase 0. Set CTIME(N) = NTIME(C) + TURN(ENTARC(C), N) + TRAVEL(N)

Step 1.4: IF CTIME(N) > BUDGET go to Step 1.2

Step 1.5: add arc N to the tentative PPA; add node j to START if j is not already in START

Step 1.6: $N = N + 1$; go to Step 1.2

Step 1.7: select a node from START that has not already been selected. If no nodes can be chosen go to Phase 2; otherwise designate the selected node as C and go to Step 1.1

PHASE 2: TEST ARCS FOR FEASIBILITY

Step 2.1: choose arc N. If N does not point to an arc in the tentative PPA, go to step 2.5

Step 2.2: Set $T^* =$ BUDGET – CTIME(N)

Step 2.3: enumerate the arcs in travel sequence along the shortest path trough the network from (and including) arc N to the travel destination. At each arc and node, subtract the appropriate turn and travel time from T^*. Continue until T^* is exhausted (no further arc can be added without making T^* negative)

Step 2.4: if the path enumerated in Step 2.3 reaches the travel destination (i.e. is incident to that node), than arc N is feasible; retain it in the PPA. If the path in Step 2.3 does not reach the travel destination, arc N is not feasible; remove it from the PPA

Step 2.5: $N = N - 1$; If $N = 0$ STOP; otherwise go to Step 2.1

This procedure results in the identification of the set of arcs that comprise the PPA given the specified travel origins and destinations and the time budget for travel.

4.4 Implementation issues

The inputs, outputs and generic procedure identified in Section 4.3 indicate several features that must be present in a GIS for the derivation of network based PPAs. These features can be broadly classified into data handling capabilities and network operations. With respect to data handling, the basic requirement of a GIS is the capability to store and manipulate topological network structure in terms of a set of uniquely identified nodes and directed arcs connecting these nodes. The arcs and nodes must also maintain data on travel times, turn times and the locations of travel origins and destinations. Each directed arc should be able to accommodate several travel times for that arc, and the analyst should have the ability to input data on travel time directly or calculate these within the GIS on the basis of an arc's length and assumed travel velocities. Similarly, the analyst should have the option of reading node turn times directly from a database or calculating these within the GIS. Additionally, each node must be able to maintain data on any travel origins and destinations coincident with the node.

This generic procedure contains two basic network operations: the calculation of all shortest paths from a particular origin up to a cumulative impedance (i.e., travel time) limit along each path (Phase 1 of the procedure) and the calculation of the shortest path from an arc to a particular destination up to a variable cumulative impedance (Phase 2 of the procedure). In addition, Phase 2 requires a test to determine whether the path from an arc to the destination is achieved within its impedance limit. To implement the generic PPA procedure, the GIS should accommodate these basic network operations in addition to the feasibility test.

Efficiency is a very critical aspect of implementing the generic PPA procedure in a GIS. Real transportation networks are typically very complex and can easily contain hundreds of nodes and arcs. As the number of data handling and network operations for any problem of non-trivial size can be substantial, the GIS must be efficient in its implementation of the algorithm. This efficiency is a function of the flexibility of the GIS command structure in adapting to the required operations and the ease with which the required commands can be combined into the overall generic algorithm. Several points can be made in this respect.

It is important that data handling and network operations are easily accessible to each other and are not separated by intervening commands as with many modular software designs. For example, it is likely that an analyst will be interested in performing sensitivity analyses with the network PPA (i.e., changing travel velocities or travel time in the network to examine the effect on accessibility). In this instance, having to re-read in the entire network data whenever a network datum is changed can greatly hamper the usefulness of the GIS implementation. Therefore, data handling should be directly and easily accessible from the network operations. Further, the network operations themselves need to be accessible to each other. The phases of the generic procedure entail two distinct network operations which must be combined in the algorithm, and a modular structure with intervening commands can greatly hamper the efficiency of the implemented procedure.

The GIS should have a sufficiently flexible command structure so that basic network operations are not only present in the purest from possible, but can also allow the parameters of the operations to be tailored to meet the analyst's needs. The generic PPA algorithm needs only very basic network operations (e.g., shortest path calculations), but requires these operations to be in a specific format. For example, the feasibility test in Phase 2 of the generic algorithm requires the calculation of the shortest paths from all tentative PPA arcs to the travel destination only until the residual time budget at each arc is exhausted. Unless the GIS allows the calculation of simultaneous multiple shortest paths (in this instance from multiple origins to a single destination), the number of individual shortest path commands required for any realistic network PPA would be enormous. Further, calculating each path only until the residual budget is exhausted and then testing for connectivity to the destination allow valuable computational savings over completely enumerating the path and then examining its cumulative impedance, as a good proportion of the arcs in the tentative PPA will typically not be feasible (although this will vary greatly depending on the location of the origin and destination pair). Thus the GIS should allow the shortest path calculation up to a variable limit.

In brief, while the generic PPA algorithm requires only very basic capabilities for data handling and network operations, it does demand a high degree of flexibility at the levels of the individual command and the overall program structure. Unfortunately, much of the current GIS technology appears to be designed for standard applications rather than for flexibility.

Consider, for example, the mainframe version 5.0 of ARC/INFO, perhaps the most popular widely available commercial GIS. This version contains a set of network related commands organized under the appropriate label of NETWORK (NETWORK User's Guide 1987). The basic data handling and network operations required in the generic procedure are nominally present in this software. ARC/INFO NETWORK can handle a set of unique nodes and directed arcs, keep records of locations within this structure and handle numerous travel times at both nodes and arcs in the network. The GIS can also perform arc allocation based on all the shortest paths from a given location, and can perform shortest path routing between origin–destination pairs. However, ARC/INFO's modular structure separates key data handling and network operations, as well as the network commands that comprise Phases 1 and 2 of the generic procedure. It also cannot perform simultaneous shortest path analysis among sets of origins and destinations or calculate these paths only up to a variable limit. A less crucial, but nevertheless a substantial, restriction is the inability of the software to overlap arc allocations to different origins within a single coverage. This can be very useful when deriving more than one PPA for a study area.

Therefore, although ARC/INFO NETWORK can meet the requirements for standard GIS applications, it is inefficient and unwieldy in meeting the more specialized needs of the network PPA procedure. Whereas ARC/INFO is certainly not representative of all GIS software, it does provide a benchmark which indicates the problems encountered by analysts who wish to use GIS technology in more specialized research and modelling.

5 Summary and conclusions

Space-time prism constructs are a powerful yet neglected approach to analysing the accessibility of individuals to an environment. With the space-time prism, accessibility can be assessed relative to spatial and temporal constraints on individual behaviour. The perspective does not require restrictive behavioural assumptions or the adoption of a particular behavioural stance — it simply addresses the behavioural possibilities of individuals given some basic constraints. The most valuable aspect of the space-time prism is that it allows the direct incorporation of considerations of accessibility into locational analysis and transportation planning.

Geographical information systems, through their ability to manipulate and analyse spatial data, can allow more widespread use of the space-time perspective in spatial modelling and analysis. In particular, the ability of a GIS to handle data on street networks can provide a realistic operational version of the potential path area, or the spatial extent of an individual's reach given spatial and temporal constrains on movement. This paper has presented an analysis of requirements and feasibility for deriving space-time prism within a GIS and has discussed some potential applications, identified the required inputs and expected outputs from a GIS and provided a generic procedure for producing a network based potential path area. It has also discussed issues related to implementing these requirements in current GIS technology. A modular structure and inflexibility of key commands and procedures can render GIS software unable to derive effectively the desired space-time prism construct, as was illustrated by an assessment of the mainframe version 5.0 of ARC/INFO.

The remarkable growth of GIS technology is well documented. As GIS becomes more widespread, it can be expected that more researchers will approach the methodology with the intent that motivated this study: the use of GIS technology to address research issues that previously could not be handled and were abandoned owing to difficulties in maintaining their unwieldy requirements. If the GIS community is to encompass these researchers, GIS software that is flexible enough to meet their variable and specific requirements is needed. As illustrated here, an analyst can approach GIS technology with a specific set of prior needs which GIS software designed for standard applications will not satisfy.

Potentially, there is a large number of researchers who could benefit from the spatial data handling and manipulation capabilities of GIS technology. The language used by many of these researchers in their modelling and analysis consists of well defined bodies of spatial and network analytical techniques and procedures. What is needed is GIS software which is sufficiently broad and flexible to accommodate these techniques.

Acknowledgements

This work was conducted as part of a Ph.D. minor in geographical information systems at the Department of Geography, Ohio State University. The author thanks Morton O'Kelly, three anonymous referees and especially Duane Marble for very helpful comments. A previous version of this paper was presented at the 1991 Annual Meeting of the Association of American Geographers in Miami, FL, U.S.A., where it was awarded first prize in the GIS Specialty Group Student Paper Competition.

References

Burns, L. D., 1979, *Transportation, Temporal, and Spatial Components of Accessibility* (Lexington, MA: D. C. Heath).

Garling, T., and Garling, E., 1988, Distance minimization in downtown pedestrian shopping. *Environment and Planning* A, **20,** 547.

Golledge, R. G., and Stimson, R. J. 1987, *Analytical Behavioral Geography* (London: Croom-Helm).

Hägerstrand, T., 1970, "What about people in regional science?" Papers of the Regional Science Association, **14,** 7.

Horowitz, J. L., 1985, Travel and location behavior: state of the art and research opportunities. *Transportation Research* A, **16A,** 441.

Janelle, D.G., and Goodchild, M. F., 1983, Transportation indicators of space-time autonomy. *Urban Geography*, **4,** 317.

Landau, U., Prashker, J. N., and Alpern, B., 1982, Evaluation of activity constrained choice sets of shopping destination choice modeling. *Transportation Research* A, **16A,** 199.
Landau, U., Prashker, J. N., and Hirsh, M., 1981, The effect of temporal constraints in household travel behavior. *Environment and Planning* A, **13,** 435.
Langran, G., 1989, A review of temporal database research and its use in GIS applications. *International Journal of Geographical Information Systems*, **3,** 215.
Lenntorp, B., 1976, *Paths in Space-Time Environments: A Time-Geographic Study of the Movement Possibilities of Individuals,* Lund Studies in Geography, Series B, Number 44.
Lenntorp, B., 1978, A time-geographic simulation model of individual activity programmes. In *Timing Space and Spacing Time Volume 2: Human Activity and Time Geography,* edited by T. Carlstein, D. Parkes and N. Thrift (London: Edward Arnold), pp. 162.
NETWORK User's Guide 1987 (Redlands, CA: Environmental Systems Research Institute).
Pred, A., 1977, The choreography of existence: comments on Hägerstrand's time-geography and its usefulness. *Economic Geography,* **53,** 207.
Sigal, C. E., Pritsker, A. A. B., and Solberg, J. J., 1980, The stochastic shortest route problem. *Operations Research*, **28,** 1122.

Modelling Accessibility Using Space-Time Prism Concepts within Geographical Information Systems: Fourteen Years On

Harvey J. Miller

Motivation

In retrospect, I would like to say that I was farsighted in my speculations about the future of geographical information systems (GIS) and transportation science, and the role of time geography in both. But in reality I was attempting to complete a Ph.D. minor in GIS at Ohio State University under the direction of Duane Marble, and I needed a research project to satisfy its requirements. At that time, someone at Ohio State, perhaps Marble, introduced me to time geography, an area that I had only been dimly aware of prior to my Ph.D. studies. The basic idea behind the paper — improving the realism of the space-time prism through GIS-based network databases and network analytical tools — seemed obvious. Marble liked the idea, as did Morton O'Kelly (my major Ph.D. advisor). After the requisite literature search, I was somewhat shocked to discover that no one had published a paper on this idea previously.

I wrote the paper, including an algorithm for calculating the network *potential path tree* (PPT; as I now call it) and an assessment of Arc/Info (as it was called at the time) as a platform for performing the calculations. I submitted the paper to *IJGIS* and was pleasantly surprised to have it accepted easily with only minor revisions. The paper also won the first Student Paper Competition Award from the GIS specialty group of the Association of American Geographers (AAG), although I discovered later that it was the *only* entry; how times have changed!

Subsequent work

After a digression into my lesser-known dissertation research on optimal spatial search behavior, I returned to the topic of time geography and GIS. Part of my motivation was the recognition that the *network time prism* concepts defined in the

1991 paper were limited with respect to the improving capabilities of GIS, availability of digital geo-referenced data and the deployment and applications of *location-aware technologies* (LATs). Another motivation was my recognition that the 1991 paper was receiving many more citations that my dissertation-related research!

In the early 1990s, I noticed a new network data model called (inappropriately) *dynamic segmentation* that allowed network attributes to be maintained at arbitrary locations within arcs. I also noted the increasing availability of address-ranged network databases and address-matching algorithms, which allowed address-indexed databases to be geo-referenced. I realized that the PPT was limited since it only provided the accessible nodes in the network, not arbitrary locations within the network. At about the same time, Okabe and Kitamura (1996) published a very nice paper that showed how to calculate market areas within networks to an arbitrary level of resolution. Using their approach, I developed a method for calculating the *potential network area* (PNA). The PNA is the true network equivalent to the potential path area (the spatial footprint of the space-time prism) in classic time geography, the PPT being a cruder surrogate. I was also able to integrate classic accessibility measures, such as spatial interaction and benefit measures, into this structure (Miller, 1999). We later extended the PPT and PNA to the case of *dynamic networks* where travel times vary not only by location, but also by time (Wu and Miller, 2001).

More recently I addressed the issues of high-resolution measurement of time-geographic entities. LATs, such as the global positioning system, have great potential for improving the collection of data on human activities in space and time. However, time geography was not ready for LATs: the literature described basic entities and relationships such as prisms and bundling only informally. The key to a formal measurement theory for time geography is temporal disaggregation. The location or spatial extent of time-geographic entities at any instant in time can be solved as compact spatial sets, or the intersection of compact spatial sets. These simple sets support evaluation of time-geographic relationships such as bundles and intersections. An ancillary benefit is a generalization to n-dimensional space rather than the strict two-dimensional space of classical time geography. The cases of direct interest in time geography are $n = 1$ (networks), 2 (planar space), or 3 (natural space), and the measurement theory can support consistent measurements and analysis of time-geographic entities across these cases (for example, a vehicle moving within and outside a transportation network) (Miller, 2005a). I also extended the measurement theory to encompass virtual interaction and tele-presence through information and computing technologies (ICTs) (Miller, 2005b).

Broader developments

My 1991 paper was simply an idea whose time had come. Several trends have emerged and converged to give it the appearance of influence.

GIScience. Integrating time into GIS beyond the simple "snapshot" metaphor had been a major concern in GIScience prior to 1991. Much progress has been achieved (Yuan, 2001). Of particular relevance for time geography is the development of transportation-oriented data models that can accommodate travel and activities

(Frihida, Marceau, and Thériault, 2002) as well as mobile objects databases (Pfoser and Jensen, 1999).

Others in the GIScience community have worked specifically on GIS design and applications in time geography. For example, Kwan and her students have developed several pragmatic algorithms for calculating feasible opportunities using GIS tools (see Kwan, 2004). O'Sullivan, Morrison, and Shearer (2001) compute space-time prisms for multimodal transportation networks. Hornsby and Egenhofer (2002) illustrate methods for modeling mobile objects over varying spatial and temporal granularities. It is also encouraging that burgeoning GIS scholars are addressing time geography and being recognized for their contributions. For example, in 2004, Scott Bridwell of the University of Utah won the AAG GIS specialty group student paper competition for his work on space-time masking in time geography, and in 2005, Hongbo Yu of the University of Tennessee won the same award for a GIS-based time geography query and visualization system.

Transportation, urban, social, and other human sciences. In addition to advances in geographic information science and technology, there is a revolution occurring in the broader sciences concerned with human phenomena such as transportation systems, urban dynamics, social networks, and epidemiology. This involves bottom-up approaches to understanding systems such as humans and the environment. Bottom-up approaches draw from complex system theory: this suggests that some phenomena are ecological rather than mechanical in nature and emerge from interactions among individual agents and the built and/or natural environment. This suggests not only that these systems are more complicated than previously supposed, but also that we cannot engineer their growth; rather, we can only influence or shape their evolution. This is not defeatist; rather, it suggests humility and the need for sophisticated, nuanced approaches to directing human systems toward efficient, equitable, and sustainable outcomes.

In transportation science, individual-level activity-based approaches to understanding and managing transportation systems have become dominant (Timmermans et al., 2002). Agent-based approaches have become prevalent in urban modeling (Benenson, 2004) and land-use/land-cover change (Parker et al., 2003). Researchers in epidemiology are discovering the value of the time-geographic perspective in analyzing the spread of infectious disease over space and time (Jacquez et al., 2005).

Research frontiers

There is still much to do in time geography and GIS. In my opinion, the following are the high-priority research frontiers.

A representation theory for time geography and activity theory. Although the time-geographic analytical theory discussed above is useful for high-resolution measurement and analysis of time-geographic entities and relationships, it is not comprehensive. As I have argued elsewhere (Miller, 2005c), there is a need for a rigorous representation theory that encompasses time geography and activity theory in transportation science. This will not be an easy task due to the diversity of entities, relationships, and events in these related but distinct fields. However, it is necessary if these two communities are to communicate more effectively, and if standardized databases and software tools are to be developed. It might also reduce the plethora of new terminology for existing time-geographic concepts that materialize as new researchers discover, or in some cases reinvent, time geography.

Visualization and knowledge discovery. Although the potential for new insights from visualizing and exploring space-time paths and prisms is promising (Kwan, 2000), there is still a need for innovative and effective methods for massive space-time activity databases. It is one thing to visualize and explore 200 or 2000 space-time paths; it is more challenging and ultimately more useful to make sense of 200,000 or 2,000,000 paths. Closer adherence to the basic principles of knowledge discovery is required (Laube et al., 2005).

Scalable techniques. A related issue to the one discussed above is a need to develop and apply scalable techniques to massive space-time activity databases. Off-the-shelf GIS tools linked though scripting languages or Visual Basic cannot handle the data volumes required; next-generation techniques are required. Time-geographic analysis can be easily decomposed into parallel computations based on tasks and/or data parallelism; computational techniques such as those based on parallel or grid computing architectures has promise for revealing new knowledge from these databases (Armstrong et al., 2005).

Multiscale linkages. Time geography has conceptual linkages from the individual to aggregate entities and relationships such as bundles, projects, or systems. However, there is still a need for analytical and modeling linkages. How do aggregate dynamics, such as traffic jams and trendy neighborhoods, emerge from individual activities? How do plans, decisions, and adjustments propagate through a system? Can we understand and predict these dynamics? Of practical importance are synoptic measures of patterns and dynamics from individual-level data and simulations.

Locational privacy. The potential for new insights and improved decision making engendered by time geography is exciting, but the possibility of unwarranted surveillance and control is equally frightening (Dobson and Fisher, 2003). Techniques and protocols are required based on the evolving concept of *location privacy* to preserve the scientific value of data without violating individual privacy and freedom.

Proof of concept. Empirical research applying time geography and GIS is suggestive; it certainly appears that a time-geographic perspective tells us something different than approaches based on aggregate spatial units and flows. But precisely what new knowledge have we gained? I would argue that dramatic and novel insights have yet to be achieved. I am extremely optimistic that this will occur, but it will require progress with respect to the research frontiers listed above.

References

Armstrong, M.P., Wang, S., and Cowles, M., 2005, Using a computational grid for geographic information analysis. *Professional Geographer*, **57**, 365–375.

Benenson, I., 2004, Agent-based modeling: From individual resident choice to urban residential dynamics. In *Spatially Integrated Social Science*, M. F. Goodchild and D. G. Janelle, Eds., Oxford, New York, 67–94.

DOBSON, J.E. AND FISHER, P.F., 2003, Geoslavery. *IEEE Technology and Society Magazine* 22(1), 47–52.

FRIHIDA, A., MARCEAU, D.J., AND THÉRIAULT, M., 2002, Spatio-temporal object-oriented data model for disaggregate travel behavior. *Transactions in GIS,* **6**, 277–294.

HORNSBY, K. AND EGENHOFER, M.J., 2002, Modeling moving objects over multiple granularities. *Annals of Mathematics and Artificial Intelligence*, **36**, 177–194.

JACQUEZ, G.M., GOOVAERTS, P., AND ROGERSON, P., 2005, Design and implementation of a space-time intelligence system for disease surveilliance. *Journal of Geographical Systems*, **7**, 7–23.

KWAN, M.-P., 2000, Interactive geovisualization of activity-travel patterns using three-dimensional geographical information systems: A methodological exploration with a large data set. *Transportation Research C*, **8**, 185–203.

KWAN, M.-P., 2004, GIS methods in time-geographic research: Geocomputation and geovisualization of human activity patterns. *Geografiska Annaler B*, **86**, 267–280.

LAUBE, P., IMFELD, S., AND WEIBEL, R., 2005, Discovering relative motion patterns in groups of moving point objects. *International Journal of Geographical Information Science*, **19**, 639–668.

MILLER, H.J., 1999, Measuring space-time accessibility benefits within transportation networks: Basic theory and computational methods, *Geographical Analysis*, **31**, 187–212.

MILLER, H.J., 2005a, A measurement theory for time geography. *Geographical Analysis*, **37**, 17–45.

MILLER, H.J., 2005b, Necessary space-time conditions for human interaction. *Environment and Planning B: Planning and Design*, **32**, 381–401.

MILLER, H.J., 2005c, What about people in geographic information science? In *Re-Presenting GIS*, P. Fisher and D. Unwin, Eds., Wiley, Chichester, 215–2242.

OKABE, A. AND KITAMURA, M., 1996, A computational method for market area analysis on a network. *Geographical Analysis*, **28**, 330–349.

O'SULLIVAN, D., MORRISON, A., AND SHEARER, J., 2000, Using desktop GIS for the investigation of accessibility by public transport: An isochrone approach. *International Journal of Geographical Information Science,* **14**, 85–104.

PARKER, D.C., MANSON, S.M., JANSSEN, M.A., HOFFMANN, M.J., AND DEADMAN, P., 2003, Multi-agent systems for the simulation of land-use and land-cover change: A review. *Annals of the Association of American Geographers,* **93**, 314–337.

PFOSER, D. AND JENSEN, C.S., 1999, Capturing the uncertainty of moving-object representations. In *Advances in Spatial Databases: 6th International Symposium* (SSD'99), Lecture Notes in Computer Science 1651, R.H. Güting, D. Papadias, F. Lochovsky, Eds., Springer, Berlin, 111–131.

TIMMERMANS, H, ARENTZE, T., AND JOH, C.-H., 2002, Analysing space-time behaviour: New approaches to old problems. *Progress in Human Geography*, **26**, 175–190.

WU, Y.-H. AND MILLER, H.J., 2001, Computational tools for measuring space-time accessibility within dynamic flow transportation networks. *Journal of Transportation and Statistics*, **4** (2/3), 1–14.

YUAN, M., 2001, Representing complex geographic phenomena in GIS. *Cartography and Geographic Information Systems*, **28**, 83–96.

International Journal of Geographical Information Systems,
1992, Vol. 6, No. 1, 31–45.

9 Geographical Information Science*

Michael F. Goodchild

Abstract. Research papers at conferences such as the European Geographical Information Systems (EGIS) and the International Symposia on Spatial Data Handling address a set of intellectual and scientific questions which go well beyond the limited technical capabilities of current technology in geographical information systems. This paper reviews the topics which might be included in a science of geographical information. Research on these fundamental issues is a better prospect for long-term survival and acceptance in the academy than the development of technical capabilities. This paper reviews the current state of research in a series of key areas and speculates on why progress has been so uneven. The final section of the paper looks to the future and to new areas of significant potential in this area of research.

1 Introduction

The geographical information system (GIS) community has come a long way in the past decade. Major research and training programmes have been established in a number of countries, new applications have been found, new products have appeared from an industry which continues to expand at a spectacular rate, dramatic improvement continues in the capabilities of platforms, and new significant data sets have become available. It is tempting to say that GIS research, and the meetings at which this research is featured, are simply a part of this much larger enthusiasm and excitement, but there ought to be more to it than that.

What, after all, is the purpose of all of this activity? Expressions such as 'spatial data handling' may describe what we do, but give no sense of why we do it. This was one of the themes behind Tomlinson's keynote address at the First International Symposium on Spatial Data Handling in Zürich in 1984 (Tomlinson 1984). The title of the conference suggests that spatial data are somehow difficult to handle, but will that always be so? It suggests a level of detachment from the data themselves, as if the U.S. Geological Survey were to send out tapes labeled with the generic warning 'handle with difficulty'. It is reminiscent of the name of the former Commission on Geographical Data Sensing and Processing of the International Geographical Union. A quick review of the titles of the papers at that or subsequent meetings should be

* Based on keynote addresses by the author at the Fourth International Symposium on Spatial Data Handling, Zürich, July 1990 (Goodchild 1990), and EGIS 91, Brussels, April 1991 (Goodchild 1991).

enough to assure anyone that their authors are concerned with much more than the mere handling and processing of data — from a U.S. perspective, that the community is more than the United Parcel Service of GIS.

Geographical information systems are sometimes accused of being technology driven, a technology in search of applications. That seems to be more true of some periods of the 25-year history of GIS than of others. For example, it is difficult to suggest that Tomlinson and the developers of the Canadian Geographical Information System (CGIS) (Tomlinson *et al.* 1976) were driven by the appallingly primitive hardware capabilities of 1965. On the other hand the prospect of a menu driven, full colour, pull-down menu raster GIS in the 386-based personal computer on one's desk has clearly sold many systems in the past few years. Technological development comes in distinct bursts, and so does the technology drive behind GIS. It may be the motivation behind the desire to handle spatial data, but it fails to explain many of the diverse research efforts being reported at meetings and in the literature.

There have also been phases when applications have driven GIS. CGIS itself was an application in search of a technology, and the drive was sufficiently strong to lead to the prototype of the first map scanner, and to numerous other technological developments (Tomlinson *et al.* 1976). McHarg had worked out the principles of the map overlay technique (McHarg 1969) long before Berry and others automated them in MAP and its derivatives (Berry 1987); school bus routing software has been around much longer than the problem's implementation in a standard GIS. But again, much of the subject matter of GIS research lies well beyond any reasonably foreseeable application.

There have also been phases when applications have driven GIS. CGIS itself was the widespread distribution of Landsat and SPOT imagery, and the availability of digital elevation models and street files in many countries have certainly led to applications well beyond those used to justify the data's compilation. TIGER, for example, appears to be spawning its own industry of updaters, repackagers and application developers, although it exists in principle only to serve the needs of the 1990 U.S. Census.

However, although the driving seat of GIS is undoubtedly crowded, I would like to deal in this paper largely with the fourth driver located apparently irrelevantly in the back seat, the 'S' word. It seems to me that there is a pressing need to recognize and develop the role of science in GIS. This is meant in two senses. The first has to do with the extent to which GIS as a field contains a legitimate set of scientific questions, the extent to which these can be expressed and the extent to which they are generic, rather than specific to particular fields of application or particular contexts. To what extent is the GIS research community driven by intellectual curiosity about the nature of GIS technology and the questions that it raises? And if GIS can be motivated by science, then what are its subfields, what are its questions, and what is its agenda? The second sense has to do with the role of GIS as a toolbox in science generally — with GIS for science rather than the science of GIS. What do we need to do to ensure that GIS, and spatial data handling technology, play their legitimate role in supporting those sciences for which geography is a significant key, or a significant source of insight, explanation and understanding?

To do this we must first establish that spatial, or rather, geographical data are unique, and that their problems cannot therefore be subsumed under some larger field. We must also establish that there are problems which are generic to all geographical data, or at least establish that it is possible to distinguish those that are from those that are not. For example, the accuracy of attributes on a choropleth map of crime statistics would seem to be very little informed by knowledge of attribute accuracy for geographical data generally, but to require instead a level of understanding of the specific problems of crime statistics. However, the accuracy of population estimates for an arbitrarily defined polygon may well be known from, or at least informed by, the general properties of the modified areal unit problem (Openshaw 1977).

2 What is unique about spatial data?

In many facilities management systems, the role of the GIS is to provide an alternative key to data, a method of access based on geographical location. In essence, a spatial database has dual keys, allowing records to be accessed either by attributes or by locations. However, dual keys are not unusual. The spatial key is distinct, as it allows operations to be defined which are not included in standard query languages. For example, it is possible to retrieve all point records lying within an arbitrary, user-defined polygon, an operation which is not defined in standard query languages such as SQL. In essence, the spatial key is multidimensional, but again multidimensional keys are known from other areas, and analogues of point in polygon retrieval can be defined for non-spatial dimensions.

What distinguishes spatial data is the fact that the spatial key is based on two continuous dimensions. It is possible to visit any location (x, y) in the real, geographical world, defined in principle with unlimited precision, and return a value for a variable, for example, topographic elevation z. Terrain is thus characterized by an infinite number of tuples $<x, y, z>$. In network applications z is defined only for locations on the network, but the number of tuples is still infinite if variation is continuous along this one-dimensional structure of links and nodes. Time series also have continuous keys, but are rarely conceived, measured or represented as continuous, and there appears to be little commonality of interest in the problems of temporal data handling. By contrast, there is ample evidence of commonality in the spatial data handling disciplines.

Many of our data models, particularly polygon networks and triangulated irregular networks (TINs), reflect an underlying view of space as continuous and the need to accommodate the user who wishes to determine z at some arbitrary and precise (x, y). One implication of this is that there exists a multiplicity of possible conceptual data models for spatial data, and that the choice between them for a given phenomenon is one of the more fundamental issues of spatial data handling.

Another distinctive feature of spatial data is what Anselin (1989) refers to as spatial dependence, the propensity for nearby locations to influence each other and to possess similar attributes. Without spatial dependence, there would be no reasonable prospect of creating even approximate views of continuous spatial variation within a discrete, finite machine. It is not uncommon for tuples which have similar values of a key to have similar values of other attributes, but the structure of spatial dependence is unusual, relying as it does on both dimensions of the (x, y) key, with similarity determined by a metric.

Finally, geographical data are distributed over the curved surface of the earth, a fact which is often forgotten in the limited study areas of many GIS projects. We have worried for centuries about how to portray the earth's surface on a flat sheet of paper, and have developed an extensive technology of map projections. However, as a result we have few methods for analyzing data on the sphere or spheroid, and know little about how to model processes on its curved surface. Moreover, we tend to have treated GIS displays as if they were virtual sheets of paper, and insisted on viewing geographical data as if they were projected to a flat surface, instead of exploiting the potential of electronic display to create views of the globe itself. We need to develop the appropriate techniques for working with the globe, and making use of solid modeling rather than conventional two-dimensional graphics, if we are to understand geographical processes at the global scale and contribute effectively to global science. We must rescue the orthographic projection from its present obscurity.

3 The content of geographical information science

Having established that geographical information has unique properties and problems, we can now review the set of generic questions which might make up a geographical information science. This can be done in a largely linear fashion, from data collection to analysis, although some themes tend to cut across this simple arrangement. However, it seems appropriate to begin this review with a disclaimer. What I present in this paper is in many ways my own view, and I would expect it to be challenged. I think my own biases will become clear in what follows. Because of the field's diversity and dynamism it is difficult, if not impossible, for any one individual to attempt a general overview. What follows is therefore almost inevitably incomplete and uneven.

Research is often identified as either pure or applied — driven by basic and innocent human curiosity or by the practical everyday needs of human society. Many GIS are a response to human needs for information management and analysis, and in that sense one might expect GIS research to be more applied than pure. However, one view of pure research is that it is research that has not yet found application; pure research is a long-term investment just as applied research is a short-term investment. From an academic perspective, pure research is often associated with higher prestige, but applied research with greater funding. I have tried to cover the full range from pure to applied, feeling that both are important to GIS. At the same time 'basic research' is the primary purpose of the U.S. National Center for Geographical Information and Analysis, and the center is very fortunate in being funded to do research the applications of which may lie years or even decades into the future.

During the design phase of the CGIS in the 1960s, it became clear that the only practical way to input the large number of maps needed would be by some form of scanning device (Tomlinson *et al.* 1976). At that time no scanner for map-sized documents existed, and it was necessary to invent one. A prototype drum scanner was built by IBM Canada and successfully tested, at what by modern standards would be regarded as vast expense. Other parts of the CGIS design team were busy inventing other, equally fundamental and now familiar solutions to technical GIS problems, such as the Morton order.

In the almost three decades of development of GIS that are now behind us, similar 'how to do it' research has produced a large number of algorithms, data structures, spatial indexing schemes and other technological solutions. Some of these are unique to GIS, but many have been reinvented in several related disciplines. The Morton order, for example, occurs in the literature of several spatial data handling fields under different names (Samet 1989), and descriptions of algorithms for finding Thiessen polygons are spread over a wide range of journals. At the same time there is a growing sense in GIS research that our emphasis has changed, as more and more of the underlying technical problems of GIS are solved. Attention has moved from primitive algorithms and data structures to the much more complex problems of database design, and the issues surrounding the use of GIS technology in real applications. The following sections identify some of these key issues.

3.1 Data collection and measurement

If spatial reality is continuous and subject to complex structures of spatial dependence, then how should it be compiled and measured? More generally, how do people perceive the real world of geographical variation, structure it and learn about it? Although many of these questions are part of the research agendas of remote sensing, photogrammetry, geodesy and cognitive psychology, the lines of demarcation are far from distinct. Should GIS or remote sensing concern the problems of transferring information from one technology to the other, and more importantly making good sense of it? Is it GIS or remote sensing if ancillary geographical information is used to improve the accuracy of classification or if an image is used to update a GIS layer? Ultimately it matters little to which of the many pigeonholes we assign each topic. There are undoubtedly substantial scientific questions here, which require a depth of understanding of the nature of spatial variation, and one person's remote sensing may well be another's geographical information science.

The process of discretization, with its implied generalization, abstraction and approximation, takes place as data are collected, interpreted or compiled, and choices are made at this stage that affect the ultimate uses of the data. When those uses change, as they have been doing with the widespread use of GIS, it may be necessary or beneficial to rethink the process of data collection. For example, with digital management and delivery of census data, is it still appropriate to conduct a census on a decennial basis? Is the traditional approach to geological field mapping the most appropriate if the eventual objective is a digital three-dimensional representation of the subsurface? How will topographic mapping change now that it is cost-effective to survey new features using the Global Positioning System? Geographical data collection is often the domain of specialists in well established disciplines, so it may be many years before these kinds of questions are investigated or answered. To date the introduction of GIS seems to have had very little effect on the process of data collection.

3.2 Data capture

Enormous strides have been made in the technology for capturing digital geographical data in the past decade, and the systems now on the market are capable of a high level of intelligence in interpreting scanned map documents. The problem remains the poor quality of the documents, and the ambiguities that are caused by aspects of map design. As a result, manual digitizing remains a widely used approach, despite its high cost, tedium, and failure to show significant improvements in efficiency. Two trends may change this situation substantially in the next few years. One is the increasing avoidance of the map document as a step in the data compilation and input process. Surveying and photogrammetry are moving away from compilation using paper maps, and the more interpretive fields such as land use, vegetation or soil mapping are likely to follow suit. The digital total station is likely to be followed by the digital plane table and perhaps even the digital field geology notebook. The other is the long recognized possibility that comparatively minor changes in a map's design can make it vastly easier to scan and interpret (Shiryaev 1987).

3.3 Spatial statistics

As spatial data are always an approximation or generalization of reality, they are full of uncertainty and inaccuracy. A change of data model or scale can introduce a loss of information, as can digitizing or scanning. Processing in a finite machine also inserts its own form of uncertainty, although this is often insignificant in relation to the errors inherent in the data themselves. Many human geographical constructs are implicitly uncertain, including spatial objects ('Indian Ocean', 'Europe') and their relationships ('in', 'across'). Whether we think of uncertainty in set theoretical terms through notions of fuzziness or in statistical terms through the calculus of probabilities, the study of spatial data uncertainty, its measurement and modeling, and the analysis of its propagation through the processes of spatial data handling are undoubtedly part of geographical information science. How should one compile an accurate representation of geographical variation for input to a database? How should one represent the uncertainty or inaccuracy present in a digital representation? How can uncertainty be propagated from database to GIS products?

Geographical data bring their own special set of problems to spatial statistics. Whereas in medical imaging the problem may be to determine the true location of objects from 'dirty' pictures (Besag 1986), in geographical images there is often no clear concept of truth, as objects are often the products of interpretation or generalization. We need much better methods of measuring and describing uncertainty, particularly in the complex spatial objects common in GIS. We need better methods for dealing with the world as a set of overlapping continua, instead of forcing the world into the mould of rigidly bounded objects. Most of the answers to these questions will have to come from spatial statistics, but geographical information specialists must provide the motivation and the examples, and define the overall objectives and constraints.

Although all geographical data are uncertain to some degree, all of the current generation of GIS follow the common practice in cartography and represent geographical objects as if their positions and attributes were perfectly known; data quality may or may not be addressed in a separate statement. The consequences of uncertainty for GIS products are never estimated. Recent research has followed several different and productive lines in attempting to address the problem of data quality. One is to match precision to accuracy. In a locational sense, this means using limited precision in data representation and processing, most often through the use of a raster whose size is determined by data accuracy. Various forms of quadtree structure have also bee used to fit locational precision to known levels of accuracy. There have been several recent papers on finite resolution processing in GIS (e.g. Franklin 1984, Dutton 1989) and finite resolution geometry is an active research area in mathematics.

Another productive approach has been to incorporate techniques from geostatistics, notably kriging, as the statistical basis of these techniques makes uncertainty explicit. We now have several useful models of digitizing error, and its consequences for estimated measures such as area (e.g., Chrisman and Yandell 1988, Keefer *et al.* 1988). Finally, there have been several successful efforts to model geographical data sets as random fields, or derivatives of random fields, and to use this approach to model uncertainty in GIS objects (e.g., Goodchild 1989). Between all of these methods, we probably now have an adequate set of models of accuracy from which to build an error-tracking GIS. However, spatial statistics is not an easy field, and many of these techniques go well beyond elementary statistics in their conceptual sophistication.

3.4 Data modeling and theories of spatial data

Data models are the logical frameworks which we use to represent geographical variation in digital databases. As each must be an approximation, the choice between alternative models constrains not only the functions available, but also the accuracy of products. Of all the developments in GIS in the past decade, perhaps the most exciting has been the proliferation of data models, and the growing literature on their relative merits. The debate over raster and vector goes back to the earliest days, but has now been joined by debates over objects, layers, the philosophy of object orientation, hierarchical models of complex objects, and the entire range of possibilities inherent in time dependence and three dimensions. Despite the interest, we still do not have a complete and rigorous framework for geographical data modeling, even in the static two-dimensional case, and without one it is difficult to see how GIS can escape the constraints imposed by specific system implementations. How much capability is being lost by forcing contemporary applications into the multilayer raster model used by many systems, or the point–line–area coverage model used by many others? This is both a pure and an applied research problem. On the one hand, we must develop a comprehensive framework for geographical data modeling, with an associated terminology, to provide the basis for standards and an ideal against which specific systems can be measured. On the other hand, an abstract framework is of little value if it does not influence practice, through implementation in the products of the vendors. Here the real issue is whether it is possible to enlarge or 'retrofit' the data model underlying an existing product, or whether any attempt to do so is doomed to cause inconsistency and incoherence.

These issues are precipitating lively discussion over the entire question of the degree to which we view, analyse, represent and model the world as discrete or continuous, as a collection of objects or a set of fields. Do we think in terms of variables with defined values everywhere in space, or of an empty space littered with possibly overlapping objects? In essence, these issues have brought the GIS debate from the comparative obscurity of internal data structures to the much more general issues of how we understand geographical variation. Everyday human experience sees a world of objects, but the science of natural processes deals more with continuous variation (Frank and Mark 1991). Thus the object oriented debate threatens to pit the New Agers against the embattled remnants of the Enlightenment, and what could be more stimulating than that?

3.5 Data structures, algorithms and processes

Many of the results of basic research which have accumulated over the past 25 years in this field of research concern internal representations of data, and the algorithms which operate on them. The quadtree (Samet 1989), band sweep algorithms for overlay (White 1977), analysis of computational complexity (Preparata and Shamos 1988) and the arc-node data structure (Peucker and Chrisman 1975) are all intellectual breakthroughs of lasting significance. Many challenging problems remain, for example in the design of efficient algorithms to minimize overposting and in other areas of cartographic design, or in developing better methods for converting between various terrain data models. Many systems now handle data through database management systems, and data structure issues have moved more and more into the realm of computer science. We seem, however, to have reached a point where all of the simpler, more generic problems have been solved, and where what remains is a set of difficult, context-specific problems. It seems clear, for example, that further advances in the conversion of terrain data models (for example, from contour to TIN) will require a much better understanding of the nature of terrain (Mark 1979), and will perhaps have to be specific to terrain type (e.g. fluvial versus glacial). There will also continue to be a need for research on efficient methods of storage and access to deal with the enormous volumes of data likely to become available in the coming decade.

3.6 Display

Geographical information systems have often been criticized for failing to give adequate attention to principles of cartographic design (Buttenfield and Mackaness 1991), or for regarding the map as a simple store of information rather than a tool for communication. If we think of the database as the truth, then a map is no more than a store, as there is often a simple correspondence between objects in the database and objects on the map. However, if the database is seen merely as an approximation of the geographical truth, then the design of output displays is critical, as it can affect the user's view of the world. Such simple things as the choice of background colour, or the contrast between adjacent polygons (McGranaghan 1991) can have a significant effect.

The capabilities of electronic display go far beyond those of conventional cartography. We need research on the design of animated displays, three dimensional display, the use of icons and metaphors in user interfaces, continuous gradation of colour and tone, zoom and browse, multiple media including voice and pointing devices, multiple windows which allow simultaneous access to spatial and temporal series of multivariate data. We need to use the electronic medium to think far beyond improvements to the design of choropleth maps. All of these are fundamental problems to a science of geographical information.

3.7 Analytical tools

A GIS is a tool for supporting a wide range of techniques of spatial analysis, including processes to create new classes of spatial objects, to analyse the locations and attributes of objects, and to model using multiple classes of objects and the relationships between them. It includes primitive geometric operations such as calculating the centroids of polygons, or building buffers around lines, as well as more complex operations such as determining the shortest path through a network. The functionality of leading products continues to grow, with no obvious end in sight.

Despite widespread recognition that analysis is central to the purpose of a GIS, the lack of integration of GIS and spatial analysis, and the comparative simplicity of the analytical functionality of many systems continues to be a major concern. In the early days of the statistical package SAS, there was a very rapid increase in the range of tests and techniques implemented in the system. Unfortunately, the same has not been true of GIS, and remarkably little progress has been made in incorporating the range of known techniques of spatial analysis into current products.

There are many reasons for this. One obvious reason is the heavy emphasis in the GIS marketplace on information management rather than analysis. The lucrative markets for GIS technology have comparatively unsophisticated needs, emphasizing simple queries and tabulations. Another is the relative obscurity of spatial analysis, a set of techniques developed in a variety of disciplines, without any clear system of codification or strong conceptual or theoretical framework. Even now it is difficult to identify more than a handful of texts (e.g., Haining 1990, Upton and Fingleton 1985). Although one might expect that GIS could provide the basis for a system of codification for spatial analysis, the poor level of current understanding of geographical data models is a major difficulty. Tomlin (1990) has made one of the few attempts to add some sort of structure or framework to the proliferation of GIS functions, which in the case of ARC/INFO is already around 10^3. We badly need a taxonomy of spatial analysis, developed perhaps from an enumerated set of data models, but going well beyond the primitive geometrical operations.

At this stage, integration of GIS and spatial analysis is proceeding slowly, in at least three different modes. Some analytical capabilities are being added directly to

GIS, for example in the recent expansion of functionality in several modules for network analysis. Some progress is being made in loosely coupled analysis, where an independent analysis module relies on a GIS for its input data, and for such functions as display. However, still missing is an effective form of tight coupling, in which data could be passed between a GIS and a spatial analysis module without loss of higher structures, such as topology, object identity, metadata, or various kinds of relationships. At present this is impossible, to a large extent because of a lack of standards for data models. Instead, coupling has to occur at a lower level, and higher structures have to be rebuilt on an arbitrary basis.

Integration between GIS and spatial analysis might also take the form of a language, whose primitive elements would represent the fundamental operations of spatial analysis. The beginnings of such a language already exist in the macro languages of many of the current generation of GIS, and in various attempts to extend SQL to spatial operations. However, all of these are specific to, and heavily dependent on limited data models, and there is remarkably little similarity between them at this time. At Santa Barbara we have been attempting to define a common language from an analysis of the languages used by a variety of current GIS, but a more satisfactory solution would begin with the conceptual framework provided by a comprehensive data model.

Another problem in integrating GIS and spatial analysis is that in the former discretization of space is explicit, whereas in many forms of spatial analysis it is often either implicit, or unspecified. Many forms of spatial analysis are written on continuous fields, and fail to deal with the uncertainties introduced by the inevitable process of discretization. For example, in GIS there can be no measure of slope that is independent of discretization, and similarly the length of an area object's boundary is dependent on its digital representation. However, slope and length commonly appear as unqualified parameters in spatial models. In this sense, the integration of GIS and spatial analysis is a two-way process, in which the inadequacies of both GIS and spatial analysis must be addressed.

Most of the current generation of GIS provide some sort of macro or script facility, allowing the user to define products from complex sequences of operations, but to invoke them with a single instruction. Although these often include the ability to construct customized environments and interfaces, they do not as yet provide tools which are specific to the needs of spatial analysis. One limited exception is Prime/Wild's ATB, a set of tools constructed on top of System/9 which allows the user to work with complex analyses, visualize their sequences and manage intermediate results. Tools like this will be needed increasingly if GIS are to move into an era of more sophisticated analysis and decision support, because it is not uncommon for relatively simple GIS products to involve processing tens of layers through similar numbers of primitive steps. We need to research methods for keeping track of data lineage and error propagation, backtracking to recover intermediate results, and preventing the user from combining operations in incorrect or meaningless ways (Lanter 1990). We also need research on ways of incorporating this sort of analysis into the GIS acquisition and planning process.

This emphasis on complex multistage analysis and the generation of products from a multilayered database seems very different from research on knowledge based systems, spatial reasoning and spatial query. One of the attractions of the GIS field is its breadth of applications, and the correspondingly extreme variety of environments for the design of user interfaces. In data modeling, the important question is not whether extended relational or object oriented models are better for geographical data, but what types of geographical data are best modeled by each approach. Similarly, the important research issue in the design of user interfaces is to determine the optimal environment for each of the many types of GIS application. What is best for a vehicle navigation system may be entirely different from what is best for a forest resource manager with a deeply seated fear of keyboards and VDUs, either colour or monochrome.

3.8 Institutional, managerial and ethical issues

Research is just beginning to appear on the issues involved in implementing and managing GIS, especially in large institutions. This is difficult research, and generalizations are not discovered easily. However, the success of several large projects in the U.S.A., and the discussions surrounding several large acquisitions by federal agencies, have created the opportunity for a number of useful case studies. Many more are needed, particularly given the importance of such research for improving the institutional environment in the future. We need a much better understanding of the processes of adoption of GIS technology and its effects on organizations; of the value of geographical information and the benefits of GIS; and of processes for utilizing geographical information in decision making. Theoretical frameworks for addressing many of these issues already exist in the relevant social science disciplines, and we need to make much more effective use of them in tackling the specific issues of GIS.

Despite the problems involved in adopting any new technology, GIS have been widely adopted in local government, utilities and resource management agencies. In fact, the introduction of GIS has had a major effect on the management of geographical information in society. At the same time there is increasing concern over the power of GIS for surveillance and invasion of privacy. The research community has a responsibility to monitor and study the more substantive aspects of the GIS phenomenon, including its significance to society as a whole. What will GIS mean to the balance of power in society? Will it be a technology available only to the empowered, or will it somehow serve to even the distribution of power? Thus far there have been remarkably few studies of the ethics of GIS.

4 Tests of commonality

The preceding sections have looked at various candidate areas for inclusion in a geographical information science. In each case there are clearly challenging scientific questions to be posed and researched. There is no reason to believe that the list is complete, or that there are not additional and substantive questions in other related areas. In each case the spatial context appears to be distinctive, although clearly it is more so in some than others. For example, we might debate whether the spatial context was distinctive in the area of decision theory, but the issue seems clear-cut for data modelling.

In the NCGIA research plan (NCGIA 1989), we argued that the absence of solutions to issues such as these constituted impediments to the effective applications of GIS technology. Other discussions of the GIS research agenda have come to similar conclusions, although with different emphases (Craig 1989, Maguire 1990, Masser 1990). Many are old issues, recognized long before the advent of GIS in fields such as cartography, geodesy and geography. Some may not be unique to GIS. For example, it is not immediately obvious that GIS technology diffuses in a fundamentally different fashion, or shows fundamentally different patterns of adoption from other technologies. Is the measurement of GIS benefits a unique problem, or an example of the more general problem of measuring the benefits of information technology? Of course these questions are in themselves research issues.

At the same time it is very important to identify those areas where GIS have created new and unique issues that are not common to other fields. In the early days of GIS, it was possible to argue that the technology was filling an existing gap, and making possible tasks that had been previously identified, but that were not easy to carry out manually. The use of GIS or suitability analysis, by overlaying layers (Tomlin 1990), mirrors the manual technique popularized by McHarg, although admittedly adding some interesting new capabilities. CGIS was justified on the grounds that the computer was a cost-effective alternative to hand measurement of overlaid areas. But GIS make it possible to do things with data that the data's gatherers may never have envisioned. GIS technology is producing radical changes in the way geographical data are collected, handled and analysed, and it will be many years before the impact of existing technology is felt, let alone the impacts of future developments.

Here are some of the issues that seem unique to GIS: how to model time-dependent geographical data; how to capture, store and process three-dimensional geographical data; how to model data for geographical distributions draped over surfaces embedded in three dimensions; how to explore such data, for example, what exploratory metaphors are useful; and how to evaluate the geographical perspective on information and processes relative to more conventional perspectives?

These are important issues for GIS, and the GIS community needs a strong commitment to research if it is going to make significant progress on them. As issues that arise within the context of GIS, they are not of major concern in other disciplines. However, at the same time the GIS community can benefit enormously from interdisciplinary research. Statisticians can make a very valuable contribution to solving the error problem in GIS, and research in cognitive psychology may be helpful in designing the cognitive aspects of user interfaces in GIS.

This argument leads naturally to a proposed definition of GIS research: research on the generic issues that surround the use of GIS technology, impede its successful implementation, or emerge from an understanding of its potential capabilities. Is this 'research about GIS' or 'research with GIS'? In a sense it is both, because these are issues that are both fundamental to the technology of GIS, and also issues that must be solved before the technology can be successfully applied. If the problems of doing research with GIS are generic, then they are best tackled as part of the GIS research agenda. However, problems that are specific to the application of GIS in a particular field clearly need to be addressed in the context of that field, and with the benefit of its expertise. Accuracy issues provide a useful example. There are aspects of the accuracy problem that span a wide range of types of geographical data, and need to be solved using generic models of uncertainty, analogous to the role played by the Gaussian distribution in the theory of measurement error. However as noted earlier, an analysis of crime data using a GIS will also raise problems of accuracy that are specific to that particular application, and need an understanding of the processes operating in criminology and in the collection of crime data if they are to be understood fully.

However, mere existence of scientific questions is far from an adequate basis for a science. Is there a commonality of interest here? Can these subfields find sufficient basis for interaction that they will develop the lasting accoutrements of a science, such as journals, societies, books and philosophers? Will researchers in these subfields behave as a group of scholars? Is there a valid analogy between the systems and science of geographical information on the one hand (tools supporting researchers) and statistical packages and statistics on the other? Statistics is a highly formalized discipline, but more technologically oriented groups can be found in such areas as exploratory data analysis, statistical visualization and applied statistics. Certainly the relationship between science and tools is stormy at times, but nevertheless vital to the success of both. The ongoing debate over the value of statistical software in teaching statistics has interesting implications for the same issue in GIS.

It may be useful to look briefly at the arguments for a commonality of interest in geographical information science, first in principle and then in practice. The field is small — rhetoric about growth in the industry aside, no one would suggest that the field of GIS is a major discipline. It is distinct, with its own reasonably unique set of questions. And it is certainly challenging and innately appealing. On the negative side, it is multidisciplinary, competing with longstanding cleavages and rivalries. It lacks a core discipline, unlike the statistical analogy, where there has been a steady growth in the number and size of academic departments for the past few decades. One of the claimants to the core, geography, has traditionally been a non-technical field, and in some areas of social geography there is a strong and fundamental antipathy to technological approaches.

In practice, commonality of interest is evident in the proliferation of GIS meetings, and we are beginning to see a supply of books and journals. However, the scientific track at GIS meetings is often small. People who attend GIS meetings need a constant supply of novelty, whether in scientific research or vendor products, and will soon desert if the supply dries up.

5 Options for the future

Looking back over nearly three decades of GIS research, it is clear that the greatest progress has been made on the best defined and easiest problems, where solutions lay in advances in the technology itself. Rapid progress was made on algorithms and data structures in the 1970s and 1980s, but many of the difficult problems of data modeling, error modeling, integration of spatial analysis and institutional and managerial issues remain. Some of these may be unsolvable: for example, there may simply be no generalities to be discovered in the process of adoption of GIS by government agencies, however easy it may be to pose the research question.

Other issues have already been solved in a pure research sense, but implementation remains a major question of applied research. In accuracy, for example, a substantial set of techniques has been defined, but the problem of moving them into actual application remains. The academic research environment is set up to pursue significant areas of research, but is generally poor at providing the means of implementation. For that we need a software industry that is tightly coupled to the research community, but able to find the resources to motivate development. More importantly, we need an education system that responds rapidly to new research and is able to build new concepts quickly into its programmes. Unfortunately, the higher education sector is too often characterized by conservatism, and it may take many years for new ideas to work themselves into the curriculum.

Research in GIS is like geographical data — the more closely one looks, the more interesting issues appear. GIS research has only begun to tackle the important issues in the research agenda. We are in an enviable position, working in a field with such strong motivation and such a strong underlying industry, and with such an interesting set of problems spanning so many disciplines and fields. I hope I have shown in this paper that the handling of spatial information with GIS technology presents a range of intellectual and scientific challenges of much greater breadth than the phrase 'spatial data handling' implies — in effect, a geographical information science. The term 'geographical' seems essential — much of what GIS research is about concerns the geographical world and our relationships with it, and the term is much richer than 'spatial'. The change in meaning of the 'S' word — from systems to science — seems to be going well, as evidenced by the success of the spatial data handling series of conferences, the move of the AutoCarto series to fully refereed papers, the new texts, subscriptions to the *International Journal of Geographical Information Systems,* and submissions of GIS papers to such established journals as *Geographical Analysis, Computers and Geosciences, Computer Vision, Graphics and Image Processing* and publications of the Regional Science Association and the IEEE.

I hope I have also shown that a strong scientific programme serves not only itself, but also the needs of industry and GIS users. GIS needs a strong scientific and intellectual component if it is to be any more than a commercial phenomenon, a short-lived flash in the technological pan. It is too easy to see current GIS as a hardware and software technology in search of applications, and to see the field of GIS as defined by the functional limits of its major vendor products. We need to move from system to science, to establish GIS as the intersection between a group of disciplines with common interests, supported by a toolbox of technology, and in turn supporting the technology through its basic research. As currently perceived, GIS sometimes seem about as close to a science as FORTRAN is to algebra.

In recent years we have seen a growing cleavage in GIS between two traditions, that of spatial information on the one hand and that of spatial analysis on the other. The spatial information tradition stresses large inventory databases, and gives geography the role of an access mechanism. The spatial analysis tradition stresses rich functionality and a range of data models, and gives geography a fundamental role in analysis and modelling. The two traditions share common data structures and algorithms, and rely on the same sources of data and hardware. However, this is not enough to convince the academy of the existence of a scientific field. To claim this we need to take a broader view, and to include data modelling accuracy, cognition, reasoning, human–computer interfaces (HCI) and visualization, and to show how these are integral parts of both traditions.

Without such arguments, the GIS field will fragment, and the GIS storm will blow itself out. Associations as fundamentally disjoint as the Association of American Geographers and AM/FM will find it impossible to justify joint sponsorship of conferences. Vendors will specialize in data input workstations, spatial analysis workstations or facility management systems, with little potential for interaction or integration. This would be tragic.

How can we ensure a lasting future for both geographical information systems and science? Disciplines are like tribes, with their own totems, symbols and membership rules, languages and social networks. The GIS tribe is currently very cohesive; it is well funded, the field is exciting and much useful research is being done. However, in the longer term the field has not done well at behaving as a science, and the academy is still doubtful about whether it needs to be taken seriously. Science is hard and places heavy obligations on its practitioners. We have been too busy, and technology has been moving too quickly. Too much of our literature is in conference proceedings, which bring fast exposure but only to limited audiences, and lack sufficient quality control. Few people have had the time to write the textbooks or to identify the intellectual core, or to publish the good examples.

I believe we ensure the future of GIS by thinking about science rather than systems, and by identifying the key scientific questions of the field and realizing their intellectual breadth. Geographical information systems are a tool for geographical information science, which will in turn lead to their eventual improvement. We need to speak to the academy, both directly and through key articles and texts, on the philosophy, methodology and foundations of the field, and by placing GIS papers in strong journals. All three communities — users, vendors and researchers — have vital and symbiotic roles to play, and we will serve all three best by playing ours in the fullest possible sense.

Acknowledgments

The National Center for Geographic Information and Analysis is supported by the National Science Foundation through grant SES 88-10917.

References

Anselin, L., 1989, What is special about spatial data? Alternative perspectives on spatial data analysis. *Technical Report 89-4* (Santa Barbara, CA: National Center for Geographic Information and Analysis).

Berry, J. K., 1987, Fundamental operations in computer-assisted map analysis. *International Journal of Geographical Information Systems,* **1**, 119–136.

Besag, J., 1986, On the statistical analysis of dirty pictures. *Journal of the Royal Statistical Society,* **B48**, 259–302.

Buttenfield, B. P., and Mackaness, W. A., 1991, Visualiztion. In *Geographical Information Systems: Principle and Applications* edited by D. J. Maguire, M. F. Goodchild and D. W. Rhind (London: Longman).

Chrisman, N. R., and Yandell, B. 1988, A model for the variance in area. *Surveying and Mapping,* **48**, 241–246.

Craig, W. J., 1989, URISA's research agenda and the NCGIA. *Journal of the Urban and Regional Information Systems Association,* **1**, 7–16.

Dutton, G., 1989, Modeling locational uncertainty via hierarchical tessellation. In *Accuracy of Spatial Databases,* edited by M. F. Goodchild and S. Gopal (London: Taylor & Francis), pp. 125–140.

Frank, A. U., and Mark, D. M., 1991, Language issues for GIS. In *Geographical Information Systems: Principles and Applications,* edited by D. J. Maguire, M. F. Goodchild and D. W. Rhind (London: Longman).

Franklin, W. R., 1984, Cartographic errors symptomatic of underlying algebra problems. *Proceedings, International Symposium on Spatial Data Handling, Zurich,* pp. 190–208.

Goodchild, M. F., 1989, Modeling error in objects and fields. In *Accuracy of Spatial Databases,* edited by M. F. Goodchild and S. Gopal (London: Taylor & Francis), pp. 107–114.

Goodchild, M. F., 1990, keynote address: spatial information science. *Proceedings, Fourth International Symposium on Spatial Data Handling, Zurich,* **1**, 13–14.

Goodchild, M. F., 1991, Keynote address: progress on the GIS research agenda. *Proceedings, EGIS 91, Brussels,* pp. 342–350.

Haining, R., 1990, *Spatial Data Analysis in the Social and Environmental Sciences* (Cambridge: Cambridge University Press).

Keefer, B. J., Smith, J. L., and Gregoire, T. G., 1988, Simulating manual digitizing error with statistical models. *Proceedings GIS/LIS 88* (Falls Church, VA: American Society of Photogrammetry and Remote Sensing/American Congress on Surveying and Mapping), pp. 475–483.

Lanter, D. P., 1990, Lineage in GIS: the problem and a solution. *Technical Paper 90-6* (Santa Barbara, CA: National Center for Geographic Information and Analysis).

Maguire, D. J., 1990, A research plan for GIS in the 1990s. *The Association for Geographic Information Yearbook 1990,* (London, Taylor & Francis), pp. 267–277.

Mark, D. M., 1979, Phenomenon-based structuring and digital terrain modeling. *GeoProcessing,* **1**, 27–36.

Masser, I., 1990, The Regional Research Laboratory initiative: an update. *The Association for Geographic Information Yearbook 1990* (London: Taylor & Francis), pp. 259–263.

McGranaghan, M., 1991, Modeling simultaneous contrast on choropleth maps. *Technical Papers, 1991, ACSM/ASPRS/Auto-Carto 10 Annual Convention, Baltimore, MD, March 25–29, 1991,* Vol. 2, pp. 231–240.

McHarg, I. L., 1969, *Design with Nature* (New York: Doubleday).

NCGIA (National Center for Geographic Information and Analysis), 1989, The research plan of the National Center for Geographic Information and Analysis) *International Journal of Geographical Information Systems,* **3**, 117–136.

Openshaw, S., 1977, A geographical solution to scale and aggregation problems in region-building, partitioning and spatial modelling. *Transactions of the Institute of British Geographers,* **2** (NS), 459–472.

Peucker, T. K., and Chrisman, N. R., 1975, Cartographic data structures. *American Cartographer,* **2**, 55–69.

Preparata, F. P., and Shamos, M. I., 1988, *Computational Geometry: An Introduction* (New York: Springer-Verlag).
Samet, H., 1989, *The Design and Analysis of Spatial Data Structures* (Reading, MA: Addison-Wesley).
Shiryaev, E. E., 1987, *Computers and the Representation of Geographical Data* (New York: Wiley).
Tomlin, C. D., 1990, *Geographic Information Systems and Cartographic Modeling* (Englewood Cliffs, New Jersey: Prentice Hall).
Tomlinson, R. F., 1984, Keynote address: geographical information systems — a new frontier. *Proceedings, International Symposium on Spatial Data Handling, Zurich,* **1**, 2–3.
Tomlinson, R. F., Calkins, H. W., and Marble, D. F., 1976, *Computer Handling of Geographical Data* (Paris: UNESCO).
Upton, G. J., and Fingleton, B., 1985, *Spatial Data Analysis by Example* (2 volumes) (New York: Wiley).
White, D., 1977, A new method of polygon overlay. *Proceedings, Advanced Study Symposium on Topological Data Structures for Geographic Information Systems, Harvard.*

Geographical Information Science: Fifteen Years Later

Michael F. Goodchild

Introduction

In early 1990 Kurt Brassel asked me if I would be willing to give a keynote at the Fourth International Symposium on Spatial Data Handling (SDH), which was being planned for July of that year. I accepted, but it was not until late March that I gave serious thought to what I would say. At the time, the U.S. National Center for Geographic Information and Analysis (NCGIA) was 18 months old, with a mandate from the National Science Foundation to pursue basic research on geographic information systems (GIS) and to promote the use of GIS within the sciences. That same year, Terry Jordan, president of the Association of American Geographers, had called GIS "nonintellectual expertise," reflecting a growing sentiment within many parts of the discipline that saw GIS as a matter of pushing the right buttons, and wondered why it was worth any more attention from geographers than, say, the hand calculator. The SDH series had evolved as the premier international meeting of the GIS research community, and it seemed to me appropriate that my keynote should focus on science, and on what it would take to bring respect for the field from the broader academy, both inside and outside geography. David Simonett, my NCGIA codirector, had argued for many years that remote sensing needed a strong emphasis on science and theory to be regarded as anything more than a bag of tricks, and was adamant that the GIS community needed to build along similar lines. It annoyed me that the premier international meeting of the field had a title that suggested that we were driven merely by the difficulty of "handling" a particular type of information — that we were merely "the United Parcel Service of GIS." I tried to capture some of these ideas in a conference proceedings paper, which I wrote one afternoon while in St. Lucia working on *Geographical Information Systems: Principles and Applications* with David Rhind and David Maguire (Maguire et al., 1991).

The paper seemed to strike a chord with the audience, and I repeated the message with some embellishments at the EGIS (European GIS) conference early in 1991. The two conference papers were eventually combined into a single paper, which appeared in *IJGIS* in 1992. To quote:

The GIS community has come a long way in the past decade. Major research and training programs have been established in a number of countries, new applications have been found, new products have appeared from an industry which continues to expand at a spectacular rate, dramatic improvement continues in the capabilities of platforms, and new and significant datasets have become available. It is tempting to say that GIS research, and the meetings at which GIS research is featured, are simply a part of this much larger enthusiasm and excitement, *but there ought to be more to it than that*. (Goodchild, 1992; emphasis added).

In the paper I deliberately played on the GIS acronym, asking whether the "S" might not usefully stand for *science* rather than *systems*, and in turn asking what such a science might encompass. Since then others have continued the wordplay, suggesting that GIStudies might be a useful term for investigations into the social context and impacts of GIS, and that GIServices should denote any remotely invokable processing capability (Longley et al., 2005).

Evidently the paper did not fully satisfy the need for a complete and lasting definition of the field. As the paper makes clear, my intent was to capture those aspects of GIS research that concerned questions of fundamental scientific significance, and that could drive a science that would eventually earn the respect of the academy — that would lead, for example, to election of GIS researchers to the U.S. National Academy of Sciences or the United Kingdom's Royal Society (the field has been successful on both counts), or to the establishment of professorships in GIS in the most prestigious institutions. I was also concerned that the field's questions be unique, and that it have internal coherence. I chose to define GIScience as "research on the generic issues that surround the use of GIS technology, impede its successful implementation, or emerge from an understanding of its potential capabilities"; Mark (2003) has provided an excellent history of the various efforts to clarify the definition of GIScience over the years. I also noted that GIScience might take two essentially distinct forms: research *about* GIS that would lead eventually to improvements in the technology, and research *with* GIS that would exploit the technology in the advancement of science. Both of these themes are clearly evident in the way the term GIScience is used today.

The concept of GIScience seems to have been adopted enthusiastically. Journals (including *IJGIS*) have been renamed, books have been published (Bishop and Schroder, 2004; Cho, 2005; Duckham et al., 2003; McMaster and Usery, 2004; Raper, 2000), a major consortium of U.S. universities have been established (the University Consortium for Geographic Information Science, www.ucgis.org), and specialist programs have appeared in academic institutions. The *systems versus science* issue has been revisited (Wright et al., 1997), and reports have been written calling for the establishment of major programs of research funding (Mark, 1999). Efforts have been made to define the *grand challenges* of the field (Longley et al., 2005; Mark, 1999), and to identify its fundamental principles (Goodchild, 2003). Problems of nomenclature will always be with us, of course, and one can find numerous terms in use that are either fully or partially equivalent: geoinformation science, spatial science, spatial information engineering, geomatics, geoinformatics, and geospatial information science to name a few. But despite the well-known caution, "Every field that needs to call itself a science probably isn't," after 15 years there seems every reason to believe that GIScience is a genuine, challenging, and fruitful area for scientific research with its own unique scientific questions and discoveries.

Reassessment: What did I miss?

In the third section of the paper, titled "The Content of GIScience," I attempted to outline what seemed to me at the time to be the major divisions of the field. They make interesting reading 15 years on:

- Data collection and measurement
- Data capture
- Spatial statistics
- Data models and theories of spatial data
- Data structures, algorithms, and processes
- Display
- Analytic tools
- Institutional, managerial, and ethical issues

While it bears strong resemblance to early statements of the NCGIA research agenda (NCGIA, 1989), it clearly misses many of the subsequent developments in the field. "Data models and theories of spatial data" is a poor apology for what subsequently became the basis for the COSIT (Conference on Spatial Information Theory) conference series, drawing heavily on work in linguistics and cognitive science, and crystallizing around the term *ontology*. The emphasis on the *tools* of analysis in the penultimate bullet, rather than on analysis itself, missed virtually all of the key developments in analytic methods of the past fifteen years. "Display" scarcely credits or anticipates what has evolved into the highly productive field of geovisualization, while "Spatial statistics" does little to convey the importance of uncertainty, which is now one of the most conspicuous specialties of GIScience, or the degree to which it has also drawn on the field of geostatistics. Comparison with the 2002 research agenda of UCGIS (www.ucgis.org) simply reinforces these observations.

This new paradigm of GIS has been recognized in numerous ways. In the UCGIS research agenda of 2002 it underlies several topics: the long-term research challenges of spatial ontologies, spatial data acquisition and integration, distributed computing, and the future of the spatial information infrastructure; and the short-term research priorities of the geospatial semantic web, geospatial data fusion, institutional aspects of spatial data infrastructures, geographic information partnering, and pervasive computing (www.ucgis.org). It has been described in many books (Peng and Zhou, 2003; Plewe, 1997), and in a growing journal literature. As the editors noted in the comments that introduced the second edition of *Geographical Information Systems* (Longley et al., 1999), the most obvious shortcoming of the first edition was its complete failure to anticipate the impact of the Internet and the Web. We continued to see a GIS as something that existed *in one place*, long after early visionaries had argued that computers would inevitably be networked and their tasks distributed. In hindsight, the case for distributing a technology that deals with the distribution of activities and phenomena over the earth's surface seems particularly obvious and compelling. Above all, however, I missed the impact of the Internet and the way in which it has massively impacted the GIScience research agenda since 1993. From the perspective of 1990, GIS was akin to a desktop *butler*, an intelligent machine for performing what its master or mistress found too tedious, imprecise, costly, or complex to do by hand. With the Internet, however, GIS became primarily a *medium*, a means of communicating one person's knowledge of the planet's surface and near-surface to others. Suddenly the sharing and dissemination of data became the subject of enormous investment in digital libraries, data warehouses, metadata standards, and geoportals (Longley and Maguire, 2005). The entire field of GIServices was unanticipated in 1990, as was the challenging research topic of interoperability (Goodchild et al., 1999), with its concern for differences of syntax and semantics. Indeed, one wonders what a GIScientist of 1990 would have made of the term *spatial web*.

Directions in GIScience

As I have argued, the GIScience of 2005 is very different from the one that prompted my keynote of 1990, and there is every reason to expect that GIScience will continue to evolve and change in the coming years. Three topics seem particularly worthy of attention at this time.

First, there are strong arguments that the focus of GIScience needs to shift from representation and analysis of the *form* of the earth's surface to a much stronger concern for the *processes* that define its dynamics (Goodchild, 2004). But while geographers have always been custodians of knowledge about form, arguably the custodians of process have been the substantive sciences of surficial geology, ecology, hydrology, epidemiology, demography, economics, and so on. A concern for process is therefore likely to change the landscape of GIScience dramatically, requiring much closer interaction with these sciences. The notion of research *with* GIS takes on different meaning, requiring that GIS be redesigned to support the process models of the sciences, rather than generic and simplistic representations of form.

Second, the past year has seen an unprecedented series of developments in the ways in which the general public interacts with GIS. Nowhere is this more evident than in the case of Google Earth and its look-alikes (Microsoft's Virtual Earth, NASA's World Wind, and so forth). Today, a child of ten can generate a flyby using a simple user interface with no more than a few minutes of instruction, a task that would previously have required a year's exposure to commercial off-the-shelf GIS in a university course. We are moving rapidly from a *concert pianist* model of GIS as a tool confined to experts, to a *child of ten* model in which the power of GIS is available to all, the obvious concerns about powerful and complex technology in the hands of naïve users notwithstanding. Google's publication of KML has empowered a creative population of hackers that dwarfs the development staffs of the major GIS software vendors.

Third, it is clear that the knowledge accumulated by the discipline of GIScience is applicable to varying degrees in any space, and not limited to the space of the earth's surface and near-surface. Much of the GIScience research agenda can be motivated equally well by other spaces, such as the three-dimensional space of the human brain, or the one-dimensional space of the human genome. At the same time, advances made in the study of other spaces may be suitable sources of cross-fertilization in GIScience. Perhaps the next decade will see a much greater degree of interaction between GIScience and the sciences of other spaces, and much more productive collaboration.

References

BISHOP, M.P. AND SCHRODER, J.F., Jr., Eds., 2004, *Geographic Information Science and Mountain Geomorphology*, Springer, New York.

CHO, G., 2005, *Geographic Information Science: Mastering the Legal Issues*, Wiley, New York.

DUCKHAM, M., GOODCHILD, M.F., AND WORBOYS, M.F., Eds., 2003, *Foundations of Geographic Information Science*, Taylor & Francis, New York.

Goodchild, M.F., 1992, Geographical information science, *International Journal of Geographical Information Science*, **6**, 31–45.

GOODCHILD, M.F., 2003, The fundamental laws of GIScience. Keynote address, Annual Assembly, University Consortium for Geographic Information Science, Monterey, CA, June .

GOODCHILD, M.F., 2004, GIScience: geography, form, and process, *Annals of the Association of American Geographers,* **94**(4), 709–714.

GOODCHILD, M.F., EGENHOFER, M.J., FEGEAS, R., AND KOTTMAN, C.A., Eds., 1999, *Interoperating Geographic Information Systems,* Kluwer, Boston.

LONGLEY, P.A., GOODCHILD, M.F., MAGUIRE, D.J., AND RHIND, D.W., 1999, *Geographical Information Systems: Principles, Techniques, Applications and Management*, Wiley, New York.

LONGLEY, P.A., GOODCHILD, M.F., MAGUIRE, D.J., AND RHIND, D.W., 2005, *Geographic Information Systems and Science*, second edition, Wiley, New York.

LONGLEY, P.A. AND MAGUIRE, D.J., Eds., 2005, Geoportals. *Computers, Environment and Urban Systems*, **29**(1), 1–85.

MAGUIRE, D.J., GOODCHILD, M.F., AND RHIND, D.W., 1991, *Geographical Information Systems: Principles and Applications*, Longman, Harlow, U.K.

MARK, D.M., Ed., 1999, *Geographic Information Science: Critical Issues in an Emerging Cross-Disciplinary Research Domain.* Workshop Report. http://www.geog.buffalo.edu/ncgia/workshopreport.html (accessed 26 September 2005).

MARK, D.M., 2003, Geographic information science: defining the field. In *Foundations of Geographic Information Science*, M. Duckham, M.F. Goodchild, and M.F. Worboys, Eds., Taylor & Francis, New York, 3–18.

MCMASTER, R.B. AND USERY, E.L., Eds., 2004, *A Research Agenda for Geographic Information Science,* CRC Press, Boca Raton, FL.

NCGIA (National Center for Geographic Information and Analysis), 1989, The research plan of the National Center for Geographic Information and Analysis. *International Journal of Geographical Information Systems*, 3(2), 117–136.

PENG, Z.R. AND TSOU, M.H., 2003, *Internet GIS: Distributed Geographic Information Services for the Internet and Wireless Networks*, Wiley, Hoboken, NJ.

PLEWE, B., 1997, *GIS Online: Information Retrieval, Mapping, and the Internet*, OnWord Press, Santa Fe, NM.

RAPER, J.F., 2000, *Multidimensional Geographic Information Science*, Taylor and Francis, New York.

WRIGHT, D.J., GOODCHILD, M.F., AND PROCTOR, J.D., 1997, Demystifying the persistent ambiguity of GIS as "tool" versus "science." *Annals of the Association of American Geographers,* 87(2), 346–362.

International Journal of Geographical Information Systems,
1993, Vol. 7, No. 4, 331–347.

10 Algorithm and Implementation Uncertainty in Viewshed Analysis

Peter F. Fisher

Abstract. In most documentation of geographical information systems (GIS) it is very rare to find details of the algorithms used in the software, but alternative formulations of the same process may derive different results. In this research several alternatives in the design of viewshed algorithms are explored. Three major features of viewshed algorithms are examined: how elevations in the digital elevation model are inferred, how viewpoint and target are represented, and the mathematical formulation of the comparison. It is found that the second of these produces the greatest variability in the viewable area (up to 50 per cent over the mean viewable area), while the last gives the least. The same test data are run in a number of different GIS implementations of the viewshed operation, and smaller, but still considerable, variability in the viewable area is observed. The study highlights three issues: the need for standards and/or empirical benchmark datasets for GIS functions; the desirability of publication of algorithms used in GIS operations; and the fallacy of the binary representation of a complex GIS product such as the viewshed.

1 Introduction

One issue of quality in GIS which has received relatively little attention is the issue of software quality and, of particular concern in the research reported here, the degree to which different GIS, which offer the same function, may actually report different outcomes. Essentially, different algorithms, which are designed to derive one particular phenomenon from the same data, or different implementations of a single algorithm, may give inconsistent results, and those inconsistencies may be large. The need for such testing is widely accepted outside the GIS community where such standards are easily testable. For instance, the working of much of the functionality of the word processor used to write this paper is easily tested, and is regularly so tested in reviews and comparative tests in the popular computer literature. Although Hemenway (1992), among others, makes an eloquent appeal for such benchmarking of GIS, only a limited amount has been done (e.g., Knaap 1992). In an effort to expand that literature, this paper focuses attention upon the commonly-available function known variously as the viewshed, viewable area, or visibility, and both algorithms and actual GIS implementations are compared.

The need for benchmarking is stated in the next section, and is followed by a brief review of previous work. Subsequently, the viewshed is defined, and two test data sets, which are used throughout the research reported, are introduced. The paper then goes on to examine two different issues with respect to the viewshed: algorithms for defining the viewshed are reviewed, the results of different versions coded by the author are examined, and the results of the viewshed operation from a number of different widely-available implementations are compared. The discussion presents a salutary lesson in the importance of benchmarking the precision of implementations of GIS functions, and culminates in the proposal that a probabilistic representation of the viewshed is more acceptable than the usual binary product.

2 The case of benchmarking

The operation of GIS is dependent upon a computer programmer either reading, understanding and implementing a published algorithm, or developing his/her own algorithm, to achieve a specified goal. If the programmer is lucky, he or she may even find public domain computer code to achieve the functionality required. For most complex GIS functions at least two alternative algorithms exist, and frequently many more can be found in the literature. Where numerous algorithms exist for the same function, they may make different initial assumptions and use different approximations in achieving the same goal. If two programmers sit down to achieve the same functionality, however, even if they adopt the same broad algorithm, it is unlikely that the code will look exactly the same. Parts of the implementation may be executed in alternative orders, and equations may be split up into variable numbers of lines of code in the implementation. If the two programmers are working on alternate computer-platforms, then the code may be identical, but still derive inconsistent outcomes owing to the different precisions of either compilers or hardware (Burrough 1986). Problems with a high level of algorithm complexity, therefore, are almost certain to have as many different, but correct answers as there are implementations. There is very clearly a need for study and evaluation of standards in the functionality of GIS, so that users may have confidence in the modeling they do (Hemenway 1992, Jordan and Star 1992).

3 Previous work

Some comparative studies of GIS algorithms do exist in the literature. Skidmore (1990) compared three of the many methods for mapping drainage networks from gridded DEMs as well as proposing his own. Lee (1991) studied four methods of converting from a Digital Elevation Models (DEM) to triangulated irregular networks (TIN) data structure, with elimination of many observations in the DEM. These are both relatively exotic operations, and only occasionally available in operational GIS.

On the other hand, both Skidmore (1989) and Kvamme (1990) have studied the precision of algorithms for extracting slope and aspect from gridded DEMs, and Srinivasan and Engel (1991) have explored how that algorithm error propagates into estimates of soil erosion. Wagner (1989) has given an exhaustive review of three polygon overlay operators.

Several discussions of raster to vector conversion algorithms have occurred in the literature (e.g., Clarke 1985, Piwowar *et al.* 1990), and, unusually, this operation is the subject of Knaap's (1992) examination of implementations. Knaap reports the rather pessimistic finding that among the eight packages tested, no two produced the same rasterization of the test vector patterns, although there are some reassuring consistencies. The need for such comparisons has been largely ignored in the drive for standards in GIS which have concentrated on data transfer (e.g., DCDSTF 1988) to the almost total exclusion of any others (but see Hemenway 1992, and Jordan and Star 1992). This paper aims to further the literature in this area by presenting the results of an exploration of both algorithms and operational implementations of the viewshed operation.

4 Experimental design and test data

4.1 The viewshed

In all implementations examined here, the viewshed function takes a gridded DEM and a single viewing point, and derives a new raster image, showing those cells which are visible from the viewing point and those which are not (coded 1 and 0, respectively). This is in conformity with the usual reporting of the viewshed. In all GIS examined here, a gridded DEM is processed. Options which exist in some of the implementations tested and others available include specification of the height of the viewer at the viewing point, the distance to which the viewer is interested, and a height of land covers in the study area which may block views (trees, walls, etc.). Less commonly it may also be possible to specify the angle of viewing (with respect to north), a height above or below which objects may not be visible, earth curvature and haze effects, although the last is usually available only for landscape visualization. In the current research the total viewable area from a specified height above a particular location over a piece of terrain unencumbered by vegetation is the desired result, this being the lowest common denominator of the different viewshed implementations examined.

Unlike some other GIS functions the viewshed is not actually verifiable in the field nor can it be logically validated, anything more than trivially, by examination of test figures, as Wagner has done for the overlay algorithm (Wagner 1989). The problem is that almost everywhere vegetation intervenes in the landscape to some height above the ground which is hard to measure with any precision. Furthermore, atmospheric refraction and earth curvature, which are both complex and rarely included in the viewshed function, cause direct lines-of-sight to be different from the actual lines viewed along. Furthermore, the database error in the DEM has a significant impact on the viewable area that can be calculated (Fisher 1991). In short, even if a precisely determined DEM were to exist with minimal vertical and horizontal errors (surveyed by GPS, for example) for a vegetation-free landscape, it is not certain that a laser determined line-of-sight between points would correspond with the viewer's line of sight. For all these reasons, the approach taken here is of comparative testing to establish the possible numerical and spatial variability in viewsheds.

4.2 The test area

The test areas are used which have been employed in other research published by this author (Fisher 1991, 1992). Each area is a 100 by 100 pixel subset of the level 1 DEM for the Coweeta Experimental Watershed in North Carolina (the Otto 7·5 minute quadrangle) (Figure 10.1). Each DEM has a single viewing point associated. Maps and histograms of the elevations of the two areas are presented in Figure 10.1, and it can be seen that one viewpoint is chosen to be relatively low in the local landscape (in a valley position), and one high in the landscape (on a hilltop). Both locations have considerable relief in the immediate vicinity of the viewing point. The testing reported here could have been achieved with any DEM, and any viewing location on that DEM.

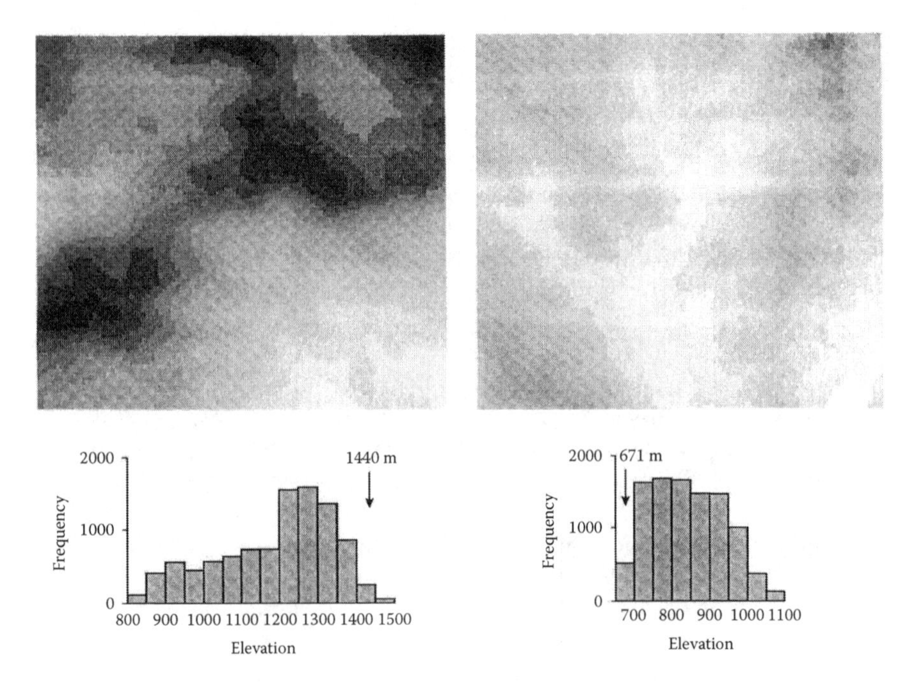

FIGURE 10.1 Shaded contour maps and histograms of the the elevations in the two DEMs used for testing. In the contour maps, higher elevations are darker, the shades do not overlap between maps, and the viewpoints are indicated as isolated black pixels. In histograms viewpoint elevations are shown by arrows.

5 Testing algorithms

5.1 The basic algorithm

Two essential steps occur in the basic algorithm addressed in the research reported here. The first detects the horizontal locations at which the line-of-sight (LOS) from the viewing point to the target intersects the grid of the DEM. The second compares the DEM elevation on the LOS. If the latter is higher, then processing can proceed to evaluate the next intersection point, otherwise the target can be declared as invisible from the viewpoint, and processing proceed to consider another target.

The method for inferring elevations within the grid and determination of the number and positions of lines for comparison with the LOS is profoundly important in this process. Similarly, the way in which the grid cell structure is interpreted with respect to the viewing point and the target, and, indeed, the whole DEM has clear importance. The second step, when the elevations are calculated, is also crucial to finding the viewshed, because this is the decision step.

Some very different algorithms have been suggested (e.g., Mark 1987, Teng and Davis 1992), but they are not explored because they yield only approximations, and because the research which is reported here makes a significant point without confusing the issue by branching into these alternatives.

5.2 Inferring elevations

From a review of the literature, and discussions with individuals implementing the viewshed operation, it has been possible to identify four different methods for inferring elevations from the basic DEM. These are either documented, suggested or implied.

The first method is to take each elevation in the DEM to be at the centre of a grid cell, and to infer continuous, linear change to the centres of each of the four neighbours of that grid cell, the DEM allowing inference of a mesh of sloping lines. The LOS from the target location is calculated to the intersection of each of the mesh lines between neighbours (Figure 10.2a), and heights compared at the target. If a second viewpoint is chosen then the mesh of elevations remain the same (Figure 10.2b). This method is specified by Yoeli (1985).

A second method, which can be carried over from the TIN data structure, is to triangulate the DEM, inferring a connection from one cell to the diagonal neighbours as well as the orthogonal neighbours. Eight lines of continuously changing elevation from each grid cell are possible. Use of all eight would mean that, at the crossing of the two diagonals within a cell, there would very likely be two elevations. This would not be acceptable, and so the triangulation is executed as illustrated in Figure 10.2c and Figure 10.2d. The characteristic of this is that in moving from one viewing point to another the elevation network to be compared changes, since the diagonals change.

Tomlin (1990) suggests the imposition of a secondary grid on the first, where new elevations are inferred equidistant between pairs of neighbours and four neighbours. On original grid lines the inferred elevations are the mean of the two neighbours,

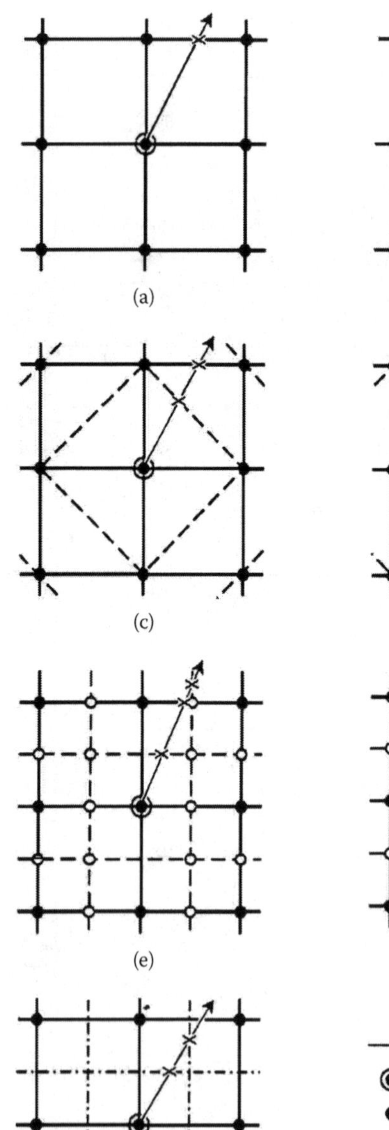

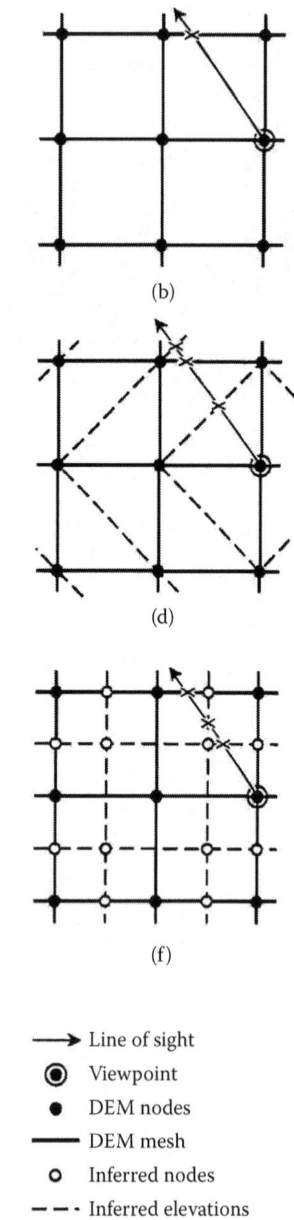

FIGURE 10.2 Alternative methods of extrapolating from the elevation values in a DEM to an inferred network of elevations, including (a) and (b) linear interpolation between grid neighbours, (c) and (d) triangulation of the grid, (e) and (f) grid constraint of the mesh, and (g) the stepped model.

while at the centers of original mesh squares they are the mean of the four neighbours (Figure 10.2e). Moving the viewing point does not change the network of elevations (Figure 10.2f).

These first three models increasingly constrain the possible viewshed, because the number of points at which the LOS to the target will be compared with the LOS to the mesh location increases before any point can be shown to be in view. It therefore follows that in any situation the viewshed is likely to be progressively smaller from the first to the third.

Finally, an elevation model may be regarded as stepped. Each elevation is applied to the whole grid cell, with four vertical faults at the edges of the cell where elevations change to the next cell (Figure 10.2g). This model is not only used in viewshed calculation (Felleman and Griffin 1990, Teng and Davis 1992) but is widely used in illustrating other algorithms for manipulating DEMs (e.g., Travis *et al.* 1975, Burrough 1986).

5.3 Results for elevation inference

Each of the four foregoing approximations of elevation models was implemented, and run for the two test viewing locations. The results are presented in Table 10.1. It can be seen that in the cases of both viewpoints, the area of the viewshed decreases

TABLE 10.1
The number of cells included in the viewsheds from the test locations. Results for the four different methods for inferring elevations discussed in the text, and illustrated in Figure 10.2, are given.

Elevation approximations	Test site 1	Test site 2
Grid	2381	2034
Triangular constraint	2312	1917
Grid constraint	1992	1442
Stepped	656	92
Frequency of cells in summed image (Figure 10.3)		
5 (Viewing point)	1	1
4	604	72
3	1409	1357
2	324	513
1	46	96
Total	2384	2038

N.B. the first entry is the same as in Tables 10.2 and 10.3.

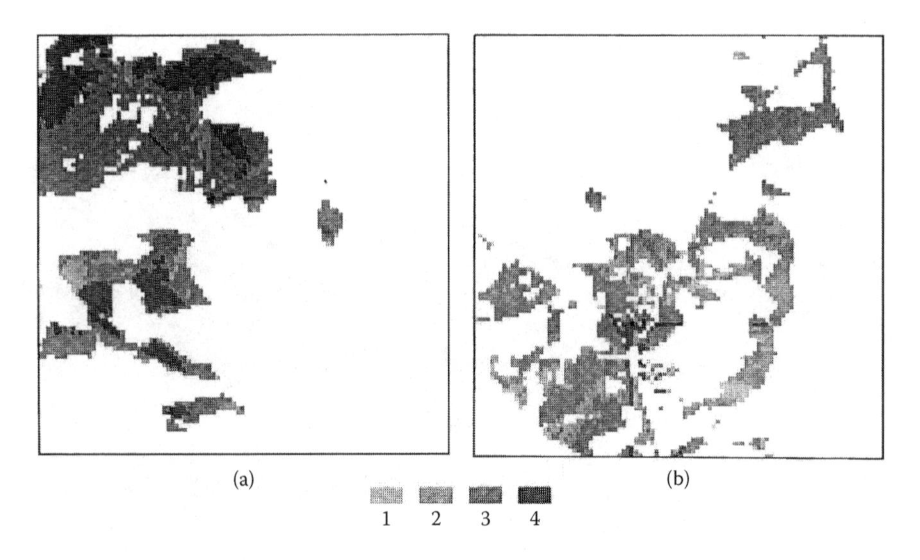

FIGURE 10.3 The pattern of grid cells included in viewsheds derived from alternative methods of inferring elevations.

from the simple grid to the stepped grid. The final step from the grid-constrained model to the stepped model is the most dramatic, but that from the triangular to the grid constrained is also large. Furthermore, the degree of change is not consistent for both viewpoints, and so may be dependent on the landscape position.

The pattern of cells included in the various viewsheds derived is presented in Figure 10.3. It can be seen that generally the patterns are nested within each other. In detail, however, this observation does not hold. Some methods have identified cells as being out-of-view, when they are in the heart of the viewable areas defined by other methods, and an examination of the frequency data presented in Table 10.1 shows that no nesting is really occurring; if it were the number of cells with the highest cell count (four in this case) should be equal to the smallest viewshed, and the total number of cells included in all viewsheds should equal the largest of the original viewsheds. In the event, the number of most frequent cells is considerably smaller than the number of cells in the smallest viewshed reported. On the other hand, the total number of cells is very nearly the same as the largest, except for a few cells. There is therefore some mixing of pixels in- and out-of-view in the single viewsheds.

5.4 Viewer and target locations

Specification of how the viewer and target locations should be treated within the viewshed operation can also cause considerable variation in the viewable area. The grid models of the DEM discussed above suggest the approximation of the viewing and target locations as nodes of the DEM mesh; although some other point may be the viewing location, it would not be logical for the target to be other than the node as representative of the raster grid cell for which visibility is to be reported. This point approximation is, however, not implicit in the raster data structure to which

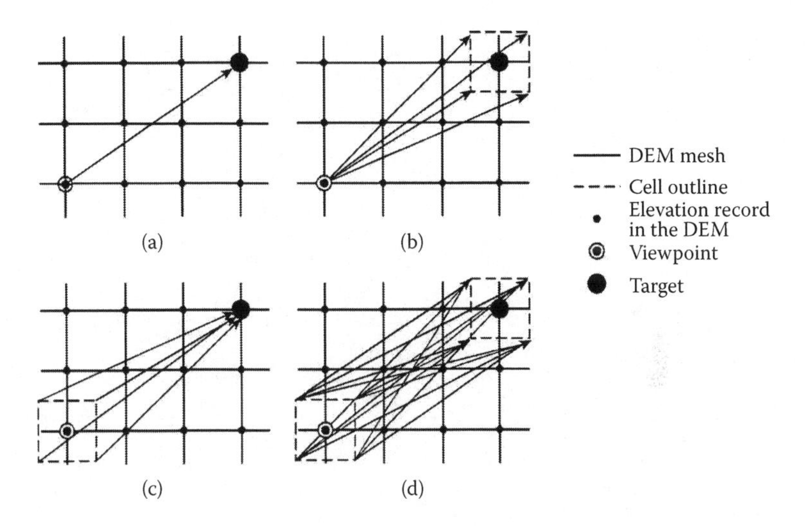

FIGURE 10.4 Alternative strategies for approximating the location of the view point and target, including (a) point-to-point, (b) point-to-cell, (c) cell-to-point, and (d) cell-to-cell.

the DEM is tied, and in which format the viewshed is reported. Instead, a grid cell approximation might be taken and the visibility to or from the whole are of the grid cell examined. The stepped approximation of elevation inference effectively recognises all cells as grid cells. Recognition of two alternative interpretations of a grid cell yields four different possible combinations of viewing and target locations: point-to-point (Figure 10.4a), point-to-cell (Figure 10.4b), cell-to-point (Figure 10.4c) and cell-to-cell (Figure 10.4d).

Exactly how the Line-of-Sight should be found for approximations involving cells is not clear, because technically every single possible viewing point on the cell should be explored, but there is an infinite number. Here it is assumed that the LOS from the corners of the viewing cell or to the corners to the target cell will determine whether a target is visible. In this case of cell-to-cell, this leads to 4^2 separate comparisons for any one target. This is of course reduced, because once the target is declared visible from one corner of the viewing cell, then it is in-view, and processing can proceed to consider the next case, but the target cell can only be declared as out-of-view after 16 LOS calculations. In all tests the simple grid method for inferring elevation is used.

5.5 Results of viewpoint and target approximation

As might be expected, the point-to-point yields the smallest viewshed and the cell-to-cell the largest (Table 10.2). The changes can be dramatic, with an increase of nearly 50 per cent being recorded in the area viewable from viewing point 1 between the point-to-point and cell-to-cell cases. The viewsheds might be expected to be nested in this case, but again this is not precisely the case, although in both cases the total number of cells with some degree of visibility in the sum image (Figure 10.5) is very close to the total for the cell-to-cell comparison.

TABLE 10.2
The number of cells included in the viewsheds from the test locations. Results for the four different methods for approximating viewer and target locations and illustrated in Figure 10.4, are given.

Viewing point	Test site 1	Test site 2
Point to point	2381	2034
Cell to point	3328	2271
Point to cell	2707	2666
Cell to cell	3970	2907
Frequency of cells in summed image (Figure 10.5)		
5 (Viewing point)	1	1
4	2350	2014
3	129	176
2	1069	545
1	457	200
Total	4006	2935

N.B. the first entry is the same as in Tables 10.1 and 10.3.

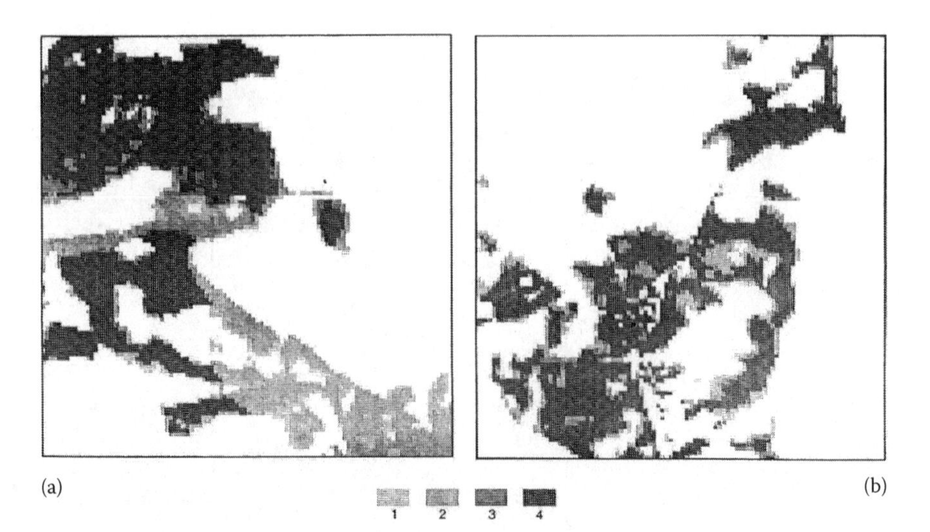

FIGURE 10.5 The pattern of grid cells included in viewsheds derived from alternative methods of viewpoint and target approximation.

Interestingly the cell-to-point and cell-to-cell analyses of the viewshed for test 1 are the only ones in all the tests reported here to identify a whole area to the bottom/centre of the image as being in view. The very straight, northeast facing edge of this area implies that it is just visible over a ridge.

5.6 *Comparing elevations*

Once the location at which the elevation should be tested is determined, there are actually several ways in which the judgment as to the relative elevations of the two lines can be made (Figure 10.6). Given that ultimately the vertical side of two right angle triangles are being compared, at least the five simple tests listed in Figure 10.6 can be evaluated to yield a correct answer, although they are, of course, all interrelated and *should* give the same answer. These five possible tests are: the lengths of the vertical lines (the relative height), the hypotenuses, or either of the two non-right angles of the triangle may be tested, as may the gradients of the line-of-sight to the target and to the interim location. The angles may be evaluated in either a trigonometric form or in degrees or radians. The interesting point is that the result should be the same whichever is tested, and usually will be so long as the difference in height is large; when the difference is small rounding errors in the compiler and/or

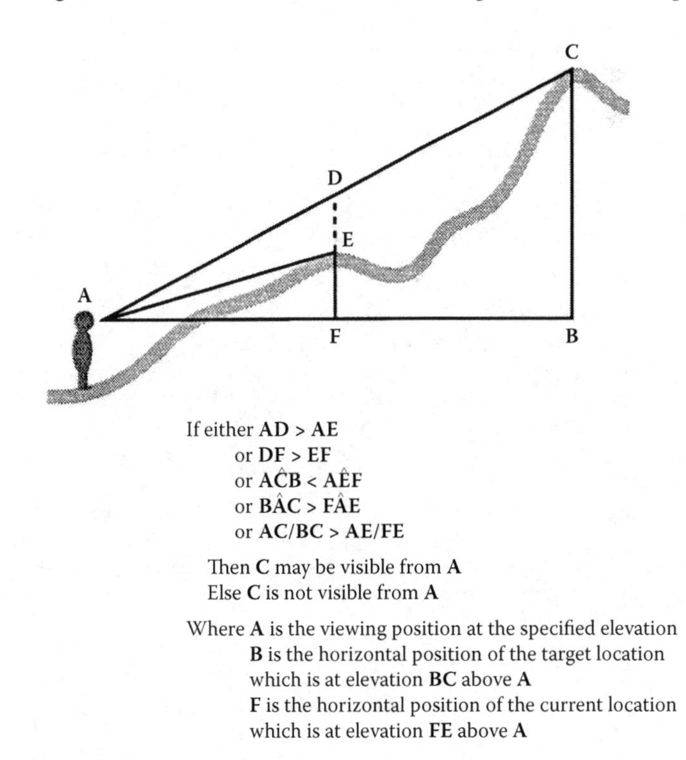

FIGURE 10.6 Five alternative methods for evaluating elevations discussed in the text when the elevation of the target is higher than the viewer. If the elevations are reserved, then the geometry is different, and the conditional statements listed are reversed.

the processor may create inconsistencies. Furthermore, when the two test heights, gradients, etc., are equal, the target should be reported as out-of-view, but it would be a simple error in coding for this not to be the case; probably an error with slight, but measurable consequences.

To test the significance of the precision, three different versions of the elevation comparison were implemented. In the first, the height at the intersection of the LOS and the grid was tested (this test was used in all the inferring comparisons above), in the second the gradient of the two lines was compared, and in the last the final result and all interim calculations were rounded to 16 bit integer values without scaling (to simulate Fortran integer arithmetic which is used in some systems to speed processing). In all cases the simple grid inference, and point-to-point viewing were used.

5.7 Results of elevation comparisons

The results are presented in Table 10.3. Differences are noted between all three methods, but they are not as large as in the preceding experiments. A consistent pattern can be discerned, however. The integer rounding always produces the largest area, and the gradient calculation yields the smallest. No particular reason is apparent for this.

Again the pattern of cells appears to show nested viewsheds (Figure 10.7), but examination of frequencies in the image shows considerable mixing in the viewsheds (Table 10.3).

TABLE 10.3
The number of cells included in the viewsheds from the test locations. Results for three of the different methods discussed in the text, and illustrated in Figure 10.7, are given.

Comparisons	**Test site 1**	**Test site 2**
Height	2381	2034
Gradient	2310	1877
Integer height	2553	2052
Frequency of cells in summed image (Figure 10.7)		
4 (Viewing point)	1	1
3	2270	1794
2	104	223
1	223	132
Total	2598	2150

N.B. the first entry is the same as in Tables 10.2 and 10.3.

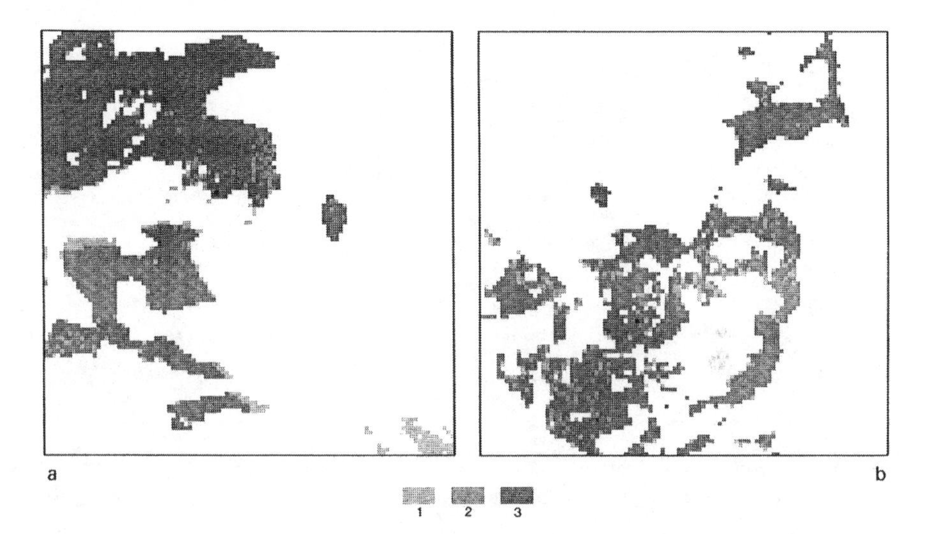

FIGURE 10.7 The pattern of grid cells included in viewsheds derived from three alternative methods of comparing elevation.

5.8 Other possible causes of variation in the viewshed

The results presented show considerable variety in the area of the viewshed derived. Although not attempted here, similar variations to those found in the third test might result from moving platforms owing to variable processor precision, or changing compilers or languages because of different methods of handling interim values in formulations, and because of different precisions in handling floating point numbers. In evaluating a single line of code some compilers store interim calculations to very high precisions (the highest allowed by the operating system or processor), lowering the precision to the level for the variable declared (integer, real, double, etc.) only at the last stage, while others limit the precision to the final precision throughout. Of course, the precision is also dependent upon the hardware, and the code on one platform (with, say, 16-bit precision) may yield a different result from another (with 32-bit precision).

6 Comparing implementations

6.1 The GIS

With the cooperation of a large number of individuals (academics, system developers, and some who are both, see list of acknowledgments) it has been possible to run the data through a number of different GIS packages all purporting to report the viewable area. Table 10.4 presents some of the characteristics of seven systems for which the viewshed was tested. Most of the systems are in the educational domain owing to the availability of software and the willingness of colleagues to give their time. Data were supplied to those testing the programs in the Idrisi ASCII format, with the DEMs and viewpoints being separate image files. Participants were asked

TABLE 10.4
Operating characteristics of the different viewshed implementations compared.

GIS software	Hardware	Viewpoint	DEM file
Idrisi	IBM PC	Grid cell	Real
OSU Map-for-the-PC	IBM PC	Grid cell	Integer
PC MAP	IBM PC	Grid cell	Integer
Map II	Macintosh	Grid cell	Integer
Grass	Sun workstation	Grid cell or Vector point	Integer
Arc/Info	Sun workstation	Vector point	Real or Integer
EPPL	IBM PC	Grid cell	Byte

to read the data into their own GIS, to execute the viewshed, forcing the program to report for the whole area, and placing the viewpoint at 2 m above the DEM surface. The resulting viewsheds were returned to the Idrisi ASCII format and combined under that GIS. Some GIS systems were excluded from the comparison because of the apparent lack of the functionality to allow export to any useful format.

The systems run on a variety of platforms, and store the DEM and viewpoint as various data types (integer, real and byte). The viewing location is specified as both grid cell and as point location (Table 10.4). In EPPL7 all data are stored in byte format, so the full elevation range in either test DEM had to be compressed to the numerical range 0–255. Since the viewer elevation also has to be in byte, it is not possible to make the viewer elevation exactly comparable with the elevation in other tests. The result is not strictly comparable with others, therefore, but it is included for completeness. In the Arc/Info Visibility operation, the viewing point is specified as a vector point, and this was digitized within Arcedit as close to the mid-point of the viewing grid cell used in the other tests as possible.

6.2 Results of implementation comparisons

The results are striking (Table 10.5, Figure 10.8). PC MAP and OSU-MAP-for-the-PC are the only two programs to derive exactly the same viewshed. The fact that the latter is based on the former, with enhancements to the user interface and functionality, means that they are almost certainly using not only the same code, but probably the same compiler. The fact that these yield identical viewable areas, which are smaller than any other in both instances, suggests that the grid constraint is imposed, which is also not surprising since it was suggested by Tomlin (1991 p. 35), the developer of PC MAP.

All other programs have produced different results, but with consistently larger areas than MAP. Assuming that all implementations are based on the LOS algorithm adopted here, it would appear that elevations are inferred by either the grid or triangulation method, and that all use point-to-point viewpoint-to-target approximation, because the areas of the viewsheds are closest. Nothing can be said about the elevation comparison method used, because the variation in the controlled experiments is less than in the implementations.

TABLE 10.5
The number of cells included in the viewsheds from the test locations. The viewshed areas calculated for the test data from the test locations are included according to 8 different GIS implementations.

Elevation approximations	Test site 1	Test site 2
Idrisi	2433	2270
OSU MAP-for-the-PC	1780	1465
PC MAP	1780	1465
MAP II	2157	2161
GRASS	2390	2156
Arc/Info visibility	2304	2174
EPPL7	2610	2263
Frequency of cells in summed images (Figure 10.8)		
8	1	1
7	1421	1283
6	288	135
5	162	239
4	351	335
3	118	173
2	150	226
1	431	493
Total	2922	2885

N.B. the first entry is the same as in Tables 10.2 and 10.3.

Only one gross error was found in the testing, and this has been reported to the developers. It causes a slight overestimate of one viewable area, and is easily identified in Figure 10.8b as a horizontal line running right across the image, through the viewpoint. The same error occurs in the viewshed for point 1, but it is not so obvious in Figure 10.8a, because the erroneously identified cells are not continuous across the image.

7 The probable viewshed

The variability of the viewshed, caused by changing minor parameters in the viewshed operation, highlights the inappropriateness of the binary representation of the viewshed. It is believed that the probable viewshed yields a more appropriate model. It recognises the Boolean nature of the viewshed, but for any location there is not a report that it is simply in or out of sight; rather a likelihood that it can be seen is given.

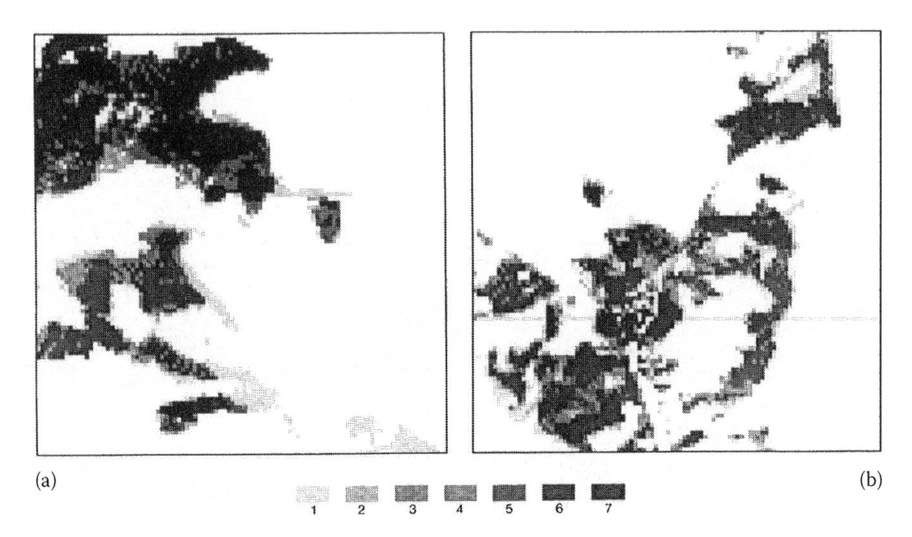

FIGURE 10.8 The pattern of grid cells included in viewsheds from the implementations tested.

Figure 10.9 shows the probable viewsheds for the test areas used throughout this paper. They have been generated by the method discussed by Fisher (1992), involving Monte Carlo simulation of error fields for the DEM, re-calculation of the viewshed, and summing of the binary viewsheds found. As in the earlier work these probable viewsheds are based on 20 simulations, using Root Mean Squared Error (RMSE) = 7 (the value reported for the DEM), and Spatial Autocorrelation (Moran's I) = 0 (a necessary assumption from the USGS error reporting).

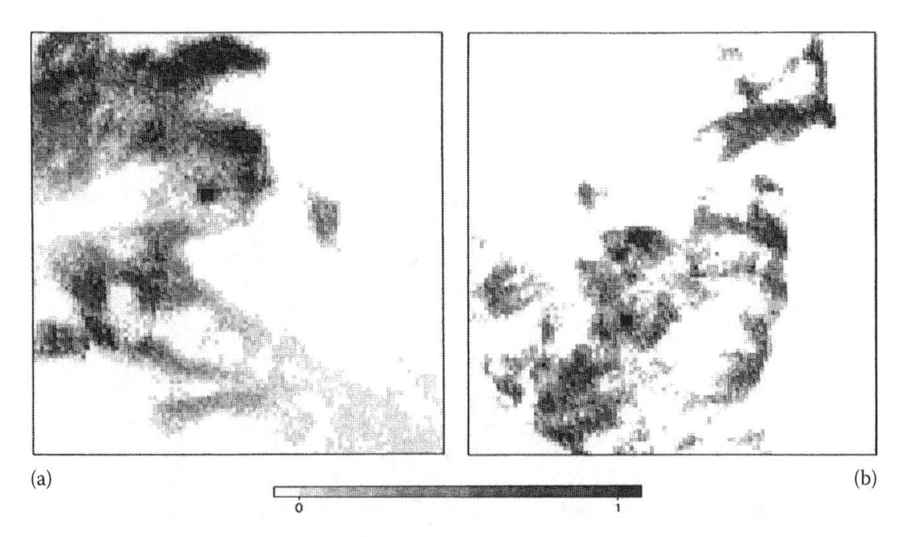

FIGURE 10.9 The patter of probabilities of cells being visible from the viewing points, as defined by repeatedly simulating error in the DEM, and adding the binary viewsheds found together (Fisher 1992).

Examination of the probable viewshed (Figure 10.9) shows that the total area which might possibly be visible includes much of those areas only included in the largest binary viewshed found from the cell-to-cell analysis. Indeed, the total area of the probable viewshed from viewpoint 1 covers 4115 cells, and from viewpoint 2 covers 3435. These are actually larger than the largest binary viewshed found here, but much of the area has only a slight probability of being visible. Therefore, although the probable viewshed is found from simulation of database error in the DEM, it can be suggested that it is broadly related to implementation variability.

It should be noted that the fuzzy viewshed is erroneously referred to by Fisher (1992), and future work will clarify the distinction between the fuzzy and probable viewshed.

8 Conclusion

From the various different algorithms for defining the viewshed which are briefly described above, and have been implemented in the course of the research described here, it has been shown that the viewable area is completely dependent on how the problem is formulated and implemented. Indeed, the ratio of the maximum to minimum viewable area is over 6 and 31, for viewpoints 1 and 2 respectively, comparing cell-to-cell to stepped, or around 2 for both test locations if the second smallest viewable area, from the grid-constrained method, is used. Furthermore, it has been shown that all actual implementations derive viewsheds which are of an area within the areal and spatial limits bounded by those coded in the course of research, but are nonetheless variable. Thus all the programs tested (bar one) seem to be 'correct'; differences result from alternative design decisions at the time that algorithms were specified by different groups. Those design decisions were undoubtedly defensible and acceptable both at the time and now. The point is that many decision stages exist for any implementation of a viewshed algorithm, and designs may diverge and cause inconsistent viewsheds to be found (unless programs share code as in PC MAP and OSU MAP-for-the-PC). Users have a right to know the algorithm used in any one version of the viewshed operation, so that they may assess how much faith can be placed on the results of an analysis. If the cell-to-cell algorithm is used, then clearly the viewshed is not so acceptable, but perhaps if point-to-point is employed using grid constraint, then considerable confidence can be placed on locations identified as being viewable actually being so. Insufficient testing has been done to establish which algorithm gives the best results in any particular circumstance, even if that were possible. Future research will attempt to examine the relative performance of the different algorithms under different landscape conditions.

The viewshed is only one of a number of complex functions contained in many GIS which have different possible algorithms. There is a very clear need for extensive comparative testing of GIS functions, as is reflected in several recent appeals for software standards (Hemenway 1992, Jordan and Star 1992). It is preferable to use simple geometric objects where the result is completely predictable, but in many cases of GIS functions (such as the viewshed) this is not appropriate. An experimental design similar to that used here, which tests the consistency in products, is a natural approach to analysing the results of implementations of these complex algorithms.

The primary and overriding conclusion of this study is that the viewshed as it is calculated is not the precise phenomenon the Boolean representation implies. Rather the viewshed is a fundamentally uncertain phenomenon within a GIS, and is simply not repeatable across a spectrum of systems. This should not be seen as a case of algorithm or implementation *error*, since all may be assumed to be justifiable, but rather a case of algorithm *uncertainty*. If the viewshed operation is not precise, the Boolean representation and any logical combinations which may be made are not acceptable. Therefore, an alternative model for the viewshed must be sought, and the probable viewshed, which can be found by Monte Carlo simulation of elevation errors (but other methods are suggested by Teng and Davis 1992), is suggested as such a model here.

This work implies no criticism of any particular GIS implementation of the viewshed operation, since all are close to the range of values of the control analyses, and none is shown to be grossly wrong. The error is a conceptual one, and is in the precision of the viewshed implied by all operators. It is believed that the results obtained here would not have differed if the DEMs had been of any natural and variable terrain on earth, and the viewing locations anywhere within the terrain (except if it were at the bottom of a hole!).

9 Invitation

Only seven GIS packages were used in this study, and those are dominantly in the academic domain. If any other GIS developers would like to receive the test data, and run their own viewshed operations on the data, the author is very willing to supply the test datasets. Anyone participating should be able to return the resulting viewshed in ASCII grid format. If enough other implementations are forthcoming, the author will undertake to submit updated versions of Table 10.5 and Figure 10.8 for publication as a note in a future issue of this journal.

Acknowledgments

This work would not have been possible without the assistance and encouragement of many individuals. The following either provided software, or ran the experimental data through particular systems, I wish to thank them all for their time: Andy Baffes, Bill Carstensen, Bill Hickin, Ravi Narasimhan, Howard Veregin, and Jim Westervelt. Among others, Nick Chrisman, Jack Dangermond, Armondo Guevera, Robert Mills, and Dana Tomlin showed interest or provided encouragement in the research by word or by letter. Kate Moore assisted with figure preparation. Finally one reviewer provided very useful and perceptive comments on the original manuscript.

References

Burrough, P. A., 1986, Principles of Geographical Information Systems for Resource Evaluation (Oxford: University Press)

Clarke, K. C., 1985, A comparative analysis of polygon to raster interpolation methods. *Photogrammetric Engineering and Remote Sensing*, 51, 575–782.

DCDSTF (Digital Cartographic Data Standard Task Force), 1988, The Proposed Standard for Digital Cartographic Data. *The American Cartographer*, **15**, 11–142.

Felleman, J., and Griffin, C., 1990, *The Role of Error in GIS-based Viewshed Determination: A Problem Analysis*. IEPP Report No. EIPP-90-2, Syracuse, New York.

Fisher, P. F., 1991, First experiments in viewshed uncertainity: the accuracy of the viewshed area. *Photogrammetric Engineering and Remote Sensing*, **57**, 1321–1327.

Fisher, P. F., 1992, First experiments in viewshed uncertainity: simulating the fuzzy viewshed. *Photogrammetric Engineering and Remote Sensing*, **58**, 345–352.

Hemenway, D., 1992, GIS observer: benchmarks and comparisons. *Photogrammetric Engineering and Remote Sensing*, **58,** 31.

Jordan, L. E., and Star, J., 1992, A call for action: Standards for the GIS community. *Photogrammetric Engineering and Remote Sensing*, **58,** 863–864.

Knaap, W. G. M. van der, 1992, The vector to raster conversion: (mis)use in geographical information systems. *International Journal of Geographical Information Systems*, **6**, 159–170.

Kvamme, K., 1990, GIS algorithms and their effects on regional archaeological analysis. In *Interpreting Space: GIS and Archaeology* edited by K. M. S. Allen, S. W. Green and E. B. W. Zubrow (Basingstoke: Taylor & Francis), 112–125.

Lee, J., 1991, Comparison of existing methods for building triangular irregular network models of terrain from grid digital elevation models. *International Journal of Geographical Information Systems*, **5,** 267–285.

Mark, D. M., 1987, Recursive algorithms for the analysis and display of digital elevation data. *Proceedings of the First Latin American Conference on Computers in Cartography, San José, Costa Rica* (Columbus, Ohio: International Geographic Union Commission on Spatial Data Sensing and Processing), pp. 375–397.

Piwowar, D. J., Ellsworth, F. L., and Dudycha, D. J., 1990, Integration of spatial data in vector and raster formats in a geographical information system environment. *International Journal of Geographical Information Systems*, **4**, 429–444.

Srinivasan, R., and Engel, B. A., 1991, Effect of slope prediction methods on slope and erosion estimates. *American Society of Agricultural Engineers, Applied Engineering in Agriculture*, **7,** 779–783.

Skidmore, A. K., 1989, A comparison of techniques for calculating gradient and aspect from a gridded digital elevation model. *International Journal of Geographical Information Systems*, **3,** 323–334.

Skidmore, A. K., Terrian position as mapped from a gridded digital elevation model. *International Journal of Geographical Information Systems*, 4, 33–49.

Teng, Y. A., and Davis, L. S., 1992, *Visibility Analysis on Digital Terrain Models and its Parallel Implementation*. Technical Report CAR-TR-625, Centre for Automation Research, University of Maryland.

Tomlin, C. D., 1991, *Cartographic Modelling and Geographic Information Systems*. (Englewood Cliffs, N. J.: Prentice Hall).

Travis, M. R., Elsner, G. H., Iverson, W. D., and Johnson, C. G., 1975, VIEWIT: *Computation of Seen Areas, Slope, and Aspect for Land-use Planning*. USDA Forest Service General Technical Report PSW-11/1975, Pacific Southwest Forest and Range Experimental Station, Berkeley, California.

Wagner, D. F., 1989, *A comparison and evaluation of three polygon overlay algorithms*. Technical Report Series TR 89-01, GIS Laboratory, The Ohio State University.

Yoeli, P., 1985, The making of intervisibility maps with computer and plotter. *Cartographica*, **22**, 88–103.

Algorithm and Implementation Uncertainty: Any Advances?

Peter F. Fisher

In writing this commentary I wish to make three points. First, I show how the original paper fits into a focussed corpus of sustained work that is perhaps incomplete, but covers an interesting set of questions with respect to the one operation, the viewshed, as an example of geographical information system (GIS) operations. Second, I point out how the main message of this particular paper has not been followed up in the subsequent years, but is still important. Finally, I highlight that admitting your mistakes is not necessarily a bad thing.

The focus

By 1991 I had done some research on the accuracy of the area that is determined by the viewshed function (Fisher, 1991; 1992). I started examining the viewshed as a result of an invitation from Mike Goodchild to attend the first initiative meeting of the NCGIA (National Center for Geographic Information and Analysis) on the accuracy of spatial databases (due to some early research I had done on uncertainty in soil information). At the meeting I listened to Stan Openshaw talking about a Monte Carlo framework for modelling the consequences of database error on analytical outcomes. Subsequently, I realised that I knew about Monte Carlo simulation and was aware of its advantages over other statistical methods, as well as some of the problems (Kiiveri, Chapter 15 in this volume). The viewshed is a function with a simple, clear, and unequivocal outcome — that area which is visible and that which is not. This, together with the relatively simple inputs, a Digital Elevation Model (DEM) and a view point, make it ideal for exploratory investigations.

I therefore did the research, using hours of PC time and small datasets, to make the simulations possible. I used a simple randomisation procedure to simulate error in the DEM-based on white noise, as well as various levels of simulated autocorrelation, to evaluate the effect of DEM error on the visible area. This was not the first time such a procedure was applied, but to the best of my knowledge it was the first time it was done with DEMs to explore the effect on an analytical derivative. Since then others have used similar approaches in various contexts (Hunter and Goodchild, 1997). My work used a simple (not to say simplistic) approach to modelling the measurement error in a DEM-based as it was based on first neighbours for determining autocorrelation. Various studies since have improved that approach using geostatistical simulation, which is more appropriate (Fisher, 1998; Holmes et al., 2000).

Error modelling was not my only interest in this track of research, but rather it was in the viewshed as a case study of a GIS operation. In subsequent work I asked whether the question a user might be asking of the viewshed was actually answered by the function (Fisher, 1996a, 1996b) and whether there was any predictability of the viewshed operation with change in grid resolution (Fisher, 1996c). I have also examined the possible use of some of these new functions and operations in landscape planning (Fisher, 1995; 1996b), but my work only scratched the surface compared with more recent work (Llobera, 2003, Chapter 20 in this volume; Rana, 2003).

The message

The viewshed is a simple outcome with simple inputs, but it is based on a huge amount of computation — multiple calculations repeated many times along each individual line of sight, which build up to give the total viewshed — and as a result, in most implementations the operation also takes a long time to run. In the first version of Idrisi running on an IBM PC XT, the operation for a relatively small dataset (200 x 200 cells) ran for up to 1 hour!

In programming the routine myself to support my research, I rapidly realised that there are a large number of design decisions to make even if the landscape is treated as composed of straight-line segments for simplicity. This means that the likelihood that every developer will have written an algorithm that does everything in the same order making all the same decisions was remote. And if they had not, then there was an unquantified potential for doubt about the outcome of the analysis grounded simply in the coding of the operation. Therefore I set out to do two things. First, I considered a number of different ways in which the landscape could be approximated and how the line of sight could be calculated. I implemented them all, and I came up with the results documented in the paper (Fisher, 1993). That much is not so terribly different from what Skidmore (1989, Chapter 5 in this volume) and a number of other people have done since for slope and aspect, and Desmet and Govers (1996) and Zhou and Liu (2002) have done for routing algorithms. What I went on to do, and what is different from much other work (especially on DEMs), was to go through the frustrating process of persuading friends who had access to GIS unavailable to me to run the viewshed operation in that GIS and send me the results.

Of course I wanted the results in digital form so I could integrate them and analyse them as I wished. A number of well-known GIS at the time (SPANS and GENASYS), could not be used in the experiment because although they were perfectly capable of reading the ASCII DEM files that would be provided, they did not have an export capability to any widely used format. The failure to implement data transfer standards probably contributed to the fact that neither system has stood the test of time. So I ended up with seven different realisations of a viewshed analysis by commercial and semicommercial software. The actual extent of inclusion and exclusion of the different implementations is more disturbing than surprising because software is being used to make fundamental decisions in planning and other areas that may effect people's quality of life or even (in military applications) their life or death.

The problem of consistency in analysis from different software products is potentially vast. The viewshed is a relatively modest function compared to many within GIS. It has relatively simple mathematical calculations with little risk of miscoding, or poor handling of floating points or rounding errors, especially when compared to more complex functions like routing or operations using statistical distributions. The fact is that very few papers have been published that systematically compare the outputs of various operations, in either GIS or Remote Sensing. Are the outputs of all maximum likelihood classifications the same? When I started the research reported as Fisher (1993) it seemed unlikely that the viewshed functions would agree and there is an equally small chance of perfect agreement among other complex operations. The only other paper I know of that does compare implementations of functions is the work of Knaap (1992) on vector to raster conversion. Perhaps comparisons have been done and no difference found; after all, such negative results do not usually find their way into the refereed literature, but it would be interesting and worth at least a technical note, if people found that a complex operation repeated across a range of software did produce identical results. I believe that the real problem with research of this type is twofold. It relates to the complexity of the software interfaces and user options, making it hard to ensure that exactly the same query is put to all software, and to the general unavailability of multiple software packages to most researchers.

In the paper, I appealed for software vendors to publish their algorithms more fully. This could be achieved either as technical documentation, refereed publication, or more helpfully, by citations to published research papers or books in the online documentation for the software. In addition, I called for a systematic and a public approach to verification of analytical outcomes. In the general image processing and statistical software industries, software developers establish publicly available datasets that are used to verify the functionality of software (Huber 2005). There seems to have been no advance in either of these areas for GIS users. Along with other appeals of the same type (Hemenway, 1992; Jordan and Star, 1992), my call seems to have fallen on deaf ears.

The mistake

In my early paper (Fisher 1992) I referred to the simulation of error in the DEM and the adding together of the multiple realisations as a way of determining the fuzzy viewshed. Unfortunately I was wrong. This is not a fuzzy viewshed — it can be interpreted as having some of the properties of such a viewshed, but they are coincidental to the process and method. It is not correct to call it a fuzzy viewshed, and in the 1993 paper I admitted that with the statement: "It should be noted that the fuzzy viewshed is erroneously referred to by Fisher (1992), and future work will clarify the distinction between the fuzzy and probable viewshed" (Fisher, 1993, p. 334; p. 221 in this volume). This expansion was given by Fisher (1994) where the distinction is made explicit. I believe that it is important to correct yourself when you realise that you are wrong in the first place, and I take this opportunity to point out the citation to the fuller explanation of the difference.

References

DESMET, P.J.J. AND GOVERS, G., 1996, Comparison of routing algorithms for digital elevation models and their implication for predicting ephemeral gullies, *International Journal of Geographical Information Systems*, **10**, 311–331.

FISHER, P.F., 1991, First experiments in viewshed uncertainty: The accuracy of the viewshed area, *Photogrammetric Engineering and Remote Sensing*, **57**, 1321–1327.

FISHER, P.F., 1992, First experiments in viewshed uncertainty: Simulating fuzzy viewsheds, *Photogrammetric Engineering and Remote Sensing*, **58**, 345–352.

FISHER, P.F., 1993, Algorithm and implementation uncertainty in viewshed analysis, *International Journal of Geographical Information Systems*, **7**, 331–347.

FISHER, P.F., 1994, Probable and fuzzy concepts of the uncertain viewshed. In *Innovations in GIS 1*, M.Worboys, Ed., Taylor & Francis, London, 161–175.

FISHER, P.F., 1995, An exploration of probable viewsheds in landscape planning, *Environment and Planning B: Planning and Design,* **22**, 527–546.

FISHER, P.F., 1996a, Reconsideration of the viewshed function in terrain modelling, *Geographical System*, **3**, 33–58.

FISHER, P.F., 1996b, Extending the applicability of viewsheds in landscape planning, *Photogrammetric Engineering and Remote Sensing,* **62**, 1297–1302.

FISHER, P.F., 1996c, Propagating effects of database generalization on the viewshed, *Transactions in GIS,* **1**, 69–81.

FISHER, P.F., 1998, Improved modelling of elevation error with geostatistics, *GeoInformatica*, **2**, 215–233.

HEMENWAY, D., 1992, GIS observer: Benchmarks and comparisons. *Photogrammetric Engineering and Remote Sensing*, **58,** 1–31.

HOLMES, K.W., CHADWICK, O.A., AND KYRIAKIDIS, P.C., 2000, Error in a USGS 30m DEM and its impact on terrain modeling, *Journal of Hydrology*, **233**, 154–73.

HUBER, D., 2005, Computer vision test images. http://www.cs.cmu.edu/~cil/v-images.html (accessed 10 November 2005).

HUNTER, G.J. AND GOODCHILD, M.F., 1997, Modeling the uncertainty of slope and aspect estimates derived from spatial databases, *Geographical Analysis,* **29**, 35–49.

JORDAN, L.E. AND STAR, J., 1992, A call for action: Standards for the GIS community, *Photogrammetric Engineering and Remote Sensing*, **58,** 6, 863–864.

LLOBERA, M., 2003, Extending GIS-based visual analysis: the concept of visualscapes, *International Journal of Geographical Information Science* **17**, 25–48.

RANA, S., 2003, Visibility analysis: Guest editorial, *Environment and Planning B: Planning and Design* **30**, 641–642.

SKIDMORE, A.K., 1989, A comparison of techniques for calculating gradient and aspect from a gridded digital elevation model, *International Journal of Geographical Information Systems,* **3**, 323–334.

VAN DER KNAPP, W.G.M., 1992, The vector to raster conversion; (mis)use of geographical information systems, *International Journal of Geographical Information Systems,* **6**, 159–170.

ZHOU, Q. AND LIU, X., 2002, Error assessment of grid-based flow routing algorithms used in hydrological models, *International Journal of Geographical Information Science,* **16**, 819–842.

International Journal of Geographical Information Systems,
1995, Vol. 9, No. 4, 359–383.

11 Development of a Geomorphological Spatial Model Using Object-Oriented Design

Jonathan F. Raper and David E. Livingstone

Abstract. This paper argues that spatial modelling within the environmental sciences is not best achieved through the low level integration of environmental models and GIS (Geographical Information Systems), but by the creation of new integrated object-oriented modelling environments. In the light of this assertion the philosophical background to representation and spatio-temporal referencing is considered in order to outline the context for the design of Oogeomorph — an object-oriented spatial modelling system for geomorphology. The paper concludes with an example of the use of Oogeomorph to represent and test a coastal geomorphological theory.

1 A perspective on GIS and environmental modelling

One of the key research issues in GIS to have emerged in the mid 1990s concerns how environmental models might be developed within a GIS framework (Goodchild *et al.* 1993). Although a considerable amount of research has now been carried out in this area (see reviews in Raper 1991, 1993) most researchers continue to adopt a series of representational compromises when formulating environmental models within a GIS. Specifically, the representational basis of the GIS is often allowed to drive the form and nature of the environmental model. This has led to the adoption, by default, of essentially geometrically-indexed methods for representing environmental models in a spatial context. These methods are based on the principle of 'planar enforcement' (space exhaustion) within a layer which does not permit the handling of overlapping features (Goodchild 1992) or temporally changing phenomena (Langran 1991). Implicitly, these methods force a segmentation of the entities being represented into separate layers whenever they interact in time or space: adopting this representational method forces compromises on most environmental modelling.

In a recent theoretical critique of this kind of work as a basis for integrating environmental modelling with GIS, the authors attempted to turn this reasoning around (Livingstone and Raper 1994). Rather than forcing the modelling into a geometrically-indexed representation, the wider question 'what structure in the environmental problem should drive the representational basis of the model?' was posed. Attempting to answer that question led to a consideration of alternative approaches to (spatial) representation. The specific aim was to identify a scheme which possessed rich semantics capable of representing a knowledge of environmental problems in a realistic spatio-temporal context. This search led to an examination of object-oriented analysis and design approaches (such as that of Booch 1991) which, in other application domains, have led to considerable progress in the development of new systems of representation and in the formulation of associated methodologies for modelling. This paper is an attempt to develop and formalize a spatio-temporal model using object-oriented techniques for a specific environmental problem in geomorphology.

There are, however, significant differences between approaches to the spatial modelling of environmental problems using geometrically-indexed and object-oriented methods. Using the former approach makes the coordinate system of the layer into the primary index of the spatial representation and dictates much of the representational structure of the environmental problem of interest. In the object-oriented approach the environmental scientist must declare the nature of the real world entities identified first: their characteristics and behaviour structure the spatial representation. Geographical Information Systems developers are not accustomed to receiving structures from application domains — the GIS normally provides them. Hence the first part of this paper consists of a review of the philosophy and methodology required to develop an object-oriented spatial representation in an environmental context. The use of this methodology to examine a coastal geomorphological problem is then discussed.

2 The philosophical background to environmental representation

Defining systems of representation based upon the identification of entities and their relationships is a process dating back to the earliest philosophers, and in many ways is the foundation of all science. However, in deciding that a defined and persistent identity can be assigned to an entity in an environmental domain involves taking certain standpoints in relation to the philosophy of science which need to be made explicit as Chapman (1977), Haines-Young and Petch (1986), Thorn (1988) and Unwin (1993) have all stressed.

Simply stating that 'many physical geographers have no difficulty accepting the traditional science paradigm' (Kirkby 1988, p. 256) leaves many questions unanswered about how entities and their behaviour are recognised and structured in environmental science. Does this view, for example, uncritically accept the deductive, logical approach to science based upon the falsification of theories — the so-called critical rationalist view expounded by Popper (1972)? If so, then environmental scientists must now recognize the wide modifications made to critical rationalism in the last two decades. These changes require the testing of research frameworks against new considerations:

- The status of theories and models.
- The process of entity identification.
- The representation of a continuous reality using discrete entities.

This review sets the process of entity identification in a wider philosophical and scientific context, since, in the object-oriented approach discussed in this paper it is the identified entities which structure the representation and not geometry.

A first consideration is the status of theories and models for research in the environmental field. Environmental theories originate from a variety of sources such as paradigms, formalized problem situations or general concepts and are often not rigorously presented using logical statements. This has led to the charge that many environmental theories cannot be considered to be true theories, since they cannot be tested (Haines-Young and Petch 1986). 'Relativist' philosophers of science such as Kuhn and Lakatos have argued that, in reality, scientists are influenced by paradigms or the objectives of research programmes when they judge the predictions of theories, rather than purely by the procedures of falsification. Since environmental theories are most often expressed as mathematical models, it is suggested by Unwin (1988) that for such environmental models to be soundly based and fully testable their predicted outcomes must be compared with real outcomes. This implies that environmental models expressed within spatial representation should be fully specified in terms of entities and their behaviour. This is not only possible but is required in the object-oriented approach illustrated in this paper.

Several general frameworks have been proposed for modelling, such as those by Ziegler (1976) and Casti (1989) which are loosely based on General Systems Theory (von Bertalanffy 1962). Most are based on logical models expressed using mathematical symbols, and, in recent times have been made finite so that they can be discretized in computer languages and handled by computer processing. Many different formulations for environmental models have been suggested, although most of them focus on process. Kemp (1993) reviews and classifies them using the Jorgensen (1990) approach in which models are distinguished by time and space behaviour, type of data, parameters and expressions used, model structure and type of mathematics used. Process models generally express theories predicting the nature of the exchange of energy or mass within systems, over time. By contrast, data models express theories predicting the structure of real world domains in terms of entities and their attributes organized in inter-related sets. One of the basic problems encountered when 'coupling' environmental models with GIS is that the former are specified as process models while the latter are specified as data models. Process and data models can be linked when implemented in object-oriented systems.

As a second consideration, and following on from the discussion of theories and models, it is necessary to examine the process by which environmental entities are given an identity. Views on this question focus on the extent to which the process is governed by theory in the abstract, i.e., whether the process is formal (logical) or whether it is governed by cognitive or linguistic constructions. Haines-Young and Petch (1986) adhere to the former view, i.e., that 'naming and classifying are only undertaken in the context of some theory' (p. 158) and observe that knowledge gained about named objects is always conjectural. This is a narrowly rationalist view: there is a considerable body of work in philosophy and cognitive science that suggests that the identification of (environmental) entities is governed by cognitive processes.

The classical view (referred to as 'objectivism' by Lakoff 1987) is that reasoning is achieved by the manipulation of abstract symbols associated with entities which exist in the world: this implies that there is a 'mind-free' world composed of interrelated things. Objectivists divided the entities in this world into categories which were defined by their shared properties and modelled them using set theory. This view of categories was modified by Rosch (1973) who suggested that categories should be characterized by prototypes or exemplars, and that other members of the category could be described in terms of their differences from the prototype. Rosch also discriminated between categorization at different levels, noting that 'basic-level' categories could be most readily defined in terms of perception, function, language and knowledge, and that most attributes of categories were stored at this level.

Objectivism has been criticised by Putnam (1981), Johnson (1987) and Lakoff (1987) who argue that it is the mind that creates the structures through which the world is understood: in a metaphysical sense, no 'mind-free' world is possible. The salient features of a mind-constructed world view include the role of the body and human culture in defining organizing concepts (called 'experientialism' by Lakoff). It is argued in experientialism that the world is perceived through 'schemas' which interface between the sensory apparatus and the mind. Johnson (1987) argues that these schemas directly constrain understanding and that they provide a means for organizing many different domains of knowledge, often through metaphor. Frank and Mark (1991) have noted that many of these schemas are spatial (e.g., container, blockage, path, surface, link, near-far, contact, centre-periphery and scale) and that others such as 'object' and 'part-whole' can be used in making spatial representations. The experientialist view has itself been modified by others who argue that the structure and architecture of the stored knowledge defines constraints in understanding (Pylyshyn 1981).

These contrasting views pose significant challenges to the identification of significant (environmental) entities. It is possible to take the view that entities in the real world are defined on the basis of attributes selected and described in a theory working from the objectivist point of view (Haines-Young and Petch 1986), or that they are singled out and defined through mind-created schemas (Johnson 1987) from the experientialist point of view. However, both views recognize the notion of natural discontinuities in the world (Lakoff 1987) and therefore converge on *means* of object identification while differing about the *origins and motivation* of the process. Both views can be used to create categorizations in the environmental domain and are examined below. The key lesson that may be learned from experientialism for the study of an environmental problem is that named forms and processes must be treated with caution and not incorporated into models uncritically. It is likely that many of the observations which have led to such classifications only reflect a minority of the spatial and temporal expressions of some phenomena.

A third consideration concerns the representation of a continuous reality using discrete entities. Although most physical theories are spatially and temporally continuous the evaluation of these theories almost always requires the existence of entities which are spatially and temporally discrete. To test such theories applied to the environment requires that ways are devised to create discrete 'abstract states' by entification at a particular granularity. Such discrete entities can either be derived from theory — such as the local minima and maxima of continuous functions, or from measurement — such as the identification of rapidly changing gradients or patterns in sampled data point.

The modelling framework proposed by Casti (1989) is based on the notion that there are an infinite set of 'states' of natural systems in the real world and that models can represent a finite subset of these named 'abstract states'. Casti defines the concept of an 'observable' as a rule for associating a real number with each 'abstract state' defined by the modeller. This is a key issue here as much spatial database development proceeds through the acquisition of datasets from 'blind' sampling exercises might be driven by instrumentation or established for reasons unrelated to the investigation in hand, and therefore could be argued to be theory-neutral. This is particularly facilitated by an information systems approach.

However, Haines-Young and Petch (1986) state that 'the view … that measurements can be taken without any theory … is entirely mistaken' (p. 167). They assert that issues of measurement scales and units force the process of measurement to make reference to theory since the form of entities that are referenced by the theory must determine the structure of any sampling. Sampling must also be decided with reference to (conjectural) knowledge and potential errors in the observations. This requires that the modeller be given the means to control the process of entification and its specific spatial expression. Nunes (1991) has described this process as the search for 'geographic individuals'. At present entities defined in an environmental model are forced to become spatially and temporally discrete in inappropriate ways by the geometrically-indexed nature of many GIS. An object-oriented approach allows measurements made on environments to become candidates for the discrete expressions predicted in theories.

In summary, it is clear that the 'traditional science paradigm' has been considerably modified in the last few decades in a number of ways. Firstly, by work questioning how theories and models are developed and tested, particularly in the environmental domain where process models must link to data models. Secondly, by studies considering how the world is understood and classified into categories, and what this dictates about the procedures for entity identification. Thirdly, by consideration of the methods by which the identified entities are discretized from measurement and how data should be used in models. These new perspectives on method now demand a more rigorous approach to modelling, classifying and discretizing when studying environmental problems. Adoption of the object-oriented approach to spatial representation recognizes these priorities and enables solutions to some of the problems of environmental models coupled to GIS to be circumvented. However, such new representations must be placed in a clear spatio-temporal context.

3 Concepts of space and time

Representation of environmental problems in models faces a challenge beyond those in many other application areas since they must specifically incorporate and manifest concepts of space and time. The nature of environmental representations has traditionally been dominated by 'timeless' geometric methods focused on two dimensional planes. However, recent theoretical developments in physics have expanded the range of concepts available and object-oriented approaches to software design have facilitated the implementation of such new ideas in spatial representations.

3.1 Physical concepts of space and time

This discussion of which concepts of space and time can be used in a representation starts from the premise that both space and time are dimensions that are defined by the entities that inhabit them and not vice versa. This premise has been discussed by Nunes (1991) and Livingstone and Raper (1994) but is best illustrated in a note by Einstein in the fifteenth edition of his book 'Relativity; the Special and the General Theory' (1952):

> 'I wished to show that space-time is not necessarily something to which one can ascribe a separate existence, independent of the actual objects of physical reality. Physical objects are not in space, but these objects are spatially extended. In this way the concept "empty space" loses its meaning' (p. vi).

A direct consequence of such a stance is that space and time must be considered relative concepts, i.e., they are determined by the nature and behaviour of the entities that 'inhabit' them (the concept of 'relative space'). This is the inverse of the situation where space and time themselves form a rigid framework which has an existence independent of the entities (the concept of 'absolute space'). Whereas an 'absolute space' approach may be suitable for an application such as land resource management, there are arguments for the use of a 'relative space' approach in the study of environmental problems.

The importance of the formulation of fundamental concepts of space and time has been noted by a number of authors, both in relation to GIS (Feuchtwanger 1989, Dikau 1990, Herring 1991, Nunes 1991, Langran 1992) and to geomorphology (Thornes and Brunsden 1977, Unwin 1993). However, Unwin (1993, p. 194) claims that 'Geographers have had a surprisingly insignificant role in contemporary philosophical debates concerning the nature of space', and that they cling to the notion of space as being an absolute reference frame of 3-dimensional Euclidean coordinate geometry. Whilst much of the work of mathematicians and physicists such as Minokowski and Einstein is relevant only at extreme scales or velocities, notions such as relative concepts of space and time are pertinent to environmental science. Unwin (1993, p. 199) has reviewed these concepts and concluded that, for geographers included, 'it thus no longer makes sense to talk of space and time; rather we should begin to examine the meaning of space-time'.

There is thus an argument for adopting an approach such as Minkowski's four dimensional extension to the three dimensional Euclidean world in environmental science (described by Einstein 1952):

$$(x1, x2, x3, x4)$$

where x4 = ict ($i = \sqrt{-1}$, c = velocity of light, t = time)

Here x1, x2, x3 represent the three spatial dimensions and x4 represents the temporal dimension. Hazelton *et al.* (1990) showed how such four dimensional representations of space-time could be constructed using measurement units of light-seconds. This four dimensional Euclidean approach can be used to implement a 'relative space' approach by the use of a set of 'worlds', where each 'world' has its own four dimensional reference system. A 'world' would therefore be a higher order concept than any metric that was used to represent each 'world'.

3.2 Spatio-temporal concepts suggested in the context of environmental science

In the study of environmental systems variations of phenomena in space and time are of fundamental importance and hence the development of a spatial representation must incorporate a physically justifiable model of space-time. Thornes and Brunsden (1977) approach this issue in geomorphology by characterizing space and time according to the constraints on measurement and developed a scheme of spatio-temporal ordering (Table 11.1). In philosophical terms the scheme is a hybrid one, since orders zero to three imply 'absolute space' while order four is compatible with 'relative space'.

However, the way that spatio-temporal processes are studied is strongly influenced by the model of space and time that is adopted. Using 'absolute space' concepts, space and time must be theoretically separated since time is unidirectional (i.e., moving forward) while spatial fluxes can be multi-directional. The widespread use of this distinction is based on a psychological perception that space and time are distinct: movement back and forward through largely unchanging environments is a reality of much urban life. The observation that a phenomenon has returned to a former location is only justifiable though if it is accepted that its state is not dependent upon temporal changes to itself or to its environment. Hence, the modelling of spatially reversible fluxes using absolute space concepts implies the use of temporal generalizations, interpolations and projections between time frames.

TABLE 11.1
Ordering of measurement in space and time (after Thornes and Brunsden 1977, p. 176).

Order	Spatial measurement	Temporal measurement
0	Serial	None
	None	Serial
1	Discrete	Discrete
2	Continuous	Discrete
	Discrete	Continuous
3	Continuous	Continuous
4		Multivariate space-time

By adopting four dimensional 'relative space' concepts multi-directional spatial fluxes are treated simply as forward movements through space-time. In other words, a flux forward and backwards through space and over time can also be seen as a move along a space-time 'spiral', i.e., since the return is not to the same time, it cannot be to the same place. This, it is argued here, is a more physically justifiable model of geomorphological systems (especially at the coast) often evolve such that whole suites of landforms with internally consistent organization move in space through time. This kind of change can only be represented by concepts of 'relative space'. Hence the assertion that Langran (1991, p. 28) makes is essentially self-limiting:

'It is useful and interesting to examine the thoughts of philosophers concerning the nature of time and reality ... the role of cartography is neither to discover nor to represent all of reality'.

It is axiomatic that environmental scientists aim to discover and represent the reality of processes and patterns in the landscape. To do this implies the use of appropriate forms of spatio-temporal representation.

3.3 Spatio-temporal concepts used in GIS

Unwin's claim about the lack of geographical input into concepts of space and time (Unwin 1993) is certainly borne out if the representations used by GIS for referencing phenomena are examined. These are mostly based upon 'absolute space' concepts where spatial relationships are inferred through two dimensional geometrical coincidences. Since spatio-temporal phenomena cannot be explicitly handled within most current GIS, various recent attempts have been made to extend and redesign GIS for this purpose.

One form of extension to current GIS designed to handle temporal change is based upon the use of multiple layers to represent 'time slices'. Livingstone and Raper (1994) examined a cross-section of recent environmental models (Farmer and Rycroft 1991) and noted that most dynamic phenomena were represented using sequences of arrays which were manipulated in raster GIS. Another form of extension is based on the projection of the spatio-temporal data into a multidimensional data space which can be indexed or partitioned in a relational database. Four alternative theoretical schemes for spatio-temporal data storage and access were qualitatively evaluated by Langran (1991). These range from object clipping approaches (which expand the number of objects unacceptably), to approaches using irregular partitions based on R-tree indexing (which reduce the efficiency of searches due to the overlapping partitions).

The redesign of GIS to represent spatio-temporal behaviour has taken various forms. One strategy has been to extend the two-dimensional raster approach to represent phenomena in space-time. O'Conaill *et al.* (1993) extended linear quadtrees to store the data in the third and fourth dimension. They adhere to the view that unified space-time representations are necessary stating that:

> '... there is no need to hardwire the human psychological view of a separate space and time into the data structure' (p. 106).

However, this approach is limited to raster data and a more general, less data-dependent approach is required for the study of environmental problems. Another strategy proposed by Peuquet and Wentz (1994) in the TEMPEST project, is to develop a temporal database where all changes to the stored spatial entities are referenced to a set of unequal ordered temporal intervals. Frank (1994) generalizes this approach, presenting a design accommodating both fully- and partially-ordered temporal intervals where the intervals need not be described in ratio scale units. A further strategy involves the use of logic-based languages to model the spatio-temporal behaviour of environmental phenomena, such as the approach of Smith (1992).

Most other approaches to the redesign of GIS to handle spatio-temporal phenomena have employed object-oriented techniques. The key design issue in object-oriented approaches is how to structure the object classes and object attributes to handle temporal change, which in turn depends on the concept of time that is to be implemented. In the 'absolute space' approach objects are made in space and time, and are bounded by events. 'Events' are defined as 'instants in time when objects are changed'. Wachowicz and Healey (1994) suggest a design in which 'events affecting objects' create 'versioned objects' such that temporally different versions of the same object can exist.

In the 'relative space' approach time is a property of the objects: time is made of objects with a spatial and temporal extent. In most 'relative space' systems using object-oriented techniques the 'objects' are geometric primitives which are time stamped. Worboys (1994) reviews several potential designs ranging from the time-stamping of composite geometry down to the time-stamping of each point. Kemp and Kowalczyk (1994) propose a design using the Zenith object management system to create a spatio-temporal data store which allows 'has-version' relationships for stored objects composed of composite geometry. They also point out that it is possible to divide attributes between levels in the object class hierarchy such that the time invariant attributes are stored at a higher level than the time variant attributes. Ramachandran *et al.* (1994) propose a design called TCObject in which objects with geometric and non-geometric attributes are given past, present and future states: the temporal reference is established using dates of birth and death for the object. In a similar approach, Hamre (1994) proposes a design based upon OMT (Rumbaugh *et al.* 1991) where a 'four dimensional dataset' object is composed of a spatio-temporal component (including a 'time of creation' temporal reference) and a non-spatio-temporal component.

It is argued here that systems based upon four dimensional 'relative space' concepts meet the requirement that representations of environmental systems should be physically justifiable in spatio-temporal terms. Development of systems based upon this design would bring considerable benefits such as the ability to execute queries of the relationships between four dimensional space-time phenomena, spatio-temporal interpolation and the facility to extract instances of phenomena from data expressed in four dimensions.

4 The organising of geomorphological concepts

The foregoing review has looked at the concepts available in general science to motivate the identification and behaviour of environmental phenomena, the methods by which such phenomena can be described and the spatio-temporal context in which they can be embedded. These perspectives and techniques offer a framework within which an environmental spatial representation can be created. However, any new representation must be created with respect to the current state of disciplinary knowledge. This section reviews the key concepts of relevance to coastal geomorphology to provide a background to the example discussed in the next section, and to illustrate the theoretical concerns which have motivated this new approach to the design of spatial representations.

Concepts used in the formation of theories in coastal geomorphology can be divided into those relating to object identification and those relating to object behaviour. Typically, object identification is concerned with the classification of coastal forms and materials, whilst object behaviour is concerned with coastal processes (primarily driven by wave, tide and wind energy). Formulation of models to express coastal theories has progressed rapidly from object identification and classification to object behaviour over the last 20 years.

A variety of simple classifications of coastal forms have been suggested including: the classic work of Johnson (1919) who divided coasts into submerged and emerged types; Bloom (1978) who suggested a subdivision of coasts into bold and low coasts; classifications of the coast based on storm- or swell-dominated wave energy regimes such as by Davies (1980); and the classification of coasts into wave-or tide-dominated environments (Carter 1988). Coastline mapping initiatives have also attempted classification according to sedimentary environments and topographic elements (for example Pearson *et al.* 1990). However, coastal mapping presents particular definitional problems as originally noted by Richardson (1961) who showed that there is a predictable relationship between the length of the coast and the scale at which the determination is made. This study has focused on low coasts consisting of storm- and wave-dominated environments. Form identification approaches rely on methods to represent the terrains being studied: these are either based on sampling or terrain interpretation.

Qualitative syntheses of basic coastal morphological observations have been carried out to develop genetic classifications. Several schemes exist for barrier island coastlines such as that by Hayes (1979) who divides coasts into micro-, meso- and macro-tidal environments. Hayes characterizes barrier island shape as similar to a 'hot dog' in microtidal environments where there is frequent washover occurring and similar to a 'drumstick' in mesotidal environments where washover is much less common. Oertel (1985) also synthesized the wide range of published descriptions of barrier islands into a general scheme stating that:

> 'the barrier island system is composed of six major coastal environments. The six environments are also required elements needed to impose the designation "barrier island" to a littoral sand body' (p. 3).

The key concept in the Oertel scheme is the 'environmental' which integrates processes, form and materials within spatially defined zones. These zones are defined over a complex suite of characteristics which are temporally variable and have boundaries which are not rigorously defined — no one factor is consistently used.

These kinds of schemes are really only useful as a set of concepts — assertions that certain features of an environment exist — and pose considerable difficulties for implementation in a GIS.

Classifications of barrier islands have also been carried out using statistical characterization. For example Williams and Leatherman (1993) assert that qualitative classifications of barrier islands suffer from the use of too few variables in the synthesis. Using six landform variables and four process variables Williams and Leatherman carried out a factor analysis for sets of variables for 120 barrier islands, testing the results with a canonical correlation. Five classes of island emerged from this process: widest; narrowest; longest; shortest; outliers. This approach is weakened by the subjectivity of variable quantification and inter-dependence among the variables.

A key conclusion to draw from this survey of coastal entity identification and classification research is that geomorphologists have long been searching for appropriate metrics to distinguish features of genetic importance in the environments under study. The lack of progress towards explanation in this area may be a function of the atemporal frame of analysis i.e., forms are being associated together at different points in evolutionary sequences or transient situations. This may be a consequence of the lack of available representational forms for spatio-temporal form observing and analysis.

Studies of coastal behaviour have dominated in recent years. Many coastal models use similar organizing concepts to those in terrestrial environments. Hence, the concept of the cycle has been applied in coastal studies to the impact of tidal cycles on landforms and processes. Oost *et al.* (in press) have suggested that the 18·6 lunar nodal cycle which controls tidal amplitudes has an impact on tidal stream velocities and therefore on the balance between wave and tide energy at tidal inlets. Models based on the concept of a cycle can, however, be criticized for their time-dependent nature. An alternative framework was provided by Hack (1960), who suggested that forms, physical properties of materials and processes were mutually adjusted in a state of dynamic equilibrium. This approach challenged the idea that spatial variations in form were the result of evolution and suggested that spatial (and temporal) variations in energy were more important. This idea has been adopted widely in geomorphological analysis through the development of process-form models.

The concept of an equilibrium coastal landform was used by Tanner (1974) to describe the balance between wave energy and coastal sediment transport. The equilibrium coastal landform concept was elaborated by the 'Australian School' of coastal geomorphologists who developed the 'beach stage' model of shoreline morphodynamics (Wright *et al.* 1979). The beach stage model identified a set of shoreline profile- and plan-form states which are associated with a set of wave and tide conditions i.e., a classic process-form model. By showing how forms reflected process regimes and how these forms oscillated between equilibrium states, this work helped to add a temporal dimension to coastal behaviour studies.

A further organizing principle in geomorphology widely used at the coast is the concept of the system adapted from the General Systems Theory (von Bertalanffy 1962). Chorley (1967) introduced the systems concept into geomorphology defining systems as components and relationships formed into a functional structure, noting that they were generally created from 'experience' or prior work. Schumm (1979) placed the study of systems in a more complete geomorphological context showing how internal system mechanisms such as thresholds and feedbacks would result in system behaviour such as complex response, dynamic equilibrium and variable periods of relaxation time for perturbed forms. Implementing models using these concepts explicitly requires sophisticated spatio-temporal concepts. Brunsden and Thornes (1979) set the study of geomorphological process-form relationships in a system context in their study of 'landscape sensitivity'. Their central contention was that characteristic landforms will develop where impulses of change arrive less frequently than the time taken for adjustment to them, when this occurs the system can develop spatially and temporally complex outputs.

Systems have been used in coastal behaviour studies as an organizing principle through the coastal cell concept. Coastal cells are transient zones of characteristic forms, processes and materials found in the nearshore zone. Formally, a coastal cell can be defined as a zone of convergent incident waves bounded on either side by points where longshore wave energy power (P_L) is at a local minimum (Carter 1988). The convergence of incident waves in this way causes systematic variations in values of P_L from a local minimum at the cell boundaries to a local maximum at the centre of the cell. Since wave crest arrivals is almost always oblique the longshore variation in P_L is using asymmetric, either side of the cell centre. Moving through the cell from the direction of the oblique wave arrival, the rate of change in P_L with distance (x) alongshore ($\delta P_L/\delta x$) reaches a maximum positive value at the point where the P_L gradient from the cell boundary towards the centre is at a maximum. Conversely $\delta P_L/\delta x$ reaches a minimum negative value where the P_L gradient from the cell centre towards the boundary is at a maximum.

May and Tanner (1973) formalized the coastal cell concept by identifying five internal points within the coastal cell as follows:

a	cell boundary	P_L = local minimum (on the side of the cell facing oblique wave arrival)
b	P_L rising	$\delta P_L/\delta x$ positive and at a local maximum
c	cell centre	P_L = local maximum
d	P_L falling	$\delta P_L/\delta x$ negative and at a local minimum
e	cell boundary	P_L = local minimum

This formalization has become known as the May and Tanner 'a, b, c …' model and is used in the worked example discussed in the final section. The 'a, b, c …' model predicts that erosion occurs from points 'a' to 'c' and deposition from points 'c' to 'e' as a consequence of the uneven longshore distribution of wave energy. Tanner and Spicola (1985) applied the 'a, b, c …' model to barrier island development showing that if the centre of the island is considered to be equivalent to point 'a' and the spits to points 'e' in two different cells, then over time the two points 'c' will migrate away from the central point 'a' leading to the erosion of the centre of the island and to accumulation on the spits. This model can be used to explain landform development on barrier islands.

However, Terwindt and Wijnberg (1991) consider that the fundamental research problem in this field is now one of identifying the appropriate scales for each problem of interest. Coastal researchers have begun to address spatio-temporal methodological questions, as it has become apparent that most classifications and models have limited persistence or scaleability in time and space. Initially research focused on fixed scales such as small, meso and large scales (de Vriend 1991). However, work has now begun to identify the appropriate spatio-temporal context within which coastal behaviour models are conceived. For example, Terwindt and Kroon (1993) use the concept of scale as an organizing principle:

> 'The selection of the scale of interest in a certain coastal problem is arbitrary. However, once selected, this scale establishes the scale of the morphodynamic entities…' (p. 193). 'There is no general spatial and temporal scale of the entities' (p. 194). '…every defined scale has its own parameterization' (p. 194).

These statements illustrate how the focus of any coastal (or environmental) problem must itself inevitably define the spatio-temporal framework for the associated representations. However, Terwindt and Kroon note that forming relationships between entities defined at different scales is problematic, whether carried out by including all the variation at the lower level in the entities at the higher level, or by the creation of new aggregate entities with new attributes. Capobianco *et al.* (1993) opt to summarize lower level variation through scaling up operations employing parameter filtering and to use more qualitative concepts (e.g., sediment diffusion) in the higher level model.

Wijnberg and Terwindt (1993) develop new spatio-temporal frames of reference by attempting to identify the dynamic boundaries of coastal phenomena (such as an offshore bar configuration) within spatio-temporal datasets. For a 27-year Dutch coastal profile dataset they observe gaps between offshore bars ('windows') moving at consistent rates through space and time. This observed behaviour indicates that the coastal system in this area is substantially conditioned by this process and that any process model would have to be linked to a representation capable of reflecting these moving, changing phenomena. Other researchers exploring spatio-temporal behaviour of coastal systems have focused on the dependence of behaviour on the Markovian inheritance of previous system states and the importance of lagged responses from forcing processes (Cowell *et al.* 1993).

A key theoretical development which has proved important in understanding such behaviour in systems is chaos theory. Phillips (1992) defines deterministic chaos as 'complex, random-like behaviour arising from the dynamics of deterministic non-linear systems' (p. 365). Chaotic systems exhibit a sensitive dependence on initial system conditions so that with alternative starting points the system may develop in radically different ways. The significance of chaos in geomorphic system is that apparently random spatio-temporal patterns may result from the behaviour of chaotic systems. Culling (1987) suggested that the existence of chaotic behaviour may make equifinality (where initially different forms converge towards a single final form) unlikely. An analysis of chaotic behaviour shows how patterns and trajectories within an *n* dimensional system phase space may be attracted towards certain configurations ('strange attractors'). Searching for chaotic behaviour by looking for system divergence from a common start point is a technique used by geomorphologists such as Slingerland (1981) and Phillips (1992).

Carter and Orford (1991) suggested that coastal forms may oscillate between various equilibrium states under the control of forcing processes, internal and external to the system. Using chaos theory, they suggested that these equilibrium states act as 'attractors' within the geomorphic system. Since the equilibrium states identified for gravel barriers in their study were associated with the level of coastal cell development, the trajectories of the forms towards the equilibrium states can be seen as a type of system-defined space-time organization. Orford *et al.* (1993) attempted to identify the time domains associated with such trajectories by studying the periods of stability and instability of a gravel barrier in Nova Scotia. Since air photos only offer periodic evidence of movement, an analysis of tide gauge evidence for the area (detrended for sea level rise) was carried out to identify the periods of forcing by tidal surge events. This analysis identified 'forcing domains' of 6 and 22 years which were correlated to periods of barrier stability and instability determined from air photos. These 'forcing domains' are in effect spatio-temporal frames of reference, defined explicitly by reference to geomorphological behaviour.

Geomorphologists studying localized behaviour formulated as mathematical models (usually in computer form) are beginning to look at the spatio-temporal framework in which the model is embedded. New studies aimed at the identification of forcing domains in space and time have been carried out. However, this kind of work would benefit from usable, spatio-temporal representations developed within a GIS context to make it possible to extend the range of representations and operations.

5 Spatial modelling in geomorphology

The discussion of (coastal) geomorphological theories in this paper has revealed a developing focus on the spatio-temporal framework in which geomorphological forms and their behaviour are understood. However, many modern GIS use static geometry as their primary index and do not appear to have the representational structure or tools to store and manipulate complex and dynamic geomorphologies. This paper sets out a new approach to spatial representation called 'OOgeomorph' which allows geomorphological models to be constructed and populated with spatio-temporal data so that observed forms and behaviour might be compared to theory.

5.1 Design principles for spatial modelling in geomorphology

Working from the previous discussion of object identification, model development, spatio-temporal referencing and geomorphological knowledge, the structure of 'OOgeomorph' is described below. To formalize the thinking behind 'OOgeomorph' a series of requirements answered by design principles are presented in order to show how the representation was developed. These statements represent the authors' view of the key steps needed to create a generic geomorphological model embedded within spatial referencing. Clearly, since this process was motivated by the study of coastal geomorphological environments alternative views of the appropriate requirements for a generic geomorphological approach are possible: at present few others, if any, have been presented.

Requirement 1

To create a testable theory about a geomorphological domain requires the discretization of explicit relationships between formalized phenomena.

Design principle 1

Relationships and phenomena forming conceptual models of geomorphological domains must be consistent with geomorphological meta-theories such as the concepts of 'cycles' or 'systems'.

Requirement 2

To discretize geomorphological phenomena requires the use of heuristics provided by the relevant meta-theory to determine the 'basic level categories' in the geomorphological knowledge domain.

Design principle 2

If using the geomorphological 'system' meta-theory the 'basic level categories' can be considered to be the processes, forms and materials found in the domain under study.

Requirement 3

To be consistent with the conjectural nature of geomorphological knowledge, the discretization of relationships between phenomena requires definitions referenced to attributes of processes, forms and materials.

Design principle 3

The representation of a phenomenon must be defined by the aggregate set of values of process, form and material attributes.

Requirement 4

To represent complex and rapidly changing geomorphological phenomena requires a physical model of space-time which is continuous and can handle any number of overlapping phenomena.

Design principle 4

The phenomena of interest must be discretized such that each representation has its own four dimensional spatio-temporal referencing and is set in a 'world' containing associated four dimensionally referenced phenomena.

Requirement 5

To permit the spatial modelling of four dimensionally referenced aggregates of process, form and material attributes requires that behaviour be incorporated into the representation of the phenomenon.

Design principle 5

The four dimensionally referenced aggregates of process, form and material attributes must have associated operations characterizing the geomorphological behaviour of the phenomenon and its relationships to other phenomena.

5.2 Using the Booch system of object-oriented design to develop OOgeomorph

Based upon the set of design principles above, a generic geomorphological modelling framework called OOgeomorph is now presented. It is constructed using the toolset for object oriented design defined by Booch (1991). Although there is a range of other approaches to object-oriented design such as Object Modelling Technique (Rumbaugh *et al.* 1991); Coad and Yourdon (1990); Information Engineering/Ptech (Martin and Odell 1992); and Objectory (Jacobsen 1992), the Booch approach has rich semantics for relationships between classes and good diagramming tools.

The basic concept of the Booch methodology and other object-oriented design approaches are those of 'class' and 'object'. A *class* represents 'types of things' and a class structure indicates 'typologic relationships between things', whereas *objects* represent 'instances of things' and an object structure indicates 'interaction relationships'. Taking an example from coastal geomorphology, a salt marsh creek is a type of thing and therefore a class — it is also a type of channel (as is a river), and as such the relationship between channel and salt marsh creek constitutes an 'is_a' inheritance relationship between classes. However, 'Big creek' is a particular (imaginary) instance of a salt marsh creek and therefore an object; it drains 'Big Marsh' (which is an instance of a marsh) and as such this is a 'using' relationship indicating object-object interaction.

In the Booch system it is possible to assign attributes to both classes and objects, although class attributes apply to all objects that are instances of the class whereas object attributes only apply to that particular instance of the class. A further concept that is introduced by Booch is that of a metaclass: a metaclass is a class that is used to define classes and is best defined as a class whose instances are themselves classes. Metaclasses can be created by a CASE (computer aided software engineering) tool module. It is intended that the implemented system will have a user interface accessible to users unfamiliar with GIS, databases or object-oriented systems.

5.3 The design of OOgeomorph

OOgeomorph has been developed to enable a geomorphologist to formulate and test theories expressed within the framework of a spatial representation. The design of OOgeomorph embraces four key concepts:

(1) *The separation of the geomorphological representation from the structure of the stored data.* The system design (Figure 11.1) envisages a GIS or other spatial database referred to as the Geomorphological Spatial Database (GSD) which is separate from, but communicating with, the geomorphological representation. The GSD could be any form of spatial database ranging in sophistication from ASCII

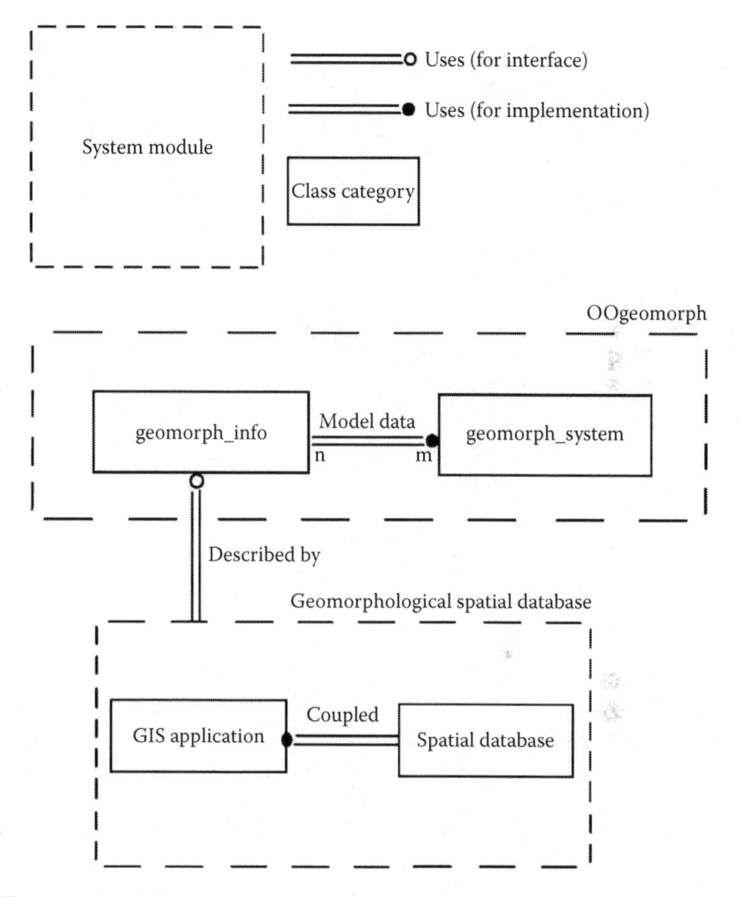

FIGURE 11.1 Overview of the system design.

files of x, y, z coordinates to a fully topologically structured GIS. This means that the low-level data structures of the stored data do not exert any influence on the structure of the geomorphological representation. This approach is consistent with the 'layered' philosophy of the Universal Geographic Information eXecutive (UGIX) system design proposed by Raper and Bundock (1993).

The geomorphological representation is formed within the OOgeomorph environment. Two distinct class structures are envisaged within OOgeomorph: first, a class structure called 'geomorph_system' to store the geomorphological representation; and secondly, a class structure called 'geomorph_info' is defined so that relevant data stored in the GSD can be matched to the geomorphological representation. The classes in 'geomorph_info' include metadata as attributes to describe the associated data and encapsulate the translation mechanisms required to extract the data from the GSD. The 'geomorph_info' classes are broken down according to the form of the data rather than the form of the geomorphological representation.

(2) *The use of geomorphological concepts to organize the structure of the representation.* Within OOgeomorph the 'geomorph_system' class structure is formulated

with reference to the design principles set out above. Since the design principles are flexible, the development of a CASE tool to implement a class structure consistent with different principles is envisaged. In the design presented here the use of concepts from the 'systems' meta-theory (design principle 1), and the assertion that the 'basic level categories' in geomorphology can be considered to be processes, forms and materials (design principle 2) are used to organize the overall structure. This does not cover all possibilities: for instance, it does not recognize a biocommunity category or a cultural category as semantic concepts at a similar organizing level to processes, forms and materials. Such concepts can be added to the structure or incorporated at a lower level. For example, the effects of colonization of communities on geomorphological phenomena may be incorporated as processes.

In the design of OOgeomorph the class 'geomorph_system' has attributes describing:

- The nature of the geomorphological system to be modelled (such as name and theories used).
- The names of the underlying CASE tools available to construct specific class structures for the processes, forms and materials recognized in the theory used.

The CASE tools developed for OOgeomorph are called mprocess, mform and mmaterial (Figure 11.2). These can be used by a geomorphologist to create specific sets of 'process', 'form' and 'material' classes relevant to the investigation or environment. Theories or conventions can be used to determine which attributes are relevant and the range of values allowed for each: OOgeomorph allows this procedure to be experimental and iterative by allowing multiple sets of the classes to be created by mprocess, mform and mmaterial.

To permit the manipulation of integrated geomorphological phenomena the collected attributes of these process, form and material classes are aggregated to create a sub-class called 'phenomenon'. This approach ensures that geomorphological phenomena are not forced into a taxonomic hierarchy (as in some GIS designs) based upon geometric descriptions of their state as a particular point in space-time, but rather that they are assembled from their process, form and material expressions. The 'phenomenon' sub-class is used to implement design principle 3, i.e., the representation of a phenomenon must be defined by the aggregate set of values or process, form and material attributes.

(3) *The assignment of four dimensional space-time referencing to geomorphological attributes.* To be consistent with design principle 4 concerning space-time referencing, all the attributes of the classes process, form and material should have a four dimensional spatial reference as required by the theory in use. Hence, the attributes of process, form and material at the first level will be referenced by a 'time of knowing' and a one, two or three dimensional geometric description: these are 'attributes of attributes'. To achieve this, all of the 'attributes of attributes' are implemented as classes in their own right at the second level and these 'classes implementing process, form or material attributes' are aggregated together under

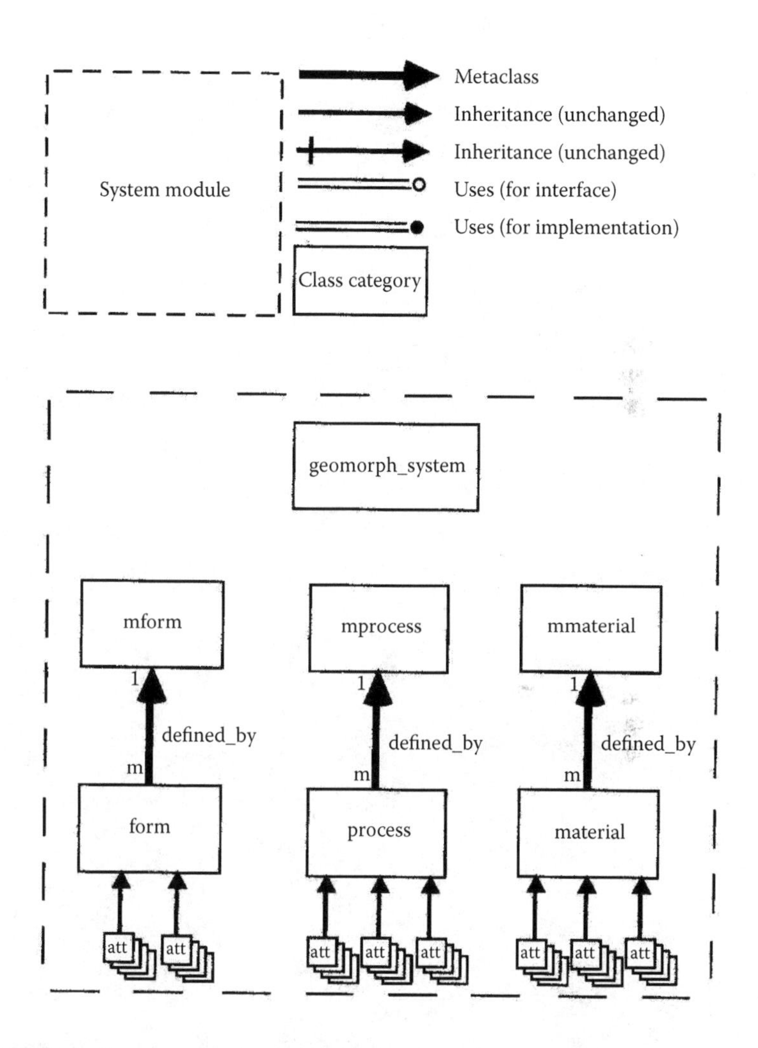

FIGURE 11.2 The class structure created below 'geomorph_system'.

each parent process, form or material attribute. This means that the 'phenomenon' sub-class stands at the third level down the hierarchy (Figure 11.3) and is created by multiple inheritance from all the 'classes implementing process, form or material attributes' at the second level.

The concept of time implemented in OOgeomorph equates to the concept of 'valid time' considered by Worboys (1994) and others. The 'time of knowing' is a point in time (recorded as a clock time and date) when each attribute of process, form and material can be known. Note that the phenomenon sub-class will inherit all the 'classes implementing process, form or material attributes' from the second level, each with its own space-time reference which may be disjunct, overlapping or identical.

In a typical scenario the mform, mprocess and mmaterial CASE tools may create classes with (say) *three* process attributes, *two* form attributes and *three* material

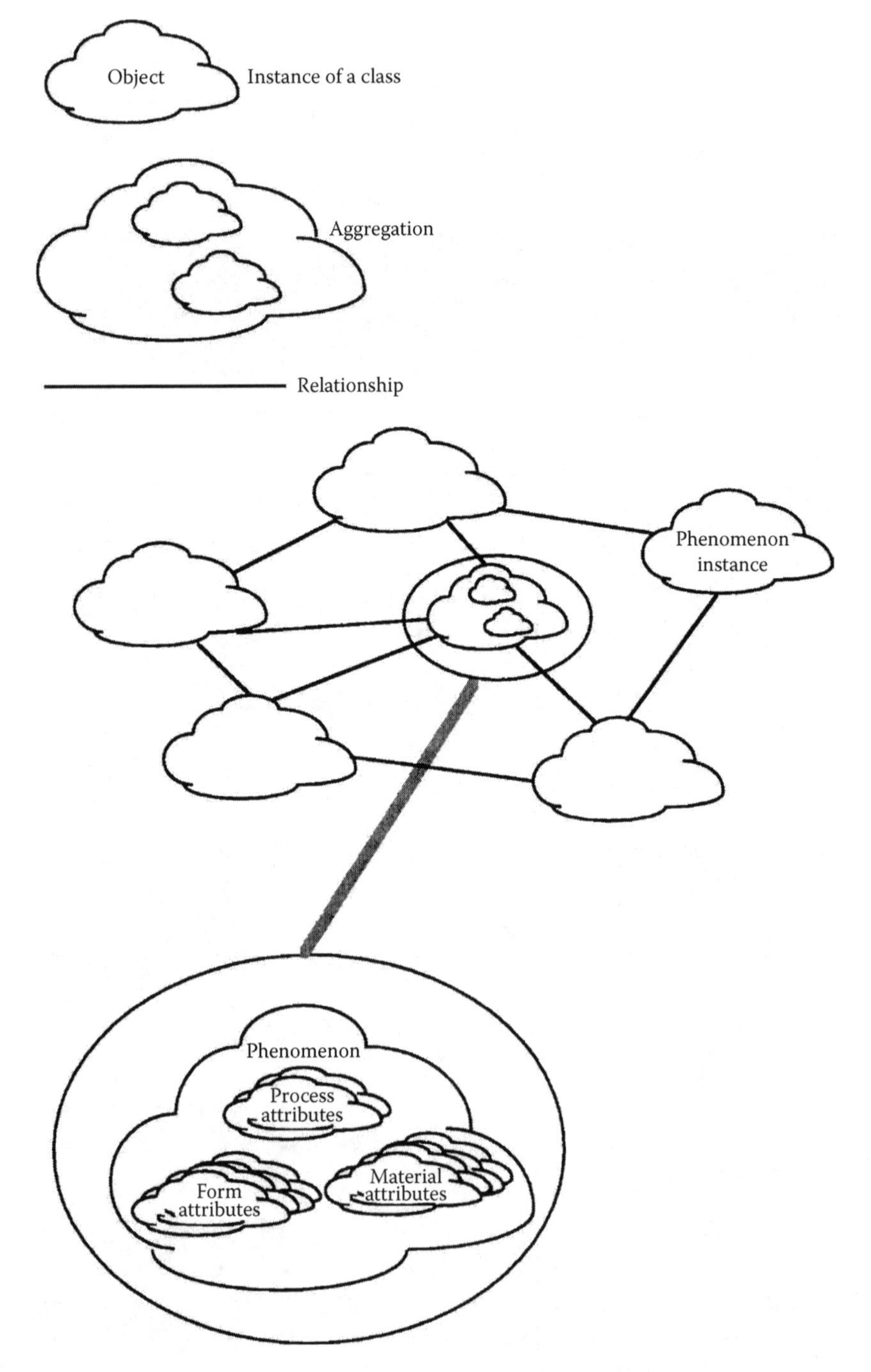

FIGURE 11.3 Generic class structure for a geomorphological phenomenon.

attributes (i.e., eight in all). If these attributes themselves have a reference identifier, a value, a 'time of knowing' and a geometric description, then there will be a total of 32 'classes implementing process, form or material attributes' required to implement this approach and these will be inherited by the phenomenon class. If the geomorphologist now populates this class structure with data by linking 'geomorph_info' to 'geomorph_system' the phenomenon sub-class will be filled with instances. Each of these can be referred to as a phenomenon instance (PI).

The spatio-temporal structure of each PI and the set of PI's in the phenomenon class is entirely in the hands of the geomorphologist, since each of the eight attributes of the process, form and material sub-classes created in the example given can have a different spatio-temporal extent: in fact this may be required by limitations in the observations available or possible. This design makes it likely that the set of PI's stored will be spatially and temporally heterogeneous ranging from highly observation-dependent forms ('over this space at this time' attributes) to infinite steady forms ('always, everywhere' attributes). An implication of this design is that PI's of widely differing 'scales' can be created and compared rendering the concept much less important. The key control over scale is, therefore, the granularity of the theory governing selection of process, form and material attributes.

It is an explicit design aim for OOgeomorph that such complex expressions be facilitated as it permits the storage together of geomorphological phenomena considered comparable as a working hypothesis, though they may be both spatially and temporally disjunct and have slightly different values over the process, form or material attributes. Any differences can then be examined using tools created to compare the PI's under objective conditions.

(4) *The assignment of behaviour expressed as mathematical models to geomorphological phenomena.* Given that the contents of the phenomenon class are likely to be heterogeneous in nature, it is necessary to implement a set of tools to operate on PI's. Such tools can be implemented as operations 'encapsulated' with the phenomenon sub-class. Generic operations would include:

- Classification operations to form subsets of the PI's based on objective classifiers.
- Generation of new PI's by substituting attribute values according to mathematical models.
- Temporal and/or spatial projection to harmonize the space-time bounds of the attributes.
- Temporal and/or spatial generalization when PI's vary slowly over space and time.

The operations can be implemented in a fourth generation language supported by the system or in a standard language such as C.

The OOgeomorph design presented above offers geomorphologists the opportunity to define rich representations which embrace both what have been termed 'data models' (defining fixed associations between data storage loci) and process models (defining the behaviour of mathematical models). Using OOgeomorph a representation can be created so that it fits with the observations available or so that it helps formalize the observations which would be required for the test of a process model. The system need not be difficult to use since the geomorphologist can use a customized user interface and need only:

(1) Identify the attributes of the classes of process, form and materials required.
(2) Specify the domains of the attributes and the appropriate space-time reference.
(3) Attach operations to the phenomenon class.
(4) Populate the system with data by importing data via 'geomorph_info'.

The geomorphologist can refine the representation by generating new class structures using mform, mprocess and mmaterial.

6 Using OOgeomorph in a coastal geomorphological study

As an illustration of how OOgeomorph can be used and to demonstrate the advantages of this approach over that of a conventional GIS, a short example of the design of an OOgeomorph class structure with its associated operations is given below. The class structure is formulated with reference to the May and Tanner (1973) theory of coastal cell development as outlined above. It has been created to test some of the hypotheses about the spatio-temporal development of coastal cells which Carter (1988), advanced, in this case with specific reference to medium term coastal development.

The procedure for formulating a spatial model for the coastal cell theory using OOgeomorph should begin with an examination of the concepts used in the theory. In keeping with design principles 1–3 above these concepts should be assigned to form, process and material categories — this is attempted in Table 11.2 (no associations along rows are intended).

To formulate these concepts within OOgeomorph requires that they each be formalized as attributes consisting of values or types with discrete four dimensional spatio-temporal referencing. The key issue in this procedure is the granularity of the representation, i.e., at what time and over what space can these attributes be expressed. In most cases the theory should provide a definition of the granularity of representation such that the temporal and spatial domains can be identified.

TABLE 11.2
Concepts used in the coastal cell theory as formulated by Carter (1988).

Form concepts	Process concepts	Material concepts
Internal points	Wave energy (E)	Unconsolidated sediment
Erosion	Wave crest approach angle (a)	
Deposition	Longshore wave energy power (P_L)	
High spring tide level	Sediment discharge (Q_S)	
Low spring tide level		

TABLE 11.3
Form, process and material attributes for the coastal cell representation in OOgeomorph.

Form attributes	Process attributes	Material attributes
Candidate point	Wave crest approach angle (a)	Sediment type
High spring tide level	Longshore wave energy power (P_L)	
Low spring tide level	Sediment discharge (Q_S)	

In this case the May and Tanner Theory provides concepts of granularity in the form of the 'internal points' of the coastal cell, which themselves represent minima and maxima of longshore wave energy power flux within the cell. To implement this representation it was decided to use measurable shoreline cross-sections at-a-time to define the granularity of the attributes. Shoreline cross-sections are normal to wave run-up and tidal fluxes and therefore offer versatile 'candidates' for the internal points. They also offer appropriate ways to discretize the other concepts in the theory such as sediment flux, angle of approach for wave crests and material properties. The concepts in Table 11.2 can be refined into the form, process and material attributes shown in Table 11.3.

The various characteristics of the attributes in Table 11.3 are implemented as a set of 'classes implementing process, form or material attributes' each with the following internal form:

Reference ID	{Char* }	a code identifying the particular project
Value of attribute	{Char* }	an alphanumeric value for the attribute
Time-of-knowing	{Char* }	clock time and date
Location	{Geom* }	a series of points defining a shoreline cross-section

When the four 'classes implementing process, form or material attributes' associated with each of the seven form, process and material attributes are aggregated together, the phenomenon class created will have 28 'attributes'. Any redundancy that this design creates, for instance multiple occurrences of the same geometry, is eliminated at the physical level of storage.

The other key aspect of the OOgeomorph design is the association of behaviour statements with the phenomenon class. In this example the operations implementing the behaviour would include:

- An operation to determine erosion and deposition changes between measuring cycles.
  This operation determines changes in the position of the beach crest for each shoreline cross-section over time, determining whether erosion or deposition had taken place. Since the cross-sections may not always be taken in the same place the operation identifies nearest neighbour cross-sections in space over time in order to make the comparisons. This operation

creates a new attribute for the phenomenon class i.e., net shoreline change, which is instantiated by the operation.

- An operation to identify internal points 'a' to 'e'.
  Since it is difficult to determine longshore wave energy power (P_L) directly (Carter 1988), this operation uses secondary evidence to identify candidate points. Approaches include those based upon wave refraction modelling and beach elevation/sediment type variation alongshore. Candidate points can also be proposed from evidence of erosion and deposition recorded in the previously created net shoreline change attribute.

- Operation to define the boundaries of features such as spits.
  This operation allows the geomorphologist to define the boundaries of features such as spits by using morphometric analysis of slope angles and other attribute values. The geometry resulting from such an operation is stored with the phenomenon class and allows vectors of movement to be computed between sets of shoreline cross sections measured at different times.

- Operations to determine the frequency of high tides in different elevational ranges.
  Given the local elevations of the highest and lowest astronomical tides this operation uses tide tables to establish the water level at any clock time and to determine the frequency with which high and low tides affect any form, limit the action of any process or cover any material.

The structuring created by OOgeomorph makes it possible to execute four dimensional 'range queries' which look for space-time coincidences. In studying the evolution of coastal cells, typical questions which take the form of a range query include:

- Over what longshore distance does an internal point move on average during a predetermined period?
- Where do rates of movement for feature boundaries defined by an operation on the PI's reach their maximum and when?
- When do the candidates for internal points differ most in longshore terms when defined by differing criteria?

All the above operations and queries can be carried out on any set of phenomenon instances defined in the study zone. The existence of this kind of system will make it possible to carry out such queries on many different phenomenon instance sets defined using different tools. The authors have been collecting the data required to instantiate the representation for the Scolt Head barrier island (in north Norfolk, England) at regular intervals over the last 5 years and are engaged in the implementation of this system.

7 Conclusions

What has been proposed in this paper reflects a major shift in thinking about the computer modelling of environmental phenomena. While the discussion is based on coastal geomorphological concepts it is hoped that the philosophical viewpoint and methodology set out in this paper will stimulate a debate about the nature of the spatial modelling systems required to solve environmental problems. Perhaps the research question posed at the start of the paper — how to integrate environmental models and GIS — is already outdated. It is argued here that the next step should be the fusion of models and spatial representations within new object-oriented environments and *not* the integration of incompatible systems which force representational compromises.

Acknowledgments

The authors gratefully acknowledge the permission of English Nature to carry out research in the National Nature Reserve at Scolt Head Island which stimulated much of this thinking and provided the example. The Wardens, Colin Campbell and Chris Everett offered much advice and help during the fieldwork. The authors are also indebted to Kim Styles, Melissa Currie, Paul Davies, Nick Fairfax, Geraldine Garner, Sian John, Simon Lewis, Clare Mellish, Tim McCarthy, Mark Smithard, Andrew Thompson and Paul Tobin for assistance with the arduous fieldwork. Birkbeck College and the Central Research Fund of the University of London gave financial support. The authors would like to thank Richard Aspinall, Charlie Bristow, Roy Haines-Young, Karen Kemp and David Unwin for stimulating conversations during the development of the ideas in this paper.

References

Bloom, A. L., 1978, *Geomorphology* (Englewood Cliffs, NJ: Prentice Hall).

Booch, G., 1991, *Objects-oriented design with applications* (Redwood City: Benjamin-Cummings).

Brunsden, D., and Thornes, J., 1979, Landscape sensitivity and change. Transactions of the Institute of British Geographers, **4**, 463–84.

Capobianco, M., de Vriend, H. J., Nicholls, R. J., and Stive, M. J. F., 1993, Behaviour-oriented models applied to long term profile evolution. *Proceedings of the Conference on Large Scale Coastal Behaviour*, St. Petersburg, FA, USA (Washington, DC: US Geological Survey) Open File Report 93–381, pp. 21–24.

Carter, R. W. G., 1988, Coastal Environments: an Introduction to the Physical Ecological and Cultural Systems of Coastlines (London: Academic Press).

Carter, R. W. G., and Orford, J., 1991, The sedimentary organisation and behaviour of drift-aligned barriers. *Proceedings of Coastal Sediments '91*, Seattle, WA, USA (Washington, DC: American Society of Civil Engineers), pp. 934–948.

Casti, J. L., 1989, *Alternate Realities: Mathematical Models of Nature and Man* (Chichester: John Wiley).

Chapman, G. P., 1977, *Human and Environmental Systems* (London: Academic Press).

Chorley, R. J., 1967, Models in geomorphology. In *Models in Geography*, edited by R. J. Chorley and P. Haggett (London: Methuen).

Coad, P., and Yourdon, E., 1990, *Object-Oriented Analysis* (Englewood Cliffs, NJ: Prentice Hall).

Cowell, P. J., Roy, P. S., and Jones, R. A., 1991, Simulation Modelling of Large Scale Coastal Behaviour. *Proceedings of Coastal Sediments '91*, Seattle, WA, USA (Washington, DC: American Society of Civil Engineers), pp. 41–44.

Culling, W. E. H., 1987, Equifinality: modern approaches to dynamical systems and their potential for geographical thought. *Transactions of the Institute of British Geographers*, **12**, 57–72.

Davis, J. L., 1980, *Geographical Variation in Coastal Development* (London: Longman).

de Vriend, H. J., 1991, G6 coastal morphodynamics, *Proceedings of Coastal Sediments '91*, Seattle, WA, USA (Washington, DC: American Society of Civil Engineers), pp. 356–370.

Dikau, R., 1990, Geomorphic landform modelling based on hierarchy theory. *Proceedings of the 4th International Symposium on Spatial Data Handling,* Zürich, Switzerland (Zürich: International Geographical Union), pp. 230–239.

Einstein, A., 1960, *Relativity: The Special and the General Theory* (London: Methuen).

Farmer, D. G., and Rycroft, M. J., 1991, *Computer Modelling in the Environmental Sciences*, (Oxford: Clarendon Press).

Feuchtwanger, M., 1989, Geographic logical database model requirements. *Proceedings of Auto-Carto 9*, Baltimore, MD, USA (Bethesda, MD: American Congress on Surveying and Mapping) pp. 599–609.

Frank, A. U., 1994, Qualitative temporal reasoning in GIS-ordered time scales. In *Advances in GIS Research. Proceedings, 6th International Symposium on Spatial Data Handling*, Edinburgh, UK, edited by T. C. Waugh and R. G. Healey (London: Taylor & Francis), pp. 410–30.

Frank, A., and Mark, D., 1991, Language issues for GIS. In *Geographic Information Systems: Principles and Applications*, edited by D. Maguire, M. Goodchild and D. Rhind (Harlow: Longman).

Goodchild, M. F., 1992, Geographical data modelling. *Computers and Geosciences*, **18**, 401–8.

Goodchild, M. F., Parks, B. O., and Steyaert, L. T., (eds), 1993, *Environmental Modelling with GIS* (Oxford; OUP).

Hack, J. T., 1960, Interpretation of erosional topography in humid tropical regions. *American Journal of Science*, **258-A**, 80–97.

Haines-Young, R., and Petch, J., 1986, *Physical Geography: its Nature and Methods* (London: Harper and Row).

Hamre, T., 1994, An object-oriented conceptual model for measured and derived data varying in 3D space and time. In *Advances in GIS Research. Proceedings, 6th International Symposium on Spatial Data Handling*, Edinburgh, UK, edited by T. C. Waugh and R. G. Healey (London: Taylor & Francis), pp. 868–81.

Hayes, M. O., 1979, Barrier island morphology as a function of tidal and wave regime. In *Barrier islands*, edited by S. P. Leatherman (New York: Academic Press), 1–29.

Hazelton, N. W. J., Leahy, F. J., and Williamson, I. P., 1990, On the design of a temporarally-referenced, 3D Geographical Information Systems: development of a four dimensional GIS. *Proceedings of GIS/LIS '90*, Anaheim, CA, USA (Bethesda, MD: American Congress on Surveying and Mapping), pp. 357–372.

Herring, J., 1991, The mathematical modelling of spatial and non-spatial information in Geographic Informations Systems, In *Cognitive and Linguistic Aspects of Geographic Space*, edited by D. M. Mark and A. U. Frank (Dordrecht: Kluwer), pp. 313–50.

Jacobsen, I., 1992, *Object-Oriented Software Engineering* (Workingham: Addison-Wesley).

Johnson, D. W., 1919, *Shore Processes and Shoreline Development* (New York: Wiley).

Johnson, M., 1987, *The Body in the Mind: the Bodily Basis of Meaning, Imagination and Reason* (Chicago: University of Chicago Press).

Jorgensen, S. E., 1990, Modelling concepts. In Modelling in Ecotoxicology. (*Developments in Environmental Modelling 16*), edited by S. E. Jorgensen (Amsterdam: Elsevier).

Kemp, K., 1993, *Environmental Modelling with GIS: a Strategy for Dealing with Spatial Continuity.* Technical report 93–3, National Center for Geographic Information and Analysis, Santa Barbara, USA.

Kemp, Z., and Kowalczyk, A., 1994, Incorporating the temporal dimension into a GIS. In *Innovations in GIS*, edited by M. Worboys (London: Taylor & Francis), pp. 89–103.

Kirkby, M., 1988, The future of modelling in physical geography. In *Remodelling Geography*, edited by W. MacMillan (Oxford: Basil Blackwell), pp. 255–272.

Lakoff, G., 1987, *Women, Fire and Dangerous Things* (Chicago: University of Chicago Press).

Langran, G., 1991, *Time in Geographical Information Systems* (London: Taylor & Francis).

Livingstone, D. E., and Raper, J. F., 1994, Modelling environmental systems with GIS: theoretical barriers to progress. In *Innovations in GIS*, edited by M. Warboys (London: Taylor & Francis), pp. 229–240.

Martin, J., and Odell, J., 1992, *Object-Oriented Analysis and Design* (Workingham: Addison Wesley).

May, J. P., and Tanner, W. F., 1973, The littoral power gradient and shoreline changes. In *Coastal Geomorphology*, edited by D. R. Coates (New York: Binghampton State University), pp. 50–99.

Nunes, J., 1991, Geographic space as a set of concrete geographic entities. In *Cognitive and Linguistic Aspects of Geographic Space*. NATO ASID 63, edited by D. M. Mark and A. U. Frank (Dordrecht: Kluwer), pp.9–34.

O'Conaill, M. A., Mason, D. C., and Bell, S. B. M., 1993, Spatiotemporal GIS techniques for environmental modelling. In *Geographical Information Handling — Research and Applications*, edited by P. M. Mather (Chichester: Wiley), pp. 103–112.

Oertel G. F., 1985, The barrier island system. *Marine Geology*, **63**, 1–18.

Oost, A. P., de Haas, H., Ijnsen, F., van den Boogert, J. M., and de Boer, P. L. (in press) The 18·6 year nodal cycle and its impact on tidal sedimentation. *Marine Geology.*

Orford, J. D. Carter, R. W. G., and McClosky, J., 1993, A method of establishing mesoscale (decadal to subdecadal) domains in coastal gravel barrier retreat rate from tide gauge analysis. *Proceedings of the Conference on Large Scale Coastal Behaviour*, St. Petersburg, FA, USA (Washington, DC: US Geological Survey) Open File Report 93–381, pp. 155–158.

Pearson, I., Funnell, B. M., and McCave, I. N., 1990, Sedimentary Environments of the Sandy Barrier Tidal Marsh Coastline on north Norfolk. Bulletin Geological Society of Norfolk, **39,** 3–44.

Peuquet, D., and Wentz, E., 1994, An approach for time-based analysis of spatio-temporal data. In *Advances in GIS research. Proceedings, 6th International Symposium on Spatial Data Handling*, Edinburgh, UK, edited by T. C. Waugh and R. G. Healey (London: Taylor & Francis), pp. 489–504.

Phillips, J. D., 1992, Qualitative chaos in geomorphic systems, with an example from wetland response to sea level rise. *Journal of Geology*, **100**, 365–74.

Popper, K., 1972, *The Logic of Scientific Discovery* (London: Hutchinson).

Putman, H., 1981, *Reason, Truth and History* (Cambridge: CUP).

Pylyshyn, Z., 1981, The imagery debate: analogue media versus tacit knowledge. *Psychological Review*, **87**, 16–45.

Ramachandran, B., MacLeod, F., and Dowers, S., 1994, Modelling temporal changes in a GIS using an object-oriented approach. In *Advances in GIS Research. Proceedings, 6th International Symposium on Spatial Data Handling*, Edinburgh, UK, edited by T. C. Waugh and R. G. Healey (London: Taylor & Francis), pp. 518–37.

Raper, J. F., 1991, Geographical Information Systems: progress report for 1990. *Progress in Physical Geography*, **15**, 108–114.

Raper, J. F., and Bundock, M. S., 1993, Development of a generic spatial language interface for GIS. In *Geographical Information Handling — Research and Applications*, edited by P. M. Mather (Chichester: Wiley), pp. 113–143.

Richardson, L. F., 1961, The problem of contiguity. General Systems yearbook, **6** 139–87.

Rosch, E., 1973, On the internal structure of perceptual and semantic categories. In *Cognitive Development and the Acquisition of Language*, edited by T. E. Moore (New York: Academic), pp. 111–144.

Rumbaugh, J., Blaha, M., Premerlani, W. E., and Lorensen, W., 1991, *Object-Oriented Modelling and Design* (Engelwood Cliffs, NJ: Prentice Hall).

Schumm, S., 1979, Geomorphic thresholds; the concept and its applications. *Transactions, Institute of British Geographers*, **4**, 485–515.

Slingerland, R., 1981, Qualitative stability analysis of geologic systems with an example from river hydraulic geometry. *Geology*, **9**, 491–93.

Smith, T. R., 1992, Towards a logic-based language for modelling and database supporting spatio-temporal domains. *Proceedings, 5th International Symposium on Spatial Data Handling, Charleston*, SC, USA (Charleston: International Geographical Union), pp. 592–601.

Tanner, W. F., 1974, Advances in near-shore physical sedimentology: a selective review. *Shore and Beach*, **42**, 1–125.

Tanner, W. F., and Spicola, J. J., 1985, The asymmetrical 'a-b-c …' model. In *Proceedings of Iceland Coastal and River Symposium*, Reykjavik, Iceland, edited by A. Sigbjarnson (Reykjavik: University of Iceland), pp. 369–87.

Terwindt, J. H. J., and Kroon, A., 1993, Theoretical concepts of parameterisation of coastal behaviour. *Proceedings of the Conference on Large Scale Coastal Behaviour*, St. Petersburg, FA, USA (Washington, DC: US Geological Survey) Open File Report 93–381, pp. 193–95.

Terwindt, J. H. J., and Wijnberg, K. M., 1991, Thoughts on large scale coastal behaviour. *Proceedings of Coastal Sediments '91*, Seattle, WA, USA (Washington, DC: American Society of Civil Engineers), pp. 1476–87.

Thorn, C. E., 1988, *Introduction to Theoretical Geomorphology* (London: Unwin Hyman).

Thornes, J., and Brunsden, D., 1977, *Geomorphology and Time* (New York: Wiley).

Unwin, T., 1993, *The Place of Geography* (London: Longman).

Unwin, D. U., 1988, Three questions about modelling is physical geography. In *Remodelling Geography*, edited by W. MacMillan (Oxford: Basil Blackwell), pp. 53–57.

von Bertalanffy, L., 1962, General system theory: a critical review. *General Systems*, **7**, 1–20.

Wachowicz, M., and Healey, R. G., 1994, Towards temporality in GIS. In *Innovations in GIS*, edited by M. Worboys (London: Taylor & Francis), pp. 105–115.

Wijnberg, K. M., and Terwindt, J. H. J., 1993, The analysis of coastal profiles for large scale coastal behaviour. *Proceedings of Large Scale Coastal Behaviour*, St. Petersburg, FA, USA, pp. 224–27.

Williams, A. T., and Leatherman, S. P., 1993, Process form relationships on USA east coast barrier islands. *Zeitschrift für Geomorphologie*, **37** (2), 179–197.

Worboys, M. F., 1994, Unifying the spatial and temporal components of geographical information. In *Advances in GIS Research*, edited by T. C. Waugh and R. G. Healey (London: Taylor & Francis), pp., 505–517.

Wright, L., Chappell, J, Thom, B., Bradshaw, M., and Cowell, P., 1979, Morphodynamics of reflective and dissipative beach and inshore systems, south Australia. *Marine Geology.*, **32**, 105–40.

Ziegler, B., 1976, *Theory of Modelling and Simulation* (New York: Wiley).

Development of a Geomorphological Spatial Model Using Object-Oriented Design: An Everyday Story of Space, Time, and GIS Folk

Jonathan F. Raper and David E. Livingstone

As it is more than 10 years since the first appearance of this paper, we now have the benefit of deep perspective on its authorship and its impact. In this commentary we want to explore both the influences on and practice of authorship, and to trace the (sometimes surprising) course of citation and implication.

In 1994 when the paper was written, Jonathan Raper was a lecturer in GIS and David Livingstone was a researcher at Birkbeck College in the University of London. At that time Birkbeck had a large research school in GIS established by the great influence and energy of David Rhind (although he had departed to the Ordnance Survey in 1992), and it played host to the ESRC-funded South East Regional Research Laboratory directed by John Shepherd. There were 15 to 20 staff members working on GIS and its applications, and the GIS bubble was at its height. At the time there was every opportunity for us to pursue application-oriented or technology-oriented GIS studies, and to direct them at the next GIS conference. With the software and hardware we had at our disposal (Arc/Info, Laser-Scan, and Smallworld on state-of-the-art Unix workstations), there was a temptation to produce these kinds of studies. In academic terms, this was the comfort zone — plenty of funds, rapid change in the field, and a local and global research community to spark ideas.

However, three things conspired to disrupt this pattern of publication for us. In retrospect, we were vulnerable to "disruption" by our respective intellectual heritages — David in astronomy and philosophy at St. Andrews, and Jonathan in geography at Cambridge. We knew that geographical information (GI) systems needed GI science, and had applauded Mike Goodchild's 1992 articulation of this (Goodchild, 1992, Chapter 9 in this volume). Thus, as we surfed the crest of the GIS innovation wave, we were looking for something more.

The first disruption was Jonathan's attendance at the workshop on cognitive and linguistic aspects of geographic space in Las Navas del Marqués in Spain in August 1990, led by David Mark and Andrew Frank, which led to the publication of a book by the same name (Mark and Frank, 1991). This workshop (funded by North Atlantic Treaty Organization [NATO]) brought together a large and very diverse group of researchers for *two whole weeks* (unheard of now). It introduced Jonathan to aspects of cognitive science and linguistics (George Lakoff and Zenon Pylyshyn spoke) that did not form part of the geographic tradition, considerably widening his horizons.

The second disruption was the acquisition of Smallworld GIS in 1990 at Birkbeck, particularly for David. Birkbeck had been the first European site for Arc/Info, which by then had been the leading GIS for almost a decade. Smallworld GIS was object oriented in its data modelling and object relational in implementation, and as such provided a massive challenge to our thinking, which up to that time had been dominated by the efficient but cartographically oriented georelational model of Arc/Info. From 1990 to 1992 we had been joined by one of the shareholders in and key developers for Smallworld, Mike Bundock, who worked part-time for the company while working at Birkbeck as a researcher on GIS user interfaces. David and Mike spent many hours discussing representational problems and the contrasting ways in which Smallworld and Arc/Info approached spatial and temporal representation. We realised from these insights that object-oriented data modelling offered an alternative way forward for those who saw limitations in the georelational model GIS.

The third disruption, and a crucial one for the paper, was the geomorphological fieldwork we were conducting at Scolt Head Island on the north Norfolk coast of England. Jonathan had been introduced to this place as an undergraduate at Cambridge by the extraordinary David Stoddart: an inspiring coastal geomorphologist and world expert on coral reefs, who had continued a coastal research project established at Cambridge in the 1920s by Alfred Steers. Unusually (then and now), Stoddart got undergraduates and Ph.D. students involved in his own research and the intellectual debates sitting around the table in the research station out on the coast were as intellectually stimulating as they were memorable. But David Stoddart had departed to Berkeley in 1988 leading to a hiatus in the 70-year research sequence, and in 1991 David and Jonathan saw the opportunity to establish a new project studying the extremely dynamic sandy spits with Charlie Bristow and Diane Horn (Raper et al., 1999). This was pure blue sky research in every sense: Jonathan's proposal for funding to do the work was turned down by NERC (Natural Environment Research Council), but the buoyant GIS short course receipts at Birkbeck meant that it could be funded in a low-key way from teaching income. It was when we tried to map the sand spits with their extraordinary changing and shifting shapes that we realised there was something deeply interesting at hand, which posed major representational challenges — challenges that object-oriented data modelling and cognitive concepts could assist with.

The major point we wanted to make in the paper was that researchers should start with structure of the problem domain and then find the data structure, and not the other way around. This seems like a mundane point in this era of richer tools, but was not really understood then, except by a few. This could clearly be facilitated by object-oriented modelling (OOM) and data structures, and so in the paper we tried to express our problem domain in Booch's OOM framework and speculate about how to implement such a model. In fact, our concept of phenomenon instance required multiple inheritance within the model, and even Smallworld was not able to implement this for spatial data types. The phenomenon instance concept also gave rise to another implication: if you had data about an environment, but did not know exactly what phenomena they constituted, then a spatio-temporal database could be constructed to hold the data and provide tools for the user to conjecture alternative sets of phenomena into existence. The raw data would then be stored along with alternative conjectures and their parameterisations.

At the time this aspect was not really picked up, and most of the citations have seen the paper as simply an early articulation of the potential of object-oriented data modelling in GIS (Peuquet, 2002). However, one reader in particular picked up the implication. In a strange episode, Doreen Massey (the assessor assigned to read papers for the Birkbeck geography department for the U.K. Research Assessment Exercise [RAE]), who would not normally have read a paper in a GIS journal, read it with interest and saw links to her own work. Once the RAE was over, and to our amazement, Doreen then wrote a paper for the leading geography journal *Transactions of the Institute of British Geographers* (*TIBG*) in which she selected the paper for some detailed commentary along with another paper by David Sugden (Sugden, 1996). The broad thesis of the commentary was that despite apparent schisms between the disciplines of physical and human geography, there were "commonalities" that could be identified in the way that practitioners were conceptualizing space, time and space-time. It was ironic that, given Pickles (1995) damning methodological critique of the apparently entrenched quantitative approach to geography typified by GIS, it was a GIS paper that should stimulate this appraisal by a leading human geographer.

The paper by Doreen Massey stimulated a debate both inside and outside the pages of *TIBG*, and this included a half-day workshop entitled Rethinking Space and Time: Transgressing the Human-physical Boundary at Portsmouth University, organised by Rob Inkpen, Giles Mohan, and David Livingstone. The debate highlighted how the notions of space and time, which geographers of all persuasions were struggling with, should not be regarded as a poor relation of the space-time theories of physicists (so-called *physics envy*), but instead should be seen as an assertion of the complex nature of geographical investigation. Contemporary papers in physical geography (Lane and Richards, 1997; Phillips, 1999) recognise this complexity in notions such as *emergence*, *self-organisation*, and *historical contingency*, which reject the classical notion of linear causality (Raper and Livingstone, 2001).

To quote from the Massey paper (1999, p. 262):

> Let me begin, however, with Raper and Livingstone's paper. This is an argument for the importance of a concept of relative space in the representation/modelling of environmental problems. "... Traditionally, the authors argue, while environmental representations have been somewhat unthinking about the concepts of space and time that they imply and necessarily incorporate, they have in fact been dominated by 'timeless' geometric methods focused on two dimensional planes" ([Raper and Livingstone, 2001], 363). Raper and Livingstone's aim is to disrupt this unthought assumption and to argue for a more self-conscious and "relative" understanding.

> The implication of this is, of course, that the GIS folk have to receive the spatio-temporal framework from the application domain, rather than, as heretofore, themselves being in a position to decide it. Second, this approach to space-time enables the conceptualization of entities themselves as a set of "worlds" (365), where each world has its own four-dimensional reference system. "Time," they write, "is a property of the objects" (366). Third, and implicit in all of this, is that for the kind of work that Raper and Livingstone are addressing, it is necessary to think not in terms of space and time separately, but in terms of a four-dimensional space-time (364). All of this was, for me, totally engrossing. It rang many bells with my own work, and that of many others, within human geography. We, too, have been struggling to understand space (and space-time) as constituted through the social, rather than as dimensions defining an arena within which the social takes place. We too have tried to consider the idea of local time-spaces, time-spaces specific to the entities with which they are mutually constitutive.

The Portsmouth meeting and the associated debate stimulated us to produce a response to the points made by Massey in the *TIBG* (Raper and Livingstone, 2001, p. 237):

> Our starting point is Massey's assertion that "the representation of space-time is itself an emergent product of the conceptualisation of the space-time entities themselves" (269). It seems to us that this phrase marks out a number of key theoretical debates:

1. The role of representation
2. The question of how entities are conceptualised
3. The framing role of space and time
4. How representation and conceptualisation are mutually constituted

In a longer-term development, this debate led Jonathan to conceive and write a monograph on the concepts of space and time used in GIS and their extension to three and four dimensions (Raper, 2000), chiefly to further explore the implications of this work. A paper that was initially conceived as addressing a data-modelling problem took us on a philosophical journey that we are still traversing and, given the opportunity, hope to continue on into the future.

References

Goodchild, M.F., 1992, Geographical information science, *International Journal of Geographical Information Systems*, **6**, 31–46.

Lane, S. and Richards, K.S., 1997, Linking river channel form and process: time, space and causality revisited, *Earth Surface Process and Landforms*, **22**, 249–260.

Mark, D.M. and Frank, A.U., Eds., 1991, *Cognitive and Linguistic Aspects of Geographic Space*, Kluwer, Dordrecht.

Massey, D., 1999, Space-time, "science," and the relationship between physical geography and human geography, *Transactions of the Institute of British Geographers*, **24**, 261–276.

Peuquet, D., 2002, *Representations of space and time*, Guilford Press, New York.

PHILLIPS, J.D., 1999, Divergence, convergence, and self organization in landscapes. *Annals of the Association of American Geographers*, **89**(3), 466–488.

PICKLES, J. (Eds.), 1995, *Ground Truth: The Social Implications of Geographic Information Systems*, Guilford Press, New York.

RAPER, J.F., 2000, *Multidimensional geographic information science*, Taylor & Francis, London.

RAPER, J.F., LIVINGSTONE, D., BRISTOW, C., AND HORN, D., 1999, Developing process-response models for spits. In *Proceedings of Coastal Sediments 1999*, Kraus, D., Ed., American Society of Civil Engineers, Reston, VA, 1755–1769.

RAPER, J.F. AND LIVINGSTONE, D.L., 2001, Spatio-temporal identity and geographic entities, *Transactions of Institute of British Geographers* 26, 237–242.

SUGDEN, D., 1996, The East Antarctic ice sheet: unstable ice or unstable ideas? *Transactions of the Institute of British Geographers*, **21**, 443–454.

International Journal of Geographical Information Systems,
1995, Vol. 9, No. 3, 251–273.

12 Integrating Geographical Information Systems and Multiple Criteria Decision-Making Methods

Piotr Jankowski

Abstract. Many spatial decision-making problems, such as site selection or land use allocation require the decision-maker to consider the impacts of choice-alternatives along multiple dimensions in order to choose the best alternative. The decision-making process, involving policy priorities, trade-offs, and uncertainties, can be aided by Multiple Criteria Decision-Making (MCDM) methods. This paper presents a framework for integrating geographical information systems (GIS) and MCDM methods. In this framework the MCDM methods are classified and matched with choice heuristics used by the decision-makers in the presence of competing alternatives and multiple evaluation criteria. Two strategies for integrating GIS with MCDM are proposed. The first strategy suggests linking GIS and MCDM techniques using a file exchange mechanism. The second strategy suggests integrating GIS and MCDM functions using a common database. The paper presents the implementation of the first strategy using PC-ARC/INFO, a file exchange module, and four different MCDM computer programs.

1 Introduction

In the early 1980s GIS software emerged commercially as a new information processing technology offering unique capabilities of automating, managing, and analyzing a variety of spatial data. From the early beginnings, dating back to the development of the Canadian Geographic Information System (CGIS) in the 1960s (Tomlinson 1988), GIS has been depicted as a decision support technology. Many applications of GIS developed over the last decade provided information necessary for the decision-making in diverse areas including natural resources management, environmental pollution and hazard control, regional planning, urban development planning, and utilities management. Cowen (1988) underscored the decision support function of GIS by describing it as a 'decision support system'. This notion of GIS was disputed, however, by other researchers who argued that the current GIS technology does not offer sufficient decision support capabilities (Densham and Rushton, 1988, Goodchild 1989, Densham and Goodchild 1989, Harris and Batty 1991).

Two perspectives on developing better decision support capabilities of GIS can be identified, one based on analytical problem solving as a centerpiece of Spatial Decision Support System (SDSS) and another based on integration of GIS and specialized analytical models. According to the first perspective, SDSS should offer modelling, optimization, and simulation functions required to generate, evaluate, recommend, and test the sensitivity or problem solution strategies. These capabilities are essential to solving semi- or poorly structured spatial decision-making problems. The model-based approach of SDSS represents an alternative and an extension of a toolkit approach embodied in the current generation of GIS software offering enough flexibility to accommodate a wide range of geometric operations at the expense of modelling, optimization, and dynamic simulation functions. The second perspective on improving the decision support capabilities focuses on the expansion of GIS descriptive, prescriptive, and predictive capabilities by integrating GIS software with other general software packages (e.g., statistical software) and with specialized analytical models such as environmental and socio-economic models (Goodchild 1987, Dangermond 1987, Burrough *et al.* 1989, Birkin *et al.* 1990, Nyerges 1993, Maidment 1993). According to this view, mapping, query, and spatial modelling functions of GIS can provide data display at different scales, preprocessing, and data input for environmental and statistical models. Different levels of integration between GIS and analytical models can be considered, starting from a loose integration where GIS and models are linked using a file exchange mechanism, through a tighter integration with a common user interface and user-transparent file exchange process, to a fully integrated GIS modelling system with shared memory and a common file structure (Fedra 1993).

Both perspectives share a common concern for providing better procedures for spatial decision-making support. Within this framework, MCDM methods received in the past little attention despite their ample potential as spatial decision support tools. A few but notable contributions on integrating of GIS and multiple criteria evaluation techniques can be found in the papers by Janssen and Rietveld (1990), Janssen and van Herwijnen (1991), and Carver (1991). A more recent and concentrated research effort at integrating GIS and MCDM methods has been undertaken by the IDRISI Project team at Clark University (Eastman *et al.* 1993).

The general objective of MCDM is to assist the decision-maker (DM) in selecting the 'best' alternative from the number of feasible choice-alternatives under the presence of multiple choice criteria and diverse criterion priorities. The problem of multicriterion (multiobjective) choice in decision-making is the paramount challenge faced by individuals, public, and private corporations. The nature of the challenge is two-fold;

1. How to identify choice alternatives satisfying the objectives of parties involved in the decision-making process.
2. How to reduce/order the set of feasible choice alternatives to identify the most preferred alternative.

The challenge of multicriterion choice can be attributed to many spatial decision-making problems involving search and location/allocation of resources (for applications of MCDM in spatial decision-making problems see Massam 1980, and Voogd 1983). These problems, often analysed in GIS, include location/site selection for: service facilities, recreational activities, retail outlets, hazardous waste disposal sites, and critical areas for specific resource management and control practices. Other examples include the selection of optimal utility routes in urban/rural areas such as powerline routes (Harris 1992), or the selection of water pipeline routes (Jakowski and Richard 1994), and land use decision-making involving trade-offs among conflicting land uses at individual parcel locations (Berry 1992).

The purpose of this paper is twofold: (1) to clarify the role of GIS and multicriteria decision-making methods in supporting spatial decision-making, and (2) to present a framework for integrating GIS with MCDM. The rest of the paper proceeds as follows. A brief exposition of the rational model of the decision-making process and the role of GIS are discussed in section two. Section three presents an overview and classification of multiple criteria decision-making techniques. The choice heuristics used by the decision-makers in the presence of competing alternatives and evaluation criteria are presented in section four. A few of these heuristics can be implemented directly in GIS using the database processing capabilities of commercial GIS software. Others, that cannot be implemented directly in GIS, can be matched with the selected MCDM techniques. The recommendations for the effective integration of GIS and MCDM are provided in section five followed by an implementation example.

2 GIS and the rational model of the decision-making process

The fundamental concept of various models of individual and organizational decision-making behaviour is *rationality*. The conceptual origins of the term rationality can be traced back to the philosophy of Rationalism which asserts the superiority of intellect over empirical experience (Encyclopaedia Britannica 1974). Simon (1957) argued for replacing the concept of rationality, built into the classical model of decision-making, with the principle of *bounded rationality*. According to this principle, individuals and organizations follow a satisficing decision-making behaviour, based on search activity, to meet certain aspiration levels rather than the optimizing behaviour aimed at finding the best decision alternative. In his more contemporary work Simon (1978, 1979) presented the concept of *procedural rationality* which means the effectiveness of decision support procedures in search of the relevant decision alternatives. The procedurally rational model of decision-making process distinguishes the following four steps that are generally appropriate for a structured approach to decision situations (McKenna 1980):

1. *Problem definition*: a discrepancy between the present state and the desired state is recognized as a need. That need is formulated as problem calling for decision.
2. *Search for alternatives and selection criteria*: the feasible alternatives (potential problem solutions) and criteria for evaluating the alternatives are established.
3. *Evaluation of alternatives*: the impacts of each alternative on every evaluation criterion are assessed.
4. *Selection of alternatives*: alternatives are ordered from the most desirable to the least desirable and either the top alternative is selected or the group of more desirable alternatives is retained for further evaluation.

The role of GIS in implementing this model in spatial decision-making has been the search for feasible alternatives (step 2). The search for feasible sites, routes, and land use allocations has been traditionally carried out using manual map overlays as part of *suitability analysis*. The basic tenets of this approach involved the specification of physical, economic, and environmental criteria important in determining the technical and social feasibilities of the project, and next the cartographical representation of these criteria on map layers. All of the layers were then overlaid using transparencies and the lightest areas represented the most desirable decision alternatives (McHarg 1969). With the advent of GIS and its capability to overlay digital maps, land suitability studies became more sophisticated. For example, Harris (1992) reported a powerline siting study that included 40 various physical, environmental, and socio-economic suitability factors.

Suitability analysis often results in generating a number of feasible alternatives. This can be due not only to more than one alternative satisfying technical suitability criteria but also due to political context of the decision making process that requires taking into account additional social and environmental criteria. Different strategies can be applied to reduce/order the set of feasible alternatives and hence, to facilitate the choice. One possible strategy suggests introducing for every decision criterion a threshold value that must be passed by an alternative in order to qualify. The threshold values can be gradually tightened until only one alternative is left. This approach, called the *reduced processing* strategy (Hong and Vogel 1991), does not set a high cognitive demand before the decision-maker since it does not require the evaluation of trade-offs among the alternatives. The reduced processing strategy can be implemented in GIS by an iterative suitability analysis resulting in step-by-step reduction of feasible alternatives. Such a procedure is, however, time consuming when the number of alternatives is large. A faster approach is to apply GIS in conjunction with one of the reduced processing MCDM techniques. In this approach the initial set of feasible alternatives is still generated in GIS but the reduction process is carried out using an MCDM technique. The final result may be visualized using the GIS display capabilities. This approach can be enhanced by integrating MCDM techniques with GIS.

Another strategy to reduce and/or order the set of GIS-generated, feasible alternatives is to include in the analysis the DM's preferences. These preferences may concern the decision criteria and performance levels of alternatives on every decision criterion. The process of quantifying the DM's preferences requires making trade-offs among the criteria and it is cognitively demanding, especially if the number of criteria is large. This strategy, called *full processing*, can also be implemented by integrating the processing capabilities of GIS with the structured decision-making approach featured by MCDM techniques.

In many instances the decision-making process involves a group of decision-makers representing different, often conflicting interests. Examples of such decision-making problems include many environmental dilemmas, such as, selection of habitat restoration sites, land use allocation, or siting of utility line corridors. In these problems, choice strategies aimed at reducing/ordering the set of GIS-generated feasible alternatives should include the consideration of multiple and conflicting viewpoints, and facilitate a movement towards consensus. These capabilities, lacking in the current GIS software, can be created by integrating GIS with MCDM techniques useful in the group decision-making context (Hwang and Lin 1987).

The following section provides a brief overview and classification of MCDM techniques that can be potentially useful for integrating with GIS.

3 Multiple criteria decision-making — An overview of the methodology

It is a widely accepted notion in the literature that MCDM consists of two categories: multiple attribute decision-making (MADM) and multiple objective decision-making (MODM) (Cohon 1978, Hwang and Masud 1979, Hwang and Yoon 1981). MADM is concerned with choice from a moderate/small size set of discrete actions (feasible alternatives), while MODM deals with the problem of design (finding a Pareto-optimal solution) in a feasible solution space bounded by the set of constraints (Colson and De Bruyn 1989). MADM is often referred to as multicriteria analysis (Teghern *et al.* 1989) or multicriteria evaluation (Nijkamp *et al.* 1990, Voogd 1983), whereas MODM is viewed as a natural extension of mathematical programming, where multiple objectives are considered simultaneously. This paper deals with MADM which involves choosing, based on the decision criteria and criterion priorities, from a moderate/small size set of alternatives; hence the term multiple criteria decision-making (MCDM) will be used in reference to MADM.

3.1 A general model of MCDM

The point of departure for any MCDM technique is the generation of the discrete set of alternatives, the formulation of the set of criteria, and the evaluation of the impact of each alternative on every criterion. Every MCDM technique shares a

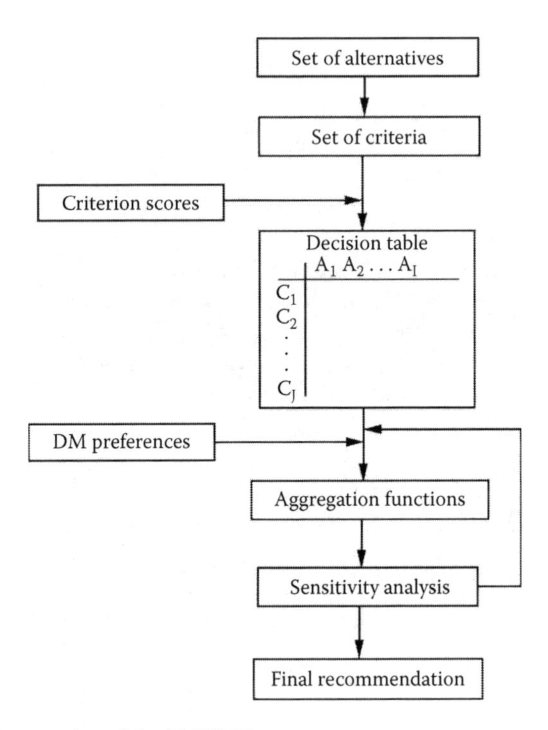

FIGURE 12.1 A general model of MCDM.

common approach called here a general model of MCDM (Figure 12.1). The estimated impacts of alternatives on every criterion, called criterion scores, are organized into a decision matrix C:

$$C = \begin{bmatrix} c_{11} \dots c_{1I} \\ \vdots \quad \vdots \\ c_{J1} \dots c_{JI} \end{bmatrix} \qquad (12.1)$$

where:

c_{JI} = is the criterion score,
J = represents criteria,
I = represents alternatives.

Depending on the MCDM technique the criterion scores may be aggregated in nonstandardized format (raw scores) or in standardized format (standardized score). Various linear and nonlinear standardization procedures exist that normalize the criterion scores so that $0 <= c_{JI} <= 1$ (Hwang and Yoon 1981, Voogd 1983, Jankowski 1989).

The second element of MCDM techniques, besides the criterion scores, is the DM's preferences. The preferences may be formulated in regard to criterion scores taking the form of cut-off values (minimum or maximum threshold) or the desired aspiration levels

(Lotfi *et al.* 1992). They may also be formulated in regard to decision criteria and expressed in a cardinal vector of normalized criterion preference weights **w** where:

$$\mathbf{w} = (w_1, w_2, \ldots, w_J), \quad \text{and} \quad 0 <= w_J <= 1 \qquad (12.2)$$

The criterion scores and the DM's preferences are processed using single or multiple aggregation functions that return a solution in the form of either:

1. One recommended alternative.
2. Reduced decision space consisting of several 'good alternatives'.
3. Ranking of alternatives from best to worst.

In many real world decision-making problems, criterion scores express predictions of impacts likely to be caused by the adoption of a given alternative and as such are prone to imprecisions of forecast and uncertainties of the future. In addition, the DM's preferences are often inconsistent, subject to shifting, and inaccuracies in determination (Bouyssou 1991). The problems of imprecision, uncertainty, and inaccurate determination are addressed in MCDM techniques by the sensitivity analysis of decision recommendations. There are two common approaches to sensitivity analysis in MCDM techniques:

1. Considering two alternatives at a time by calculating changes in weight values and changes in criterion scores required to bring the two alternatives to an equal position in the final ranking (Nagel and Long 1989).
2. Considering all alternatives taking part in the evaluation process and calculating changes in their ranking positions as the result of changing criterion scores and criterion weights.

3.2 Classification of MCDM techniques

The MCDM techniques can be classified according to the level of cognitive processing demanded from the DM and the method of aggregating criterion scores and the DM's priorities (Figure 12.2).

According to the cognitive processing level two classes of MCDM techniques can be distinguished: *compensatory* and *noncompensatory* (Hwang and Yoon 1981). The compensatory approach is based on the assumption that the high performance of an alternative achieved on one or more criteria can compensate for the weak performance of the same alternative on other criteria. In a compensatory MCDM technique the high score of an alternative achieved on one criterion is traded off, according to the DM's preference structure, for the low score received on another criterion. This means that a low score of a given alternative on a high-preference criterion may be compensated by a high score on another high preference criterion, or it may require high scores on two low-preference criteria to offset a low score on one high-preference criterion. The compensatory approach is cognitively demanding since it requires the DM to specify criterion priorities expressed as cardinal weights or priority functions.

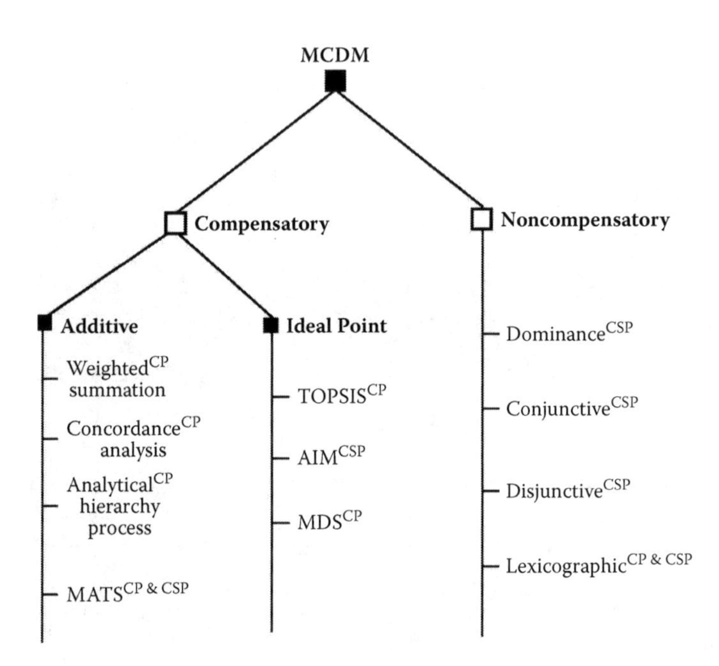

FIGURE 12.2 Classification of MCDM techniques. The superscripts denote criteria priorities (CP) and criterion score priorities (CSP). The techniques listed in the figure constitute a limited representation of the full set of MCDM techniques. For a more complete account of MCDM techniques see Ozernoy (1992).

Under the noncompensatory approach a low criterion score for an alternative cannot be offset by another criterion's high score. The alternatives are compared along the set of criteria without making intra-criterion trade-offs. The noncompensatory approach is cognitively less demanding than the compensatory approach since it requires, at the most, the ordinal ranking of criteria based on the DM's priorities.

3.2.1 Compensatory MCDM techniques

The class of compensatory MCDM techniques can be divided further, according to the method of aggregating criterion scores and the DM's priorities, into two subclasses;

1. *Additive* techniques.
2. Techniques based on the *ideal point* approach.

Within both subclasses the MCDM techniques can be further distinguished based on the approach to formulating the DM's priorities. The two approaches included in Figure 12.2 are: priorities regarding the decision criteria (criteria priorities or CP) and priorities regarding the value levels of criterion scores (criterion score priorities or CSP).

In the additive techniques the criterion scores are standardized to enable the inter-criterion trade-offs and to allow the comparison of the alternative performance on a common scale. The total score for each alternative is calculated by multiplying the criterion weight by the criterion performance score. The alternative with the highest score is recommended as the top choice. *Weighted Summation* (Voogd 1983) is probably the best known representative of the subclass. The basic form of the weighted summation technique can be depicted in this matrix notation:

$$\begin{bmatrix} s_1 \\ \vdots \\ s_I \end{bmatrix} = \begin{bmatrix} c_{11} \ldots c_{J1} \\ \vdots \quad\quad \vdots \\ c_{1I} \ldots c_{JI} \end{bmatrix} \times \begin{bmatrix} w_1 \\ \vdots \\ w_J \end{bmatrix} \tag{12.3}$$

where:

s_I = appraisal score for alternative I, and the criterion scores matrix is the transpose of the decision matrix C

The assumptions behind this technique are that the weights are cardinal, the criterion scores are determined on the interval/ratio scale, and data is aggregated by addition.

The *Concordance Analysis,* which is the most common technique based on pairwise comparison, determines the ranking of alternatives by means of pairwise comparison of alternatives. The comparison is based on calculating the concordance measure which represents the degree of dominance of alternative i over alternative i' for all the criteria for which i is equal or better than i', and the discordance measure which represents the degree of dominance of alternative i' over alternative i for all the criteria for which i' is better than i (Nijkamp and van Delft 1977, Nijkamp *et al.* 1990). The calculations of concordance and discordance measures are carried out in concordance analysis for every pair of alternatives. Based on these measures the differences between alternatives are quantified and a final score is calculated for every alternative. The final score is then used to rank the alternatives from best to worst.

Other MCDM techniques belonging to this subclass include the Analytical Hierarchy Process–AHP (Saaty 1980) which uses a hierarchical structure of criteria and both additive transformation function and pairwise comparison of criteria to establish criterion weights, and the Multi-Attribute Tradeoff System–MATS (Brown *et al.* 1986) in which the DM's utility functions are derived for every criterion and criterion weights are calculated based on inter-criteria trade-offs. The utility functions are used to derive utility values for every criterion score. The utility values are then combined with criterion weights, using the additive function, to obtain the total score for every alternative.

In the techniques based on the concept of an ideal point the DM is asked to locate in n-dimensional space his/her ideal solution, i.e., to specify the most desirable value for each decision criterion. Then, the distance between the ideal solution and each considered alternative is measured using a Euclidean or a non-linear distance measure in order to arrive at the ranking of alternatives.

Technique for Order Preference by Similarity to Ideal Solution — TOPSIS (Hwang and Yoon 1981), Aspiration-level Interactive Method — AIM (Lotfi *et al.* 1992), and Multi-Dimensional Scaling — MDS (Voogd 1983) are three examples of techniques based on the intuitive concept that the chosen alternative should be as close as possible to the ideal solution and as far away as possible from the worst solution. The two techniques: TOPSIS and MDS use the criterion preference approach to representing the DM's preferences. The AIM technique uses the criterion score preferences approach in which the DM determines his/her levels of aspiration for different criterion score values.

3.2.2 Noncompensatory MCDM techniques

A common facet of the noncompensatory techniques is the stepwise reduction of the set of alternatives without trading off their deficiencies along some evaluation criteria for their strengths along other criteria (Hwang and Yoon 1981, Minch and Sanders 1986). The differences among these techniques result from the following different rules of elimination:

(*a*) *The Dominance technique*: the elimination is based entirely on criterion scores; an alternative is dominated, and hence eliminated from further consideration, if there is another alternative which is better on one or more criteria and is equal on the remaining criteria.

(*b*) *The Conjunctive method*: every criterion has a minimum cut-off value specified by the DM. Those alternatives that fail a cut-off value on any of the evaluation criteria are eliminated.

(*c*) *The Disjunctive method*: also uses cut-off values but in this method an alternative is eliminated if it fails to exceed a minimum cut-off value on at least one of the evaluation criteria.

(*d*) *The Lexicographic method*: requires that the evaluation criteria be ranked from the most important to the least important. The alternatives are compared first on the most important criterion and the alternative with the highest criterion performance score wins; all the other alternatives are eliminated. If more than one alternative is left the evaluation continues on the second most important criterion, third, etc. The lexicographic technique uses both the DM's criterion score preferences (CP) to rank the criteria and the DM's criterion score preferences (CSP) to compare the alternatives.

In general, the noncompensatory MCDM techniques require from the DM the reduced level of cognitive processing and hence, can be useful in decision making situations where the DM cannot or is unwilling to formulate his/her preferences. The disadvantage of these techniques is a potential for recommending an inferior alternative due to their reduced processing strategy.

4 Choice strategies and selection of MCDM techniques

Hong and Vogel (1991) identified five different choice strategies used by decision-makers that can be matched with characteristics of different MCDM techniques. These choice strategies include:

1. *Screening of absolute rejects*: elimination of clearly dominated alternatives as the first step before any further choice deliberation,
2. *Satisficing principle*: the DM will consider all the alternatives that satisfy conjunctively or disjunctively the minimum performance levels,
3. *First-reject*: the DM wants to use exclusively the conjunctive elimination rule to reject all the alternatives that do not attain minimum threshold values,
4. *Stepwise elimination*: The DM narrows down the choice, re-evaluating the set of remaining alternatives every time one of the alternatives is eliminated,
5. *Generation of linear ordering*: the DM wants to generate a ranking of alternatives from the most preferred to the least preferred one.

The first four choice strategies can be implemented using exclusively the non-compensatory MCDM techniques. The last strategy (generation of linear ordering) requires the full processing approach and hence, can be implemented using the compensatory MCDM techniques.

Below, the five choice strategies are matched with the MCDM techniques presented in Figure 12.2.

absolute rejects =⇒ [Dominance]
satisfying principle =⇒ [Conjunctive, Disjunctive]
first-reject =⇒ [Conjunctive]
stepwise elimination =⇒ [Lexicographic]
generation of linear orderings:
  with criteria preferences =⇒ [Weighted Summation, Concordance Analysis, AHP, TOPSIS, MDS]
  with criterion score preferences =⇒ [AIM]
  with criterion preferences and criterion score preferences =⇒ [MATS]

The three variants of the 'generation of linear orderings' choice strategy were distinguished based on three different approaches to representing the DM's priorities in the MCDM techniques represented in Figure 12.2.

There exists empirical evidence that decision-makers use different choice strategies at different stages of the decision-making process (Wright and Barbour 1977). Hence, one can assume that different sequences of choice strategies can be used in decision-making situations concerning spatial decision-making problems. For example, the sequence of {first-reject, generation of linear orderings} could be used in a site selection problem if the DM desires first to drop the clearly inferior alternatives and next to rank the 'good' alternatives from most to least desirable. This situation is illustrated with the following route selection problem described in Jankowski and Richard (1994).

The Seattle Water Department (SWD 1988) studied various alternatives for selecting a route for a new section of a primary water transmission line for the City of Seattle and its purveyors in King Country, Washington (Figure 12.3).

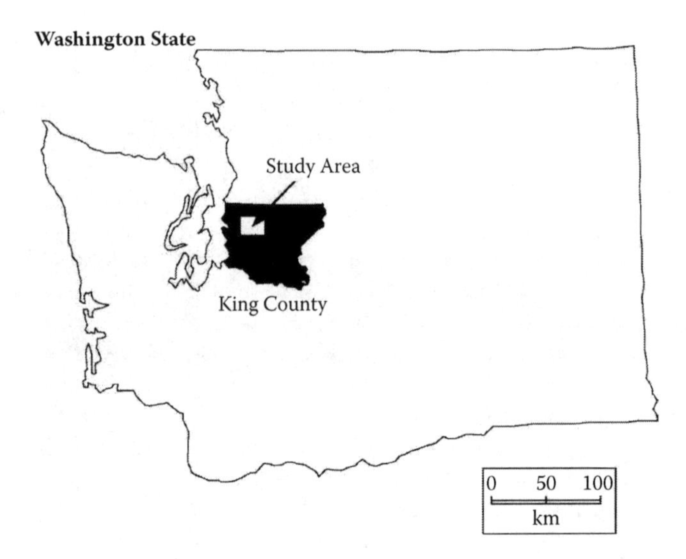

FIGURE 12.3 Study area for a water transmission line in King County, Washington.

The proposed alternatives were all located within the environmentally sensitive Snoqualmie River Valley region, making this an appropriate example of the multi-criteria decision problem. In the initial study, conducted by the Seattle Water Department, the alternatives were identified with a manual, suitability mapping approach. In the verification study conducted by Richard (1992) a GIS-based approach was used. The decision criteria included: total cost (TOTALCOST) estimated for each route alternative, the amount of public right-of-way (ROWACRES) falling within the alternative right-of-way, the reliability criteria including the normal daily traffic volume of roads, which fall with and parallel the alternatives' rights-of-way (VEH_DAY), erosion hazard areas (ERSACRES), landslide hazard areas (LNDACRES), seismic hazard areas (SEIACRES), and the environmental criteria including the area of wetlands (WETACRES) and the length of stream segments (STRMLEN) falling within the alternatives' rights-of-way. The additional criterion included in the analysis was the length of alternative (ALTLEN).

The conjunctive selection rule, representing the 'first reject' strategy, was applied to reduce the initial set of pipeline route alternatives. The choice rule included a combination of technical, land use, topographic, geological, and economic constraints that had to be satisfied by the alternatives. The six alternatives that passed the conjunctive selection rule are listed in Table 12.1 together with the decision criteria and criterion scores.

The six pipeline route alternatives can be ordered from best to worst applying the choice strategy 'generation of linear orderings'. One variant of this strategy that uses criterion utility functions for expressing criterion score preferences and weights for representing preferences on criteria can be implemented by the Multi-Attribute Utility Tradeoff System technique — MATS.

TABLE 12.1
Six conjunctively selected alternatives with the decision criteria and criterion scores.

Criteria	Units	Alternative Names					
		ALT1	ALT2	ALT3	ALT4	ALT5	ALT6
ROWACRES	[acres]	70.27	53.25	15.28	34.54	35.85	29.78
ERSACRES	[acres]	6.24	10.05	13.76	13.53	12.03	9.36
SEIACRES	[acres]	15.70	12.76	15.00	16.12	22.93	16.28
LNDACRES	[acres]	4.91	4.03	5.22	4.78	13.85	3.44
WETACRES	[acres]	5.71	5.07	1.09	0.86	6.52	6.91
STRMLEN	(m)	502	206	1347	883	675	191
VEH_DAY	(cars)	8200	8200	4900	6300	7240	8200
ALTLEN	(km)	12.38	11.83	25.57	9.88	37.05	8.1
TOTALCOST	[$ millions]	27.1	25.2	23.6	24.4	24.7	15.6

MATS, implemented as a stand-alone, interactive computer program that runs on DOS-based microcomputers (Brown *et al.* 1986), requires the DM to enter first decision criteria and to define a numeric scale for every criterion by entering maximum and minimum values. Next, with the help of the program's interrogation/specification mode, the utility functions are derived by the user for every criterion. A criterion utility function describes how much utility is received from each value of the criterion score. The utility values are normalized and range in the interval <0, 1>.

Following the specification of criterion score preferences through the criterion utility functions the user is asked to define the preferences regarding the decision criteria. The decision criteria preferences are derived through the series of trade-off questions in which the user is asked to evaluate the relative importance of one criterion versus the other criterion. The decision criteria preferences are then quantified into standardized weights that sum to 1.0. In the next step the user enters the names of decision alternatives and the criterion scores. MATS uses the following aggregation function to calculate the final score for each decision alternative:

$$S_i = \sum_{j=1}^{J} Uf_j(c_{ji})w_j \tag{12.4}$$

where:

S_i = final score for alternative i.
c_{ji} = criterion score for criterion j and alternative i
Uf_j = utility function of criterion j
w_j = cardinal weight of criterion j.

In the water pipeline route selection problem the following criterion weights, consistent with preferences stated by a citizen advisory committee (Richard 1992), were used:

Criterion	Weight
ROWACRES	0·130
ERSACRES	0·057
SEIACRES	0·057
LNDACRES	0·057
WETACRES	0·053
STRMLEN	0·027
VEH_DAY	0·053
ALTLEN	0·284
TOTALCOST	0·284

The ranking of six alternatives based on standardized final score values calculated by MATS is presented below:

Alternative	Final score
ALT6	0·821
ALT4	0·638
ALT2	0·572
ALT3	0·543
ALT1	0·478
ALT5	0·270

The sensitivity analysis of the solution, facilitated by MATS, revealed that the alternative route ALT6 was firm in the first position. It would take a significant improvement in the total cost of the second-ranked route (ALT4), or a simultaneous improvement in values of at least four other criteria, in order for ALT4 to tie ALT6.

Another variant of the decision strategy 'generation of linear orderings' incorporates the DM's priorities only in regard to decision criteria. This variant can be implemented using, among others, an MCDM Weighted Summation technique. In this example the Best Choice program (Nagel and Long 1989) was used to rank the alternatives ALT1 through ALT6. Best Choice requires the user to enter names of alternatives, decision criteria, the measurement units of criteria, and weights for each criterion. The criterion scores for each alternative are then entered into a decision matrix with alternatives represented by rows and criteria by columns. The program analyzes data by converting scores to percentages in order to deal with the multidimensionality of data, and a summary score is computed for each alternative. Negative (cost) criteria are handled simply by placing a negative sign in front of the criterion scores, or by taking the reciprocal of the criterion scores. The sensitivity analysis in Best Choice allows the user to see how much the criterion score or the criterion weight must change to alter the results of the analysis.

The ranking of six alternatives based on final score values calculated by Best Choice, using the same criterion weights as in MATS, is presented below:

Alternative	Final Score (%)
ALT6	25·27
ALT4	18·03
ALT3	16·59
ALT2	15·26
ALT1	14·17
ALT5	10·90

The final ranking obtained from Best Choice is very similar to the ranking obtained from MATS. This can be explained by the same type of aggregation function used in both programs, and by the fact that utility functions derived in MATS were nearly linear. The final ranking might have been quite different if the utility functions were concave or convex.

5 Strategies for integrating GIS and MCDM

The choice of MCDM techniques and an approach to integration with GIS depends on the type of data model used in GIS. In a raster-based GIS each individual cell is regarded as a choice alternative and hence, it is a candidate for evaluation. The number of cells, in the majority of raster maps, makes it impractical to use computationally intensive MCDM techniques based on pairwise comparisons of all alternatives (i.e., concordance analysis). In this case one can advocate weighted summation as the useful and practical MCDM technique for raster GIS. The multiple criteria evaluation, with weighted summation integrated into a raster GIS, becomes then a two-step procedure where a suitability map is developed first, followed by the rank-ordering of cells.

In a vector-based GIS the multiple criteria evaluation can also be carried out as a two-step procedure, where the first step involves creating a suitability map and the second step rank-ordering the alternatives represented by points, lines, or polygons. The difference between the multiple criteria evaluation in vector GIS and raster GIS is a relatively small number of suitable alternatives obtained from step one in a vector system in comparison to a number of raster cells. This makes possible integrating vector-based GIS and MCDM techniques other than weighted summation.

Nyerges (1993) distinguished *loose* and *tight coupling* as two approaches to interfacing GIS and computer environmental models. These two approaches can be followed when integrating GIS with MCDM technique. In the remainder of this section the conceptual basis of loose and tight coupling of GIS and MCDM are outlined followed by an example of implementing the loose coupling approach in a vector-based GIS.

5.1 *The loose coupling strategy*

The main idea of loose coupling approach is to facilitate to integration of GIS and MCDM techniques using a file exchange mechanism. The assumption behind this strategy is that MCDM techniques already exist in the form of stand-alone computer programs, which is true for many of the techniques reviewed in §3 (Lotfi and Teich 1991, Radcliff 1986). Each of these programs comes with its customized interface and the user is faced with different interfaces when running a sequence of different MCDM techniques. The results of the decision analysis may be sent to GIS for display and spatial visualization.

The integration of GIS and MCDM in the loose coupling architecture is based on linking three modules:

- GIS module
- MCDM techniques module, and
- File Exchange Module (see Figure 12.4).

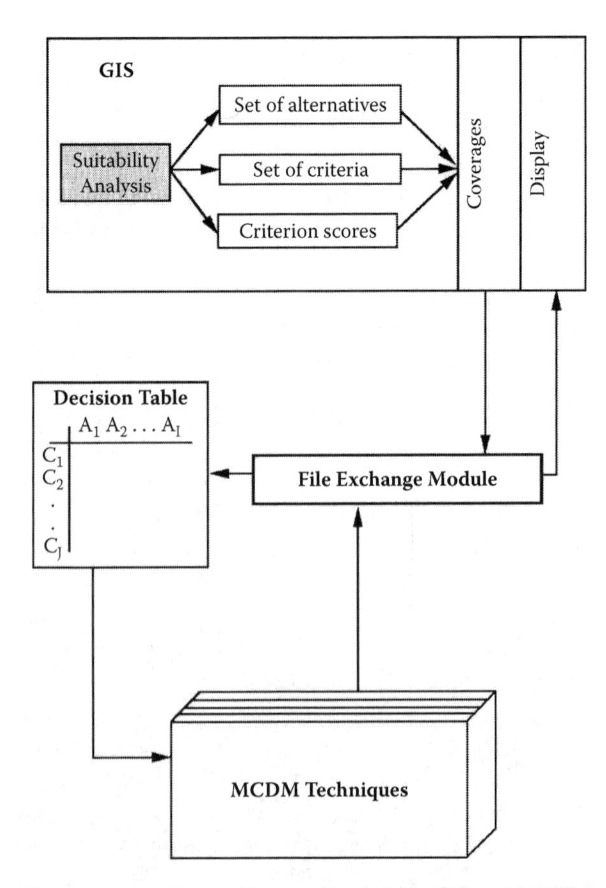

FIGURE 12.4 The loose coupling architecture for linking GIS and MCDM.

The GIS module is used for generating the set of spatial decision alternatives (e.g., sites, routes, land uses) in the course of performing land suitability analysis. The alternatives can be represented in two different ways: (1) each alternative is represented by a separate GIS map coverage, (2) each alternative is represented by one record or multiple records in one composite coverage. The set of decision criteria consists of map coverage attributes. The coverage attribute values serve as the criterion scores.

According to the loose coupling strategy the land suitability analysis is run through the GIS user interface. The results of analysis including the set of alternatives, the set of decision criteria, and criterion scores can be represented in a two-dimensional decision table with criteria in rows, decision alternatives in columns, and table elements as criterion scores. The decision table is generated as an ASCII-format file in the File Exchange module. The File Exchange module consists of programs that extract the GIS coverage attributes, create the decision table, and change its format to different input file formats required by MCDM programs.

5.2 *The tight coupling strategy*

The tight coupling strategy differs from the loose coupling by using multiple criteria evaluation functions fully integrated into GIS, a shared data base, and a common user interface (Figure 12.5).

Under this approach the GIS evaluation functions facilitate the spatial decision-making with multiple criteria. The functions include: generation of decision table,

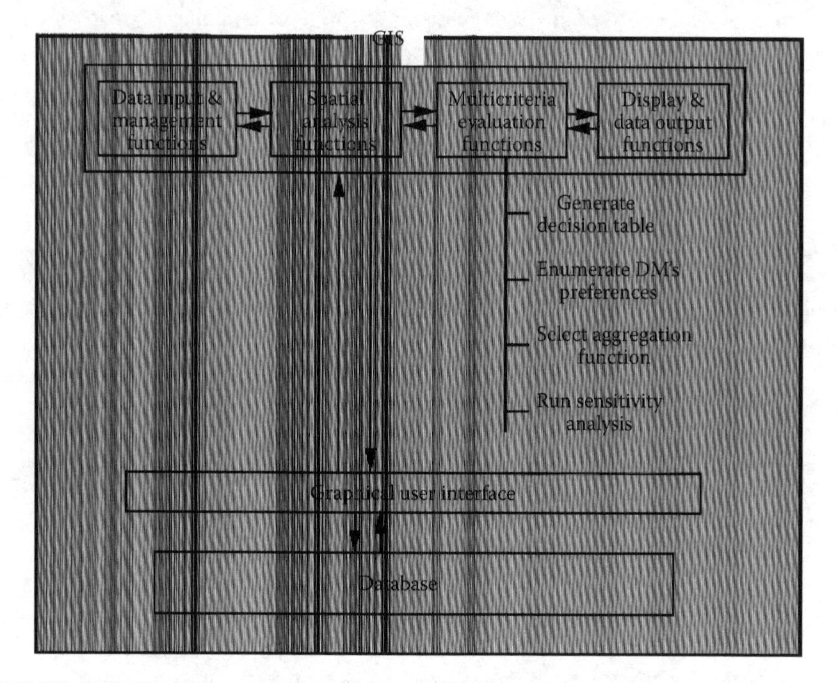

FIGURE 12.5 The tight coupling architecture for linking GIS and MCDM.

enumeration of decision-maker's preferences, selection of aggregation function, and sensitivity analysis. These functions represent operations common to many MCDM techniques. However, instead of accessing a specific MCDM technique, like in the loose coupling approach one can select a function from a common GIS user interface. This makes the multiple criteria evaluation functions a part of the GIS toolbox.

The evaluation functions contain specific operations that can be selected by the user. For example, the function enumerating DM's preferences can be carried out by: (1) rating (Voogd 1983), (2) pairwise comparison of criteria using a 9-point scale (Saaty 1980), and (3) inter-criteria trade-offs based on attribute values (Brown *et al.* 1986). The function 'select aggregation function' offers the selection of: (1) weighted summation, (2) outranking based on pairwise comparison, (3) additive utility function, and (4) ideal point distance function. The function 'sensitivity analysis' allows one to select: (1) dynamic sensitivity option in which the user can see the change in final scores of alternatives as one increases and decreases the weight of any criterion, and (2) threshold analysis option which calculates new values for criterion scores and weights that would have to be adopted in order to rank any two selected alternatives (e.g., the winner and the runner-up) equally.

The GIS database module serves as the repository of choice alternatives (sites, routes, land use areas) whose spatial features are represented by points, lines, and areas. The specification of alternatives and evaluation criteria is made by the user from the system's interface. The data extraction into the decision table is performed transparently.

The GIS interface, under the proposed design, will facilitate the map views of alternatives and their criteria. The user will be able to select any number of criteria and up to two alternatives that will be displayed in two simultaneous map views. In the case of selecting two alternatives the monitor's display will be split into two halves facilitating the pairwise comparison of alternatives on selected criteria. The quantitative comparisons can be further enhanced by displaying tables with criterion scores representing the performance of alternatives on selected criteria, and charts (pie, bar, line and spider web charts). The qualitative comparisons can be facilitated by changing the scale of a map view to better assess spatial relationships, combining map views with hypertext, photographic images, sound, and video animation. The comparative mode of visualizing the spatial properties of alternatives can help in formulating the decision-maker's preferences and trade-offs articulated in criterion weights.

5.3 Implementation of the loose coupling strategy

The integration of a vector-based GIS and MCDM was developed using the loose coupling strategy outlined in §5.1. The integrated system has three components: a GIS module, MCDM module, and file exchange module (Figure 12.6).

The GIS module is comprised of the PC-ARC/INFO® 3.4D (commercial GIS software from ESRI, Redlands, CA) that runs on DOS-based microcomputers. PC-ARC/INFO is used for land suitability analysis that results in generating feasible alternatives. These are the alternatives satisfying physical and technical constraints imposed on ARC/INFO coverage attributes. The current implementation requires that every alternative be represented by one ARC/INFO coverage.

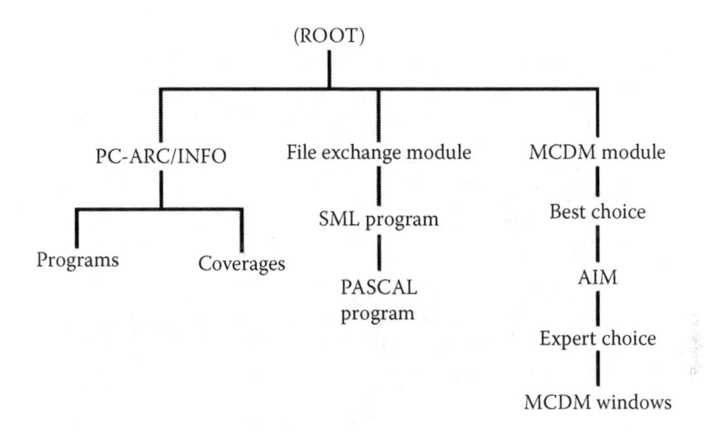

FIGURE 12.6 The directory structure of the prototype implementation of loose coupling strategy.

The MCDM module is a collection of four stand-alone DOS-based computer programs implementing different MCDM technique included in Figure 12.1. These programs are:

1. Best Choice: the interactive computer program based on the weighted summation technique (Nagel and Long 1989). Best Choice can accept up to 15 alternatives and 15 decision criteria. The DM's preferences on decision criteria are represented by cardinal weights.
2. Aspiration-Level Interactive Method — AIM 3.0: the interactive computer program based on the ideal-point approach (Lotfi *et al.* 1992). The maximum size of the decision table accepted by AIM is 50 alternatives and 10 criteria. The DM's preferences are represented in AIM by the aspiration levels on criterion scores.
3. Expert Choice 8.0: the interactive computer program based on Analytic Hierarchy Process — AHP (Saaty 1990). The program uses the hierarchical structure of criteria and pairwise comparisons among the criteria to establish criterion weights. The additive transformation function of the weighted summation approach is used to calculate a final score for each alternative. There is no limit on the size of decision problem accepted by Expert Choice. The DM's preferences on criteria are represented by cardinal weights.
4. MCDM-Windows: the interactive computer program that includes three different MCDM techniques: Conjunctive, Disjunctive, and TOPSIS. The conjunctive and disjunctive techniques are based on cutoff values as the decision rule for eliminating inferior alternatives, and TOPSIS is based on the ideal point approach. The program was developed by the author and his students as part of the GIS-MCDM integration project. There is no limit on the size of the decision table accepted by MCDM-Windows. In conjunctive and disjunctive techniques the DM's preferences on criterion

scores are represented explicitly by cutoff values. The TOPSIS technique represents the DM's preferences on criteria in the form of cardinal weights.

The file exchange module is comprised of two programs: one that generates the decision table and the other that changes its format to input file formats of the four MCDM programs. An input file containing the decision table can then be opened from a selected MCDM program by the user. The generation of a decision table is carried out by an interactive program written in SML (Simple Macro Language of PC-ARC/INFO). The formatting of the decision table file is carried out by a program written in Turbo PASCAL version 6.0.

5.3.1 Operation of the file exchange module

After preparing the decision alternatives in PC-ARC/INFO and storing them in separate coverages the user can access the file exchange module. The user is presented first with a dialogue menu asking for the name of a file that will store the decision table and then with another menu asking to select the decision alternatives from the PC-ARC/INFO coverage list (Figure 12.7).

In response to the selected alternatives the SML program presents the user with another menu containing the choice of three different formats for attributes (Figure 12.8).

Under this specification an attribute value extracted from a coverage is the sum of all record values for this attribute. If an alternative is represented by a point coverage then the extracted attribute value will be represented only by one record since one point represents one alternative. If an alternative is represented by either a line or a polygon coverage then the extracted attribute value will be represented

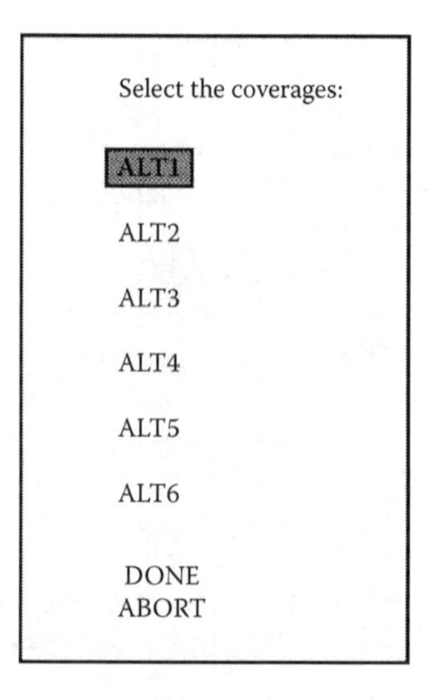

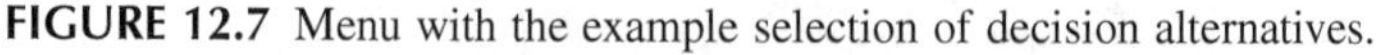
FIGURE 12.7 Menu with the example selection of decision alternatives.

Formating the attribute information ...

Attribute numbers: 1	
Which attribute format will be used?	
SUM	SUM (attribute k)
RATIO	SUM (attribute k)/SUM (attribute n) * 100
SUM PRODUCT	SUM (attribute value * constant)
DONE	
ABORT	

FIGURE 12.8 Menu with the attribute formats.

by the sum of line or polygon records, since a line or an area may be comprised of multiple line segments or multiple polygons. The choice of formats for the extracted attribute values include: (1) the sum of all records for a given attribute, (2) the percentage ratio of two attributes, and (3) the product of attribute value and the user-specified constant value.

Following the selection of the format for the first attribute the program extracts and displays the attribute names of the selected alternatives. It is assumed that all attributes representing the decision criteria are shared by the alternatives. After the user selects one attribute its values are extracted from the selected alternatives (coverages) and written into an ASCII file representing the decision table. This step, including the selection of the attribute format, is then repeated for other attributes chosen by the user as the decision criteria.

After the decision table is completed the user is presented with the final selection menu (Figure 12.9).

BEST CHOICE
AIM
EXPERT CHOICE
MCDM-Windows
DONE
ABORT

FIGURE 12.9 Menu with MCDM file formats.

The selection made in this menu is carried out by the PASCAL program that changes the original format of the decision table to the input file format of the selected MCDM program. The decision table can then be accessed from the given MCDM program as its input file.

6 Summary and conclusion

Two perspectives on developing better decision support capabilities of GIS have been identified in this paper: one based on analytical problem solving as a central function of a Spatial Decision Support System (SDSS) and another based on the integration of GIS and other specialized software. Both perspectives share a common goal of providing better procedures for spatial decision-making support. Within this framework, MCDM techniques offer ample potential as spatial decision support tools. The purpose of using MCDM is to assist the decision maker in selecting the best alternative from the number of feasible choice-alternatives under the presence of multiple choice criteria and diverse criterion priorities.

Every MCDM technique shares a common approach called a general model of MCDM. According to this model the decision-making problem is structured into the set of alternatives and the set of criteria. The performance of each alternative on every criterion is represented in the two-dimensional decision table by evaluation scores. These scores combined with the decision-maker's preferences are processed by aggregation functions that return a solution in the form of either one recommended alternative, or a reduced decision space consisting of several good alternatives, or a ranking of alternatives from best to worst. The general model of MCDM parallels the procedurally rational model of decision-making. The four steps of the procedurally rational model: problem definition, search for alternatives and selection criteria, evaluation of alternatives, and selecting an alternative, correspond to the following steps in the general model of MCDM: deriving the set of alternatives and the set of criteria, formulating the decision table, aggregating the data in the decision table and performing sensitivity analysis, and making the final recommendation.

The role of GIS in implementing the procedurally-rational model of decision-making in land use allocation, site, and route selection problems is foremost the search for suitable alternatives, but also helping the decision-makers to assign priority weights to decision criteria, evaluate the suitable alternatives, and visualize the results of choice. The search often results in the selection of a number of alternatives that satisfy the minimum threshold values. Further reduction of the set of admissible alternatives and the selection of the best alternative often requires the use of MCDM techniques. Hence, the improvement of GIS decision support capabilities can be achieved by introducing the multiple criteria decision-making techniques into the GIS context.

The MCDM can be classified according the level of cognitive processing required from the DM into compensatory and noncompensatory techniques, and according to the method of aggregating criterion scores and the DM's priorities into additive, based on the ideal point approach, pairwise comparison, and sequential elimination techniques.

The multiple criteria evaluation in raster GIS involves evaluating a large number of individual cells. This makes the integration of some computationally more intensive MCDM techniques with raster GIS impractical. In vector-based GIS the multiple criteria evaluation results in a relatively small number of suitable alternatives, in comparison to a number of raster cells. This makes feasible integrating a number of different MCDM techniques, besides the weighted summation technique, with vector-based GIS.

A few of the MCDM techniques can be implemented directly in GIS using the operators of the database query language. Others, however, can be implemented more efficiently using external programs and integrating these programs with GIS. Two strategies for integrating GIS and MCDM are proposed. The first strategy called here the loose coupling strategy suggests linking GIS and MCDM techniques using a file exchange mechanism. The second strategy called the tight coupling strategy suggests linking GIS and MCDM techniques using a shared database.

The framework presented in this paper provides guidelines for implementing the loose coupling of a vector-based GIS and MCDM. The implementation of the tight coupling strategy requires further work. More research is also needed on the topics of sensitivity analysis in an integrated GIS — MCDM system and facilitating group decision-making in the GIS — MCDM context.

Acknowledgments

The author would like to thank two anonymous reviewers for their thoughtful comments and suggestions in preparing the final version of the paper.

References

Berry, J. K., 1992, GIS resolves land use conflicts: A case study. *1993 International GIS Sourcebook* (Fort Collins, CO: GIS World), pp. 248–253.

Birkin, M., Clarke, G., and Wilson, A., 1990, Elements of a model-based GIS for the evaluation of urban policy. In *Geographic Information Systems: Developments and Applications*, edited by L. Worrall (London: Belhaven Press), pp. 133–162.

Bouyssou, D., 1991, Modelling inaccurate determination, uncertainty, imprecision using multiple criteria. In *Multiple criteria decision support. Proceedings of the International Workshop held in Helsinki, Finland, 7–11 August 1989*, edited by P. Korhonen, A. Lewandowski and J. Wallenius (Berlin: Springer-Verlag), pp. 78–87.

Brown, C. A., Stinson, D. P., and Grant, R. W., 1986. *Multi-attribute tradeoff system: Personal computer version user's manual.* (Denver: Public and Social Evaluation Office, Division of Planning Technical Services, Engineering Research Center, Bureau of Reclamation).

Burrough, P. A., van Deursen, W., and Heuvelink, G., 1988, Linking spatial process models and GIS: a marriage of convenience of a blossoming partnership? In *Proceedings, GIS/LIS '88*, vol. 2, San Antonio, TX (Bethesda, MD: American Congress on Surveying and Mapping), pp. 598–607.

Carver, S. J., 1991, Integrating multi-criteria evaluation with geographical information systems. *International Journal of Geographical Information Systems,* **5**, 321–339.

Cohon, J. L., 1978, *Multiobjective programming and planning.* (New York: Academic Press).

Colson, G., and De Bruyn, C., 1989, Models and methods in multiple objectives decision making. *Mathematical and Computer Modelling*, **12**, 1201–1211.

Cowen, D., 1988, GIS versus CAD versus CAD versus DBMS: what are the differences? *Photogrammetric Engineering and Remote Sensing*, **54**, 1551–1555.

Dangermond, J., 1987, The maturing of GIS and a new age for geographic information modelling systems (GIMS), In *Proceedings, International Geographic Information Systems Symposium: The Research Agenda*, vol. 2 (Arlington, VA: NASA), pp. 55–66.

Densham, P. J., and Goodchild, M. F., 1989, Spatial decision support systems: a research agenda. In *Proceedings, GIS/LIS '89, Orlando, FL*, vol. 2 (Bethesda, MD: American Congress on Surveying and Mapping), pp. 707–716.

Densham, P. J., and Rushton, G. R., 1998, Decision support systems for locational planning. In *Behavioral Modeling in Geography and Planning*, edited by R. G. Colledge, and H. J. P. Timmermans (New York: Croom Helm), pp. 56–90.

Encyclopedia Britannica, 1974, vol. 15 (University of Chicago), p. 527.Eastman, J. R., Kyem, P. A. K., and Toledano, J., 1993, A procedure for mutli-objective decision making in GIS under conditions of competing objectives. *Proceedings, EGIS '93*, pp. 438–447.

Fedra, K., 1993, Environmental modeling and GIS. In *Environmental Modeling with GIS*, edited by M. Goodchild, B. Parks and L. Steyaert (Oxford: University Press), pp. 35–50.

Goodchild, M. F., 1987, A spatial analytical perspective on geographical information systems. *International Journal of Geographical Information Systems*, **1**, 327–334.

Goodchild, M. F., 1989, Geographica information systems and market research. In *Papers and Proceedings of Applied Geography Conferences*, **12**, 1–8.

Harris, T. M., 1992, Balancing economic and energy development with environmental cost: a GIS approach. Presented at the *Association of American Geographers 88th Annual Meeting*, 18–22 April, 1992, San Diego, CA. Copy of the abstract available from the Association of American Geographers.

Harris, B., and Batty, M., 1991, *Locational models, geographic information and planning support systems*. Paper presented at the *38th North American Meeting of the Regional Science Association*, 7–10 November, New Orleans, LA. Copy of the abstract available from the Regional Science Association.

Hong, I. B., and Vogel, D. R., 1991, Data and model management in a generalized MCDM-DSS. *Decision Science*, **22**, 1–25.

Hwang, C. L., and Lin, M. J., 1987, *Group decision making under multiple criteria* (Berlin, Germany: Springer-verlag).

Hwang, C. L., and Yoon, K., 1981, *Multiple attribute decision making methods and applications: A state of the art survey* (Berlin: Springer-Verlag).

Hwang, C. L., and Masud, A. S. M., 1979, *Multiple objective decision making methods and applications: a state of the art survey* (Berlin: Springer-Verlag).

Jankowski, P., 1989, Mixed data multicriteria evaluation for regional planning: A systematic approach to the decision-making process. *Environment and Planning A*, **21**, 349–362.

Jankowski, P., and Richard, L., 1994, Integration of GIS-based suitability analysis and multicriteria evaluation in a spatial decision support system for route selection. *Environment and Planning B*, **21**, 323–340.

Janssen., R., and van Herwijnen, M., 1991, Graphical decision support applied to decisions changing the use of agricultural land. In *Multiple criteria decision support. Proceedings of the International Workshop held in Helsinki, Finland, 7–11 August, 1989*, edited by P. Korhonen, A. Lewandowski and J. Wallenius (Berlin: Springer-Verlag), pp. 293–302.

Janssen, R., and Rietveld, P., 1990, Multicriteria analysis and GIS: an application to agricultural land use in The Netherlands. In *Geographical Information Systems and Regional Planning*, edited by H. J. Scholten and J. C. Stilwell (Dordrecht: Kluwer).

Lotfi, V., Stewart, T. J., and Zionts, S., 1992, An aspiration-level interactive model for multiple criteria decision making. *Computer Operations Research*, **19**, 671–681.

Lotfi, V., and Teich, J. E., 1991, Multicriteria decision making using personal computes. In *Multiple criteria decision support. Proceedings of the International Workshop held in Helsinki, Finalnd, 7–11 August, 1989*, edited by P. Korhonen, A. Lewandowski and J. Wallenius (Berlin: Springer-Verlag), pp. 152–158.

Maidment, D., 1993, GIS and hydrological modeling. In *Environmental Modeling within GIS*, edited by M. Goodchild, B. Parks and L. Steyaert (Oxford: University Press), pp. 147–167.

Massam, B. H., 1980, *Spatial Search* (Oxford: Pergamon Press).

McHarg, I. L., 1969, *Design with nature* (Garden City, NJ: John Wiley & Sons).

McKenna, C. K., 1980, *Quantitative methods for public decision making* (New York: McGraw-Hill).

Minch, R. P., and Sanders, G. L., 1986, Computerized information systems supporting multicriteria decision making. *Decision Sciences*, **17**, 395–413.

Nagel, S. S., and Long, J., 1989, *Evaluation analysis with microcomputers* (Greenwich, CN: Jai Press).

Nijkamp, P., Rietveld, P., and Voogd, H., 1990, *Multicriteria Evaluation* (Amsterdam: North Holland).

Nijkamp, P., and van Delft, A., 1977, *Multi-criteria analysis and regional decision-making. Studies in Applied Regional Science*, **8** (Amsterdam: Leiden).

Nyerges, T. L., 1993, Understanding the scope of GIS: its relationship to environmental modeling. In *Environmental Modeling within GIS*, edited by M. Goodchild, B. Parks and L. Steyaert (Oxford: University Press), pp. 75–93.

Ozernoy, V. M., 1992, Choosing the 'best' multiple criteria decision-making method. *Information Systems and Operational Research*, **30**, 159–171.

Radcliff, B., 1986, Multi-criteria decision making: a survey of software. *Social Science Microcomputer Review*, **4**, 38–55.

Richard, L., 1992, Integration of geographic information system analysis and multicriteria evaluation in a spatial decision support system for a pipeline route selection study in King County, Washington. MSc. Thesis (Moscow: Department of Geography, University of Idaho).

Saaty, T. L., 1990, *Multicriteria Decision Making: The Analytic Hierarchy Process*, (Pittsburgh: Expert Choice, Inc.).

Saaty, T. L., 1980, *The analytic hierarchy process: planning, priority setting, resource allocation* (New York: McGraw-Hill).

Seattle Water Department (SWD), 1988, *Tolt Eastside Supply No. 2 (TESSL) No. 2, Phase III Draft Route Selection Study* (Seattle, WA: Seattle Water Department).

Simon, H., 1979, Rational decision making in business organizations. *American Economic Revue*, **69**, 493–513.

Simon, H., 1978, Rationality as process and as product of thought. *American Economic Revue*, **68**, 1–16.

Simon, H., 1957, *Models of economic man* (New York: John Wiley & Sons).

Teghem, J., Delhaye, C., and Kunsch, P. L., 1989, An interactive decision support system (IDSS) for multicriteria decision aid. *Mathematical and Computer Modelling,* **12**, 1311–1320.

Tomlinson, R. F., 1988, The impact of the transition from analogue to digital cartographic representation. *The American Cartographer*, **15**, 249–261.

Voogd, H., 1983, *Multicriteria evaluation for urban and regional planning* (London: Pion).

Wright, P., and Barbour, F., 1977, Phased decision strategies; sequels to an initial screening. *TIMS Studies in the Management Science*, **6**, 91–109.

Integrating Geographical Information Systems and Multiple Criteria Decision-Making Methods: Ten Years After

Piotr Jankowski

Paper roots

The interest of geographers in extending decision-support capabilities of geographic information systems (GIS) traces back to the fields of operations research and planning. In the 1960s, operations researchers began revising the concept of rational decision-making expressed by a single objective, and introduced the idea of multiple objectives as the bases for decision aiding (Thiriez and Zionts, 1976). Planners (Voogd, 1983) along with regional scientists (Nijkamp and van Delft, 1977) employed multiple-criteria analysis for evaluation of local and regional development alternatives and were among the first to advance the idea of combining multiple-criteria evaluation with the map overlay method. This was the background that stimulated writing of the paper "Integrating geographical information systems and multiple criteria decision-making methods."

GIS-based multiple criteria analysis: Now and then

By 1995, the year the paper was published, research on GIS-based multiple-criteria analysis (GIS-MCA) was already generating interest in the GIScience community. Early papers (Janssen and Rietveld, 1990; Carver, 1991; Banai, 1993; Pereira and Duckstein, 1993) introduced the philosophy of GIS-MCA, areas of application, and solutions for integrating GIS with MCA techniques. Thanks to the research initiatives on the topics of spatial decision support systems, collaborative spatial decision-making, and GIS and society, organized by the National Center for Geographic Information and Analysis between 1990 and 1998, GIS-MCA gained wider recognition. The question one may ask today is: Was GIS-MCA a short-lived foray of geographers into operations research/management science or has it established itself as a viable component of GIScience? An indicator that can help answer the question is a growing trend in the number of scholarly publications in refereed journals. Jacek Malczewski compiled an extensive bibliography of 317 papers on the topic of GIS-MCA published in refereed journals between 1990 and 2004 (http://publish.uwo.ca/~jmalczew/gis-mcda.htm). A quick analysis of the bibliography shows that after a slow start between 1990 and 1995 with only 26 papers published the publication rate tripled every four to five years with 75 papers published between 1996 and 1999, and 216 papers published between 2000 and 2004. These publications account for much of the research on GIS-Multi-Criteria Evaluation (MCE) in theoretical, methodological, and application domains. Before discussing briefly the developments in these three domains, we should acknowledge that the vast majority of work on GIS-MCE has been focused on methods and applications neglecting theoretical questions about the foundations of GIS-MCE, validity of assumptions behind the methods, and the role of spatially grounded information in decision-making processes.

Theoretical developments

One area of research that emerged in the last ten years has been directed at understanding the dynamics of decision processes supported by GIS-MCE. The research purpose has been to develop knowledge about how and when to deploy GIS-MCE tools leading to higher quality and better outcomes of decision processes. In this regard participatory decision processes involving diverse actors have been of particular interest. Empirical studies of such decision processes supported by GIS-MCE included laboratory experiments (Jankowski and Nyerges, 2001), field experiments (Nyerges et al., in press) and field studies (Kyem, 2004). For example, the study conducted by Jankowski and Nyerges (2001) revealed the need to develop improved integration between visualization and multiple-criteria analysis tools over what was available at the time. The study showed that users of GIS-MCE tools tended to separate the use of visualization tools from MCE tools, employing the former during a decision problem exploration phase and the latter during the analytical phase. Thus the potential synergistic effect of using visualization tools in concert with MCE tools was never achieved.

Methodological developments

Much of work in the last ten years has been devoted to key methodological areas of MCE including: (1) generating decision alternatives and evaluation criteria; (2) eliciting stakeholder preferences; (3) combining outcomes of decision alternatives with criterion preferences; (4) testing the stability of MCE results in light of potential uncertainties; and (5) integrating MCE with GIS software.

Methods proposed for generating spatial decision alternatives ranged from GIS-based suitability analysis to single or multiple-objective optimization models formulated using integer (often binary) programming and solved with exact or heuristic approaches including simulated annealing and genetic algorithms. Some of these methods addressed explicitly spatial constraints such as contiguity and adjacency of feasible location alternatives (Aerts and Heuveling, 2002; Aerts et al., 2003).

Techniques for eliciting stakeholder preferences can be grouped into two categories: techniques leading to calculation of criterion weights, and techniques facilitating the expression of preferences without criterion weights. The former category includes techniques such as ranking, rating, and pairwise comparison (Malczewski, 1999), which have predominated in applications of GIS-MCE. The latter category is an attempt to overcome a problem of dealing with nonlinear weights. Criterion weights represent a linear preference function allowing compensation (that is, a poor performance of an alternative on a decision criterion can be compensated by a good performance on another decision criterion). True compensation, however, may require nonlinear weights, hence a nonlinear preference function, which is difficult to elicit. Techniques in this category rely on a user's evaluation of criterion tradeoffs leading to calculation of aspiration levels instead of weights (Jankowski et al., 2001).

Techniques for combining outcomes of spatial-decision alternatives with criterion preferences have been aimed at simplifying the decision space by ordering the decision options from the highest scoring to the lowest scoring, or by selecting a subset of non-dominated decision options. Such a monotonic order or a subset of nondominated options can be useful both in situations with a few decision alternatives and in situations with a large number of alternatives. Techniques from this category used in GIS-MCE included simple additive functions such as weighted summation and its generalized form called ordered weighted averaging, goal-based procedures, and decision rules based on outranking relations (Joerin et al., 2001).

Approaches suggested for dealing with uncertainty, indetermination, and vagueness in spatial decision situations included probabilistic functions for estimating alternative outcomes, sensitivity analysis of criterion weights using a deterministic approach where a single criterion weight is changed by a certain amount and a stochastic approach where a criterion weight variance is the result of Monte Carlo simulation, and the use of fuzzy sets to represent stakeholder evaluation of decision option outcomes.

Architectures for integrating MCE techniques with GIS software have been another area of work featured in the GIS-MCE literature. Three primary architectures have been used: (1) loose integration, in which MCE software is linked with GIS software through a file exchange mechanism, (2) tight integration, in which MCE and GIS software share a common user interface and data management mechanism, and (3) full integration, in which either the MCE routines are ported into GIS or GIS functions are embedded in MCE software. Today at least two GIS software packages including IDRISI (Eastman et al., 1995) and CommonGIS (Andrienko and Andrienko, 1999; and Chapter 18 in this volume) feature MCE functionalities fully integrated with GIS.

Application areas

Most of the GIS-MCE papers published since 1995 included examples of methods applied in some application domain context. The application context ranged from hypothetical problem and hypothetical data, through hypothetical problem and real-world data to real-world problem with real-world data. Published reports of studies representing the last situation have been far fewer than the reports representing the first two. Various application domains featured urban and regional planning, transportation, natural resources, natural hazards, recreation and tourism, real estate, and a broad domain of environmental applications. The breadth of application domains demonstrates the potential of GIS-MCE for spatial decision aiding in a variety of practical situations. Making GIS-MCE more relevant to real-world problems is a grand challenge for the field requiring better integration among three research domains: theory, methods, and application areas.

Future directions

It is instructive to review the evolution of intended users of GIS-MCE tools over the last 15 years. In the early 1990s it was an analyst assisting a decision maker. Since the late 1990s an analyst has been joined by a group of stakeholders creating a collective decision-maker. In the last few years the concept of stakeholders has expanded to embrace a potentially large number of participants representing various social groups. Will GIS-MCE be in a position to aid such a diverse group of users, and, if so, which areas of the field require future research efforts?

The vast majority of GIS-MCE prototype applications published in the literature dealt with structured problems, in which evaluation criteria were already known. In order to deal with similarly structured problems in future participatory processes, stakeholder information needs must be known. Keeney (1992) argued that the anticipation of stakeholder information needs during a decision process requires the knowledge of stakeholder values understood as deeply held beliefs and moral convictions affecting the stakeholder behavior in a decision situation. The practical problem is how to elicit stakeholder values. Hence, one of research challenges is to develop robust methods of eliciting stakeholder values, knowledge of which would allow incorporation of relevant evaluation criteria in GIS-MCE applications. Relevant evaluation criteria, which reflect stakeholder concerns about consequences of alternatives under consideration, might be the key factor in GIS-MCE gaining a broader acceptance outside the academic community.

Other challenges for future work on GIS-MCE include:

- Developing robust methods of accounting for imprecision, uncertainty, and ill-determination in human judgment and in estimates of outcomes of decision alternatives.
- Methods of facilitating meaningful participation of diverse stakeholder groups in local governance.

Addressing the former will require methods and tools for discerning MCE-relevant information from spatial characteristics of decision problems. Such information can be discerned from maps integrated with other information visualization tools (charts, tables, animations). Until now only a few attempts have been made to use maps as tools for structuring information in MCE (Andrienko et al., 2003). Maps can and should play a more prominent role in brainstorming, exploration, and analysis of alternatives for solving decision problems. The uncertainty inherent in human judgment can be partially accounted for by sensitivity analysis. Some progress in developing methods of analyzing the sensitivity of stakeholder judgments in spatial decision situations has been made (Feick and Hall, 2004). More work still needs to be done in order to develop robust sensitivity analysis techniques accounting for shifts in the ranking order of decision alternatives, not only in response to shifting criterion weights, but also in response to the propagation of uncertainty in weights and in estimates of decision option outcomes.

Addressing the latter challenge will require positioning GIS-MCE in a broader context of *analytic-deliberative* decision processes. Research in this area shows that meaningful public participation is possible, and decision outcomes are improved (National Research Council, 1996; Renn et al., 1997). The analytic component provides technical information that ensures broad-based, competent perspectives are treated. The deliberative component provides an opportunity to interactively give voice to choices about values, alternatives, and recommendations. The research challenge here lies in developing usable integrations of deliberative and analytic components that would facilitate their broad use. Internet-based GIS-MCE portals supporting analytic-deliberative processes might be one way to facilitate meaningful participation in large groups, while holding down the cost to all groups that wish to participate.

References

Andrienko, G. and Andrienko, N.,1999, Interactive maps for visual data exploration, *International Journal of Geographical Information Science*, 13(4), 355–374.

Andrienko, G, Andrienko, N., and Jankowski, P., 2003, Building spatial decision support tools for individuals and groups, *Journal of Decision Systems*, 12(2), 193–208.

Aerts, J.C.J.H., Eisinger, E., Heuvelink, G.B.M., and Stewart, T.J., 2003, Using linear integer programming for multi-site land-use allocation, *Geographical Analysis*, 35(2), 148–169.

Aerts J.C.J.H. and Heuvelink, G.B.M., 2002, Using simulated annealing for resource allocation, *International Journal of Geographical Information Science*, 16(6), 571–587.

Banai, R., 1993, Fuzziness in geographical information systems: Contribution from the analytic hierarchy process, *International Journal of Geographical Information Systems*, 7(4), 315–329.

Carver, S.J., 1991, Integrating multi-criteria evaluation with geographical information systems, *International Journal of Geographical Information Systems*, 5(3), 321–339.

Eastman J.R., Jin, W.G., Kyem, P., and Toledano, J., 1995, Raster procedures for multicriteria multiobjective decisions, *Photogrammetric Engineering and Remote Sensing*, 61(5), 539–47.

Feick, R.D. and Hall, G.B., 2004, A method for examining the spatial dimension of multicriteria weight sensitivity, *International Journal of Geographical Information Science*, 18(8), 815–840.

Jankowski, P. and Nyerges, T., 2001, GIS-supported collaborative decision making: Results of an experiment. *Annals of the Association of American Geographers*, 91(1), 48–70.

Jankowski, P., Andrienko, N., and Andrienko, G., 2001, Map-centred exploratory approach to multiple criteria spatial decision making, *International Journal of Geographical Information Science*, 15(2), 101–127.

Janssen, R. and Rietveld, L., 1990, Multicriteria analysis in GIS: An application to agricultural land use in The Netherlands. In *Geographical Information Systems and Regional Planning*, H.J. Scholten and J.C. Stilwell, Eds., Kluwer, Dodrecht, 129–139.

Joerin F., Theriault, M., and Musy, A., 2001, Using GIS and outranking multicriteria analysis for land-use suitability assessment, *International Journal of Geographical Information Science*, 15(2), 153–174.

Keeney, R.L., 1992, *Value-Focused Thinking*, Harvard University Press, Cambridge, MA.

Kyem, P.A.K., 2004, On intractable conflicts and participatory GIS applications: The search for consensus amidst competing claims and institutional demands, *Annals of the Association of American Geographers*, 94 (1), 37–57.

Malczewski, J., 1999, *GIS and Multicriteria Decision Analysis*, Wiley, New York.

National Research Council, 1996, *Understanding Risk: Informing Decisions in a Democratic Society*. National Academy Press, Washington, D.C.

Nijkamp, P. and van Delft, A., 1977, *Multi-Criteria Analysis and Regional Decision Making*, Studies in Applied Regional Science, 8, Leiden, Amsterdam.

Nyerges, T., Jankowski, P., Tuthill, D., and Ramsey, K., in press, Participatory GIS support for collaborative water resource decision making: Results of a field experiment, *Annals of the Association of American Geographers*.

Pereira, J.M.C. and Duckstein, L., 1993, A multiple criteria decision-making approach to GIS-based land suitability evaluation, *International Journal of Geographical Information Systems*, 7(5), 407–424.

Renn, O., Blattel-Mink, B., and Kastenholz, H., 1997, Discursive methods in environmental decision making, *Business Strategy and the Environment*, 6, 218–231.

Thiriez, H. and Zionts, S., 1976, *Multiple Criteria Decision Making*, Springer-Verlag, Berlin.

Voogd, H., 1983. *Multicriteria Evaluation for Urban and Regional Planning*, Pion, Amsterdam.

International Journal of Geographical Information Systems,
1996, Vol. 10, No. 5, 605–627.

13 The Geography of Parameter Space: An Investigation of Spatial Non-Stationarity

A. Stewart Fotheringham, Martin E. Charlton, and Chris Brunsdon

1 Introduction

In line with the current information revolution, geographers increasingly make a distinction between **the analysis of spatial data** and **spatial data analysis**. While both types of analyses investigate data having geographical co-ordinates, the former effectively ignores the geographical component and treats the data as if they were aspatial (*inter alia*, Flowerdew and Green 1994), whereas the latter makes use of the geographical component to explore explicitly spatial aspects of the data (*inter alia*, Anselin 1988, Besag 1986). It is this latter characteristic which makes spatial data analysis such a rich field for exploratory investigation. To clarify the distinction, suppose one is interested in the relation between an attribute Y and a set of attributes $X_1, X_2 \ldots X_N$ where the data all have spatial co-ordinates. A frequently used procedure to examine relations in such a situation is to regress Y on the set of X variables using the full data set. That is, observations from all the spatial units are used to produce a 'global' set of parameter estimates that relate the variable Y to each of the X's. In following such a procedure, a great deal of spatial information is lost in that the relations being examined may in fact exhibit significant spatial variation which is not uncovered in the estimation of the global parameter estimates. This variation is referred to as **spatial non-stationarity**.

In this paper, we propose an exploratory method of spatial data analysis which allows spatial variations in relations to be examined visually. It is useful to note at this stage that exploratory analysis is of two types: pre-confirmatory, which focusses on data accuracy, and post-confirmatory, which focusses on model accuracy. However, neither type of exploratory analysis is exclusively data or model-orientated. Pre-confirmatory exploratory analysis can be used to suggest relations to be included within a modelling framework and post-confirmatory analysis can be useful in highlighting unusual aspects of data. The exploratory analysis described here is post-confirmatory and we are concerned with examining differences between a 'global' calibration of a model and many different 'local' calibrations. In particular, we are interested in exploring spatial variations in model calibration results. We would argue that this method has the following advantages over more conventional analysis:

1. It allows greater insights into the nature and accuracy of the data being examined;
2. It provides a more detailed understanding of the nature of relations and their variation over space;
3. It demonstrates the possible naiveté of conventional approaches to data analysis that often ignore spatial non-stationarity; and
4. It allows a more detailed comparison of the relative performances of different types of analysis or different models.

2 Visualization of spatial non-stationarity

Consider a region divided into **R** rows and **C** columns in which a typical analysis is undertaken so that **R** by **C** observations are used to produce a single set of parameter estimates (the analysis of spatial data or **global** approach). An alternative technique involves placing a moving window of dimensions **r** by **c** (where **r** < **R** and **c** < **C**) over every cell of the matrix for which the window can be centred without any part of it lying outside the region. For every placement of the moving window the relation between *Y* and each of the *X*'s is estimated so that a set of parameter estimates for each relation is obtained (see Figure 13.1). These sets of parameter estimates have spatial co-ordinates and can therefore be mapped to show how the relation between *Y* and a particular *X* varies over space (a spatial data analysis or **local** approach).

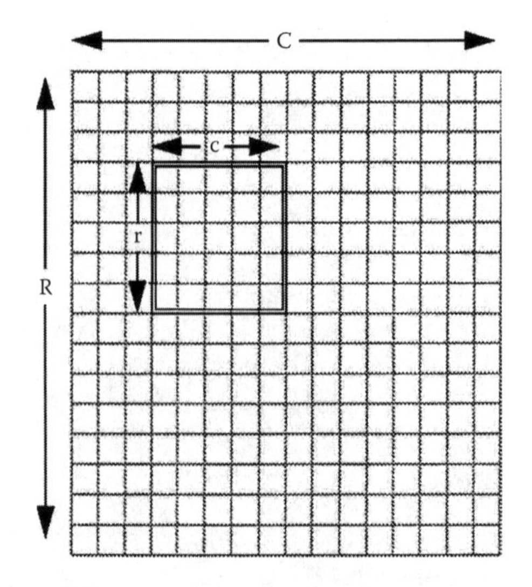

FIGURE 13.1 Moving window parameter estimation.

The distinction between the terms **global** and **local** as used here is similar to that between the analysis of spatial data and spatial data analysis. Obviously a spatial goodness-of-fit statistic and other diagnostics can be mapped in the same way. The key outputs then are two- and three-dimensional surfaces of parameter estimates, their t-values and an R-square statistic. Two-dimensional plots are also presented to compare the relative performance of global and local calibration methods.

It is noted at the outset that the detection of spatial non-stationarity depends on a subjective impression of the roughness of a three-dimensional surface of parameter estimates although work is currently in progress to produce a more objective comparative procedure. Clearly, if spatial non-stationarity does not exist, the parameter surface depicting spatial variations in relations will be smooth and increasing spatial non-stationarity will be indicated by increasing divergence from this smooth surface.

It is further noted that a square window is being employed and that other shapes of windows could be used. Perhaps the most intuitive window shape would be circular so that cells are included that are within a pre-specified distance from the centre of the window. However, a square window was chosen as being more representative of regions that are often defined for data analysis. It is more likely, for example, that a square or rectangular subset of that data is used to calibrate a model — rarely are study regions circular. As an example, the **global** region which forms the basis of this analysis and which is described below, is rectangular.

The technique proposed is similar to other developments in spatial data analysis such as those related to the modifiable areal unit problem by, *inter alia*, Fotheringham and Wong (1991) where the sensitivity of analytical results to variations in the spatial units for which data are observed is examined. Both studies utilize GIS to examine spatial aspects of the sensitivity of model performance. However, whereas the investigations of the modifiable areal unit problem keep the study area boundary fixed and examine the performance of an analytical method when the study space is divided in different ways, the technique described here in effect varies the study region so that different subsets of the data are used to calibrate a model. The technique is also part of a growing field of analytical techniques that provide spatial disaggregations of more traditional global statistics. Other examples include the development of a localised spatial association statistic by Getis and Ord (1991), the generation of variogram clouds and pocket plots by Cressie (1991), and the spatial expansion of parameter estimates by Jones and Casetti (1992) and by Fotheringham and Pitts (1995).

3 An example using OLS and bi-square regression

The basic aim of this example is to examine the presence of spatial non-stationarity in the calibration of a regression model. The model attempts to connect people and the environment by investigating the relations between population density and a series of attributes of the physical landscape, namely: elevation, land cover and soil type. The data on these variables are all taken from an area in north-east Scotland and are described more fully below.

3.1 Data

The data refer to an area 38·45 km by 51·2 km in the north-east of Scotland which is described in detail in Aspinall and Veitch (1993) and consist of a population density measure as the dependent variable, and ground elevation, land cover, and soil type as the independent variables. The object of this example is to provide a demonstration of the visualization of spatial non-stationarity rather than to model the socio-physical processes affecting the settlement pattern in the study area, but in the interests of having a reasonably plausible model on which to work, it might be expected that the distribution of population is related to land cover, elevation, and (perhaps more speculatively) the underlying soil type. The models are based on a 1 km raster.

Counts of residential population for each census enumeration district in the study area were extracted from the 1991 Census of Population together with the co-ordinates of the centroid of each enumeration district. The count data were aggregated to the 1 km raster on an 'all or nothing' basis — the raster is relatively crude by comparison with that used by Martin (1989) in allocating centroid based census data to a raster. The density for a given grid square I is calculated as:

$$POPDEN_i = \sum_j (\text{population}_j / \sqrt{((x_j - x_i)^2 + (y_j - y_i)^2)}) \tag{13.1}$$

where (x_j, y_j) is the co-ordinate of the lower left-hand co-ordinate of grid square j.

The elevation raster was resampled from a 1:250 000 scale digital terrain model, originally sampled every 100 m. The land cover data were extracted from a 50 m raster coverage, captured at 1:50 000 — the data were reclassed into 'woodland', 'grassland' and 'other', each being expressed as a percentage of the land area in a 1 km output raster. The soils data were extracted from 100 m raster data, captured at 1:250 000 scale and reclassed into 'podsols' and 'other soil types; in the output raster, the percentage of the land area that is composed of podsols is recorded.

3.2 OLS regression

The above data were standardised by transforming them into Z-scores and these were used to calibrate the following model:

$$POPDEN = \alpha + \beta_1 EL + \beta_2 WOOD + \beta_3 GRASS + \beta_4 PODS \tag{13.2}$$

such that global estimates of the four slope parameters as well as a global goodness-of-fit indicator, the R-squared statistic, were obtained. The globally fitted model is:

$$\begin{aligned} POPDEN = 0 - &\underset{(16\cdot4)}{0\cdot43EL} - \underset{(2\cdot2)}{0\cdot05WOOD} - \underset{(4\cdot1)}{0\cdot11GRASS} + \underset{(4\cdot8)}{0\cdot10\,PODS} \end{aligned} \tag{13.3}$$

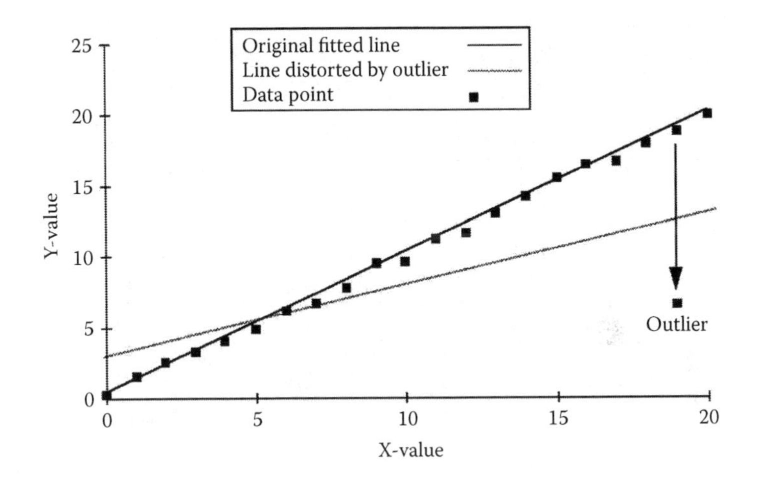

FIGURE 13.2 The effect of outliers on least squares regression.

with an R-squared value of 0·26. The values in parentheses are the absolute values of the t-statistics associated with each parameter estimate.

The results suggest that although there remains a considerable amount of variance to be explained, there do seem to be some significant relations between population density and features of the physical landscape. Population density decreases significantly as elevation rises, as the proportion of both woodland and grassland rises, and increases significantly in areas where podsols are found. A not unreasonable interpretation of these would be that in this part of Scotland, settlements are found primarily in low-lying areas which are cleared of woodland and grassland and which have fairly rich podsolic soils conducive to agriculture. We explore these interpretations in more detail below.

3.3 Bi-square regression

One of the major problems with ordinary least squares regression is its sensitivity to outlying observations as shown in Figure 13.2 and as described by many authors (*inter alia* Unwin and Wrigley 1987, Wrigley 1983). Although a general trend is followed by most observations, there is one exception that unduly influences the position of the regression line.

Here the difficulty can be overcome by manual intervention, simply by excluding the outlying case from the regression analysis. However, this is only possible when the data can be visualised. If the regression model had three or more independent variables, outliers would have to be identified in Euclidean space of at least four dimensions. For this reason, some automated controlling for outliers needs to be adopted.

In its crudest form, this can be thought of as zero-weighting the offending observations. These can be identified in a two-stage process:

1. Fit the model on all the data and compute the residuals.
2. Determine outliers by looking at the values of the residuals. Re-fit the model excluding these.

This seems a reasonable approach, but suffers from two main setbacks. Firstly, the threshold of residual size beyond which observations are deemed to be outliers is not specified. Secondly, the weighting makes a quantum leap from full weighting to absolute exclusion of the observation at this threshold. Thus, the weighting in the second stage can be thought of as a step function of r, the size of residual, and k, the threshold.

In the bi-weighted approach to outlier resistant regression, the cut-off is less drastic (Coleman *et al.* 1980). Weighting of the observations gradually falls off with increasing r, and finally becomes zero smoothly at the cut off point:

$$w(r) = \left(1 - \left(\frac{r}{k}\right)^2\right)^2 \qquad \text{if } r \le k \tag{13.4}$$

$$w(r) = 0 \text{ otherwise}$$

This is illustrated in Figure 13.3. The question still arises as to the value of k although a good experimental value for k has been found to be five times the median absolute value of the residuals. Finally, although the methodology suggested so far suggests only two iterations (i.e., fit regression, re-weight and fit again), the bi-square method can be iterated several times until the estimates of the regression coefficients converge.

In this instance, the bi-square methodology is applied to the moving window regression analysis alongside the ordinary least squares methodology. Due to the relatively small size of the regression windows (49 pixels), the effect of an outlying

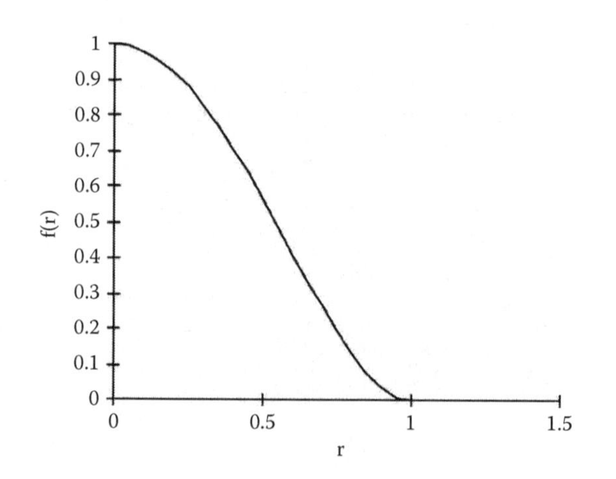

FIGURE 13.3 The weighting function in bi-square regression.

observation could be quite prominent on the estimates of the regression parameters. It is intended that this method will identify any outlying pixels and reduce any distortion that these may cause. In the case when the results from the two analyses do not differ drastically, this will at least reassure us that the ordinary least squares methodology was not affected by outliers.

The result of calibrating the global bi-square regression model with standardized data is:

$$POPDEN = -0 \cdot 08 - 0 \cdot 37\,EL - 0 \cdot 03\,WOOD - 0 \cdot 07\,GRASS + 0 \cdot 03\,PODS \quad (13.5)$$

which is similar to the OLS version although there are differences caused by the reduction in the influence of extreme values. It is not possible to report values of t-statistics for the parameter estimates because the standard errors reported in the bi-square calibration are unreliable because the weights are dependent on the observations and are therefore random variables. The R-squared value is unreliable for the same reason and should not be compared with that from the OLS regression.

4 Localised results

4.1 Localised OLS results

To obtain localised results, a 7 by 7 window was placed over each pixel which provides 49 data points for the model calibration at each pixel location. This was the smallest square window with which we felt comfortable as a basis for analysis. The data for each set of 49 pixels are locally standardised prior to model calibration. As described above, the typical output of this technique for each of the four slope parameters in the model is a three-dimensional surface as shown in Figure 13.4. A two-dimensional representation is also shown as an alternative view of the data. Clearly there are variations in the parameter estimate over the region although it is not clear how important these variations might be. In an attempt to overcome this problem we have chosen to depict surfaces of the parameter estimates divided by their standard errors, t-statistic surfaces, which slow more clearly the variations in results that can be obtained by using different parts of the data to calibrate the model. These t-statistic surfaces are shown in Figure 13.5 through Figure 13.8 for the parameters associated with the variables *elevation*, *woodland*, *grassland*, and *podsols*, respectively. The interpretation of these figures is that each point on the surface indicates the t-statistic associated with the parameter estimate that would be obtained *if a 7 by 7 matrix of cells around that point constituted the sample from which the estimate was derived.*

As an example, consider the interpretation of Figure 13.5 which depicts the degree of non-stationarity for the parameter associated with elevation. The regions of *upland* depicted in white are those areas for which the local parameter estimate is significantly positive. The regions shaded in light-grey, the *hillsides*, are where the parameter estimate is not significantly different from zero, and the *valley bottoms* shaded in dark-grey depict areas where the parameter estimate is significantly negative. So whereas the global regression result suggests a significant negative relation

FIGURE 13.4 OLS model elevation parameter.

between population density and elevation, there are a number of localised areas where population density **increases** significantly as elevation increases. The two-dimensional map, onto which the local road pattern has been added, suggests that the areas where a significant positive relation between population density and elevation exists are close to roads which in this area tend to hug lower ground. This could suggest an aversion to very low-lying ground which may be difficult to cultivate or which may be prone to flooding. Whatever the explanation, the results

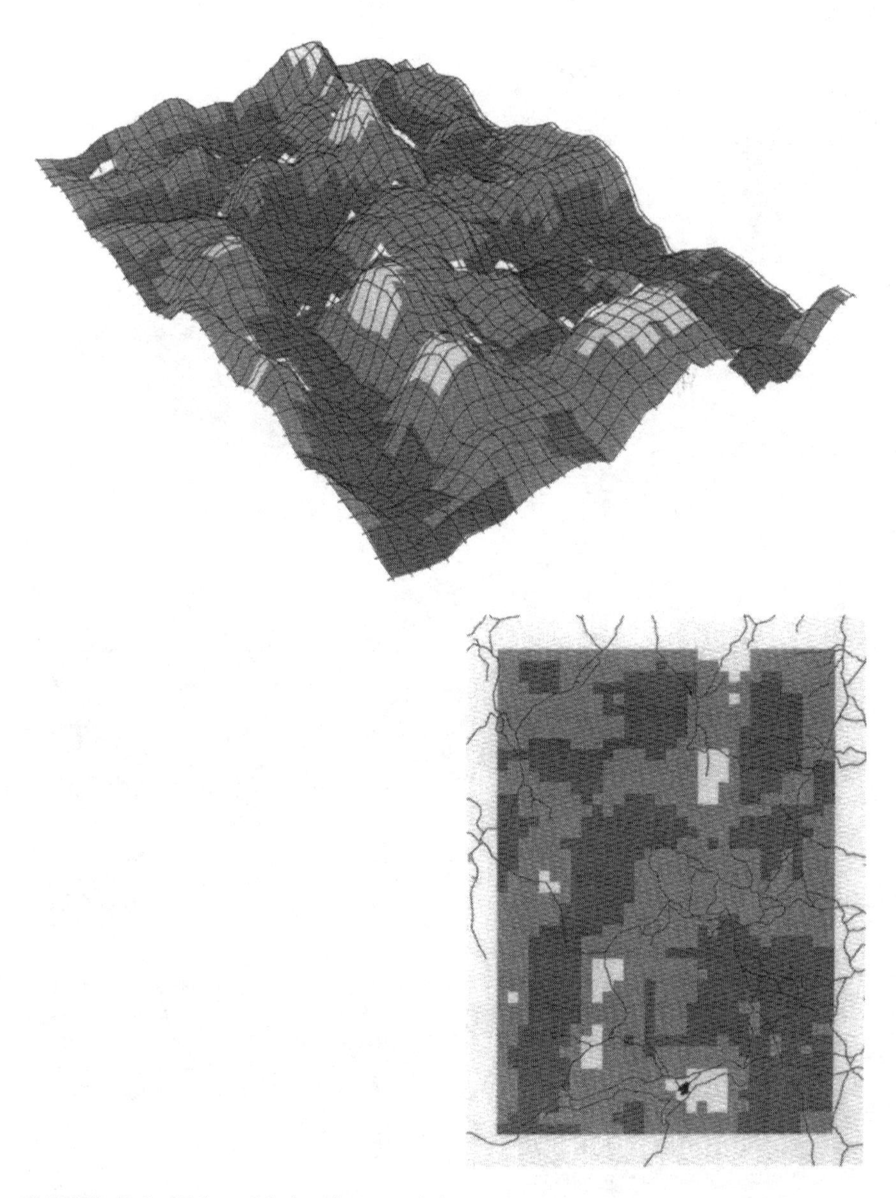

FIGURE 13.5 OLS model elevation *t*-statistic.

are striking in their depiction of the extent of spatial non-stationarity which is completely hidden in the traditional modelling approach.

Because the vertical exaggeration is held constant throughout, Figure 13.6 depicts a similar, although slightly less dramatic, picture for the woodland parameter. Use of the global data set would lead to the conclusion that there is a significant negative relation between population density and the amount of woodland although most of the local data sets would lead one to conclude that there is no significant

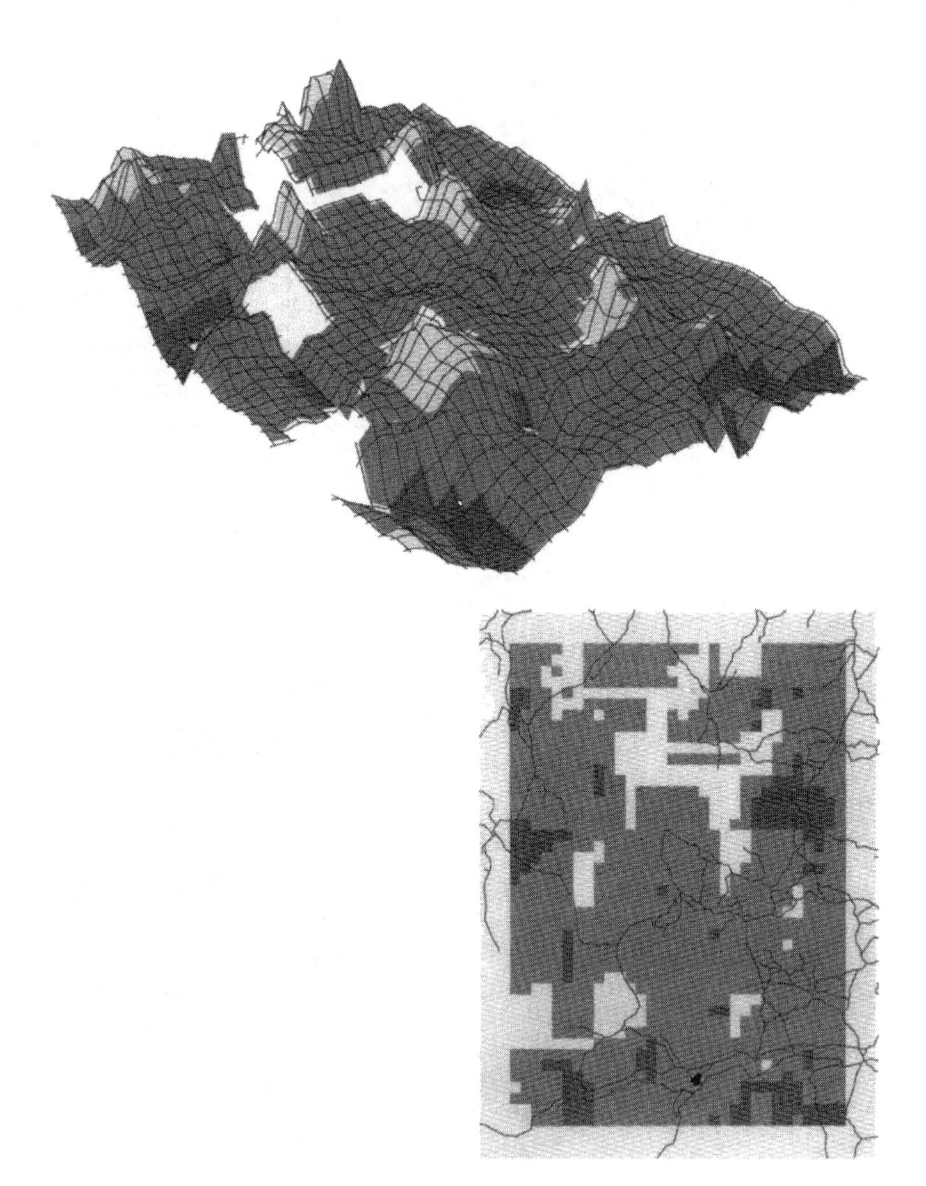

FIGURE 13.6 OLS model woodland *t*-statistic.

relation. This apparent discrepancy is probably caused by the large number of degrees of freedom in the former although it should be noted that the global estimate is just significant at the 95 per cent confidence level. There are a number of local data sets which would lead to the conclusion that there is a significant positive relation. The pixels for which no results are reported on this surface are those for which the data did not allow the local estimation of the parameter; this was because the variable was constant within the window.

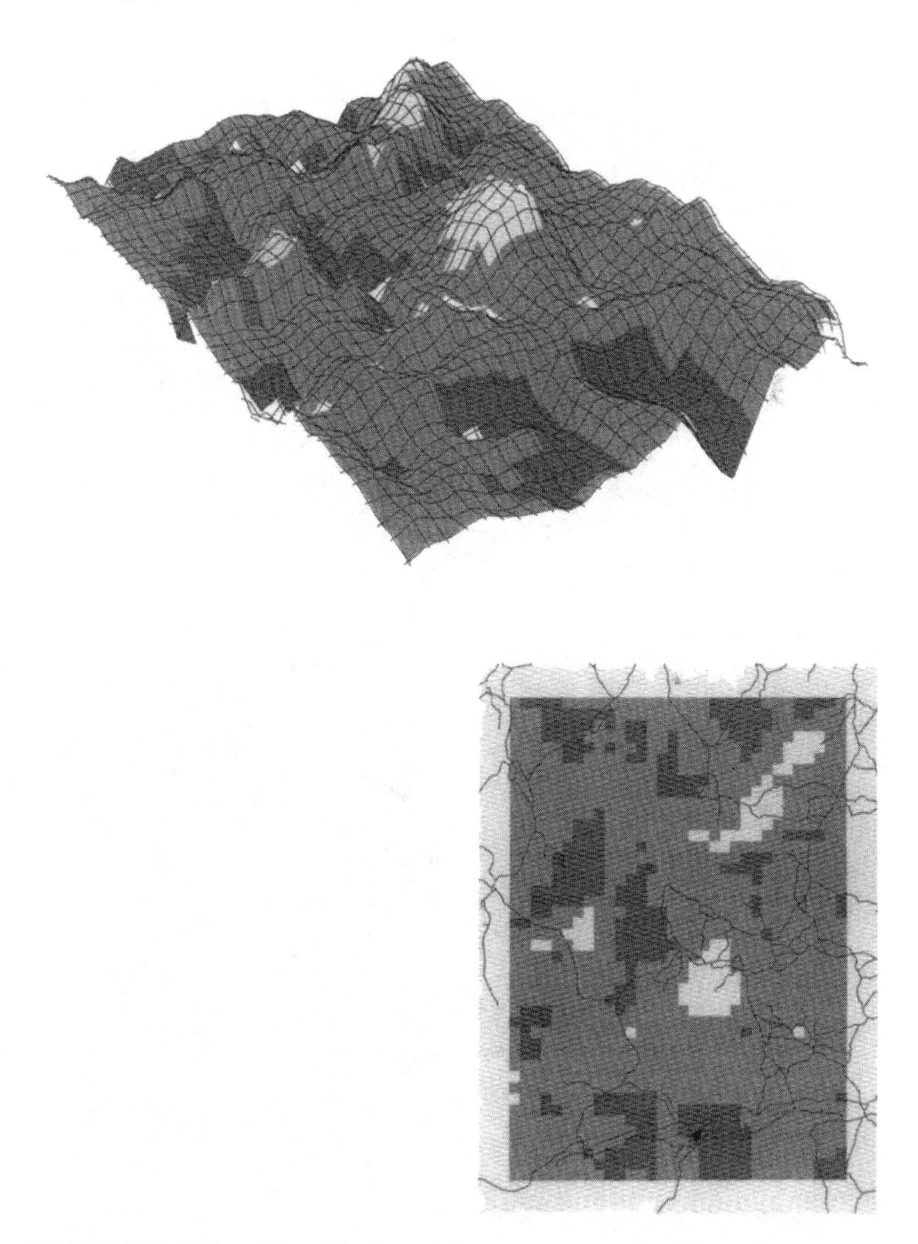

FIGURE 13.7 OLS model grassland t-statistic.

The grassland estimates in Figure 13.7 again suggest that one would reach the conclusion of no significant relation using many of the local data sets, in contrast to the globally significant negative relation. However, there are regions where the local data lead one to conclude the relation is significantly positive and others that lead one to conclude it is significantly negative. It needs more familiarity with the local geography to understand why the relations should show this degree of variation.

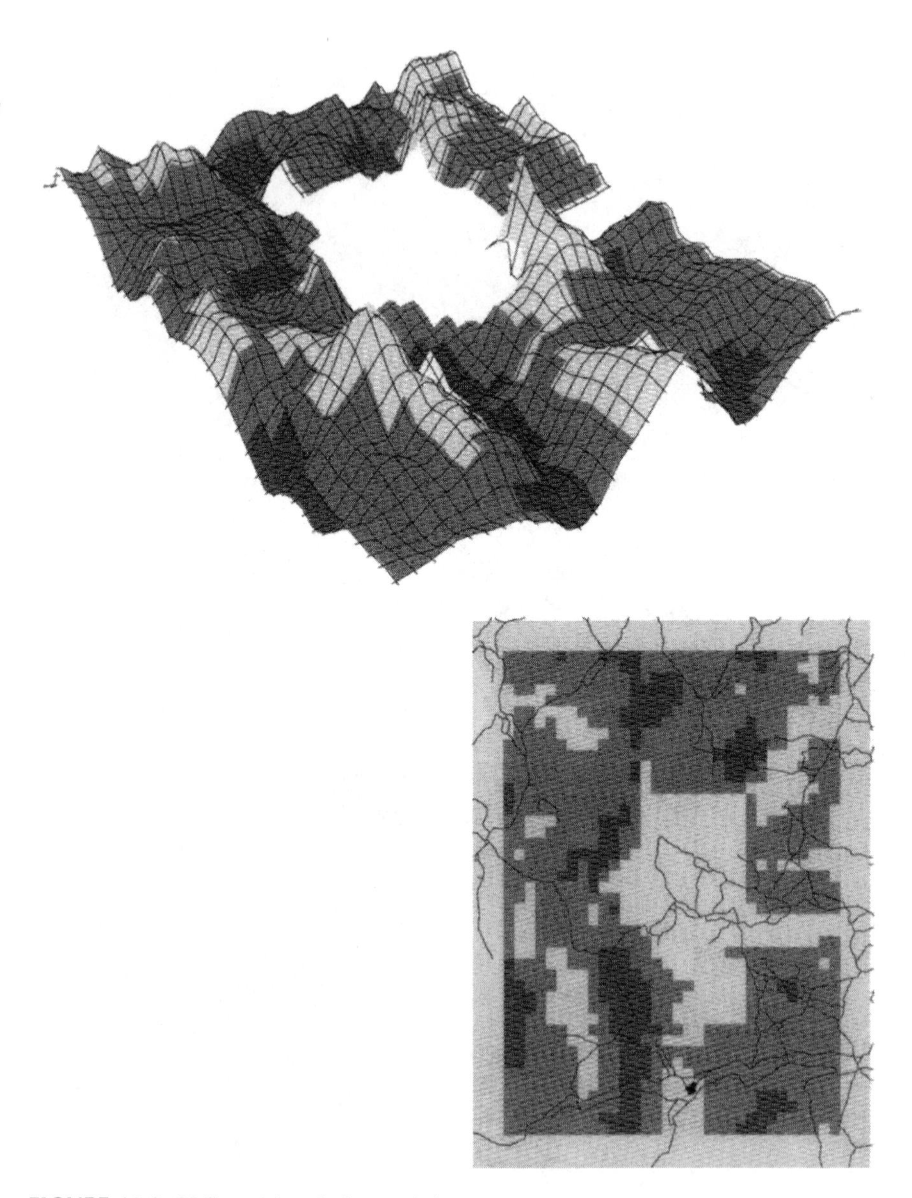

FIGURE 13.8 OLS model podsols *t*-statistic.

Figure 13.8 is dominated by a large area of missing values where the variable was zero (that is the soils were not podsolic) for all 49 pixels within the window. On the remainder of the surface there are substantial areas where the local data lead to the conclusion of a significant negative relation in contrast to the globally reported positive one.

The goodness-of-fit, measured by the *R*-squared statistic, of the local model is depicted in Figure 13.9 which uses a continuous grey scale to show spatial variations in the degree to which the model fits the data. There are clearly identifiable hills

FIGURE 13.9 OLS model goodness-of-fit.

and valleys in this surface. The model performance ranges from explaining only 1 per cent of the variance of the population density data in some parts of the region to explaining over 80 per cent in others. The globally fitted model explains 26 per cent of the variance. Again, the large spatial variation in goodness-of-fit suggests that spatial non-stationarity exists and that there are interesting fits to the data in some parts of the region, prompting questions regarding both the data and the model which need detailed local knowledge, perhaps from subsequent fieldwork, to answer. The surface therefore provides a good example of the use of this technique in encouraging a more detailed consideration of the data and the relations being examined than would happen in the conventional, global type of analysis currently the *modus operandi* of most research.

4.2 Localised bi-square results

Figure 13.10 through Figure 13.13 show the *t*-statistic surfaces produced by the bi-square calibration method. Given that this calibration method is designed to reduce the effects of outliers in the data, there will obviously be some differences between these surfaces and those produced by OLS. For reasons discussed earlier, any significance testing based on the bi-square technique is strictly exploratory. However, the differences are not particularly great and the bi-square surfaces depict a similar degree of spatial non-stationarity to the OLS ones and so will not be discussed further.

The *R*-squared surface shown in Figure 13.14 again is similar to that derived from the OLS calibration method although the absolute values cannot be directly compared to those produced by OLS because of the difference in the two calibration methods.

The similarity of the two sets of results usefully demonstrates that the OLS regression is not being unduly biased by the presence of outliers. The methodology also shows how the comparison can be used to assess relative model performance because if the results of the localised calibration were quite different, then it may be possible to make statements about which model is better able to capture spatial variations in relations.

4.3 A comparison of global, local OLS and local bi-square results

A final set of results depicts the relative performances of the three estimation techniques in replicating the observed population density surface. In Figure 13.15, three sets of scatterplots show the relations between the observed data on the horizontal axis and the global predicted values (Figure 13.15a), the local OLS predicted values (Figure 13.15b) and the local bi-square predicted values (Figure 13.15c). It is clear, and only to be expected, that the local regressions outperform the global one in replicating the data since the former allow the mean of the predicted values to vary across the study area. Perhaps more interesting is the extent to which the global regression results fail to replicate the spatial pattern of the original data as shown in Figure 13.16. Population density in this part of NE Scotland shows a marked trend with highest values towards the southeast and lowest values towards the northwest. The two locally derived estimated surfaces both succeed in picking up this trend, because of the moving mean, but the global regression model performs poorly and almost misses the trend entirely.

5 Discussion

This technique for the exploration and visualization of spatial non-stationarity can be used on any spatial data set; however, the number of spatial units should be large

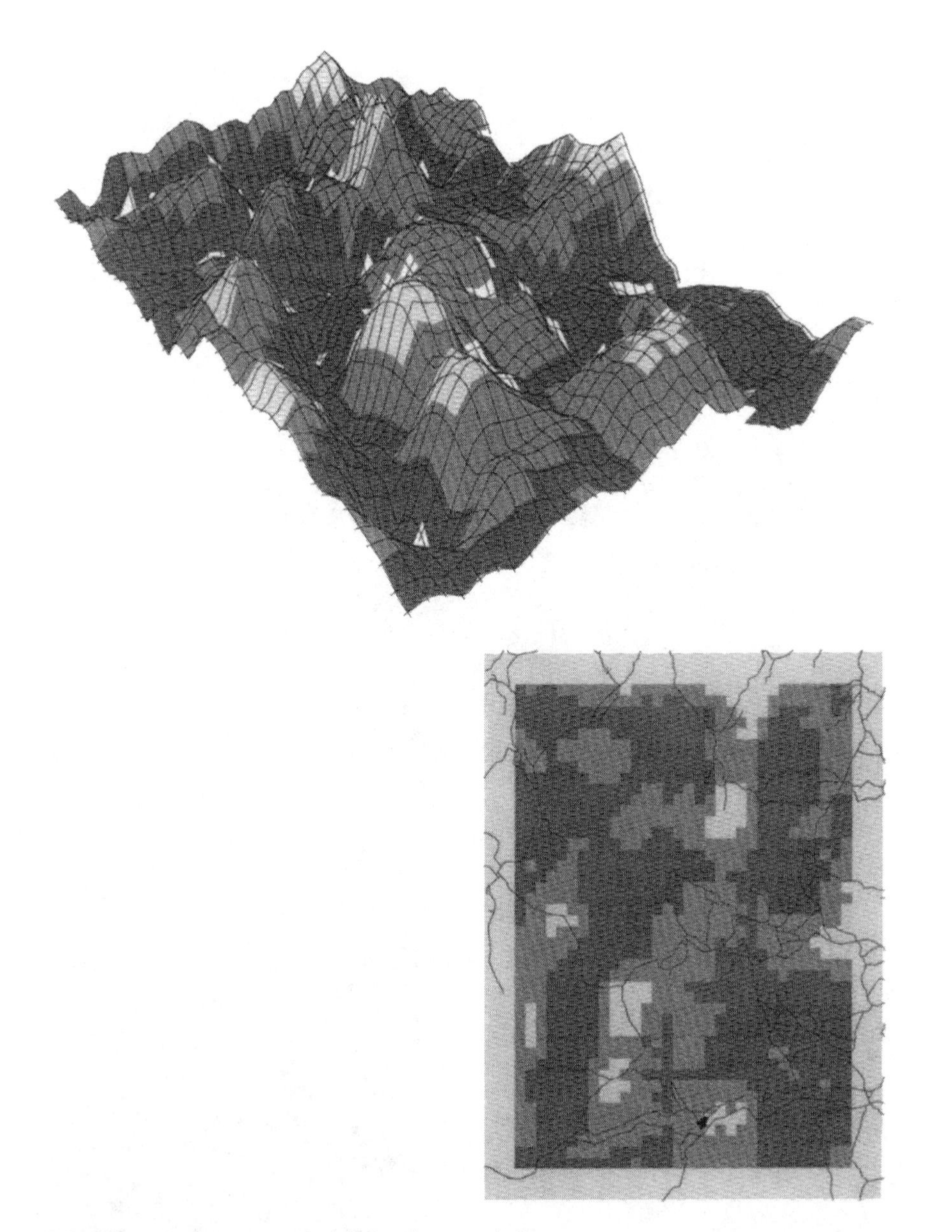

FIGURE 13.10 Bi-square model elevation t-statistic.

enough to ensure that those units excluded from the analysis, due to the windowing, form a reasonably small percentage of the total. The technique is easier to apply to raster data due to the uniform nature of the recording units but there is no conceptual reason to restrict its use to this type of data. The technique could be applied to irregular polygons although some care would have to be taken in defining the window and this is currently under investigation.

FIGURE 13.11 Bi-square model woodland *t*-statistic.

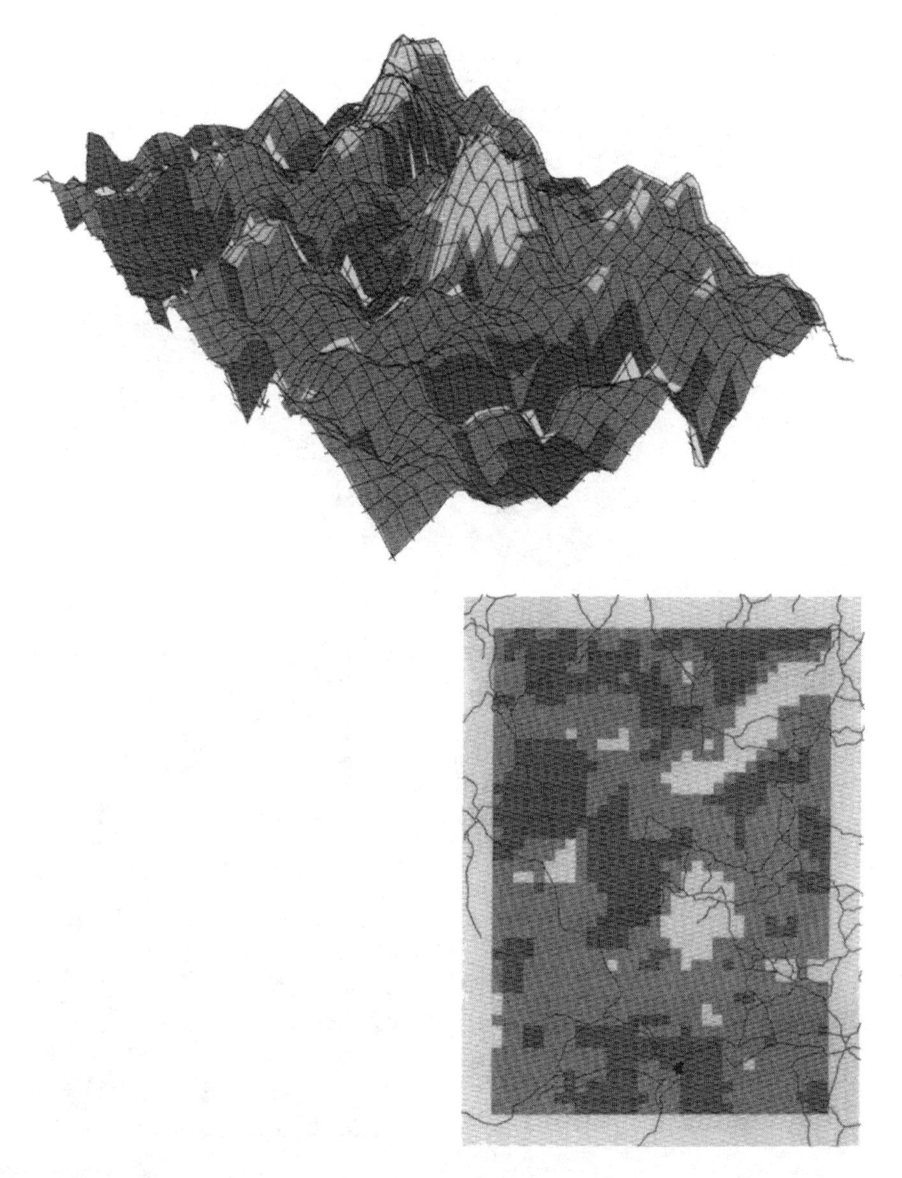

FIGURE 13.12 Bi-square model grassland *t*-statistic.

Given that a key output from this technique is a set of three-dimensional surfaces depicting, *inter alia*, spatial variations in a set of parameter estimates, judging the success of the technique is heavily visual and subjective. Consider, for example, a situation in which there was negligible spatial non-stationarity in a relation depicted by a parameter estimated through regression. The three-dimensional surface would then be a flat plain. Increasing deviations from this plain, as hills and valleys and

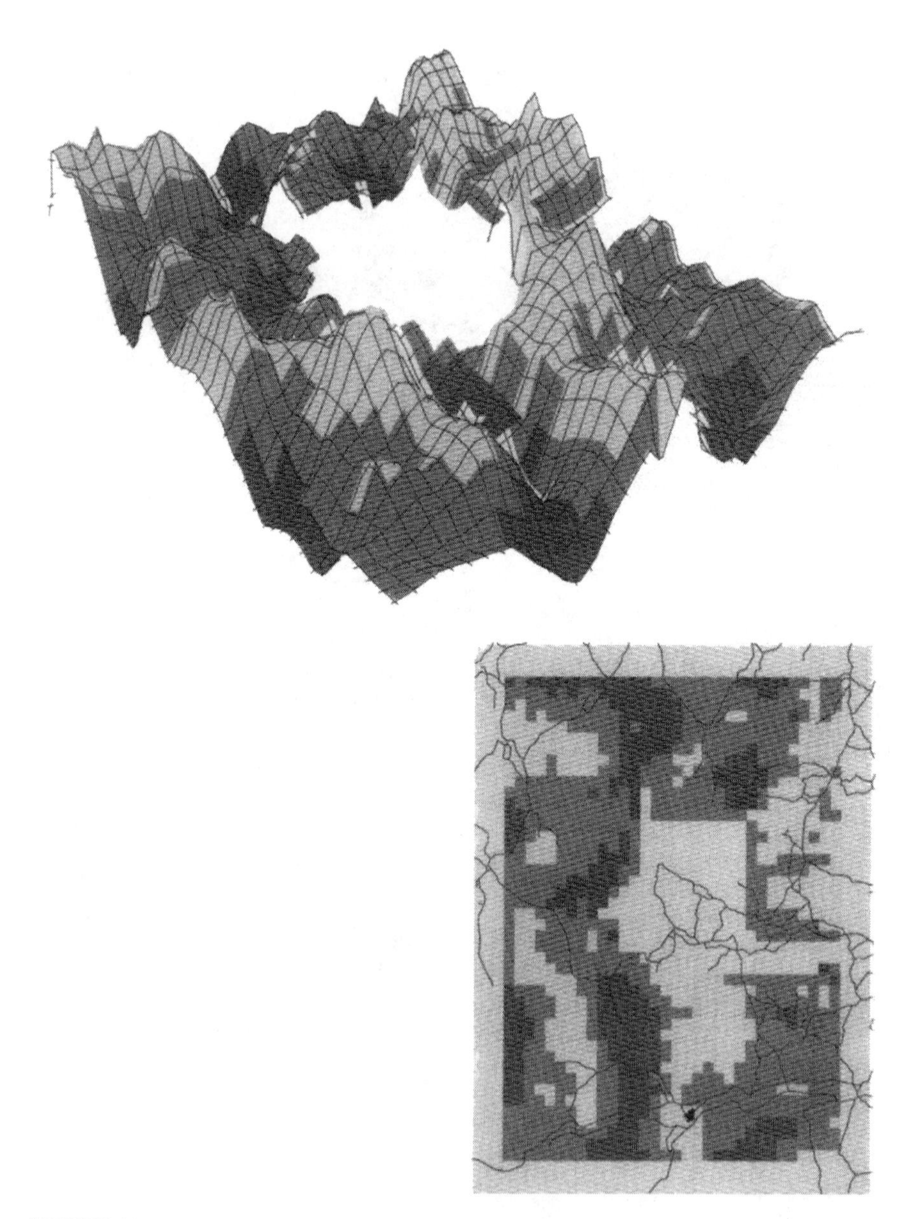

FIGURE 13.13 Bi-square model podsols t-statistic.

other features evolve, indicate increasing degrees of spatial non-stationarity. Consequently, if the result from this technique is a relatively flat surface, then one can infer that spatial non-stationarity is negligible which increases confidence in the **global** result as being a reasonable representation of a more general relation. However, if the output is a highly convoluted surface, then clearly the use of a **global** estimate is suspect.

FIGURE 13.14 Bi-square model goodness-of-fit.

The technique should be used for any modelling application where there is a suspicion of spatial non-stationarity. Although we have used a relatively simple linear model, with the justification that this is representative of the type of model used in many applications, the technique we describe can be used whenever it is useful to compare a **global** model with a series of **local** methods. For instance, the model in question could just as easily be a non-parametric regression model (Hastie and Tibshirani 1990) which could be examined with a global data set and with many different subsets of the data to allow comparisons similar to those made above. We would argue that any globally-fitted model may miss important spatial variations in

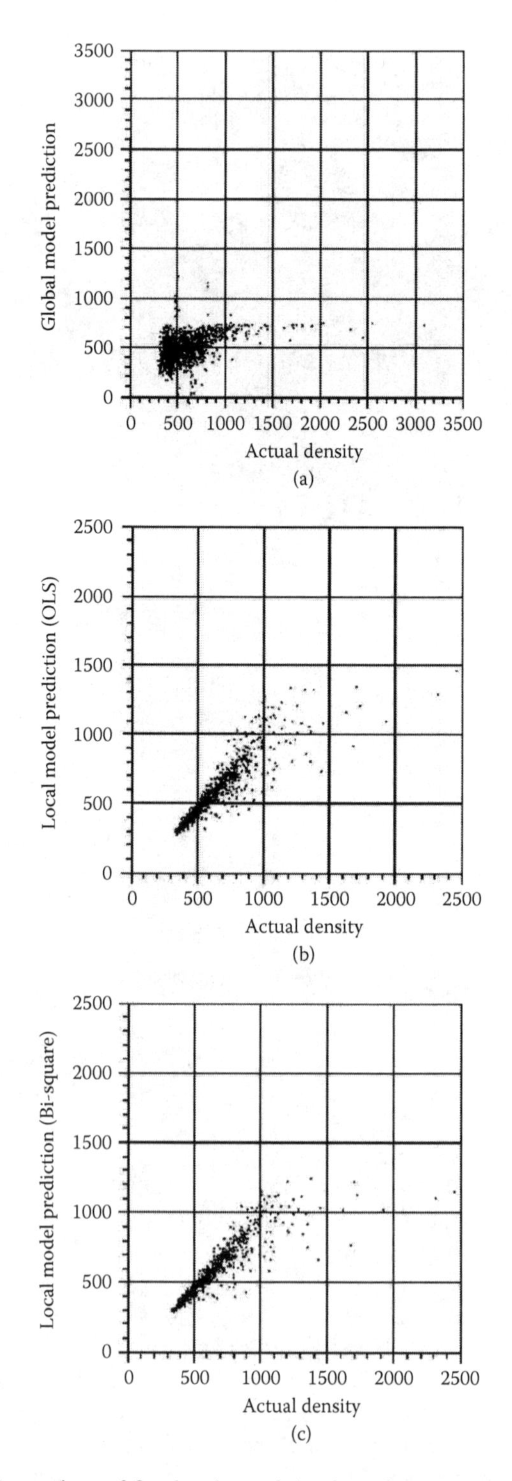

FIGURE 13.15 Comparison of fitted and actual density values.

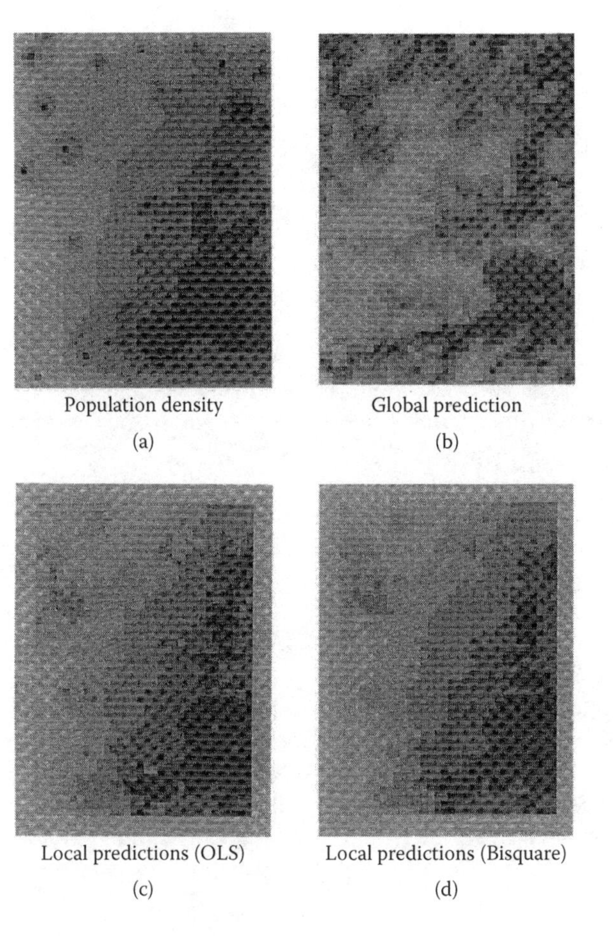

FIGURE 13.16 Spatial comparison of fitted and actual values.

relations and that this technique will provide greater insights into the data being used for calibration as well as the nature of the relations being measured.

Given the emphasis on visualization and the need to reference spatial co-ordinates, the technique should be applied through a GIS. It is not necessary to conduct the model calibrations within a GIS (and in the example described above the calibrations were undertaken outside a GIS) although the spatial referencing and graphics capabilities of a GIS make it almost a necessity in interpreting results. Hearnshaw and Unwin (1994), and particularly the chapters by Bracken (1994) and Gatrell (1994) therein, provide many other examples of the use of the visualization capabilities of a GIS for exploratory spatial analysis.

6 Reservations

We have the following reservations regarding the technique although we feel these are not sufficient to outweight the benefits.

1. The main output is a set of three-dimensional surfaces which may be difficult to interpret. In some places, we use two-dimensional surfaces to depict information although the parameter results are best seen in three-dimensions to get a sense of the variations within the area. Of course, any three-dimensional plot is subject to the vertical exaggeration used in the construction of the surface.
2. Although all the original data are used in the construction of the three-dimensional surfaces, not all the original cells of the region can be depicted because the outlying cells have to be 'stripped off' to provide consistency in the windowing. For example, if a 7 by 7 window is used on a raster image 500 by 500, the dimensions of the resulting three-dimensional surface will not be 494 by 494. In most cases this will not present a problem **unless** the extreme margins of the region are of interest or the study area is small.
3. It is much easier to apply the technique to raster data where the cell size is uniform. Where cells are of unequal size, the definition of the window will be more difficult and the numbers of data points within the window may vary. However, neither of these problems should prohibit the use of the technique.
4. To avoid the problem of multiple hypothesis testing, the surfaces of t-values should be used primarily as exploratory tools and care should be taken on drawing inferences about the parameter estimates. The surfaces indicate those parts of the study area where particular conclusions would be reached regarding a relation **if** local data were used in the calibration of the model. The surfaces therefore prompt some interesting questions about the data and the relations being examined that would not otherwise get asked.

7 Extensions

In this research we have adopted a conservative posture by moving one step at a time. To this point we have explored non-stationarity in a very common model format, linear regression, which is available in many GISs. However, it is a model that has *several* potential problems and given that one of the aims of this technique is to allow more detailed examination of any globally applied model, as well as to investigate spatial non-stationarity and data accuracy, an obvious extension would be to use the technique to evaluate modelling approaches designed explicitly to capture spatial effects. We intend to explore the capabilities of spatial lag and spatial error models in dealing with spatial non-stationarity. Would the parameter surfaces shown above be smoother if the estimates were drawn from either of these models capturing aspects of spatial dependency?

A further extension would be to apply the technique to a set of irregular polygons so that the optimal windowing technique for such a surface could be explored. In such an application it is likely that the window would be irregular and this could lead to a spatially adaptive windowing technique whereby windows of different magnitudes were allowed around each point in order to maximize some criterion, perhaps related to goodness-of-fit. It would also be interesting to examine the modifiable areal unit problem in the context of spatial non-stationarity: to what extent would the results change as the moving window changed in size? Would windows of different sizes identify non-stationarity at different scales? Presumably the surface will become smoother as window size increases and there may well be some optimal size of windowing that allows spatial trends in relations to be most clearly seen.

A final extension of the results would be to produce a more objective method of deciding when spatial non-stationarity is a serious problem. Clearly a three-dimensional parameter surface that is flat is indicative of no spatial non-stationarity and increasing deviations from this plain, as hills and valleys and other features evolve, indicate increasing degrees of spatial non-stationarity. It would be useful to have an objective means of deciding whether the degree of spatial non-stationarity invalidated the use of a global model. Such a test may be possible based on explained variances and would be similar to the F-test used in ordinary regression.

8 Summary

The classical use of regression with spatial data is to calibrate a model, spatial or aspatial, to produce a global set of parameter estimate which are implicitly assumed to represent relations that are constant across the space from which the data are drawn. This paper provides a relatively simple technique for examining this assumption and in the example above the results indicate that the assumption is questionable. Given the results provided here, for example, how much faith could one place on any extrapolation of the global results beyond this particular set of polygons in north-east Scotland?

The methodology of depicting spatial non-stationarity visually is part of a general trend towards exploratory analysis particularly in the use of spatial data. Geographers have long been embarrassed by their data because of the difficulties arising from properties such as spatial dependency and spatial non-stationarity. These problems are increasingly seen as the outcomes of data which are rich in information and which provide the catalyst for increasingly imaginative types of analysis. It is our hope that this and subsequent studies will advance the understanding of the behaviour of models in the context of spatial data analysis.

References

Anselin, L., 1988, *Spatial Econometrics: Methods and Models*, (Dordrecht: Kluwer Academic).

Aspinall, R. J., and Veitch, N., 1993, Habitat mapping from satellite imagery and wildlife survey data using a Bayesian modelling procedure in a GIS. *Photogrammetric Engineering and Remote Sensing,* **59**, 537–543.

Besag, J. E., 1986, On the statistical analysis of dirty pictures. *Journal of the Royal Statistical Society B*, **48**, 259–279.

Bracken, I., 1994, Towards improved visualization of socio-economic data. In *Visualization in GIS* edited by H. Hearnshaw and D. J. Unwin (Chichester: Wiley), pp. 76–84.

Coleman, D., Holland, P., Kaden, N., Klima, V., and Peters, S. C., 1980, A system of subroutines for iteratively re-weighted least-square computations. *ACM Transactions on Mathematical Software*, **6**, 327–336.

Cressie, N. A. C., 1991, *Statistics for Spatial Data* (New York: Wiley).

Flowerdew, R., and Green, M., 1994, Areal interpolation and types of data. In *Spatial analysis and GIS* edited by A. S. Fotheringham and P. A. Rogerson (London: Taylor & Francis), pp. 121–145.

Fotheringham, A. S., and Pitts, T., 1995, Directional variation in distance-decay. *Environment and Planning A*, **27**, 715–729.

Fotheringham, A. S., and Wong, D. W., 1991, The modifiable areal unit problem and multivariate analysis. *Environment and Planning A*, **23**, 1025–1044.

Gatrell, A., 1994, Density estimation and the visualization of point patterns. In *Visualization in GIS* edited by H. Hearnshaw and D. J. Unwin (Chichester: Wiley), pp. 65–78.

Getis, A., and Ord, J. K., 1991, The analysis of spatial association by use of distance statistics. *Geographical Analysis*, **23**, 189–206.

Hastie, T. J., and Tibshirani, R. J., 1990, *Generalized Additive Models* (London: Chapman & Hall).

Hearnshaw, H., and Unwin, D. J., 1994, *Visualization in GIS* (Chichester: Wiley).

Jones, J. P. III, and Casetti, E., 1992, *Applications of the Expansion Method* (London: Routledge).

Martin, D., 1989, Mapping population data from zone centroid locations. *Transactions of the Institute of British Geographers*, **14**, 90–97.

Unwin, D. J., and Wrigley, N., 1987, Control point distribution in trend surface modelling revisited: an application of the concept of leverage. *Transactions of the Institute of British Geographers*, **12**, 147–160.

Wrigley, N., 1983, Quantitative methods: on data and diagnostics. *Progress in Human Geography*, **7**, 567–577.

The Geography of Parameter Space: Ten Years On

A. Stewart Fotheringham, Martin E. Charlton, and Chris Brunsdon

Background

The germ of the idea for the research reported in our 1996 paper (Fotheringham et al.) came originally from two earlier papers by Fotheringham and Rogerson (1993) also in this journal (but obviously not a classic!) and Fotheringham and Charlton (1994), which outlined a series of important research topics in geographical information system (GIS)-based spatial analysis, one of which concerned spatial variations in model performance. The topic, however, was mentioned in passing and perhaps little would have happened to develop this idea further were it not for a workshop hosted by Richard Aspinall at the Maccaulay Institute in Aberdeen in 1994. Richard set a group of researchers the task of investigating the same data set from different angles. The data consisted of red deer counts in parts of the Scottish Highlands along with various environmental measurements at these locations. Our "angle" on this was to investigate whether the relationship between red deer counts and environmental factors varied over space. In the end, the data were far too sparse for this purpose and we cheated and used a different data set, which served our purpose much better! However, the nugget of the idea in these two earlier papers, allied to the empirical task we set ourselves for the Aberdeen workshop of investigating potential spatial variations in relationship, led directly to the research reported in the *IJGIS* paper. This turned out to be one of our earliest attempts to formulate what became known as geographically weighted regression (GWR) and the catalyst for a rapidly expanding research area.

Our intention was to investigate the stability of multiple linear regression models over geographical space, and in the paper we proposed investigating this using a rather crude method — a square window that "scanned" geographical data, fitting models only to the subset of the data that fell in the window. The window was centred in turn on each point in a rectangular grid, providing a series of calibrated models, each associated with a particular grid point. This gave a pixel-based view of geographical trends in the regression parameters, allowing any spatial patterns to be mapped in a raster-based GIS. This paper was thus highly influential in our later refinement of GWR, where we weighted data in a continuous manner (rather than the discrete approach in the paper) in order to examine possible spatial variations in relationships (Brunsdon et al., 1996; 1998; Fotheringham et al., 1997; 1998). The investigation of spatial nonstationarity of relationships was timely, given the concomitant development of the field of local statistics around this time (Fotheringham, 1997).

The development of GWR, which our paper was certainly one of the earliest examples, was quite different from most other local statistical measures such as Openshaw et al.'s (1987; and Chapter 2 in this volume) geographical analysis machine (GAM) and various scan statistics (Kulldorf, 1997) because rather than being a univariate statistical procedure, each parameter in a linear regression model linking a response variable y with a vector of predictor variables $\boldsymbol{x} = (x_1, x_2, \ldots, x_p)$ is regarded as a function of location in space (u,v). Thus a global regression model (Model 1) of the form

$$E(y) = b_0 + b_1\, x_1 + \ldots + b_p\, x_p \tag{13.6}$$

becomes a geographically varying model (Model 2)

$$E(y) = b_0(u,v) + b_1(u,v)\, x_1 + \ldots + b_p(u,v)\, x_p \tag{13.7}$$

The pixel-based view of parameters in the original paper, when cast into a GWR framework, equates to the estimation and visualisation of each of the continuous parameter surfaces $b_k(u,v)$. The method for estimating these surfaces in GWR borrows from the "local likelihood" ideas of Cleveland (1979). To estimate the value of $b_k(u,v)$ at some specific point (u',v') one assumes that in the locality of that point one can approximate model (2) with model (1). By only considering observed data geographically close to (u',v'), one can then calibrate model (1) to obtain an estimate of $b_k(u',v')$. The "only considering observed data geographically close to (u',v')" is achieved by weighting the log-likelihoods of all of the observed values of $\boldsymbol{x}$ and y using a decreasing kernel function of their distance from (u',v'), and then finding the regression coefficients that maximise the sum of these log-likelihoods. For a regression of type (2) with normally distributed y-values, this is equivalent to a weighted least squares approach with the weights being a kernel function centred around (u',v'). Typical kernel functions now used are the bisquare:

$$\mathrm{pos}\{[1 - (d/h)^2]^2\} \tag{13.8}$$

and the Gaussian

$$\exp(-d^2/2h^2) \tag{13.9}$$

where d denotes distance from (u',v'), h is a smoothing parameter controlling how close to (u',v') data must be in order to influence the estimate of $b_k(u',v')$ (more commonly known as the bandwidth), and pos{.} denotes that negative values inside the curly brackets are set to zero.

All of this may seem a digression, until one realises that the moving rectangular window parameter estimation used in the original *IJGIS* paper is almost identical although the kernel is binary (i.e., data are weighted with either 1 or 0), and rather than being radially symmetrical, it is box-shaped. The paper therefore formed an important bridge between our original concern with spatial instability in regression models and the continuous GWR methodology developed in subsequent papers to address this issue.

Subsequent theoretical developments

Although the key ideas of this paper were exploratory, there were some considerations of statistical inference — in particular the pseudo-F statistic was explored. Further work in this area led to a more theoretical, statistical, inference-based approach. In particular, a number of hypothesis tests associated with spatial nonstationarity have been developed. Essentially these test a null hypothesis of the global model (1) above against an alternative of model (2). Some of these are very similar to the pseudo-F statistic suggested in the original *IJGIS* paper. A nonparametric test is achieved by considering a randomisation approach in Brunsdon et al. (1996), while Leung proposes a theoretically derived test based on the assumption of Gaussian errors (Leung et al., 2000). Other approaches to model testing and selection have also been proposed — in particular, use has been made of the corrected Akaike Information Criterion (Akaike, 1981; Hurvich et al., 1998; Fotheringham et al., 2002).

There have been other developments in terms of the modelling aspect of the paper. In addition to work on the spatial nonstationarity of OLS regression models, subsequent GW models for Poisson and logistic regression have been derived. Although in its infancy, work has also taken place on "mixed" GWR models, where some parameters are allowed to vary as functions of (u,v) while others remain fixed. An example of these developments may be seen in Nakaya et al. (2005) where a mixed Poisson GWR model is used to model premature mortality rates in the Tokyo metropolitan area.

Finally, as well as regression modelling, the idea of moving-kernel-based local statistics has been applied to other areas — in particular, local descriptive statistics such as standard deviation, skewness, and correlation. One of the more complex of these approaches is that of *geographically weighted principal components analysis.* All of this seems a long way from the material in this early paper, but essentially all of these applications draw on the original idea of using a moving window to explore geographical changes in the relationships between a number of variables and this is the key issue introduced here and probably why the paper is fairly heavily cited.

Diffusion of the research

If we take this paper as laying the foundations for GWR, then the technique has now been around for a decade. During that time its take up has followed a classic logistic diffusion curve in which we appear to be still on the steep part of the slope. Published papers using GWR, or referring to it, now run into the hundreds, with applications seemingly in all areas where spatial data are used. We have also published around ten further GWR-related papers, either extending the theoretical framework or demonstrating empirical applications. These publications culminated in a book on GWR published in 2002 (Fotheringham et al., 2002), which remains an important source of knowledge for those interested in this area. We have also written specialised software for calibrating GWR models (currently the software is in its third version with a fourth imminent), which has been distributed across a wide range of countries and application areas. We have received about 500 requests for the software from academics and nonacademics ranging from such diverse entities as the Southwestern Museum of Archaeology to the Buenos Aeries Police Department! It is now installed in about a dozen computer labs (that we know about!) and has been bought by several private sector firms. We have also been asked to conduct workshops devoted to GWR in various countries. These workshops range from one day to a week and have been highly successful in publicising the use of the technique and also its limitations. Perhaps what has been most surprising and most satisfying to us has been the adoption of GWR outside geography and the traditional fields of GI Science. It would seem to have caught a wave of diffusion of spatial statistical techniques and GIS to communities beyond the traditional heartlands of GI Science. While many geographers are turning their backs on anything remotely quantitative or technical, researchers analysing data sets in other academic areas and outside academia are beginning to realise that these data often have a locational component, and that they therefore need specialised technologies and methods in order to understand them and analyse them successfully. It is gratifying, therefore, for us to be associated with a technique that had its origins in *IJGIS* but has spread well beyond this journal's traditional readership. We thank the perceptive editor and reviewers who handled the original submission.

References

AKAIKE, H., 1981, Likelihood of a model and information criteria, *Journal of Econometrics,* **16**, 3–14.

BRUNSDON, C., FOTHERINGHAM, A.S., AND CHARLTON, M., 1996. Geographically weighted regression: A method for exploring spatial nonstationarity, *Geographical Analysis*, **28**, 281–298.

BRUNSDON, C., FOTHERINGHAM, A.S., AND CHARLTON, M., 1998, Geographically weighted regression — modelling spatial nonstationarity, *Statistician*, **47**(3), 431–443.

CLEVELAND, W.S., 1979, Robust locally weighted regression and smoothing scatterplots, *Journal of the American Statistical Association*, **74**, 829–836.

FOTHERINGHAM, A.S., 1997, Trends in quantitative geography I: stressing the local, *Progress in Human Geography,* **21**, 88–96.

FOTHERINGHAM, A.S., BRUNSDON, C., AND CHARLTON, M., 1998. Geographically weighted regression: A natural evolution of the expansion method for spatial data analysis, *Environment and Planning A*, 30, 1905–1927.

FOTHERINGHAM, A.S., BRUNSDON, C., AND CHARLTON, M., 2002, *Geographically Weighted Regression: The Analysis of Spatially Varying Relationships*, Wiley, Chichester.

FOTHERINGHAM, A.S. AND CHARLTON, M., 1994, GIS and exploratory spatial data analysis: An overview of some research issues, *Geographical Systems*, **1**, 315–327.

FOTHERINGHAM, A.S., CHARLTON, M.E., AND BRUNSDON, C., 1996, The geography of parameter space: An investigation into spatial non-stationarity, *International Journal of Geographic Information Systems*, **10**, 605–627.

FOTHERINGHAM, A.S., CHARLTON, M.E., AND BRUNSDON, C., 1997, Two techniques for exploring non-stationarity in geographical data, *Journal of Geographical Systems*, **4**, 59–82.

FOTHERINGHAM, A.S. AND ROGERSON, P.A., 1993, GIS and spatial analytical problems, *International Journal of Geographic Information Systems*, **7**(1), 3–19.

HURVICH, C.M., SIMONOFF, J.S., AND TSAI, C.-L., 1998, Smoothing parameter selection in nonparametric regression using an improved Akaike Information Criterion, *Journal of the Royal Statistical Society Series B*, **60**, 271–293.

KULLDORF, M., 1997, A spatial scan statistic, *Communications in Statistics: Theory and Methods*, **26**, 1481–1496.

LEUNG, Y., MEI, C.-L., AND ZHANG, W.-X., 2000, Statistical tests for spatial nonstationarity based on the geographically weighted regression model, *Environment and Planning A*, **32**, 9–32.

NAKAYA, T., FOTHERINGHAM, A.S., BRUNSDON, C., AND CHARLTON, M., 2005, Geographically weighted Poisson regression for disease association mapping, *Statistics in Medicine,* **24**, 2695–2718.

OPENSHAW, S., CHARLTON, M., WYMER, C., AND CRAFT, A.W., 1987, A Mark I Geographical Analysis Machine for the automated analysis of point data sets, *International Journal of Geographical Information Systems*, **1**, 359–377.

International Journal of Geographical Information Systems, 1996, Vol. 10, No. 3, 269–290.

14 Qualitative Spatial Reasoning: Cardinal Directions as an Example

Andrew U. Frank

Abstract. Geographers use spatial reasoning extensively in large-scale spaces, i.e., spaces that cannot be seen or understood from a single point of view. Spatial reasoning differentiates several spatial relations, e.g., topological or metric relations, and is typically formalized using a Cartesian coordinate system and vector algebra. This quantitative processing of information is clearly different from the ways humans draw conclusions about spatial relations. Formalized qualitative reasoning processes are shown to be a necessary part of Spatial Expert Systems and Geographical Information Systems.

Addressing a subset of the total problem, namely reasoning with cardinal directions, a completely qualitative method, without resource to analytical procedures, is introduced and a method for its formal comparison with quantitative formula is defined. The focus is on the analysis of cardinal directions and their properties. An algebraic method is used to formalize the meaning of directions. The standard directional symbols (*N, W,* etc.) are supplemented with a symbol corresponding to an undetermined direction between points too close to each other which greatly increases the power of the inference rules. Two specific systems to determine and reason with cardinal directions are discussed in some detail.

From this example and some other previous work, a comprehensive set of research steps is laid out, following a mathematically based taxonomy. It includes the extension of distance and direction reasoning to extended objects and the definitions of other metric relations that characterize situations when objects are not disjoined. The conclusions compare such an approach with other concepts.

1 Introduction

Qualitative spatial reasoning is widely used by humans to understand, analyze, and draw conclusions about the spatial environment when the information available is in qualitative form, as in the case of text documents. Tobler and Wineberg (1971), for example, tried to reconstruct spatial locations of historic places from scant descriptions in old documents. Verbal information about locations of place can leave certain aspects imprecise and humans deduce information from such descriptions (for example descriptions of location in natural science collections (McGranaghan

1991). The use of cardinal directions is typical for reasoning in geographical or large-scale space, but other spatial relations are equally important.

Most formalizations or implementations of spatial reasoning rely on the Euclidean geometry and the Cartesian coordinate system. There is a clear need for a fully qualitative system of spatial reasoning, combining topological and metric relations. In qualitative reasoning, a situation is characterized by variables which *'can only take a small, predetermined number of values'* (de Kleer and Brown 1985, p. 116) and the interference rules use these values and not numerical quantities approximating them. It is clear that the qualitative approach loses some precision, but simplifies reasoning and allows deductions when precise information is not available.

The work presented here is part of a larger effort to understand how space and spatial situations are described and explained. Within the research initiative 2, 'Languages of Spatial Relations' of the National Center for Geographic Information and Analysis (NCGIA 1989) a need for multiple formal descriptions of spatial reasoning — both quantitative-analytical and qualitative — became evident (Mark *et al.* 1989, Mark and Frank 1990, Frank and Mark 1991, Frank 1990a, 1992).

Three special arguments for research in qualitative spatial reasoning serve as examples for numerous others:

1. The formalization required for the implementation in Geographic Information System (GIS), or generally for spatial information systems.
2. The interpretation of spatial relations expressed in natural language.
3. The comparison of the semantics of spatial terms in different languages.

Formalizing spatial reasoning is a precondition for the implementation of any spatial data processing method. Most methods currently used for spatial reasoning translate the problem for the quantitative to the qualitative realm and use analytical geometry for the solution (Dutta 1990, Webster 1990). This is not always an appropriate solution. The treatment of the uncertainty inherent in qualitative spatial descriptions causes problems that cannot be overcome easily in methods translating the problem to the quantitative realm (McDermott and Davis 1984). For this reason, the construction of expert systems which deal with space and spatial problems has been recognized as being difficult (Bobrow *et al.* 1986). Abstract, non-coordinate-based methods are necessary for the planning of the execution of spatial queries and their optimization. This is most urgent for intelligent spatial query languages for Geographical Information Systems (Egenhofer 1991). Specifically, algebraic properties of the operations, primarily commutativity and associativity, determine the optimization steps possible.

It has been suggested that user interfaces currently available in Geographical Information Systems are strongly influenced by Anglo-Saxon concepts which are difficult to translate to users from a different cultural or linguistic background (Campari in press, Mark *et al.* 1989 b). This problem will increase in importance, as GIS becomes widely used outside of the Anglo-Saxon world and may in the long run limit their usefulness to address the pressing needs of resource management and spatial planning in third world countries. A necessary, but not sufficient step to address this problem is to identify and formalize the concepts used in spatial reasoning, so one can compare methods used in one language with concepts present in another one (Campari and Frank in press, Campari and Frank 1993).

The structure of this paper is as follows. The next section describes large-scale space and discusses in some detail reasoning with cardinal directions as an example of spatial inference in a rule based system. Section 3 outlines desirable properties of reasoning with cardinal directions from geometric intuition. Section 4 describes alternative approaches to reason about cardinal directions based on the concepts of angular directions and projections. Section 5 outlines previous approaches to spatial reasoning and links the different perspectives. Section 6 lays out a comprehensive set of research steps which extends distance and direction reasoning to arbitrary objects and shows how to search for additional metric relations that characterize the situations when objects are not disjoint. The conclusion summarizes the results and briefly discusses alternative guiding principles for research in spatial reasoning.

2 Reasoning and cardinal directions in geographical space

Spatial reasoning is commonplace and defines the conceptualization of a situation as spatial. The details of spatial reasoning, the specific inference rules, depend on the type of space to which it applies. Conceptualization of space does not always lead to the customary Euclidean geometry, but depends on the situation. Zubin differentiates four different spaces, covering the range from the small objects in an area which can be seen at a single glance, to the large areas, where one accumulates knowledge about the space by moving through it (Mark *et al.* 1989 a, Couclelis 1992, Montello 1993). Similar differentiations have been made previously; for example, Kuipers and Levit (1990, p. 208) define large-scale space as

> 'a space whose structure is at a significantly larger scale than the observations available at an instant. Thus, to learn the large-scale structure of the space, the traveler must necessarily build a cognitive map of the environment by integrating observations over extended periods of time, inferring spatial structure from perceptions and the effects of actions.'

Geographical Information Systems, and geography and cartography in general, deal with this large-scale space, and this justifies the focus of the paper. Moreover, cardinal directions presented as an extensive example, are almost exclusively used in large-scale space. Reports of statements like the pencils are in the upper, west drawer — which is said to be used in some mid-west families — or a description as

> CHRISTY: He gave a drive with the scythe, and I gave a leap to the east. Then I turned around with my back to the north, . . . (Synge 1941, *The Playboy of the Western World,* p. 132)

which apply cardinal directions to smaller spaces, seems strange or quaint to most.

Cartographic rendering of large-scale spaces in reduced scale transfers the situation to small-scale space and makes concepts from small-scale spatial reasoning available for the analysis of geographical space. An extreme example of applying small-scale space concepts to large-scale space are maps that compare the area of two states by superimposing one onto the other, moving states in space as one would move books on a table. One can, therefore, not limit investigations to a single space and the spatial relations valid in it, but consider the transfer of concepts of spatial reasoning from one to another.

A standard approach to spatial reasoning is to translate the problem into analytical geometry and to use quantitative methods for its solution. Many problems can be conveniently reformulated as optimization with a set of constraints, e.g., location of a resource and shortest path. A similar approach of mapping a situation to coordinate space, has been applied to understand spatial reference in natural language text (Herskovits 1986, Nirenburg and Raskin 1987, Retz-Schmidt 1987). A special problem is posed by the inherent uncertainties in these descriptions and their translation into an analytical format. McDermott (1984) introduced a method using 'fuzz' and in (Dutta 1988, Dutta 1990, Retz-Schmidt 1987) fuzzy logic (Zadeh 1974) is used to combine such approximately metric data.

A qualitative approach to spatial reasoning does not rely on a coordinate plane and does not attempt to map all information into this framework. It can deal with imprecise data and, therefore, yields less precise results than the quantitative approach (Freksa 1991). This is highly desirable (Kuipers 1983, NCGIA 1989), because

1. precision is not always desirable;
2. precise, quantitative data is not always available.

Verbal descriptions have the advantage that they need not be metrically precise, just sufficient for the task intended. Imprecise descriptions are necessary in query languages where one specifies some property that the requested data should have. Consider, for example, the query "find all factories about 3 miles southeast of town A and northwest of town B". The depiction in Figure 14.1 is oversimplified and inaccurate, while the visualization of Figure 14.2 is too complex to show the desired information only.

In large-scale space several spatial relations are important. Pullar and Egenhofer (1988) classified spatial relations in:

1. *direction* relations that describe order in space (e.g. *north, northeast*),
2. *topological* relations that describe neighborhood and incidence (e.g., *disjoint*),
3. *comparative* or *ordinal* relations that describe inclusion or preference (e.g., *in, at*),
4. *distance* relations such as *far* and *near* and
5. *fuzzy* relations such as *next to* and *close*.

This paper concentrates on reasoning with cardinal directions as a major example. The specific problem addressed in this example is the following: Given the facts that San Francisco is west of St. Louis and Baltimore is east of St. Louis, one can

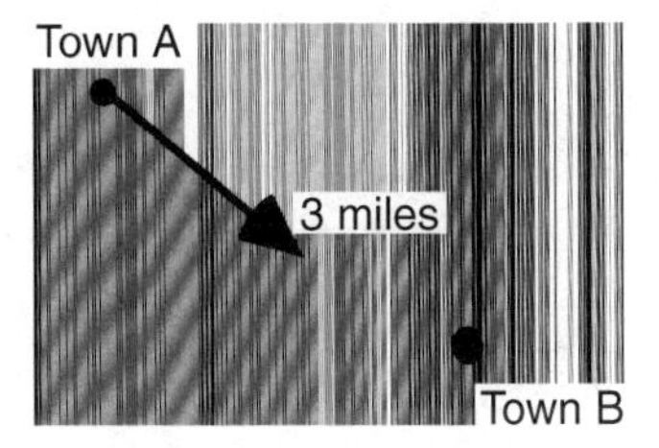

FIGURE 14.1 Overspecific visualization.

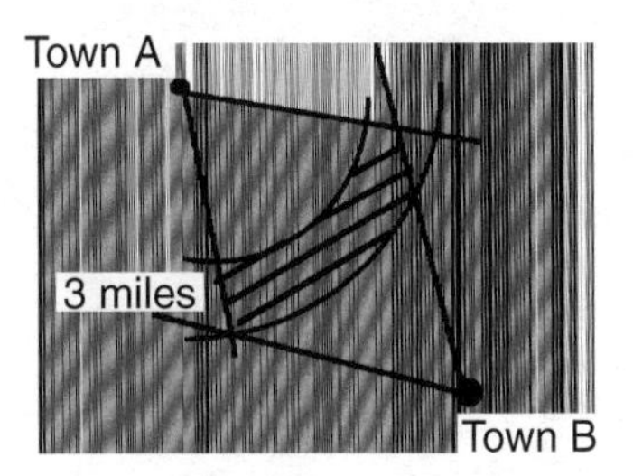

FIGURE 14.2 Complex visualization.

reduce the direction from San Francisco to Baltimore by the following chain of deductions:

1. Use 'San Francisco is west of St. Louis' and 'Baltimore is east of St. Louis', to establish a sequence of directions San Francisco-St. Louis-Baltimore.
2. Deduce 'St. Louis is west of Baltimore' from 'Baltimore is east of St. Louis'.
3. Use the concept of transitivity; 'San Francisco is west of St. Louis' and 'St. Louis is west of Baltimore', thus conclude 'San Francisco is west of Baltimore'.

This paper formalizes such rules and makes them available for inclusion in an expert system.

One is tempted to apply first-order predicate calculus to spatial reasoning with directions:

> 'The direction relation NORTH. From the transitive property of NORTH one can conclude that if A is NORTH of B and B is NORTH of C then A must be NORTH of C as well (Mark *et al.* 1989 a).'

The description of directional relationships between points in the plane can be formulated as propositions 'A is north of B' or 'north (A, B)'. Given a set of propositions, one can then deduce other relative positions as the induced set of spatial constraints (Dutta 1990). Following an algebraic concept, one does not concentrate on directional relations between points, but attempts to find rules for the manipulation of the directional symbols themselves, when combined by operators. An algebra is defined by a set of symbols that are manipulated (here the directional symbols like *N, S, E, W*), a set of operations and axioms that define the outcome of the operations.

3 Desirable properties of cardinal directions

The intuitive properties of cardinal directions are described in the form of an algebra with two operations applicable to direction symbols:

1. The reversing of the direction of travel (*inverse*).
2. The composition of the direction symbol of two consecutive segments of a path (*composition*).

The operational meaning of cardinal direction is captured by a set of formal axioms. These axioms describe the desired properties of cardinal directions and are similar to vector algebra.

3.1 Cardinal directions

Direction is a binary function that maps two points (P_1, P_2) in the plane onto a symbolic direction d. The specific directional symbols available depend on the system of directions used, e.g., $D_4 = \{N, E, S, W\}$ *or* more extensive $D_8 = \{N, NE, E, SE, S, SW, W, NW\}$. The introduction of a special symbol 0 (for zero) avoids the limitations that direction is only meaningful if the two points are different. In algebra this symbol (usually called *identity*) simplifies the inference rules and increases the deduction power. Its meaning here is that 'two points are too close to determine a direction between them'.

3.2 Reversing direction

Cardinal directions depend on the direction of travel. If a direction is given for a line segment between points P_1 and P_2, the direction from P_2 to P_1 can be deduced (Freeman 1975). This operation is called *inverse* (Figure 14.3), with

$$inv\ (dir\ (P_1, P_2) = dir\ (P_2, P_1) \quad \text{and} \quad inv(inv\ (dir\ (P_1, P_2))) = dir\ (P_1, P_2). \quad (14.1)$$

3.3 Composition

The second operation combines the directions of two contiguous line segments, such that the end point of the first direction is the start point of the second one (Figure 14.4).

$$d_1 \infty d_2 = d_3 \quad \text{means} \quad dir\ (P_1, P_2) \infty dir\ (P_2, P_3) = dir\ (P_1, P_3) \qquad (14.2)$$

Associativity: Compositions of more than two directions should be independent of the order in which they are combined:

$$d_1 \infty (d_2 \infty d_3) = (d_1 \infty d_2) \infty d_3 = d_1 \infty d_2 \infty d_3 \quad (associative\ law) \quad (14.3)$$

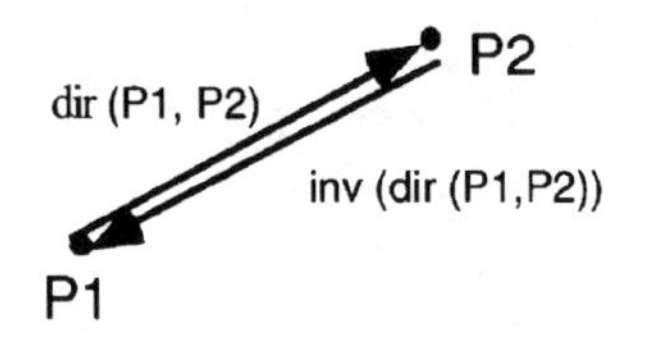

FIGURE 14.3 Inverse.

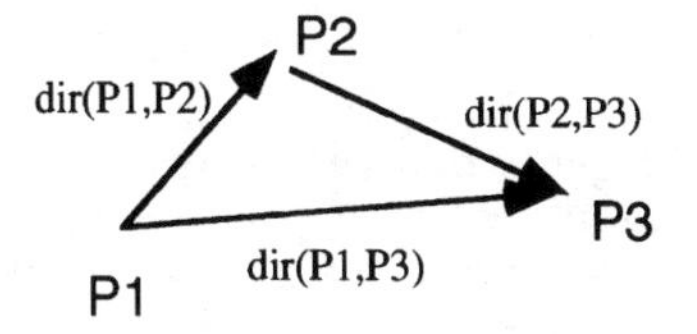

FIGURE 14.4 Composition.

Identity: Addition of the direction from a point to itself, $dir\ (P_1,\ P_1) = 0$, to any other direction does not change it.

Algebraic definition of inverse: In algebra, an inverse to a binary operation is defined such that a value, combined with its inverse, results in the identity value. Geometrically the inverse is the line segment that combines with a given other line segment to lead back to the start (Figure 14.5).

$$inv\ (d) \infty d = 0 \quad \text{and} \quad d \infty inv\ (d) = 0 \tag{14.4}$$

Computing the composition of two directions, where one is the inverse of the other, is in the general case an approximation. The degree of error depends on the definition of 0 and the difference in the distance between the points — if they are the same, the inference rule is exact. This represents a chain of reasoning like: New York is east of San Francisco, San Francisco is west of Baltimore, thus New York is too close to Baltimore (in the frame of reference of the continent) to determine the direction. In general, one observes that deduction based on directions alone is only applicable if all the distances are of the same order of magnitude.

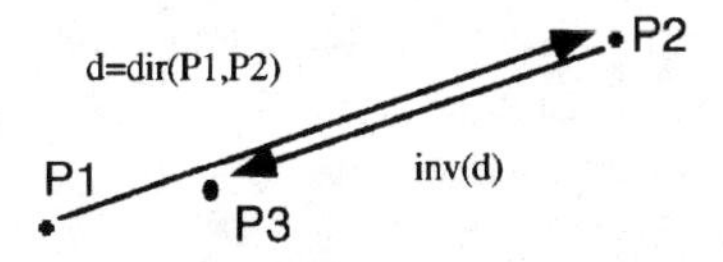

FIGURE 14.5 $d \infty inv\ (d) = 0$.

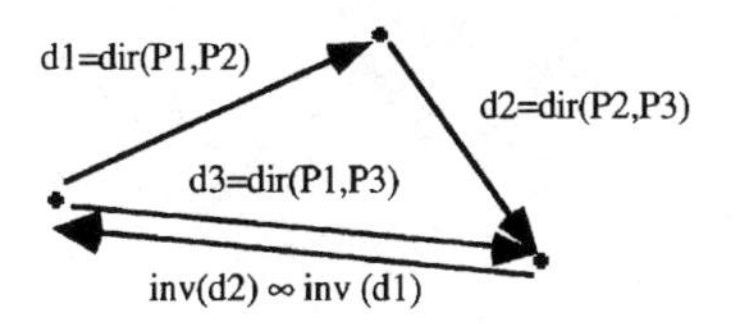

FIGURE 14.6 inv (d_3) = inv (d_1) ∞ inv (d_2).

The inverse is also used to compute the completion of a path to another given one (Figure 14.6) and one finds that a combined path is piece-wise reversible $inv\ (d_1 \infty d_2) = inv\ (d_2) \infty inv\ (d_1)$.

Idempotent: If one combines two line segments with the same direction, one expects that the result maintains the same direction. This is transitivity for a direction relation:

$$from\ d\ (A, B) \quad and \quad d\ (B, C)\ follows\ d\ (A, C)$$

3.4 Definition of Euclidean exact reasoning

These rules for qualitative spatial reasoning are compared with the quantitative methods using analytical geometry, as, for example, vector addition. The following definition gives a precise framework for such a comparison. A qualitative rule is called *Euclidean exact,* if the result of applying the rule is the same as if the data was translated to analytical geometry and applied to the equivalent functions, i.e., if there is a homomorphism. Otherwise, it is called *Euclidean approximate* (the symbol + denotes vector addition)

$$dir\ ((P_1, P_2) \infty dir\ (P_2, P_3) = dir\ (P_1, P_2) + (P_2, P_3)) \tag{14.5}$$

3.5 Summary of properties of cardinal directions

The basic rules for cardinal directions can be grouped into Group properties (6–8) and desirable properties of directions (9–13). The operations of inverse and composition are:

(a) The composition operation is associative (6).
(b) The direction between a point and itself is a special symbol 0, called *identity* (9) and (7).
(c) The combination of a direction and its inverse is 0 (cancellation rule) (8).
(d) The direction between one point and another is the *inverse* of the direction between the second point and the first (10).
(e) Combining two equal directions results in the same direction (*idempotent,* transitivity for direction relation) (11).
(f) The composition can be inverted (12).
(g) Composition is piece-wise invertible (13).

$$d_1 \infty (d_2 \infty d_3) = (d_1 \infty d_2) \infty d_3 \tag{14.6}$$

$$d \infty 0 = 0 \infty d = d \tag{14.7}$$

$$d \infty inv\ (d) = 0 \tag{14.8}$$

$$dir\ (P_1, P_1) = 0 \tag{14.9}$$

$$dir\ (P_1, P_1) = inv\ (dir\ (P_2, P_1)) \tag{14.10}$$

$$d \infty (d) = d \tag{14.11}$$

for any a, b, in D exist unique x in D such that

$$a \infty x = b \quad \text{and} \quad x \infty a = b \tag{14.12}$$

$$inv\ (a \infty b) = inv\ (a) \infty inv\ (b) \tag{14.13}$$

Several of the properties of directions are similar to properties of algebraic groups or follow immediately from them. Unfortunately, the idempotent rule (transitivity for direction relation) (3) or the cancellation rule (3′) contradict the remaining postulates. Searching for an inverse x for any $d \infty x = 0$, both $x = d$ (suing (3)) or $x = 0$ (suing (3′)) are possible, which contradicts the uniqueness of x in (4). A contradiction results also with associativity (1′), computing $d \infty$ d ∞ *inv (d)*, which yields either $d \infty$ *inv (d)* $= d \infty 0 = d$, depending how parentheses are set.

The customary method to avoid these problems is to define the composition operation to yield a set of values (not a single value as above) and to include in the resulting set all values that might apply. The cancellation rule (3′) then becomes $d \infty$ *inv (d)* = *{d,* 0*, inv (d)}*, because depending on the distances involved, any of these values may be correct.

4 Two examples of systems of cardinal directions

The differences are minor and are probably a reflection of the inherent vagueness of the concept of directions (Herskovits 1986, p. 192). Following the concept of prototype values in radial categories (Lakeoff 1987), the most likely value is selected and the reasoning rule labelled as Euclidean approximate. The results obtained are exact if the distances involved are all the same (or similar), which is the case for certain situations of navigation of large-scale spaces, e.g., walking in an inner city environment, where the block faces are comparable in length.

Two examples of systems of cardinal directions are studied, both using the same set of eight directional symbols (plus the identity). One is based on cone-shaped (or triangular) areas of acceptance, the other is based on projections. These two semantics for cardinal direction can be related to Jackendoff's principles of centrality, necessity and typicality (Jackendoff 1983, p. 121) as pointed out by Peuquet (1988).

4.1 Directions in eight or more cones

Cardinal directions relate the angular direction between a position and a destination to some directions fixed in space. An angular direction is assigned to the nearest named direction which results in cone-shaped areas for which a symbolic direction is applicable. This model has the property that 'the area of acceptance for any given direction increases with distance' (Peuquet and Zhan 1987, p. 66, with additional references) and is sometimes called 'triangular'.

TABLE 14.1
Composition of cone-shaped directions (lower case denotes Euclidean approximate inference).

	N	NE	E	SE	S	SW	W	NW	0
N	N	n	ne	0	0	0	nw	nw	N
NE	n	ne	ne	e	0	0	0	n	NE
E	ne	ne	E	e	se	0	0	0	E
SE	0	e	e	SE	se	S	0	0	SE
S	0	0	se	se	S	S	sw	w	S
SW	0	0	0	s	s	SW	sw	w	SW
W	nw	0	0	0	sw	sw	W	w	W
NW	n	n	0	0	0	w	W	NW	NW
0	N	NE	E	SE	S	SW	W	NW	0

The set of directional symbols for this system is

$$V_9 = \{N, NE, E, SE, S, SW, W, NW, 0\}.$$

and a turn of an eighth anti-clockwise is defined as:

$$e(N) = NE, \quad e(NE) = E, \quad e(E) = SE, \ldots, \quad e(NW) = N, \quad e(0) = 0.$$

Eight turns of 45° being the identity function. The inverse is defined as four eighth turns $inv\ (d) = e^4(d)$. The rules for composition of directions are (2′), (3) and the approximation rule (3′). This allows deduction for about one third of possible composition. A set of averaging rules, which allow the composition of directional values which are apart by 1 or 2 eighth turns, is necessary to complete the system. For example *N* is combined with *NE* is yield *N*.

$$e(e(d)) \infty d = e(d), \quad e(d) \infty d = d, \quad e(d) \infty inv\ (d) = 0, \quad d \infty e(d) = d, \quad \text{etc.}$$

In this system, from all the 81 pairs of values (64 for the subset without 0) compositions can be inferred, but most of them only approximately (Table 14.1).

Directions so defined do not fulfill all the requirements, because they violate the associative property; for example

$$(S \infty N) \infty E = 0 \infty E = E \quad \text{but} \quad S \infty (N \infty E) = S \infty NE = 0.$$

4.2 *Cardinal directions based on projections*

4.2.1 Directions in four half-planes

One can define four directions such that they are pair-wise opposites (Peuquet and Zhan 1987, p. 66) and each pair divides the plane into two half-planes. The direction

FIGURE 14.7 Two sets of half-planes.

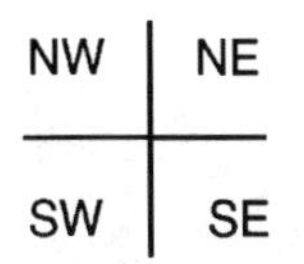

FIGURE 14.8 Directions defined by half-planes.

operation assigns for each pair of points a composition of two directions, e.g., south and east for a total of four different directions (Figures 14.7 and 14.8).

Another justification for this definition of cardinal direction is found in the structure geographic longitude and latitude impose on the globe. Cone-shaped directions better represent the direction of 'going toward', whereas the 'half-planes' better represent the relative position of points on the Earth. Frequently, the two coincide.

In this system, the two projections can be dealt with individually. When the two projections in N-S and E-W are combined to form a single system with the directional symbols:

$$V_4 = \{NE, NW, SE, SW\}$$

only the trivial cases ($NE \infty NE = NE$, etc.) can be resolved.

4.2.2 Projection-based directions with neutral zone

If points which are near to due north (or west, east, south) are not assigned a second direction, i.e., one does not decide whether or not such a point is more east or west, one effectively divides the plane in nine regions (Figure 14.9), a central neutral area,

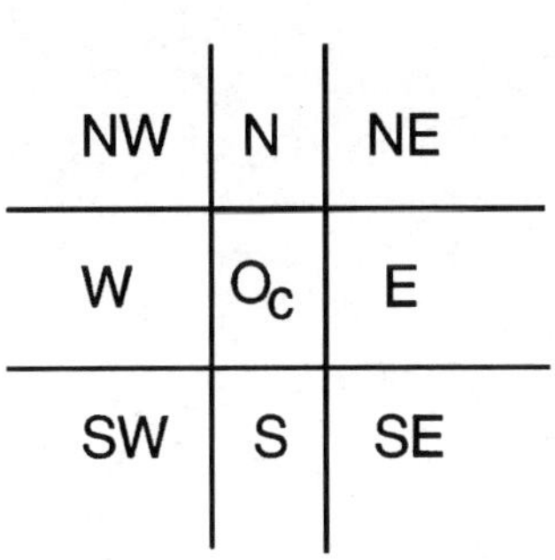

NW	N	NE
W	O_C	E
SW	S	SE

FIGURE 14.9 Directions with neutral zone.

four regions where only one direction applies and four regions where two are used. For N-S the three values for direction d_{ns} are $\{N, P, S\}$ and for the E-W direction the values are d_{ew} $\{E, Q, W\}$.

It is important to note that the width of the 'neutral zone' is not defined *a priori.* Its size is effectively decided when the directional values are assigned and a decision is made that P_2 is north (not northwest or northeast) of P_1. The algebra deals only with the directional symbols, not how they are assigned. The system assumes that these decisions are consistently made in order to determine if a deduction rule is Euclidean exact or not.

In one projection, selecting $d_{ns} = \{N, P, S\}$ as an example, the inverse operation is defined as $inv\ (N) = S$, $inv\ (S) = N$, $inv\ (P) = P$. For the composition the rules (3), (2′) and (3′) are required. The two projections in N-S and E-W are then combined to form a single system, in which for each line segment one of nine compositions of directions are assigned.

$$V_9 = \{NE, NQ, NW, PE, PW, PQ, SE, SQ, SW\}$$

This can be simplified to yield the customary cardinal directions, if P and Q are eliminated (replacing PQ by 0).

$$V_9 = \{NE, N, NW, E, W, 0, SE, S, W\}$$

If any of the results for the two projections is approximate, the total result is considered approximate reasoning.

The inverse is defined as the combination of the inverse of the projection, and the composition is the combination of the compositions of each projection. Using the three rules the values for each composition are computed (Table 14.2).

TABLE 14.2
Composition of cone-shaped directions (lower case denotes Euclidean approximate inference).

	N	NE	E	SE	S	SW	W	NW	0
N	N	N	NE	e	0	w	NW	NW	N
NE	NE	NE	NE	e	e	0	n	n	NE
E	NE	NE	E	SE	SE	s	0	n	E
SE	e	e	SE	SE	SE	s	s	0	SE
S	0	e	SE	SE	S	SW	SW	w	S
SW	w	0	s	s	SW	SW	SW	w	SW
W	NW	n	0	s	SW	SW	W	NW	W
NW	NW	n	n	0	w	w	NW	NW	NW
0	N	NE	E	SE	S	SW	W	NW	0

4.3 *Assessment*

The power of the eight direction cone-shaped and the four half-planes based on the directional systems are similar. Each system uses nine directional symbols, eight cone-shaped directions plus identity on the one hand, the Cartesian product of three values (two directional symbols and one identity symbol) for each projection on the other hand. The same rules are used to build both systems, but the cone-shaped system has to be completed with a number of 'averaging rules'.

Introducing the identity symbol 0 increases the number of deductions in both cases considerably. Only 8 out of 64 compositions could be resolved for cone-shaped directions and only trivial cases can be resolved for the half-plane based system. The systems with identity allow conclusions for any pair of input values (81 different pairs) at least approximately. The comparison reveals, that the projection-based directions result in less cases of approximation results than the cone-shaped (32 versus 56) and that it yields less often the value 0 (9 versus 25). Considering the definite values (other than 0) deduced, differences appear only for results where the 'averaging rules' are used for the cone-shaped directions, and the values differ by one eighth turn only. In conclusion, the projection-based system is producing more exact and more definite results.

An implementation of these rules and comparison of the computed combinations with the exact value was done and confirms the theoretical results. Comparing all possible 10^6 compositions in a grid of 10 by 10 points (with a neutral zone of three for the projection-based directions) shows that the results for the projection-based directions are correct in 50 per cent of cases and for cone-shaped directions in only 25 per cent. The result 0 is the outcome of 18 per cent of all cases for the projection-based, but 61 per cent for the cone-shaped directions. The direction-based system with an extended neutral zone is a quarter turn off in only 2 per cent of all cases, otherwise the deviation from the correct result is never more than one eighth of a turn, namely in 13 per cent of all cases for cone-shaped and 26 per cent for projection-based direction systems. In summary, the projection-based system of directions produces a result that is within 45° in 80 per cent of all cases and otherwise the value 0.

5 Spatial reasoning studies

Cardinal directions are similar to the four directions 'Front', 'Back', 'Left', and 'Right', which one encounters in spatial references (Herskovits 1986, Retz-Schmidt 1987). Although cardinal directions are prominent in geography, humans often use local reference frames, based on either their own position and heading or an implied frame for the object of interest. It is then necessary to convert distance and direction expressions valid in one frame of reference to the expression valid in another frame (McDermott gives a solution based on coordinate values including some 'fuzz' (1984)). In a qualitative reasoning system, composition is used.

The object in Figure 14.10 is in front of Building B which is to the right of Building A that the viewer is facing. Using composition of direction relations (Table

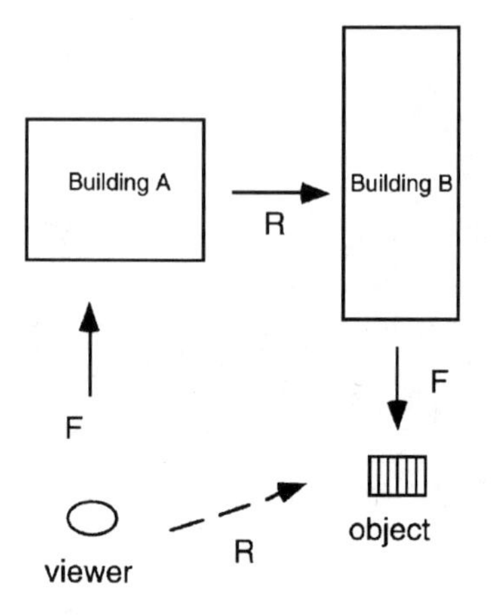

FIGURE 14.10 Reference frame example.

14.2) we can conclude that the object is to the right of the viewer using the following chain of inferences:

$$c1 = F \ (\textit{front})$$
$$c2 = R \ (\textit{right})$$
$$c'3 = F, \textit{ with } ref3 = e^{**}4$$
$$(\textit{the reference frame of Building B is 4 eights turned})$$
$$c1 \infty c2 = F \infty R = FR$$
$$c3 = e^{**}\, 4(F) = B$$
$$c1 \infty c2\ c3 = FR \infty B = R \ (\textit{using for example } NE \infty S = E \textit{ in table 2}).$$

Conclusion: the object is to the right of the viewer, which is correct.

Artificial intelligence has found the formalization of common sense knowledge a formidable task (Lenat *et al.* 1990). McCarthy (1975) proposed a subdivision of endeavours and singled out spatial reasoning as an important task, prevalent in many other kinds of reasoning. Spatial reasoning has been further subdivided in different approaches, mostly concentrating on the aspects necessary to resolve some specific tasks.

Most relevant for geographical or large-scale space is Kuipers' work, which primarily dealt with aspects of knowledge acquisition for wayfinding in large-scale space and to a lesser degree with spatial reasoning or a restricted sense (Kuipers 1982, 1983, 1990a, 1990b, Kuipers and Levit 1990). McDermott is more concerned with navigation in a complex environment consisting of buildings, etc., and deductions of routes that permit physical passage (McDermott 1980, McDermott and Davis 1984). Davis deals with the acquisition of geographical (i.e., large-scale) spatial information for navigation purposes (Davis 1986).

Others have primarily studied vision, and concentrated on aspects of understanding images for which spatial reasoning is necessary. An entirely qualitative approach is the work on *symbolic projections*. It translates exact metric information, primarily about objects in pictures, into a qualitative form (Chang *et al.* 1990). Segments in pictures are projected vertically and horizontally, and their order of appearance is encoded in two strings. Spatial reasoning, especially spatial queries, is executed as fast substring searches (Chang *et al.* 1987). Holmes and Jungert (1992) have demonstrated how symbolic projections can be applied to knowledge-based route planning in digitised maps. Papadias and Sellis (1993) have combined the concepts of symbolic projections and topological reasoning in one representation framework.

Related to composition of spatial relations is the problem of *consistency checking* in spatial constraint networks. A spatial constraint network is a qualitative graph-based description of a scene, where the nodes represent objects and the arcs sets of spatial relations corresponding to disjunctions of possible relations between objects. Inserting a new relation between two objects in the network affects not only the two objects, but because of composition, the insertion might yield additional constraints between other objects in the network *(constraint propagation)*. Studies of constraint propagation and consistency checking in networks of spatial relations include those by Guesgen (1989) and Hernández (1993).

In cognitive science and psychology the study of spatial relations and their use in human reasoning has been a topic of interest for many years. Stevens and Coupe (1978) observed distortions in the recall of direction relations between cities in Northern America and they attributed these findings to a hierarchical representation of space. Hirtle and Jonides (1985) observed similar distortions when people reason about distance relations. Twersky (1993) proposed *cognitive maps* (i.e., map-like mental constructs which can be mentally inspected), *cognitive collages* (i.e., thematic overlays of multimedia from different points of view) and *spatial mental models* (i.e., representations which capture spatial relations among distinct objects without preserving metric information) as alternative mental representations that depend on the reasoning tasks to be performed. Huttenlocker (1968) suggested that people construct spatial representations in order to draw conclusions about spatial or non-spatial tasks.

Several computational models and representation schemes have been proposed to deal with aspects of spatial reasoning. Glasgow and Papadias (1992), for instance, incorporate hierarchical spatial representations that can be used to reason with spatial relations in their representation scheme of *computational imagery.* Randell *et al.* (1992) proposed a theory for topological reasoning expressed in a many-sorted logic. Levine (1978) developed a semantic network where the arcs encode spatial relations such as left, inside, etc. Myers and Konolige (1992) designed a hybrid representation system that integrates both deductive and non-deductive spatial reasoning. A study of several systems that have been used to represent and reason with spatial relations can be found in Papadias and Kavouras (1994). Although the different systems can represent the same information about a given domain (they can be made *information equivalent*), they are not *computationally equivalent* since the efficiency of the reasoning mechanisms is not the same. Identical tasks may involve different algorithmic solutions and consequently have different complexities in distinct representational systems. A discussion about *informational* and *computational equivalence* of spatial representations can be found in (Larkin and Simon 1987).

6 Research envisioned

This section describes a comprehensive set of investigations to formalize spatial reasoning with large-scale space. The description links different aspects of spatial reasoning and discusses their relations. The research effort concentrates on the formalization of spatial reasoning using methods similar to those used above to reason about cardinal directions. Its goals are more modest than those of the researchers interested in navigation or vision, intending only to formalize inference rules relevant for spatial relations in large-scale space. This is similar to efforts of Dutta (1990). Despite these restrictions, a sufficiently complex problem remains to be investigated with results of utility for GIS.

Following the classical definition by Felix Klein, geometry is the branch of mathematics that investigates properties of configurations which remain invariant under a group of transformations (Blumenthal and Menger 1970). One differentiates the topological relations like ‘inside’, ‘connected’, ‘bounding’, etc., which remain invariant under homomorphic transformations, from metric relations, like ‘distance’, ‘direction’, etc., which remain invariant only under the much more restricted group of rotations and translations.

Initial work investigating topological structures (Egenhofer 1989, Egenhofer and Herring 1990, Egenhofer and Franzosa 1991) and metric relations, specifically distance and cardinal directions (Frank 1990, 1991) indicates that topological relations are the first level qualification which is further described by metric relations. This reflects the fact that topological relations are invariant under a much larger group of transformations than metric relations. Topological relations can be observed directly and characterized with a value on a nominal scale.

A method to characterize topological relations was proposed by Egenhofer and Herring (1990). It uses ideas from Allen (1983), initially developed for reasoning about time intervals, and applies them to two dimensional space. It introduces a simple characterization of topological relations in terms of three primitives, namely the test if the intersection of boundary, interior and complement of the two figures is empty.

The plan is stressing a stepwise extension from the known, qualitative reasoning with cardinal directions and topological relations. Each step is directed towards maximum separation of concerns — assuming that only the most simple problems can be solved (a tendency that can be generally observed in AI research). The results are then combined to yield the complex systems used by human beings — the complexity being the result of the combination and not present in the single step.

6.1 Distance and directions between points embedded in two-dimensional space

Distance and cardinal directions are two very widely used metric relations in large-scale space. Indeed, cardinal directions are only used in large-scale space. The prototypical case for the metric relations are distance and direction between points in the plane, i.e., co-dimension 2 (0 D points embedded in two-dimensional space).

6.1.1 Formalization of qualitative reasoning with distances

Preservation of metric properties is typical for our experience with physical, solid objects in small scale space (Adler *et al.* 1965). They can be moved but preserve length, angles etc. and Euclidean geometry can be seen as the ideal formalization of these concepts. In real physical space not all the assumed properties of Euclidean geometry are fulfilled. For example, one cannot insert between any two (physically embodied) points another one. Furthermore, movements back and forth over the same distance do not exactly cancel.

Similarly, visual perception cannot distinguish between two points that are sufficiently close (Roberts and Suppes 1967, Zeeman 1962). *Measurement theory* provides a theoretical base when the problems of limited acuity (the just noticeable difference) are taken into account (Krantz *et al.* 1971, Scott and Suppes 1958). Based on problems with observations of economic utility, Luce (1956, p. 181) has defined semi-orders as

'Let S be a set and $<$ and $\sim$ be two binary relations defined over S. (M, $\sim$) is a semi-ordering of S if for every a, b, c, and d in S the following *axioms hold:*

$S1$. exactly one of $a < b$, $b < a$, or $a \sim b$ obtains,
$S2$. $a \sim a$.
$S3$. $a < b$, $b \sim c$, $c < d$ imply $a < d$.
S4 $a < b$, $b < c$, $b \sim d$ imply not both $a \sim d$ and $c \sim d$'.

These observations can be visualized using graphs (Roberts 1969, 1971) and lead to tolerance geometry (Robert 1973). It will be necessary to consider these results when defining qualitative metric relations, using expressions like 'near', 'far' or 'very far' (Frank 1991).

6.1.2 Integrating reasoning about distances and directions into a single set of inference rules

Combining distance and direction reasoning, one becomes aware of the limitations of separating distance and direction reasoning in individual chains of inference. The integration of distance and direction inferences increases the precision of the results. Separating distance and direction information and applying independent rules for the inference on distance and direction, ignores the potential influence between distance and direction inference and reveals the shortcomings of both distance and direction reasoning. The result of far north composed with near south is far north (and not 0 as would result from rule (3′) above).

It is possible to construct tables for the addition of qualitative distance symbols under the assumption that the two paths are anti-parallel (i.e., parallel, but in the inverse direction). Other tables can be constructed if the two paths form a right angle, or one or three quarter turns apart. Such distance reasoning rules take into account the angle between the two paths, computed as the (symbolic) difference between the qualitative directions. The same concept is applied to the direction reasoning. Hong (1994) produced tables similar to table 2 indexed by the difference in length of the paths (computed from the qualitative distance information available).

This method improves the precision, without having to deal with the composition of all possible inputs (Mukerjee 1990). It takes advantage of the invariance of composition of paths under rotation. In lieu of a table for the results of all input values, only a small set of smaller tables is necessary.

6.2 Combining reasoning about distance and direction relations with topological relations

A valid criticism to formalized spatial reasoning, primarily if using a coordinate plane, is, that it attempts to map all information given into a single, uniform spatial frame (McDermott and Davis 1984). This does not compare with the reasoning methods of human beings, which resolves spatial relations in groups (Stevens and Coupe 1978). It is intuitively clear that spatial reasoning for a trip from home to office can be separated in several steps, starting with navigating inside one's home, moving with the car through the street network, moving inside the office building, etc.

In a first instance, the 'inside' topological relation is used, because it allows for propagation of spatial relations from the container to the parts contained. For example, if the distance between Maine and California is known to be about 4000 miles, one may conclude that the distance between any point in Maine and a point in California is in the same order of magnitude. Further the inside relation is transitive (A inside B and B inside C includes A inside C).

6.2.1 Hierarchically regionalized space

A multi-level hierarchical structure of areas each containing smaller areas is an extreme simplification but leads to a straightforward formalization. A similar simplification has been used successfully by Christaller (1966) for his central place theory. In a hierarchically regionalized space, reasoning about distances and cardinal directions between points is enormously simplified. Papadias and Sellis (1992) have shown how inference of direction relations can be achieved using hierarchical representations of space, while Car and Frank (1994) developed algorithms for hierarchical wayfinding. The errors produced in hierarchical systems resemble the ones observed in humans performing similar operations.

6.2.2 Generally partially ordered regionalized space

In general, the inclusion relation orders regions (i.e., two dimensional areas) partially, but does not automatically produce a strict hierarchy or even a lattice (Kainz 1988, Saalfeld 1985). The rules for reasoning about distances and directions developed for the hierarchical case must be extended to cope with situations, where a point is contained in multiple regions (town, school district, watershed, etc.), which do partially overlap. Papadias *et al.* (1994) have dealt with aggregation hierarchies where a geographical entity may belong to more than one parent and observed that the improper specification of multiple hierarchies can lead to the inheritance of inconsistent relations.

6.3 Generalize distance and direction relations to extended objects

Distance and direction relations between extended objects, i.e., objects, which are not point-like, need to be defined as an extension of metric relations between points. Humans ask questions like, what is the direction from Maine to Canada and expect an answer of 'north' (indeed, from Maine, Canada is also east and west, and there are some points in Maine where Canada is also found to the south). Peuquet and Zhan (1987) developed algorithms for the determination of direction relations based on a 'visual interpretation', but their determinations do not always correspond with others' judgment. Papadias and Sellis (in press) have also dealt with direction relations between arbitrary objects represented by their minimum bounding rectangles. Their method is projection-based, unlike Peuquet's method which is based on the concept of cone-directions.

Distances and directions are easily definable only for extended objects which are disjoint — i.e., the topological relations provide the first level of classification of spatial relations, and the metric relations a second level of characterization, with distance and directions only applicable for disjoint objects (different from the approach by Peuquet and Zhan 1987).

Using the results from the formalization of distance and direction between point-like objects and the solution to extending topological relations to higher co-dimensions, distance and direction relations between extended objects must be formalized, explaining distances between lines and points, two lines, lines and areas ('how far is the town of Dixmont from the Interstate'), etc.

6.4 Finding other metric relations to characterize relations of objects

Not only the situation of topologically disjoint objects can be further characterized with a metric value. Cardinal direction is also used for objects which are inside others. For example, Aroostook County (a county in the state of Maine) is said to be in the northern part of Maine — characterizing the topological relation of 'inside' additionally with a direction term. This is a different use of a cardinal direction than between two points.

Generalizing this example, the following device helps to discover additional metric relations. Imaging two objects, first quite distant from each other (Figure 14.11a) characterized by disjoint and a distance value. The smaller object approaches the other and 'meets' (Figure 14.11b), characterized by the topological relation alone, which implies distance 0. Moving the smaller object further, the objects overlap (Figure 14.11c), which could be characterized by a metric value, indicating the degree of overlap. Then the objects meet from the inside for a moment (Figure 14.11d), before one is inside the other (Figure 14.11e).

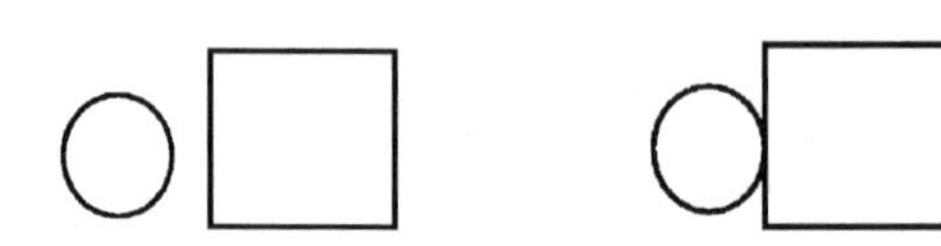

FIGURE 14.11(a) Disjoint. **FIGURE 14.11(b)** Meet.

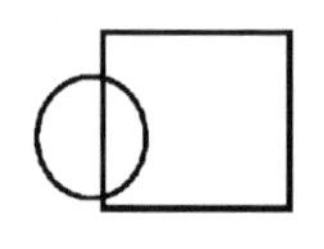

FIGURE 14.11(c) Overlap.

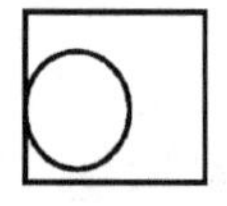

FIGURE 14.11(d) Covered.

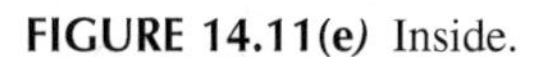

FIGURE 14.11(e) Inside.

Egenhofer and Al-Taha (1992) studied the gradual changes of topological relations for several cases of transformations involving translations, rotations, expansions and reductions. Some topological relations are 'sharp' in the sense that they apply only for a single moment when an object moves, whereas others are 'wide' and apply for some range of situations, which can be further characterized by a metric value. In English one can say

Kansas is further inside the United States than West Virginia.
Object A is 'barely' overlapping B in figure y.

Both expressions indicating that a metric characterization of the topological relation is appropriate.

The topological relations which are further characterized by a metric relation are, in first line, 'overlap' and 'inside' (but also 'meet along boundary' — characterized by the length of the common boundary). Possible candidates for a metric relation would be the relation of area inside and outside for overlap and a similar measure to determine if A is more inside than B.

Qualitative spatial reasoning is prevalent in human thinking about space and spatial situations. Its formalization almost always uses Euclidean geometry and the Cartesian coordinate system. This is admittedly not adequate to model human cognitive structure, but other solutions are not currently known. Spatial reasoning is crucial for progress in Geographical Information Systems, primarily, but not only, for the design and the programming of spatial query languages. Spatial relations depend on the type of space, for example cardinal directions are only used in large-scale space. The focus in this paper is on large-scale or geographical space, but the problems identified, the approaches, the methods and more of the results are meaningful for other types of spaces.

7 Conclusion

This paper introduced a system for qualitative spatial reasoning with cardinal distances from an algebraic point of view. Two operations, *inverse* and *composition* are applied to direction symbols and their meanings are formalized with a set of axioms. Three requirements, that directions should fulfill, are

1. The direction from a point to itself is a special value, meaning 'too close to determine a direction'.
2. Every direction has an inverse, namely the direction from the end point to the start point of the line segment.
3. The composition of two line segments with the same direction results in a line segment with the same direction.

In order to evaluate the expressive power of our qualitative reasoning approach, the notion of 'Euclidean exact' was defined using a homomorphism. A deduction rule is called 'Euclidean exact' if it produces the same results as Euclidean geometry operations would. Two different specific systems for cardinal directions, one based on cone-shaped regions, the other based on projections, are explored. They produce very similar outputs, but the projection-based cardinal directions yield more often results which are 'Euclidean exact'.

Cardinal directions are used as an example to show the advantages of the algebraic approach over the traditional propositional one. A series of research steps is laid out, which will lead to a comprehensive set of formalized spatial relations, that interact in a controlled way. Several guiding principles for the organization of such work are possible. The taxonomy of space from (Couclelis and Gale 1986) and the properties of the dominant operations could be used. A linguistic approach would primarily investigate which spatial relations are used in which context (e.g., cardinal directions are typical for large-scale spaces).

The guideline selected is based on traditional divisions in mathematics. The characterization of a spatial situation is done first by assessing the topological relations and then the metric properties and the same idea is followed to organize the investigations envisioned. A series of steps are listed, each addressing a specific researchable problem:

1. Extension of topological relations to points, lines and areas (co-dimension > 0),
2. Qualitative reasoning with distances, and integration of reasoning about distances and directions,
3. Reasoning in regionalized spaces, supporting distance and direction reasoning using the topological relation 'inside',
4. Generalize distance and direction relations to extended (non-point) objects, and
5. Find additional metric relations which characterize topological relations other than 'disjoint'.

The guiding expectation is, that the combination of different methods of qualitative spatial reasoning will provide the accuracy expected, an accuracy that cannot be found in any single method in isolation. The lead example is reasoning about distances and directions in isolation, which leads to approximate rules like cancellation (3′), which are obviously often inappropriate. However, the combination of qualitative distance and direction reasoning improves precision of inferences enormously.

Acknowledgments

Cooperation with Max Egenhofer was instrumental to shape these ideas. Discussions with David Mark, Matt McGranaghan, Werner Kuhn, and many others helped to clarify the concepts and their presentation. Dimitris Papadias helped with the revision of the paper and contributed valuable extensions. Part of this work was done while the author was with the National Center for Geographic Information and Analysis (NCGIA) and the Department of Surveying Engineering University of Maine at Orono. Funding from NSF for the NCGIA under grant SEC 88-10917 and from Intergraph Corporation is gratefully acknowledged.

References

Adler, D., Bazin, M., and Schiffer, M., 1965, *Introduction of General Relativity* (New York: McGraw-Hill).

Allen, J. F., 1983, Maintaining knowledge about temporal intervals, *Communications of the ACM,* **26,** 832–843.

Blumenthal, L. M., and Menger, K., 1970, *Studies in Geometry* (San Francisco: Freeman and Company).

Bobrow, D. G., Mittal, S., and Stefik, M., J., 1986, Expert systems: perils and promise, *Communications of the ACM,* **29,** 880–894.

Campari, I., 1995, Some notes on GIS: The relationship between their practical applications and their theoretical evolution. In *Spatial Languages,* edited by D. M. Mark and A. U. Frank (Amsterdam: Kluwer).

Campari, I., and Frank, A. U., 1993 Cultural differences in GIS: A basic approach. *Proceedings of the Fourth European GIS Conference (EGIS '93),* Geneva (Utrecht: EGIS Foundation), pp. 1–10.

Campari, I., and Frank, A. U., in press, Cultural aspects and cultural differences in GIS. In *Human Computer Interaction for GIS,* edited by T. Nyerges (Dordrecht: Kluwer), pp. 249–266.

Car, A., and Frank, A., 1994, General principles of hierarchical spatial reasoning: the case of wayfinding. In Advances in GIS — *Proceedings of the 6th International Symposium on Spatial Data Handling* (London: Taylor Francis).

Chang, S. K., Shi, Q. Y., and Yan, C. W., 1987, Iconic indexing by 2-D string. *IEEE Transactions of Pattern Analysis and Machine Intelligence,* **9,** 413–428.

Chang, S.-K., Jungert, E., and Li, Y., 1990, The design of pictorial databases upon the theory of symbolic projection. In *Design and Implementation of large Spatial Databases,* edited by A. Buchmann, O. Günther, T. R. Smith and Y.-F. Wang (New York, NY: Springer-Verlag), pp. 303–324.

Christaller, W., 1966 *Central Places in Southern Germany* (Englewood Cliffs, NJ: Prentice Hall).

Couclelis, H., and Gale, N., 1986, space and spaces. *Geografiska Analer,* **68B,** 1–12.

Couclelis, H., 1992, People manipulate objects (but cultivate fields): beyond the raster-vector debate in GIS. In *Proceedings of the International Conference GIS — From Space to Territory: Theories and Methods of Spatio-Temporal Reasoning in Geographic Space. Pisa, Italy,* edited by A. U. Frank, I. Campari and U. Formentini (Berlin: Springer-Verlag), pp. 65–77.

Davis, E., 1986, Representing and Acquiring Geographic Knowledge. Research Notes in Artificial Intelligence. Los Altos, CA: Morgan Kaufmann.

de Kleer, J., and Brown, J. S., 1985, A qualitative physics based on confluence. In *Formal Theories of the Commonsense World,* edited by J. R. Hobbs and R. C. Moore (Norwood, NJ: Ablex Publishing Corp.) pp. 109-184.

Dutta, S., 1988, Approximate spatial reasoning. In *Proceeding of First International Conference on Industrial and Engineering Applications of Artificial Intelligence and Expert Systems* (Tullahoma, Tennessee: ACM Press), pp. 126–140.

Dutta, S., 1990, Qualitative spatial reasoning: a semi-quantitative approach using fuzzy logic In *Design and Implementation of Large Spatial Database,* edited by A. Buchmann O. Günther, T. R. Smith and Y.-F. Wang (New York, NY: Springer-Verlag) pp. 345–361.

Egenhofer, M., and Herring, J. R., 1990. A mathematical framework for the definition topological relationships. In *Proceedings of the 4th International Symposium on Spatial Data Handling,* edited by K. Brassel (Zurich, Switzerland: International Geographic Union IGU, Commission on Geographic Information Systems), pp. 803–813.

Egenhofer, M. J., 1989, A formal definition of binary topological relationships. In *Proceeding of the Third International Conference on Foundations of Data Organization and Algorithms (FODO),* edited by W. Litwin and H.-J. Schek (Berlin: Springer-Verlag) pp. 457–472.

Egenhofer, M. J., 1991, Interacting with Geographic Information Systems via spatial queries Journal of Visual Languages and Computing, **1,** 389–413.

Egenhofer, M. J., and Franzosa, R., 1991, Point-set topological spatial relations International Journal of Geographical Information Systems, **5,** 161–174.

Egenhofer, M. J., and Al-Taha, K., 1992, Reasoning about gradual changes of topologic reasoning. In *Proceedings of the International Conference GIS — From Space to Territory: Theories and Methods of Spatio-Temporal Reasoning in Geographic Space Pisa, Italy,* edited by A. U. Frank, I. Campari, and U. Formentini (Berlin: Springer-Verlag), pp. 196–219.

Frank, A. U., 1990, Qualitative spatial reasoning about cardinal directions. In *Proceedings of Autocarto* 10 (Bethesda, M.D.: ACSM/ASPRS).

Frank, A. U., 1992, Spatial concepts, geometric data models and data structures. *Computers and Geosciences,* **18,** 409–419.

Frank, A. U., 1991, Qualitative spatial reasoning about distances and directions in geographic space. *Journal of Visual Languages and Computing,* **3,** 343–371.

Frank, A. U., and Mark, D. M., 1991, Language issues for geographical information systems. In *Geographic Information Systems: Principles and Applications,* edited by D. Maguire, D. Rhind and M. Goodchild (London: Longman), pp. 147–163.

Freeman, J., 1975, The modelling of spatial relations. *Computer Graphics and Image Processing,* **4,** 156–171.

Freksa, Ch., 1991, Qualitative spatial reasoning. In *Spatial Languages,* edited by D. M. Mark and A. U. Frank (Amsterdam: Kluwer), pp. 361–372.

Glasgow, J. I., and Papadias, D., 1992, Computational imagery. *Cognitive Science,* **16,** 355–394.

Güsgen, H. W., 1989, Spatial reasoning based on Allen's temporal logic. TR-89-049 (Berkeley CA.: International Computer Science Institute).

Hernández, D., 1993, Maintaining qualitative spatial knowledge. In *Proceedings of the European Conference on Spatial Information Theory COSIT, Elba, Italy,* edited by A. U. Frank and I. Campari (Berlin: Springer-Verlag), pp. 36–53.

Herskovits, A., 1986, *Language and Spatial Cognition — An Interdisciplinary Study of the Propositions in English.* Studies in Natural Language Processing (Cambridge, U.K: Cambridge University Press).

Hirtle, S., and Jonides, J., 1985, Evidence of hierarchies in cognitive maps. *Memory and Cognition*, **13,** 208–217.

Holmes, P. D., and Jungert, E., 1992, Symbolic and geometric connectivity graph methods for route planning. In digitized maps. *I.E.E.E. Transactions on Pattern Analysis and Machine Intelligence,* **PAMI-14,** 549–565.

Hong, J.-H., 1994, Qualitative distance and direction reasoning in geographic space. Ph.D. Thesis, Department of Surveying Engineering, University of Maine, Orono.

Huttenlocker, J., 1968, Constructing spatial images: a strategy in reasoning. *Psychological Review*, **4,** 277–299.

Jackendoff, R., 1983, *Semantics and Cognition* (Cambridge, Mass.: MIT Press).

Kainz, W., 1988, Application of lattice theory to geography. In *Proceedings of Third International Symposium on Spatial Data Handling*, edited by D. F. Marble (Sydney Australia: International Geographical Union, Commission on Geographical Data Sensing and Processing), pp. 135–142.

Krantz, D. H., Luce, R. D., Suppes, P., and Tversky, A., 1971, *Foundations of Measurement* (New York, NY: Academic Press).

Kuipers, B., 1982, The 'map in the head' metaphor. *Environment and Behaviour,* **14,** 202–220.

Kuipers, B., 1983, The cognitive map: could it have been any other way? In *Spatial Orientation*, edited by H. L. Pick and L. P. Acredolo (New York, NY: Plenum Press), pp. 345–359.

Kuipers, B., 1990 a, Commonsense knowledge of space: learning from experience. In *Advances in Spatial Reasoning,* edited by S.-S. Chen (Norwood, NJ: Ablex Publishing Corp.), pp. 199–206.

Kuipers, B., 1990 b, Modeling spatial knowledge. In *Advances in Spatial Reasoning*, edited by S.-S. Chen (Norwood, NJ: Ablex Publishing Corp.), pp. 171–198.

Kuipers, B., and Levit, T. S., 1990, Navigation and mapping in large-scale space. In *Advances in Spatial Reasoning*, edited by S.-S. Chen (Norwood, NJ: Ablex Publishing Corp.), pp. 207–251.

Lakoff, G., 1987, *Women, Fire, and Dangerous Things: What Categories Reveal About the Mind* (Chicago, IL: University of Chicago Press).

Larkin, J., and Simon, H. A., 1987, Why a diagram is (sometimes) worth ten thousand words. *Cognitive Science*, **11,** 65–99.

Lenat, D. G., Guha, R. V., Pittman, K., Pratt, D., and Shepherd, M., 1990, Cyc: Toward programs with common sense. *Communications of the ACM,* **33,** 30–49.

Levine, M., 1978, a knowledge-based computer vision system. In *Computer Vision Systems*, edited by A. Hanson, and E. Riseman (New York, NY: Academic Press), pp. 355–352.

Luce, R. D., 1956, Semiorders and a theory of utility discrimination. *Econometric,* **24,** 178–191.

Mark, D. M., and Frank, A. U., 1990, Experimental and formal representations of geographic space and spatial relations. Technical Report (Santa Barbara: NCGIA).

Mark, D. M., Frank, A. U., Egenhofer, M. J., Freundschuh, S. M., McGranaghan, M., and White, R. M., 1989a, Languages of spatial relations: initiative two specialist meeting pepor. Technical Report 89-2, National Central for Geographic Information and Analysis.

Mark, D. M., Gould, M. D., and Nunes, J., 1989b, Spatial Language and Geographic Information Systems: cross linguistic issues. In *II Conferencia Latinoamericana sobre la Tecnologia de los Sistemas de Informacion Geografica*, edited by R. Ponte, A. Guevara and M. Lyew (Merida, Venezuela: Universidad de los Ardes), pp. 105–130.

McCarthy, J., 1977, Epistomological problems of artificial intelligence. In *Proceedings of the International Joint Conference on Artificial Intelligence (IJCAI-77)* (Cambridge, MA), pp. 1038–1044.

McDermott, D., 1980. A theory of metric spatial inference. In *First National Conference on Artificial Intelligence of the American Association for Artificial Intelligence*, pp. 246–248.

McDermott, D., and Davis, E., 1984, Planning routes through uncertain territory. *Artificial Intelligence,* **22,** 107–156.

McGranaghan, M., 1991. Schema and object matching as a basis for interpreting textual specifications of geographical location. In *Spatial Languages,* edited by D. M. Mark and A. U. Frank (Amsterdam: Kluwer), pp. 387–402.

Montello, D., 1993, Scale and multiple psychologies of space. In *Proceedings of the European Conference of Spatial Information Theory COSIT, Elba, Italy, September*, edited by A. U. Frank and I. Campari (Berlin: Springer-Verlag), pp. 312–321.

Mukerje, A., 1990, *A Qualitative Model for Space.* TAMU-90-005 (Houston, TX: Texas A&M University).

NCGIA, 1989, The U.S. National Center for Geographic Information and Analysis: an overview of the agenda for research and education. *International Journal of Geographical Information Systems,* **2,** 117–136.

Myers, K. L., and Konolige, K., 1992, Reasoning with analogical representations. In *Proceedings of the 3rd (KR'92) International Conference on Principles of Knowledge Representation and Reasoning* (Los Altos, CA.: Morgan-Kaufman).

Nirenburg, S., and Raskin, V., 1987, Dealing with space in natural language processing. In *Spatial Reasoning and Multi-Sensor Fusion,* edited by A. Kak and S.-S. Chen (Los Altos, CA.: Morgan-Kaufmann Publishers), pp. 361–370.

Papadias, D., Frank, A., and Koubarakis, M., 1994, Constraint-based reasoning in geographic databases: the case of symbolic arrays. In *Proceedings of the second ICLP Workshop on Deductive Databases, Italy*

Papadias, D., and Kavouras, M., 1994, Acquiring, representing and processing spatial relations. In *Proceedings of the 6th International Symposium on Spatial Data Handling* (London: Taylor Francis).

Papadias, D., and Sellis, T., 1992, Spatial reasoning using symbolic arrays. In *Proceedings of the International Conference GIS — From Space of Territory: Theories and Methods of Spatio-Temporal Reasoning in Geographic Space, Pisa, Italy,* edited by A. U. Frank, I. Campari and U. Formentini (Berlin: Springer-Verlag), pp. 153–161.

Papadias, D., and Sellis, T., 1993, The semantics of relations in 2D space using representative points: spatial indexes. In *Proceedings of the European Conference on Spatial Information Theory, COSIT, Elba, Italy,* edited by A. U. Frank and I. Campari (Berlin: Springer-Verlag), pp. 234–247.

Papadias, D., and Sellis, T., 1994, On the qualitative representation of spatial knowledge in two-dimensional space. *Very Large Data Bases Journal* (Special Issue on Spatial Databases), pp. 479–516.

Peuquet, D., and Zhan, C.-X., 1987, An algorithm to determine the directional relationship between arbitrarily-shaped polygons in a plane. *Pattern Recognition,* **20,** 65–74.

Peuquet, D. J., 1988, Toward the definition and use of complex spatial relationships. In *Third International Symposium on Spatial Data Handling,* edited by D. F. Marble (Sydney Australia: International Geographical Union, Commission on Geographical Data Sensing and Processing), pp. 211–224.

Pullar, D. V., and Egenhofer, M. J., 1988, Towards the defaction and use of topological relations among spatial objects. In *Proceedings of the Third International Symposium on Spatial Data Handling* (Columbus: International Geographical Union), pp. 225–242.

Randell, D., Cui, Z., and Cohn, A., 1992, A spatial logic based on regions and connection. In the *Proceedings of the 3rd International Conference on Principles of Knowledge Representation and Reasoning (KR'92) (St. Charles, IL: Morgan-Kaufman),* pp. 165–176.

Retz-Schmidt, G., 1987, Deitic and intrinsic use of spatial propositions: a multidisciplinary comparison. In *Spatial Reasoning and Multi-Sensor Fusion,* edited by A. Kak and S.-S. Chen (Pleasan Run Resort, St. Charles, IL: Morgan-Kaufmann), pp. 371–380.

Robert, F. S., 1973, Tolerance geometry. *Notre Dame Journal of Formal Logic,* **14,** 68–76.

Roberts, F. S., 1969, Indifference graphs. In *Proof Techniques in Graph Theory,* edited by F. Harary (New York: Academic Press), pp. 139–146.

Roberts, F. S., 1971, On the compatibility between a graph and a simple order. *Journal of Combinatorial Theory,* **11,** 28–38.

Roberts, F. S., and Suppes, P., 1967, some problems in the geometry of visual perception. *Synthese,* **17,** 173–201.

Saalfeld, A., 1985, Lattice structures in geography: The quadtree and related hierarchical data structures. *Computing Surveys*, **16,** 482–489.

Scott, D., and Suppes, P., 1958, Foundational aspects of theories of measurements. *Journal of Symbolic Logic,* **23,** 113–128.

Stevens, A., and Coupe, P., 1978, Distortions in judged spatial relations. *Cognitive Psychology* **10,** 422–437.

Synge, J. M., 1941, *Plays, Poems and Prose* (London: Dent).

Tobler, W. R., and Wineberg, S., 1971, A Cappadocian speculation. *Nature,* **231,** 39–42.

Tversky, B., 1993, Cognitive maps, cognitive collages and spatial mental models. In *Proceedings of the European Conference of Spatial Information Theory, COSIT, Eiba, Italy,* edited by A. U. Frank and I. Campari (Berlin: Springer-Verlag), pp. 14–24.

Webster, B., 1990, Rule-based spatial search. *International Journal Geographical Information Systems,* **4,** 241–260.

Zadeh, L. A., 1974, Fuzzy logic and its application to approximate reasoning. In *Information Processing* (North-Holland Publishing Company).

Zeeman, E. C., 1962, The Topology of the brain and visual perception. In *The Topology of 3-Manifolds,* edited by M. K. Fort (Englewood Cliffs, NJ: Prentice Hall), pp. 240–256.

Twenty Years of Reasoning with Spatial Relations

Andrew U. Frank

Early investigation in spatial relations

I started looking at spatial relations as a consequence of my previous interest in the logic-based programming language Prolog (Clocksin and Mellish, 1981) and my implementation of a spatial extension and query facility to a database (Frank, 1984, 1985, 1986; Frank and Egenhofer, 1990). Prolog is based on predicate calculus; facts are expressed as logical relations between objects and can be combined with Boolean operators (OR, IF_THEN). This approach promised to make results from artificial intelligence research useful for building user interfaces for geographical information systems (GIS) (Frank, Robinson, and Blaze, 1986a, 1986b; Frank and Robinson, 1987; Robinson and Frank, 1987). Using Prolog we defined, for example, geomorphologic terms and built a program to identify such objects in a digital terrain model (Frank et al., 1986). Peuquet published an article at that time on methods of determining the spatial relation between extended objects (Peuquet and Zhan, 1987) as part of building an intelligent GIS using LISP.

We and others were influenced by Allen's investigation of temporal relations (Allen, 1983) that called for the application of similar methods to spatial problems. Combining logic with algebra, we turned to relation algebra (Schröder, 1890, 1966; Maddux, 1991) with the *composition* of two relations defined as:

$$R\ (a,b) \text{ and } S\ (b,c) = R;S\ (a,c).$$

In this framework the composition of direction relation terms can be explored, but it could equally well be applied to topological relations or distance terms. Composition tables give a very compact and powerful method for describing how to reason with relations; they were initially introduced to temporal reasoning by Freksa (1990) and to spatial reasoning by (Hernández, 1990, 1991).

What I see as the contribution of the article

The intention of research in spatial relations was to find symbolic, qualitative and not quantitative, approaches to spatial reasoning. We wanted to be able to reason spatially without the use of coordinate systems. Symbolic projections (Chang et al., 1987; Jungert and Chang, 1989) were first approaches for qualitative spatial reasoning, but the most important contribution to defining qualitative spatial relations is certainly the formalization of topological relations in the dissertation Max Egenhofer wrote under my supervision (1989), and the essentially similar definitions by Randel, Cohn, and Cui (Randell and Cohn, 1989). It has led to a large number of extensions.

My paper, "Qualitative spatial reasoning: Cardinal direction as an example," summarized research on metric spatial relations previously reported at a conference in 1991 (Frank, 1991) and another journal article (Frank, 1992), and added new results and a detailed research agenda.

I believe the major contributions were:

- The comparison of cone- and projection-based cardinal directions, and the introduction of a neutral zone for cardinal directions. Peuquet (1988) suggested two methods of defining cardinal directions: cones and projections. Previous papers used direction on a bipartite scheme (north-south, east-west) or a subdivision of the compass in four, eight, or more directions. Figure 14.9 in the paper introduced a neutral element in direction reasoning — the direction between two points too close to be differentiated. This is a usual approach in algebra, comparable to the unit elements for addition ($a + 0 = a$) or multiplication ($a * 1 = a$). A comparison of these methods indicated that the projection-based system, using a neutral zone, led more often to a meaningful result for composition.
- Approximate qualitative reasoning: The usual approach to composition of relations is such that the result is a single relation, a conjunction of possible relations or the universal relation when no more informative results can be obtained. I deviated from this conservative approach: in cases where several possibilities result from a composition, the most likely one was selected. These approximation results were marked in the composition tables with lower-case letters. We found that approximations were better for a projection-based model than a cone-based model of cardinal directions.

Continuation of this research

The size of the neutral zone was left free to adapt to context. An immediate application is to cardinal directions between extended objects: the ground object determines by its size the neutral zone, which divides space in nine tiles. Goyal, in a Ph.D. thesis under Egenhofer's supervision, classified the location of a figure object by observing how much area of the figure object fell into each tile (Figure 14.12). This gives a direction-relation matrix (Goyal and Egenhofer, 2000; Goyal, 2000) that characterizes the relative position of an extended figure object with respect to an extended ground object; it does not yield a single, often counter-intuitive, direction (as Peuquet's algorithm does).

The idea of adapting the neutral zone to the ground object is not only useful when considering a figure object outside of the ground object, but also for objects inside a ground object (Liu et al., 2005) (Figure 14.13). One can see these two proposals as combinations of topological relations (disjoint, inside) with cardinal directions.

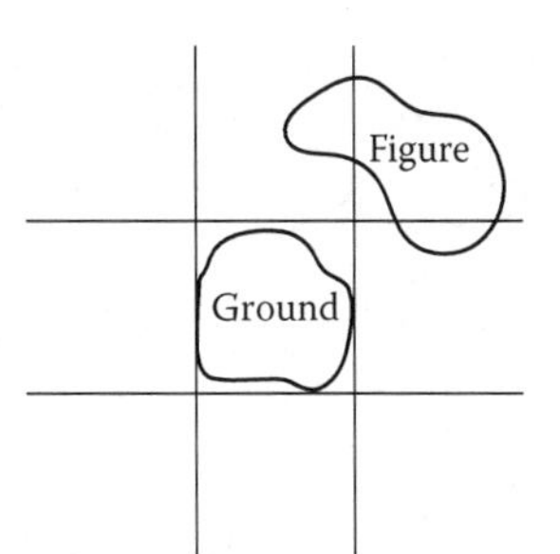

FIGURE 14.12

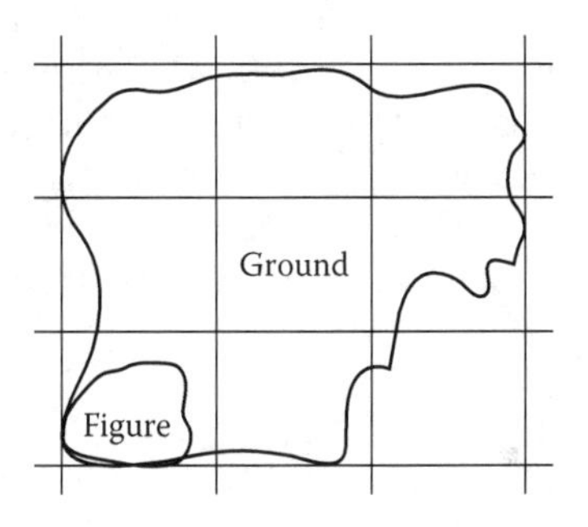

FIGURE 14.13

I pointed out in the article that some relations are *sharp*, they are valid only for exactly one configuration (e.g., meet), others are *wide*, valid for a range of situations (e.g., overlap or disjoint); this was later expanded by Galton (1997) to a theory of dominance of relation.

The major focus of my effort to combine qualitative reasoning with distances and directions was to find an optimal combination. I hoped that for some clever definition of symbolic distances, the quality of symbolic reasoning would improve. In two Ph.D. theses under Egenhofer's supervision, these ideas were further explored by Sharma (Sharma et al., 1994; Sharma and Flewelling, 1995; Sharma, 1996) and Hong (Hong, et al., 1995; Hong, 1994). They showed promising results but were not fully conclusive.

The initial focus was the definition of spatial relation terms because a need for a formal definition of spatial relations was evident in the various proposals for spatial query languages (Egenhofer, 1991). Later the focus shifted to the terminology humans use to describe spatial relations. A number of experiments with human subjects have been reported — for example, on topological relations (Egenhofer and Mark, 1995), on distance (Robinson et al., 1986; Robinson et al., 1987; Worboys, 2001), and on internal cardinal directions (Liu et al., 2005).

Spatial relations were, in my view in 1990, a special case of conceptual image schemata (Lakoff, 1987) such as CONTAINER, SUPPORT, CYCLE, and so forth. Formalization using relation composition tables was successful for topological, directional, distance relations, and some others, but not for all spatial relations. I later tried to identify the properties of spatial relations in geographic space and analyzed the conceptual image schemata PATH and CENTER-PERIPHERY. I identified *entailments* of statements (that is, what, given a context, a listener could derive from a factual statement) and used these to construct an algebra (Frank and Raubal, 1999). I believe today that the semantics of spatial relations for tabletop and for geographic space are, at least for some cases, different, and hence researchers should take into account the influence of context.

Why did we study relations and not entities?

Recently David Mark and I were discussing classification of geographic landscape terms, like valley, *cañon*, peninsula, etc., and he asked me, "Why did we [in the 1990s] place so much emphasis on spatial *relations*, and little or none on the spatial entities that would participate in those relations?" There was indeed a period in my research where my focus was on spatial relations; the paper reprinted here is the last publication of this effort and in a way a summary of it.

In my Ph.D. research (Frank, 1983), the focus was on classification of objects following the database tradition of describing entity classes with attributes (Chen, 1976), and on methods to retrieve data based on spatial location. Spatial relations were cursorily used to differentiate different spatial situations. A spatial query language was designed (Frank, 1982) and implemented (Egenhofer, 1984), but spatial relation terms played a minor role if any.

My focus changed slowly to formalization and implementation of how human beings conceived space and the objects in it. Boyle (Smith et al., 1983) identified the need for a formal *spatial theory*, under which title I got a National Science Foundation research project funded in 1986. In the programmatic article, "Towards a Spatial Theory" (Frank 1987), I laid out a research agenda that included definitions of "terms" for spatial relations and mentioned the difference between mathematically defined concepts and the "naïve geometry" of the GIS users. David Mark brought a cognitive and linguistic viewpoint to geography, influenced by Lakoff (1987). This seemed complementary to the formal-mathematical focus Max Egenhofer and I were following in Maine and influenced the NCGIA (National Center for Geographical Information and Analysis) Research Initiative 2, "Languages for Spatial Relations" (NCGIA 1989). A two-week-long meeting in Las Navas in 1990 was financed by NATO (Mark and Frank, 1991) where the first papers on spatial relations appeared (Hernández, 1990; Freksa, 1991).

Returning to David Mark's question: Why did we investigate spatial relations and not geographic objects? *My* answer is that we selected spatial relations because we had the mathematical tools to approach the problem: predicate calculus and composition tables from relation algebra. The results demonstrate that these tools were appropriate to approach the question and led to valuable insight and the development of widely applicable theories.

Object classes like dam, watershed, lake, pond, etc. were not in our focus then (Rugg et al., 1997). There were practical efforts in producing terminology lists by national mapping agencies, but without formal definitions of the terms. Mark (1993) listed factors influencing the choice of terminology in different languages: in English, the difference between a *lake* and a *pond* is mostly size, but in French, *etang*, which translates roughly to *pond*, includes much of what in English would be called a *lagoon*. With the methods then used for the description of entity classes in database schema, such differences could not be captured. Object orientation was promising (Egenhofer and Frank, 1989; Worboys et al., 1990 and Chapter 6 in this volume; Egenhofer and Frank, 1992), but inheritance, especially multiple inheritance, which would be necessary to deal with the problems described by Mark, was not well understood. I could not see how the object-orientation theory (Guttag et al., 1985; Ehrich et al., 1989) we used for the analysis of user interfaces (Kuhn and Frank, 1991), could be used to differentiate, for example, the pair *lac/etang* and compare with the difference in the pair *lake/pond*.

More than 15 years later, a number of new tools should prompt us to go back to the question David Mark asked. Modeling of object classes and spatial relations has progressed and the combination, together with modeling of operations, could be applied fruitfully. The combination of the formalism of relation algebra and specification languages can be achieved using category theory, combining the category of relations with the category of functions. Incidentally, John Herring had very early on pointed to category theory as an important formal approach (Herring, 1989; Herring et al. 1990; Herring, 1991), but we used only relation algebra, which is the category of relations, and it took 10 years for me to partially understand the importance of his suggestion (Frank, 1999).

Conclusion

Forbus et al. (1987) made the pessimistic conjecture that no powerful spatial reasoning without coordinates (or other methods of representing metric aspects of space) was possible. This conjecture does not jibe with the observation of human performance, where spatial deduction from purely qualitative data is routinely done. A recent survey of qualitative spatial reasoning (Cohn and Hazarika, 2001) lists a large number of formal systems that also casts doubt on this *poverty conjecture*.

The insight that research focus is influenced by the tools available is not novel, but is instructive in hindsight. I consider the results of the research on spatial relations a success, not the least because we focused on the part of the enormously large problem that we could tackle with the tools we had. Understanding new tools, and selecting tools and problems that fit, is the art of scientific research.

Research progresses in small steps and few papers contribute lasting ideas. I hope this paper has made the useful contribution of a neutral element for cardinal directions.

Acknowledgments

Many of my colleagues and students have helped me to advance to this point. Particular thanks go to David Mark for asking the right question over more than 20 years of scientific cooperation.

References

ALLEN, J.F., 1983, Maintaining knowledge about temporal intervals, *Communications of the ACM*, **26**(11), 832–843.

CHANG, S.K. AND SHI, AND C.-W. YAN Q.-Y., 1987, Iconic indexing by 2-D strings, *IEEE Transactions on Pattern Analysis and Machine Intelligence PAMI*, **9**(3), 413–428.

CHEN, P.P.S., 1976, The entity-relationship model — toward a unified view of data, *ACM Transactions on Database Systems*, **1**(1), 9–36.

CLOCKSIN, W.F. AND MELLISH, C.S., 1981, *Programming in Prolog*, Springer-Verlag, Berlin.

COHN, A.G. AND HAZARIKA, S.M., 2001, Qualitative spatial representation and reasoning: An overview, *Fundamenta Informaticae*, **46**(1–2), 1–29.

EGENHOFER, M.J., 1984, *Implementation of MAPQUERY, a Query Language for Land Information Systems* (in German), Institute for Geodesy and Photogrammetry, Swiss Federal Institute of Technology, (ETH), Zurich, Switzerland.

EGENHOFER, M.J., 1989, *Spatial Query Languages*, Ph.D. thesis, University of Maine.

EGENHOFER, M.J., 1991, Deficiencies of SQL as a GIS Query Language. In *Cognitive and Linguistic Aspects of Geographic Space: An Introduction,* Mark, D.M. and Frank, A.U., Eds., Kluwer Academic Publishers, Dordrecht, 477–492.

EGENHOFER, M.J. AND FRANK, A.U., 1989, Why object-oriented software engineering techniques are necessary for GIS. In *Proceedings of the International Geographic Information Systems (IGIS) Symposium*, Baltimore, MD.

EGENHOFER, M.J. AND FRANK, A.U., 1992, Object oriented modeling for GIS, *Journal of the Urban and Regional Information Systems Association*, **4**(2), 3–19.

EGENHOFER, M.J. AND MARK, D.M., 1995, Modelling conceptual neighbourhoods of topological line-region relations, *International Journal for Geographical Information Systems*, **9**(5), 555–565.

EHRICH, H.-D., GOGOLLA, M., AND LIPECK, U.W., 1989, *Algebraische Spezifikation abstrakter Datentypen*, B.G. Teubner, Stuttgart.

FORBUS, K., NIELSON, P., AND FALTINGS B., 1987, Qualitative kinematics: A framework. In *Proceedings of the International Joint Conference on AI (IJCAI-87)*, Milan, Italy, M. Kaufmann, Los Altos, 430–436.

FRANK, A.U., 1982, MAPQUERY: Database query language for retrieval of geometric data and its graphical representation, *ACM SIGGRAPH*, **16**(3), 199–207.

FRANK, A., 1983, *Datenstrukturen für landinformationssysteme — Semantische, topologische und räumliche beziehungen in daten der geo-wissenschaften*. Institut für Geodäsie und Photogrammetrie, ETH Zürich.

FRANK, A.U., 1984, Extending a network database with Prolog. In *Proceedings of the First International Workshop on Expert Database Systems,* Kerschberg, L., Ed., Kiawah Islands, SC, 665–676.

FRANK, A.U., 1985, *Combining a Network Database with Logic Programming*. University of Maine at Orono, Surveying Program, Report No. 45.

FRANK, A.U., 1986, LOBSTER: Combining database management and artificial intelligence techniques to manage land information. In *Proceedings of the XVIII International Congress of the International Federation of Surveyors (FIG)*, Toronto, Ontario, Canada, 14.

FRANK, A.U., 1987, Towards a spatial theory. In *Proceedings of the International Geographic Information Systems (IGIS) Symposium: The Research Agenda*, Crystal City, VA, NASA, Greenbelt, MA, 215–27

FRANK, A.U., 1991, Qualitative spatial reasoning about cardinal directions. In *Proceedings of Auto-Carto 10*, American Congress on Surveying and Mapping, Bethesda, MD, 148–167.

FRANK, A.U., 1992, Qualitative spatial reasoning about distances and directions in geographic space, *Journal of Visual Languages and Computing*, **3**, 343–371.

FRANK, A. U., 1999, One step up the abstraction ladder: Combining algebras — from functional pieces to a whole. In *Spatial Information Theory — Cognitive and Computational Foundations of Geographic Information Science (COSIT'99)*, Stade, Germany, Lecture Notes in Computer Science 1661, Freksa, C. and Mark, D.M., Eds., Springer-Verlag, Berlin, 95–107.

FRANK, A.U. AND EGENHOFER, M.J., 1990, LOBSTER: Combining AI and database techniques for GIS, *Photogrammetric Engineering & Remote Sensing*, **56**(6), 919–926.

FRANK, A. U., PALMER, B., AND ROBINSON, V., 1986, Formal methods for accurate definition of some fundamental terms in physical geography. In *Proceedings of the Second International Symposium on Spatial Data Handling*, International Geographical Union, Seattle, Washington, 583–599.

FRANK, A.U. AND RAUBAL, M., 1999, Formal specifications of image schemata — A step to interoperability in geographic information systems, *Spatial Cognition and Computation* **1**(1), 67–101.

FRANK, A.U., ROBINSON, V., AND Blaze, M., 1986a, Expert systems and geographic information systems: Review and prospects, *ASCE Journal of Surveying Engineering* **112**(2), 119–130.

FRANK, A.U., ROBINSON, V., AND BLAZE, M., 1986b, An introduction to expert systems, *ASCE Journal of Surveying Engineering*, **112**(2), 109–118.

FRANK, A.U. AND ROBINSON, V., 1987, Expert systems for geographic information systems, *Photogrammetric Engineering and Remote Sensing,* **53**(10), 1435–1441.

FREKSA, C., 1990, *Temporal Reasoning based on Semi-Intervals*, International Computer Science Institute, Berkeley, CA.

FREKSA, C., 1991, Qualitative spatial reasoning. In *Cognitive and Linguistic Aspects of Geographic Space: An Introduction*, Mark, D.M. and Frank, A.U., Eds., Kluwer Academic Press, Dordrecht, 361–372.

GALTON, A., 1997, Continuous change in spatial regions. In *Spatial Information Theory — A Theoretical Basis for GIS (COSIT'97)*, Lecture Notes in Computer Science 1329, Hirtle, S.C. and Frank, A.U., Eds., 1–14, Springer-Verlag, Berlin.

GOYAL, R., 2000, *Similarity Assessment for Cardinal Directions between Extended Spatial Objects*, Ph.D. thesis, University of Maine.

GOYAL, R.K. AND EGENHOFER, M.J., 2000, Consistent queries over cardinal directions across different levels of detail. In *Proceedings of the 11th International Workshop on Database and Expert Systems Applications*, Tjoa, A.M., Wagner, R., and Al-Zobaidie, A., Eds., IEEE Press, Los Alamitos, 876–880.

GUTTAG, J.V., HORNING, J.J., AND WING, J.M., 1985, *Larch in Five Easy Pieces*, Digital Equipment Corporation, Systems Research Center.

HERNÁNDEZ, D., 1990, *Relative Representation of Spatial Knowledge: The 2-D Case*, FRG Technische Universität München, Munich.

HERNÁNDEZ, D., 1991, Relative representation of spatial knowledge: The 2-D case. In *Cognitive and Linguistic Aspects of Geographic Space: An Introduction*, Mark, D.M. and Frank, A.U., Eds., Kluwer Academic, Dordrecht, 373–386.

Herring, J.R., 1989, The category model of spatial paradigms. In *Languages of Spatial Relations: NCGIA Initiative 2 Specialist Meeting Report,* Mark, D.M., Frank, A.U., Egenhofer, M.J., Freundschuh, S.M., McGranaghan, M., and White, R.M., Eds., National Center for Geographic Information and Analysis, Goleta, CA, 47–51.

Herring, J.R., 1991, The mathematical modeling of spatial and non-spatial information in geographic information systems. In *Cognitive and Linguistic Aspects of Geographic Space: An Introduction*, Mark, D.M. and Frank, A.U., Eds., Kluwer Academic, Dordrecht, 313–350.

Herring, J., Egenhofer, M. J., and Frank, A.U., 1990, Using category theory to model GIS applications. In *Proceedings of the 4th International Symposium on Spatial Data Handling*, Vol. 2, International Geographical Union, Zurich, 820–829.

Hong, J.-H., 1994, *Qualitative Distance and Direction Reasoning in Geographic Space*, Ph.D. thesis, University of Maine.

Hong, J.-H., Egenhofer, M.J., and Frank A.U., 1995, On the robustness of qualitative distance- and direction-reasoning. In *Proceedings of Auto-Carto 12*, American Congress on Surveying and Mapping, Bethesda, MD, 301–310.

Jungert, E. and Chang, S.K., 1989, An algebra for symbolic image manipulation and transformation. In *Visual Database Systems*, Kunii, T.L., Ed., North-Holland, Amsterdam, 301–317.

Kuhn, W. and Frank, A.U., 1991, A formalization of metaphors and image-schemas in user interfaces. In *Cognitive and Linguistic Aspects of Geographic Space: An Introduction*, Mark, D.M. and Frank, A.U., Eds., Kluwer Academic Publishers, Dordrecht, 419–434.

Lakoff, G., 1987, *Women, Fire, and Dangerous Things: What Categories Reveal About the Mind*, University of Chicago Press, Chicago, IL.

Liu, Y., Wang, X., Jin, X., and Wu, L., 2005, On internal cardinal direction relations. In *Spatial Information Theory (COSIT'05)*, Lecture Notes in Computer Science 3693, Cohn, A.G. and Mark, D.M., Eds., Springer-Verlag, Berlin, 283–299.

Maddux, R., 1991, The origin of relation algebras in the development and axiomatization of the calculus of relations, *Studia Logica*, **50**(3–4), 421–455.

Mark, D.M., 1993, Toward a theoretical framework for geographic entity types. In *Spatial Information Theory: A Theoretical Basis for GIS (COSIT'93),* Lecture Notes in Computer Science 716, Frank, A.U. and Campari, I., Eds., Springer-Verlag, Berlin, 270–283.

Mark, D.M. and Frank, A.U., Eds., 1991, *Cognitive and Linguistic Aspects of Geographic Space*, Kluwer Academic Publishers, Dordrecht.

NCGIA, 1989, The research plan of the National Center for Geographic Information and Analysis, *International Journal of Geographical Information Systems*, **3**(2), 117–136.

Peuquet, D. J., 1988, Toward the definition and use of complex spatial relations. In *Proceedings of the Third International Symposium on Spatial Data Handling*, International Geographical Union, Sydney, Australia, 211–224.

Peuquet, D. and Zhan, C.-X., 1987, An algorithm to determine the directional relationship between arbitrarily-shaped polygons in a plane, *Pattern Recognition*, **20**(1), 65–74.

Randell, D. and Cohn, A., 1989, Modelling topological and metrical properties of physical processes. In *Proceedings of First International Conference on the Principles of Knowledge Representation and Reasoning*, Morgan-Kaufmann, Los Altos, 357–368.

Robinson, V. B., Blaze, M., and Thongs, D., 1986, Representation and acquisition of a natural language relation for spatial information retrieval. In *Proceedings of the 2nd International Symposium on Spatial Data Handling*, International Geographical Union, Seattle, 472–487.

ROBINSON, V. AND FRANK, A.U., 1987, Expert systems for geographic information systems, *Photogrammetric Engineering and Remote Sensing*, **53**(10), 1435–1441.

RUGG, R.M., EGENHOFER, M., AND KUHN, W., 1997, Formalizing behavior of geographic feature types, *Geographical Systems*, 4(2), 159–180.

SCHRÖDER, E., 1890, *Vorlesungen über die Algebra der Logik (Exakte Logik)*, Teubner, Leipzig.

SCHRÖDER, E., 1966, *Algebra der Logik, I–III*, reprint.

SHARMA, J., 1996, *Integrated Spatial Reasoning in Geographic Information Systems*, PhD thesis, University of Maine.

SHARMA, J. AND FLEWELLING, D.M., 1995, Inferences from combined knowledge about topology and directions. In *Advances in Spatial Databases*, Lecture Notes in Computer Science 951, Egenhofer, M.J. and Herring, J.R., Eds., Springer-Verlag, Berlin, 279–291.

SHARMA, J., FLEWELLING, D.M., AND EGENHOFER, M.J., 1994, A qualitative spatial reasoner. In *Advances in GIS Research: Proceedings of the 6th International Symposium on Spatial Data Handling*, Waugh, T.C. and Healey, R.G., Eds., Taylor & Francis, London, 665–681.

SMITH, L. K., BOYLE, A.R., DANGERMOND, J., MARBLE, D., SIMONETT, D.S., AND TOMLINSON, R.F., 1983, *Final Report of a Conference on the Review and Synthesis of Problems and Directions for Large Scale Geographic Information System Development*, NASA Report, available from ESRI, Redlands CA.

WORBOYS, M., 2001, Modelling changes and events in dynamic spatial systems with reference to socio-economic units. In *Life and Motion of Socio-Economic Units*, Frank, A.U., Raper, J., and Cheylan, J.-P., Eds., Taylor & Francis, London, 129–138.

WORBOYS, M.F., HEARNSHAW, H.M., AND MAGUIRE, D.J., 1990, Object-oriented data modelling for spatial databases, *International Journal of Geographical Information Systems*, **4**(4), 369–383.

International Journal of Geographical Information Science, 1997, Vol. 11, No. 1, 33–52.

15 Assessing, Representing, and Transmitting Positional Uncertainty in Maps

Harri T. Kiiveri

Abstract. This paper considers a model for positional uncertainty in maps with applications in geographical information systems. Using the model it is shown how uncertainty in map boundaries can be assessed, represented and transmitted through set operations such as intersection (map overlaying). Expressions for uncertainty in lengths, perimeters and areas calculated from maps are given. Simulation of maps and other quantities of interest derived from them is also discussed. The ideas are illustrated with examples.

1 Introduction

The collection and storage of vast amounts of spatially related data in geographical information systems (GIS) (Maguire *et al.* 1991) has far outstripped our ability to assess their reliability. For example, large areas of the Earth's surface are mapped with little chance of substantial ground truthing. Once entered into a GIS, calculations such as map overlays or intersections are carried out assuming that the data are accurate when it is known that this is not the case. There is a need for methods for assessing, representing, and transmitting uncertainty through calculations with maps so that decision makers have some idea of the reliability of their information.

The results of this paper have application for decision making in areas such as town planning, resource management, environmental monitoring, and mineral exploration, to name a few.

Some of the issues involved in assessing, representing and transmitting uncertainty in spatial data bases are discussed in Goodchild and Gopal (1989), whilst existing work on error propagation can be found in Heuvelink *et al.* (1989), Haining and Arbia (1993), and Wesseling and Heuvelink (1993). This existing work deals primarily with the propagation of quantitative attribute errors through calculations. It does not deal with positional uncertainty and associated uncertainty in the computation of lengths, perimeters and areas which are considered here. (However, see Keefer *et al.* 1991.) In this paper our focus is on models which allow us to make probability statements about the presence of various attributes (map classes) at given points on the ground, and which allow assessment of uncertainty in location of boundaries, points and lines. The use of probability as compared to fuzzy sets (Altman 1994), to address this problem appears to be a new approach.

Data stored in a GIS can be divided into roughly three types. Raster data, for which attributes are recorded for each picture element or pixel on a rectangular grid; vector data for which features on the ground are approximated by sequences of line segments, and point or site data consisting of the co-ordinates of locations. The results in this paper apply to all three data types.

Section 2 considers a statistical model for maps which postulates random 'distortions'. This model is applicable for vector, raster and point site data. Section 3 considers the assessment of probabilities of attributes at specified locations for this model. Calculation of uncertainties in lengths, perimeters, set operations such as overlaying, and areas of map features are considered in Section 4, whilst the simulation of maps and the results of map operations is considered in Section 5. Methods for obtaining model parameters are considered in Section 6 and the paper concludes with a discussion of extensions to the model and areas for further work in Section 7.

2 Distortion models

In this section, we explore a model for describing uncertainty in spatial data, particularly map data where boundaries can be considered to be uncertain. For convenience, we take the map to be on the unit square $S = [0, 1] \times [0, 1]$ and consider the map to be a real-valued function on S i.e. each area or map category gives rise to a different value of a (step) function. In practice a simple location shift and scaling is all that is required to achieve this. Note that in the following vectors and matrices will be in bold.

The model is as follows. Given a true map $g_T(\mathbf{u})$ for $\mathbf{u} \in S$, the observed map $g_0(\mathbf{u})$ is assumed to be:

$$g_0(\mathbf{u}) = g_T(\mathbf{u} + \delta(\mathbf{u})) \tag{15.1}$$

where $\delta(\mathbf{u})$ is a (smooth) random perturbation of $\mathbf{u}$ and $\mathbf{u} + \delta\mathbf{d}(\mathbf{u}) \in \mathrm{S}$. In other words, the value we observe at $\mathbf{u}$ is in fact the true value at a position displaced from $\mathbf{u}$ by an amount $\delta(\mathbf{u})$. Equation (1) can be reformulated as:

$$g_0(\mathbf{u}) = g_T(\mathbf{P}(\mathbf{u})) \tag{15.2}$$

where P maps S onto S. Assuming that $\mathbf{Q} = \mathbf{P}^{-1}$ exists, (15.2) can also be expressed as:

$$g_T(\mathbf{v}) = g_0(\mathbf{Q}(\mathbf{v})) \tag{15.3}$$

where $\mathbf{u} = \mathbf{P}^{-1}(\mathbf{v})$. Since $\mathbf{Q}(\mathbf{v}) = \mathbf{v} + (\mathbf{Q}(\mathbf{v}) - \mathbf{v})$, (15.3) can be written as:

$$g_T(\mathbf{v}) = g_0(\mathbf{v} + \boldsymbol{\delta}^*(\mathbf{v})) \tag{15.4}$$

so that the true value at location $\mathbf{v} \in S$ is in fact the observed value at a point displaced from $\mathbf{v}$ by $\boldsymbol{\delta}^*(\mathbf{v})$.

Note that this model assumes that the **net** effect of all sources of error is a 'smooth' or rubbersheet distortion of the true map. We do not model each error source separately here. The smoothness guarantees that continuous lines are not broken and that points inside a region always stay inside after distortion, i.e. topology is preserved.

We postulate a model for **P** in (15.2) in which the displacement of **u** is modelled as a random linear combination of basis functions. Specifically, we write $\mathbf{P}(\mathbf{u}) = (P_1(\mathbf{u}), P_2(\mathbf{u}))^T$ where $\mathbf{u}^T = (x, y)$ and:

$$P_1(\mathbf{u}) = x + \sum_i \alpha_i e_i(\mathbf{u}) = x + \boldsymbol{\alpha}^T \mathbf{e} \tag{15.5}$$

and:

$$P_2(\mathbf{u}) = y + \sum_j \beta_j f_j(\mathbf{u}) = y + \boldsymbol{\beta}^T f \tag{15.6}$$

In (15.5) and (15.6), the vectors $\boldsymbol{\alpha}$, $\boldsymbol{\beta}$, **e**, **f** have corresponding components a_i, b_j, e_i, f_j. Displacements in the x and y directions (i.e., the 'columns' and 'rows' of the square respectively) are *independent*.

Following Amit *et al.* (1991), one choice for the basis functions e_i and f_j in (15.5) and (15.6) is:

$$e_i(\mathbf{u}) = \frac{2 \sin \pi n x \cos \pi m y}{\pi^2 (n^2 + m^2)} \tag{15.7}$$

and:

$$f_j(\mathbf{u}) = \frac{2 \cos \pi m' x \sin \pi n' y}{\pi^2 ((n')^2 + (m')^2)} \tag{15.8}$$

where i and j are multi-indices with $(0, 1) \le i = (m, n) \le (M, N)$ and $(0, 1) \le \mathrm{j} = (\mathrm{m}', \mathrm{n}') \le (\mathrm{M}, \mathrm{N})$ component wise. It follows that $\boldsymbol{\alpha}$, $\boldsymbol{\beta}$, **e**, **f** have $p = N(M + 1)$ non zero components. Note that in (15.7) and (15.8), $e_i(x, y) = f_i(y, x)$. Also note that for this choice of basis functions, the edge of the square maps onto itself under **P**. Another way of stating this is to notice that the displacement orthogonal to the edge is always zero on the edge of the square. These boundary conditions ensure that realizations of the model are (square) maps. We will comment further on this later.

We might attempt to specify $\boldsymbol{\alpha}$ and $\boldsymbol{\beta}$ in (15.5) and (15.6) as $N(0,\sigma_\alpha^2 I)$ and $N(0,\sigma_\beta^2 I)$ respectively. However, anticipating the need to have **P** invertible we specify the above normal distributions for $\boldsymbol{\alpha}$ and $\boldsymbol{\beta}$ truncated to the sets of the form $\boldsymbol{\alpha}^T \boldsymbol{\alpha} \le \mathrm{k}^2$

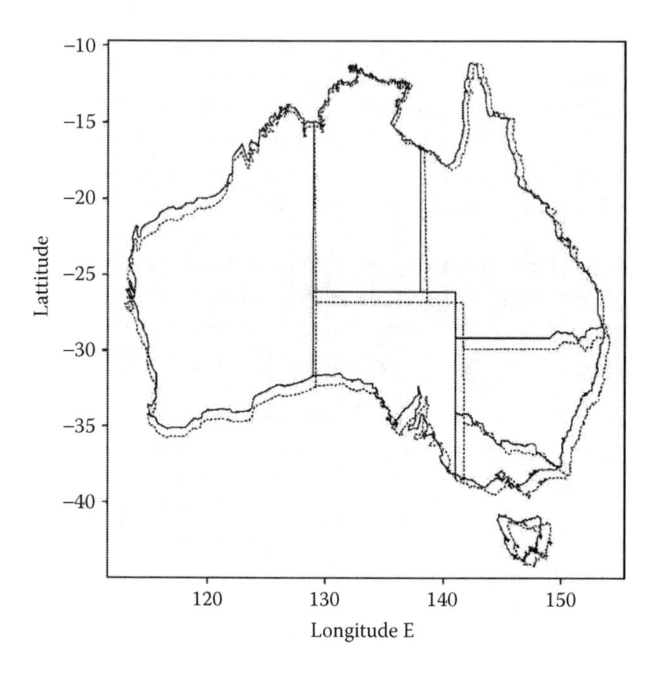

FIGURE 15.1 Map of Australia and a simulated distortion.

σ_α^2 and $\boldsymbol{\beta}^T \boldsymbol{\beta} \leq k^2 \sigma_\beta^2$. In practice, we know g_0, the observed map in (15.3). However, we do not know $\boldsymbol{\alpha}$ or $\boldsymbol{\beta}$, so our model gives a distribution for the possible 'true' maps g_T.

We have chosen the above specification for illustrative purposes since it results in a simple two parameter model, with σ_α and σ_β being directly related to the maximum displacement or 'distortion' in x and y directions respectively.

As a referee has pointed out, we could use the trigonometric functions in (15.7) and (15.8) without the weights and have diagonal covariance matrices for $\boldsymbol{\alpha}$ and $\boldsymbol{\beta}$. This increases the number of parameters, however. More work is required on identifying other useful classes of basis functions, and we will discuss this later.

Some example realizations of (15.2) are given in Figures 15.1 and 15.4.

2.1 Invertibility of *P*

The mapping $\mathbf{P}$ as defined by (15.5) and (15.6) has derivative $\mathbf{J}_{\alpha\beta} = \{\partial \mathbf{P}/\partial \mathbf{v}\}$ given by:

$$\mathbf{J}_{\alpha\beta} = \begin{bmatrix} 1 + \boldsymbol{\alpha}^T \mathbf{e}_x & \boldsymbol{\beta}^T \mathbf{f}_x \\ \boldsymbol{\alpha}^T \mathbf{e}_y & 1 + \boldsymbol{\beta}^T \mathbf{f}_y \end{bmatrix} \tag{15.9}$$

where $\mathbf{e}_x = \partial/\partial x\, \mathbf{e}$, $\mathbf{e}_y = \partial/\partial y\, \mathbf{e}$ and $\mathbf{f}_x$ and $\mathbf{f}_y$ are defined similarly. For $\mathbf{P}$ to be invertible, and for purposes of evaluating integrals later, we require:

$$|\mathbf{J}_{\alpha\beta}| = \det\{\mathbf{J}_{\alpha\beta}\} = ((1+\boldsymbol{\alpha}^T\mathbf{e}_x)(1+\boldsymbol{\beta}^T\mathbf{f}_y) - \boldsymbol{\alpha}^T\mathbf{e}_y\boldsymbol{\beta}^T\mathbf{f}_x) > 0 \qquad (15.10)$$

on the unit square S. Note that for $\boldsymbol{\alpha} = \boldsymbol{\beta} = 0$, this is certainly the case. Hence by continuity we can find a neighbourhood of the form $\boldsymbol{\alpha}^T\boldsymbol{\alpha} \leq k^2\,\sigma_\alpha^2$ and $\boldsymbol{\beta}^T\boldsymbol{\beta} \leq k^2\,\sigma_\beta^2$ for $|\mathbf{J}_{\alpha\beta}| > 0$. We will usually choose k^2 to be the 95 or 99 per cent point of the chi-squared distribution on p degrees of freedom. Conditions on k, σ_α and σ_β which $|\mathbf{J}_{\alpha\beta}| > 0$ are given in the Appendix. Typically we would expect σ_α and σ_β to be $<<$ 0·1, i.e., less than 10 per cent of the total row or column length. All the following calculations will assume use of truncated distributions.

The computation of $\mathbf{Q} = \mathbf{P}^{-1}$ is considered in the Appendix. However, at this point we note that a remarkably good approximation to $\mathbf{Q}$ is given by $\mathbf{Q}(\mathbf{v}) = 2\mathbf{v} - \mathbf{P}(\mathbf{v})$ when σ_α and σ_β are small, say less than 0·1. For example, if $\sigma_\alpha = \sigma_\beta = 0{\cdot}05$ then the error in using $\mathbf{Q}$ for a 500×500 array of image picture elements (pixels) is the order of half a pixel.

2.2 *Mean and covariance matrix of displacements*

Let $\mathbf{u}_1$ and $\mathbf{u}_2$ be two points in the unit square. Then the displacements $\delta i = \mathbf{P}(\mathbf{u}_i) - \mathbf{u}_i$ have mean zero and covariance matrix Σ given by:

$$\Sigma = c(k,p)\begin{bmatrix} \mathbf{e}_1{}^T\mathbf{e}_1\sigma_\alpha^2 & 0 & \mathbf{e}_1{}^T\mathbf{e}_2\sigma_\alpha^2 & 0 \\ & \mathbf{f}_1{}^T\mathbf{f}_1\sigma_\beta^2 & 0 & \mathbf{f}_1{}^T\mathbf{f}_2\sigma_\beta^2 \\ & \text{symmetric} & \mathbf{e}_2{}^T\mathbf{e}_2\sigma_\alpha^2 & 0 \\ & & & \mathbf{f}_2{}^T\mathbf{f}_2\sigma_\beta^2 \end{bmatrix} \qquad (15.11)$$

where $c(k, p) = \Pr\{\chi^2_{p+2} \leq k^2\} / \Pr\{\chi^2_p \leq k^2\}$, $\mathbf{e}_1 = \mathbf{e}(\mathbf{u}_1)$, $\mathbf{e}_2 = \mathbf{e}(\mathbf{u}_2)$ and $\mathbf{f}_1$ and $\mathbf{f}_2$ are defined similarly. See appendix 2. Note that the constant $c(k, p)$ will usually be close to one.

For the basis functions (15.7) and (15.8), the function $\mathbf{e}_1{}^T\mathbf{e}_1$ has a maximum at $x = 1/2$ and $y = 0$ or 1 in which case it is:

$$\left.\begin{aligned} \max(\mathbf{e}_1{}^T\mathbf{e}_1) &= \frac{4}{\pi^4}\left[\sum_{n=1}^{N}\frac{\sin^2\frac{n\pi}{2}}{n^4} + \sum_{m,n=1}^{M,N}\frac{\sin^2\frac{n\pi}{2}}{(n^2+m^2)^2}\right] \\ &\approx (0.24)^2 \quad (\text{independent of } M, N) \end{aligned}\right\} \qquad (15.12)$$

Similar results hold for the other diagonal terms. Hence if we wish to model maximum errors of the order of Δx and Δy in x and y respectively, assuming $c(k, p) \approx 1$ we could take ($q = 2$ or 3):

$$\left.\begin{array}{l}\max(q\sqrt{\mathbf{e}_1{}^T\mathbf{e}_1\sigma_\alpha}) \approx \Delta x \\ \quad \text{i.e., } \sigma_\alpha \approx \Delta x/(0.24q) \text{ or } \sigma_\alpha \approx 4\Delta x/q \text{ and similarly } \sigma_\beta \approx 4\Delta y/q.\end{array}\right\} \quad (15.13)$$

3 Evaluating and representing uncertainty

In this section we assume that, M, N, σ_α and σ_β are known. Since we observe g_0, but not the α, β defining the random displacements, we can make probability statements about true values according to:

$$\Pr\{g_T(\mathbf{v}) = l\} = \Pr\{g_0(\mathbf{Q}(\mathbf{v})) = l\}$$

Using the approximation $\mathbf{Q}(\mathbf{v}) = 2\mathbf{v} - \mathbf{P}(\mathbf{v})$, this becomes:

$$= \Pr\{g_0\ (\mathbf{v} - \delta(\mathbf{v}) = l\}.$$

By approximating the distribution of $\delta(\mathbf{v})$ (see the Appendix equations (15.A.4) (15.A.8)) this probability is:

$$= \Pr\{\mathbf{Z} \in A_1\ |\mathbf{Z} \sim N_T\ (\mathbf{v}, \Sigma(\mathbf{v}))\} \quad (15.14)$$

where A_1 is the observed region with $g_0\ (\mathbf{v}) = l$ for $\mathbf{v} \in A_i$, N_T denotes the truncated normal distribution and:

$$\Sigma(\mathbf{v}) = d(k, p)\begin{bmatrix} \mathbf{e}(\mathbf{v})^T\mathbf{e}(\mathbf{v})\sigma_\alpha^2 & 0 \\ 0 & \mathbf{f}(\mathbf{v})^T\mathbf{f}(\mathbf{v})\sigma_\beta^2 \end{bmatrix} \quad (15.15)$$

with $d(k, p) = \{\Pr(\chi_p^2 \leq k^2)/\Pr(\chi_{p-1}^2 \leq k^2)\}$. The truncation region is $\{(\mathbf{Z}_1, \mathbf{Z}_2)\colon |\mathbf{Z}_1| \leq k(\mathbf{e}^T\ \mathbf{e})^{1/2}\ \sigma_\alpha, |\mathbf{Z}_2| \leq k(\mathbf{f}^T\ \mathbf{f})^{1/2}\ \sigma_\beta\}$ (see equations (15.5) and (15.6) and Appendix 2). For practical computations we only consider the part of the distribution within ±3 standard deviations of the mean along each axis.

To illustrate how equation (15.14) is calculated, consider Figure 15.2. In this figure, observed map category 1 corresponds to the region A_1 defined by the diagonal lines. There are also five points $\mathbf{v}_1$, $\mathbf{v}_2$, ..., $\mathbf{v}_5$ (at the centre of rectangles) for which we will compute $\Pr\{g_T\ (\mathbf{v}_i) = 1\}$, i.e., the probability that the true map at $\mathbf{v} = (x_i, y_i)$ has category 1. Around each point is a rectangle B_i defined by $\{(x, y)\colon |x - x_i| \leq k\sigma_1\ (\mathbf{v}_i), |y - y_i| \leq k\sigma_2\ (\mathbf{v}_i)\}$ where $\sigma_1(\mathbf{v}_i)$, $\sigma_2(\mathbf{v}_i)$ are the diagonals of equation (15.15) evaluated at $\mathbf{v}_i$. Note that the size of the rectangles depends on the location of the point, i.e., on the variance of the x and y distortions at that point.

To compute the probability that a particular point is actually in category 1, the truncated normal distribution on the corresponding rectangle needs to be integrated over that part of the rectangle intersecting A_1. In particular, this probability is zero

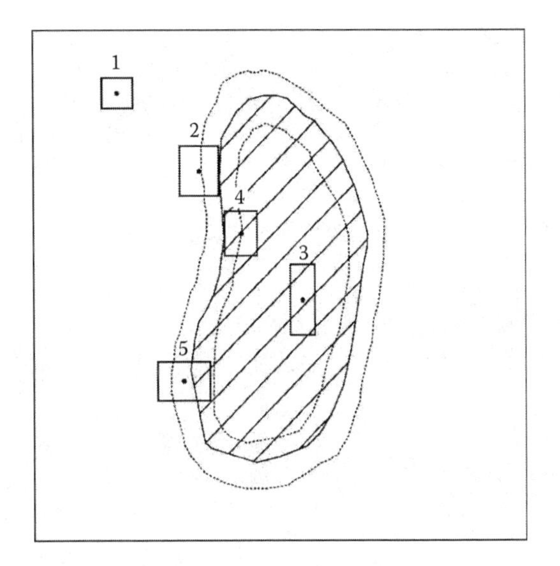

FIGURE 15.2 Example illustrating the calculation of uncertainty.

for $\mathbf{v}_1$ and $\mathbf{v}_2$, one for $\mathbf{v}_3$ and $\mathbf{v}_4$ and some value <1/3 for $\mathbf{v}_5$. Computationally it is only necessary to do calculations in the region between the broken lines in Figure 15.2.

An easy way to implement the calculation of the integrals is to rasterize the data, i.e., define a grid and approximate the regions by collections of squares or rectangles.

The effect of the model is to define a fuzzy boundary of varying thickness around the observed boundary. The use of truncated distributions results in probabilities of 1 for points in the centre of a large region and probabilities of zero for points a sufficient distance away from the region under consideration.

Note that if the variances $\sigma_1(\mathbf{v})$ and $\sigma_2(\mathbf{v})$ were constant, the calculation of probabilities could be done by convolving the (binary) raster image for a particular class with a Gaussian filter either directly or using fast Fourier transforms.

To visually represent uncertainty, a map is digitized into raster form and a grey scale image constructed for each map category or area. The probability of membership in a particular category is then calculated at the centres of each pixel and each displayed with a shade of grey depending on the probability (e.g., zero = black, 1 = white).

Colour maps for three map categories simultaneously could be displayed by assigning the probabilities of each category to different colour guns.

3.1 Transmitting uncertainty through set operations

In this section we consider two map layers with uncertainty specified by (15.3), i.e.

$$g_T(\mathbf{v}) = g_0(\mathbf{Q}_1\ (\mathbf{v}))\ \text{and}\ h_T(\mathbf{v}) = h_0\ (\mathbf{Q}_2\ (\mathbf{v}))$$

where g_0 and h_0 are the observed maps and $\mathbf{Q}_1$ and $\mathbf{Q}_2$ are defined by (15.5) and (15.6) with random variables $(\boldsymbol{\alpha}_1{}^T, \boldsymbol{\beta}_1{}^T)^T$ and $(\boldsymbol{\alpha}_2{}^T, \boldsymbol{\beta}_2{}^T)^T$ respectively. Note that M, N need not be the same for $\mathbf{Q}_1$ and $\mathbf{Q}_2$, and $\mathbf{Q}_1$ and $\mathbf{Q}_2$ are independent since there is no *a priori* reason for believing that the distortions in two different maps are correlated.

We assume that Pr $\{g_T(\mathbf{v}) = l\}$ and $\Pr\{h_T(\mathbf{v}) = m\}$ have been calculated for all categories I and m.

The following formulae are obvious,

1. *Intersections:*

$$\Pr\{g_T(\mathbf{v}) = l \text{ and } h_T(\mathbf{v}) = m\} = \Pr\{g_T(\mathbf{v}) = l\}\ \Pr\{h_T(\mathbf{v}) = m\}$$

2. *Unions:*

$$\Pr\{g_T(\mathbf{v}) = l \text{ or } h_T(\mathbf{v}) = m\} = \Pr\{g_T(\mathbf{v}) = l\} + \Pr\{h_T(\mathbf{v}) = m\} - \Pr\{g_T(\mathbf{v}) = l\}\ \Pr\{h_T(\mathbf{v}) = m\}$$

3. *Complements:*

$$\Pr\{g_T(\mathbf{v}) \neq l\} = 1 - \Pr\{g_T(\mathbf{v}) = l\}$$

Hence, given probabilities as calculated in Section 3, uncertainty can be assigned to the results of set operations by simple additions and multiplications of probabilities at each point. For raster data this corresponds to pixel-wise operations.

3.2 An example

To illustrate the ideas above consider the following example. A construction site needs to be selected in a given region and it must have the properties that:

1. The depth to bedrock is suitable.
2. The soil type is suitable.
3. The slope of the ground is suitable.

Maps are available for determining (15.1), (15.2) and (15.3) above. The first row of Figure 15.3a,b,c gives these maps indicating where depth to bedrock, soil type and slope are individually suitable for construction. In each map these regions are in black. Intersecting or overlaying these three maps gives the map (g) in the lower left hand corner of Figure 15.3. These areas in black denote constructions sites deemed to be suitable.

Choosing $M = N = 9$ and $\sigma_\alpha = \sigma_\beta = 0.05$, the second row of Figure 15.3 gives the corresponding maps allowing for uncertainty. Note the grey scale for the probabilities (white = 0 and black = 1). When the uncertainty maps are overlayed, the result is the map in the lower right-hand corner Figure 15.3h. Two observations can be made. Firstly this map indicates possible areas in the lower left corner (albeit of

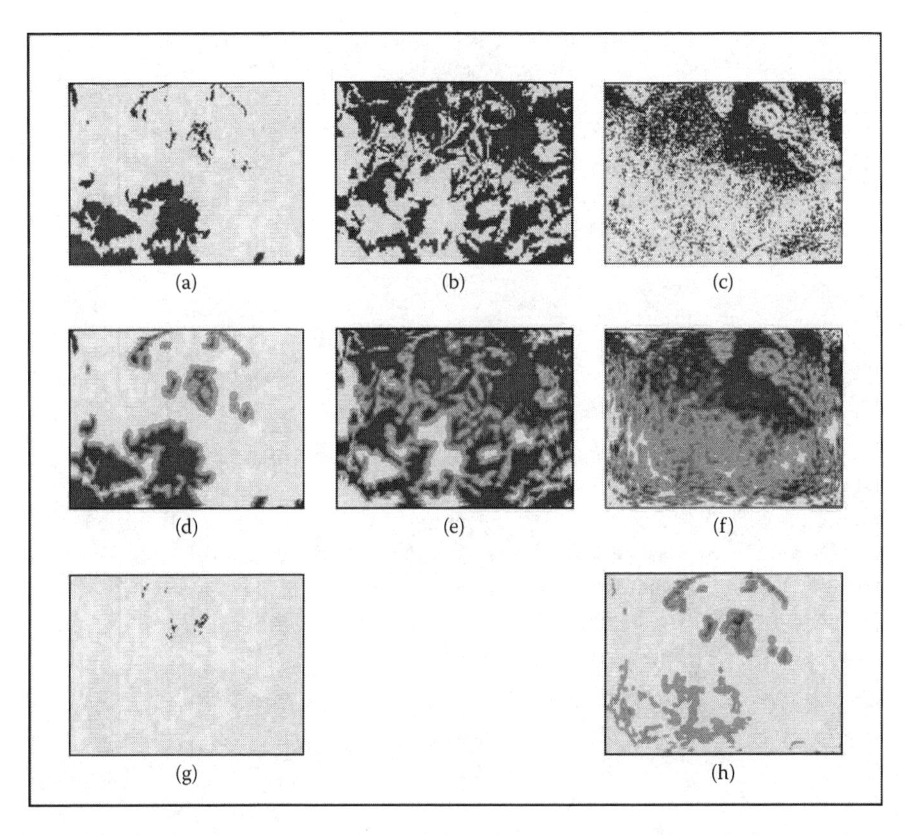

FIGURE 15.3 Example illustrating transmission of uncertainty through overlaying operations.

low probability) not given in the map ignoring uncertainty 15.3g. Secondly areas which originally appeared to have all three attributes have a maximum probability of 0·7 of having all three attributes. It can be seen that decision making ignoring uncertainty might neglect potentially useful areas and also mistakenly choose areas not having the attributes of interest.

Note that the effect of differing uncertainties in each of the maps could be studied by re-running the analyses. For example, the effect on the final result of reducing the error in one map could be assessed.

Uncertainty in map intersection using attribute (as compared to locational) errors is considered in Heuvelink and Burrough (1993).

4 Uncertainty in lengths and areas

Given data in a GIS, a common operation involves the calculation of lengths or perimeters of features in a map, fence lines, roads, or property boundaries. Below we consider the effect of uncertainty in point locations on these calculations. We will primarily be concerned with computing (approximate) means and variances, however in Section 5 on simulation we consider approximating the actual distribution of these quantities.

Although the results in this section are derived for points in the unit square S, in practise the region of interest will usually be a rectangle R at a substantial distance from the origin. Note that for all $\mathbf{v} \in R$ we can find a diagonal matrix Δ and a vector $\mathbf{m}$ such that for $\mathbf{v} \in R$, $\Delta^{-1}(\mathbf{v} - \mathbf{m})$ belongs to S. The appropriate distortion mapping $\tilde{\mathbf{P}}$ on R is then defined by:

$$\tilde{\mathbf{P}}(\mathbf{u}) = \mathbf{m} + \Delta\mathbf{P}((\Delta^{-1}\ (\mathbf{u} - \mathbf{m}))$$

where $\mathbf{P}$ is a distortion mapping on S, with inverse:

$$\tilde{\mathbf{Q}}(\mathbf{u}) = \mathbf{m} + \Delta\mathbf{Q}((\Delta^{-1}\ (\mathbf{u} - \mathbf{m}))$$

4.1 Uncertainty in lengths

In this section we consider the calculation of lengths. It should be remembered that lengths are scale dependent, e.g., coastlines, and what we are really doing is computing the length of a feature as approximated by a given series of line segments.

Consider a feature (e.g., property boundary) defined (approximately) by the collection of line segments joining the points $\mathbf{u}_0, \mathbf{u}_1, \ldots, \mathbf{u}_n$ ($\mathbf{u}_0 \neq \mathbf{u}_n$), then according to our model the true location of the points $\mathbf{u}_i$ is given by $\mathbf{v}_i = \mathbf{P}(\mathbf{u}_i)$ and the 'true' length of the feature is:

$$\left.\begin{aligned} d = d(\mathbf{v}_0, \mathbf{v}_1, \ldots, \mathbf{v}_n) &= \sum_{i=1}^{n} \{(\mathbf{v}_i - \mathbf{v}_{i-1})^T (\mathbf{v}_i - \mathbf{v}_{i-1})\}^{1/2} \\ &= \sum_{i=1}^{n} \left\| \mathbf{v}_i - \mathbf{v}_{i-1} \right\| \end{aligned}\right\} \tag{15.16}$$

Define $\mathbf{u} = (\mathbf{u}_0^T, \mathbf{u}_1^T, \ldots, \mathbf{u}_n^T)^T$ and similarly $\mathbf{v}$. Then by a Taylor series expansion we have:

$$d(\mathbf{v}) = d(\mathbf{u}) + \frac{\partial d^T}{\partial \mathbf{u}}(\mathbf{v} - \mathbf{u}) + \frac{1}{2}(\mathbf{v} - \mathbf{u})^T \frac{\partial^2 d}{\partial \mathbf{u}^2}(\mathbf{v} - \mathbf{u}) \tag{15.17}$$

Now the expected value of $\mathbf{v}$, $E\{\mathbf{v}\}$ is $\mathbf{u}$ and the variance $V\{\mathbf{v}\}$ of $\mathbf{v}$ is $\Sigma(\mathbf{u})$ where the i, jth block Σij (2×2) of $\Sigma(\mathbf{u})$ is:

$$\sum_{ij} = c(k, p) \begin{bmatrix} \sigma_{1ij} & 0 \\ 0 & \sigma_{2ij} \end{bmatrix} \tag{15.18}$$

with:

$$\sigma_{1ij} = \mathbf{e}(\mathbf{u}_i)^T \mathbf{e}(\mathbf{u}_j)\sigma_\alpha^2 \quad \text{and} \quad \sigma_{2ij} = \mathbf{f}(\mathbf{u}_i)^T \mathbf{f}(\mathbf{u}_j)\sigma_\beta^2$$

and $c(k, p)$ defined as in (15.11).

Ignoring the last term of (15.17) and taking the expectation of both sides we obtain the first order approximation:

$$E\{d(\mathbf{v})\}=d(\mathbf{u}) \tag{15.19}$$

The variance $V\{d(\mathbf{v})\}$ is:

$$\left.\begin{aligned} V\{d(\mathbf{v})\} &= \left(\frac{\partial d}{\partial \mathbf{u}}\right)^T \Sigma(\mathbf{u}) \left(\frac{\partial d}{\partial \mathbf{u}}\right) \\ &= c(k,p)\sum_{r=1}^{2}\sum_{k.l=0}^{n} \frac{\partial d}{\partial u_{rk}}\frac{\partial d}{\partial u_{rl}}\sigma_{rkl} \end{aligned}\right\} \tag{15.20}$$

where $\mathbf{u}_k=(u_{1k}, u_{2k})^T$. Note that from (15.19) the expected distance is approximately equal to the observed distance.

In a similar way we can obtain second order approximations to the mean and variance:

$$\left.\begin{aligned} E\{d(\mathbf{v})\} &= d(\mathbf{u}) + \frac{1}{2}\ \text{trace}\left\{\frac{\partial^2 d}{\partial \mathbf{u}^2}\Sigma(\mathbf{u})\right\} \\ &= d(\mathbf{u}) + c(k,p)\sum_{r=1}^{2}\sum_{ij}\frac{\partial^2 d}{\partial u_{ri}\partial u_{rj}}\sigma_{rij} \end{aligned}\right\} \tag{15.21}$$

Define, for $r=1, 2$

$$\tau_{rij} = \frac{\partial^2 d}{\partial u_{ri}\partial u_{rj}}\{(v_{ri}-u_{ri})(v_{rj}-u_{rj}) - c(k,p)\sigma_{rij}\}$$

and using the independence of $\mathbf{v}_{1i}$ and $\mathbf{v}_{2k}$ for all i and k we get the second order variance approximation

$$\left.\begin{aligned} V\{d(\mathbf{v})\} &= V_l + \frac{1}{4}\sum_{r=1}^{2}\sum_{ijlm} E\{\tau_{rij}\tau_{rlm}\} \\ &= V_l + \frac{1}{4}\sum_{r=1}^{2}\sum_{ijlm}\{b(\sigma_{ril}\sigma_{rjm}+\sigma_{rim}\sigma_{rjl}) + (b-c^2)\sigma_{ij}\sigma_{lm})\frac{\partial^2 d}{\partial u_{ri}\partial u_{rj}}\frac{\partial^2 d}{\partial u_{rl}\partial u_{rm}} \end{aligned}\right\} \tag{15.22}$$

where $b = b(k, p) = \Pr\{\chi^2_{p+4} \leq k^2\}/\Pr\{\chi^2_p \leq k^2\}$, c is defined as in (15.11), and V_1, denotes the right-hand side of (15.20). In a slightly different context similar formulae are given by Heuvelink *et al.* (1989).

For computing perimeters ($\mathbf{u}_0 = \mathbf{u}_n$) the above formulae still apply with the modification that d is a function of $\mathbf{u}_0, \mathbf{u}_1, \ldots, \mathbf{u}_{n-1}$. Simulations suggest that the first order approximation is quite good and that the second order corrections can usually be ignored. The above results for means and variances of d apply for $\mathbf{u}, \mathbf{v} \in R$ provided Σ_{ij} in (15.18) is multiplied by Δ^2. Also note that the derivative of the distortion mapping over R is related to the derivative over S by:

$$\partial\mathbf{P}/\partial\mathbf{u} = \partial\mathbf{P}/\partial\mathbf{z},\mathbf{z} = \Delta^{-1}(\mathbf{u} - \mathbf{m})$$

4.2 Uncertainty in areas

Another calculation common to GIS users is the computation of areas from maps. For example, the total area of a particular soil type, vegetation cover or property holding. Given uncertain area boundaries, below we compute the corresponding uncertainty in the calculated area as measured by its variance.

Consider an observed map class C defined by $\{\mathbf{u}: g_0(\mathbf{u}) = c\}$. Then according to our model the true region is $\mathbf{P}(C)$ for some random distortion mapping $\mathbf{P}$. This follows since $g_0(u) = c$ iff $g_T(\mathbf{P}(\mathbf{u})) = c$. Now, the change of variables integration formula (Lang 1971, p. 421) shows that under appropriate conditions the true area $m(\mathbf{P}(C))$ is given by:

$$\mu(P(C) = \int_c \det(\mathbf{J}_{\alpha\beta})d_{\mathbf{u}} \tag{15.23}$$

where $\mathbf{J}_{\alpha\beta}$is defined by (15.9) and is positive on S.

A simple calculation gives:

$$\det(\mathbf{J}_{\alpha\beta}) = 1 + \mathbf{e}_x^T\boldsymbol{\alpha} + \mathbf{f}_y^T\boldsymbol{\beta} + \boldsymbol{\alpha}^T(\mathbf{e}_x\mathbf{f}_y^T - \mathbf{e}_y\mathbf{f}_x^T)\boldsymbol{\beta} \tag{15.24}$$

Evaluating (15.23) we get:

$$\mu(\mathbf{P}(C)) = \mu(C) + \mathbf{a}^T\boldsymbol{\alpha} + \mathbf{b}^T\boldsymbol{\beta} + \boldsymbol{\alpha}^T(\mathbf{A} - \mathbf{B})\boldsymbol{\beta} \tag{15.25}$$

where μ(C) is the *observed* area of C,

$$\left.\begin{array}{ll} a = \int_c \mathbf{e}_x, & b = \int_c \mathbf{f}_x^T \\ \mathrm{A} = \int_c \mathbf{e}_x\mathbf{f}_x^T, & \mathrm{B} = \int_c \mathbf{e}_x\mathbf{f}_x^T \end{array}\right\} \tag{15.26}$$

and the integrals of vectors and matrices are defined componentwise. Note that expression (15.25) is actually a quadratic form in $(\boldsymbol{\alpha}^T\boldsymbol{\beta}^T)^T$. For the basis functions

(15.7) and (15.8), the integrals in (15.26) can be evaluated explicitly when C is a square or rectangle. The integrals over more general regions can be easily approximated by sums of integrals over squares or rectangles. This is particularly straightforward for raster maps, where the integrals can be calculated over strips of pixels.

Using (15.25) and the distribution properties of $\boldsymbol{\alpha}$ and $\boldsymbol{\beta}$ it follows that:

$$E\{\mu(\mathbf{P}(C))\} = \mu(C) \tag{15.27}$$

i.e., we expect the true area of region R to be equal to the observed area. The variance is:

$$V\{\mu(\mathbf{P}(C))\} = c((\mathbf{a}^T\mathbf{a}\alpha_\alpha^2 + \mathbf{b}^T\mathbf{b}\sigma_\beta^2) + c^2\alpha_\alpha^2\sigma_\beta^2 \text{ trace } \{(\mathbf{A}-\mathbf{B})^T(\mathbf{A}-\mathbf{B})\} \tag{15.28}$$

where c is defined in (15.11).

For areas on the general rectangle R, if δ_1 and δ_2 denote the diagonals of Δ, then corresponding areas in S are multiplied by $\delta_1\delta_2$ and variances by $(\delta_1\delta_2)^2$ to obtain the appropriate results on R.

A formulae for the area of a polygon (Stolk and Ettershank 1987) can also be used to get an explicit expression for the variance of an area. This information also enables the computation of the variance of the area of intersection of polygons from different maps.

4.3 An example

The black lines in Figure 15.4 are farm paddock and block boundaries digitized from a satellite image. (The grey lines are a realization from the distortion model, see Section 5.3.) The region is 4·6 km × 7·5 km in size.

Given the observed map (i.e. the black lines in Figure 15.4) we can make inferences about the 'true' lengths of boundary lines and the true areas of each region as follows.

For illustrative purposes we take $M = N = 3$ and $\sigma_\alpha = 0{\cdot}016$, $\sigma_\beta = 0{\cdot}026$ which translates to a maximum error in row and column co-ordinates of 2 pixels or 60m.

We also assume that the normal distributions for α and β are truncated at the 95 per cent level. Applying (15.19) and (15.20) to the lines marked one to four in Figure 15.4 gives the results in Table 15.1. Note that line L5 denotes the perimeter of the region A2 whilst the remaining lines are defined between the points (nodes) where three lines meet.

Similarly, computing (15.27) and (15.28) we obtain Table 15.2 below.

To compute the integrals required in (15.28) the region was divided into 926 by 1512 $5m^2$ pixels and areas approximated accordingly. Of course, the effect of different grid size could be studied easily.

The actual distributions of lengths and areas were also simulated and this will be mentioned in the next section.

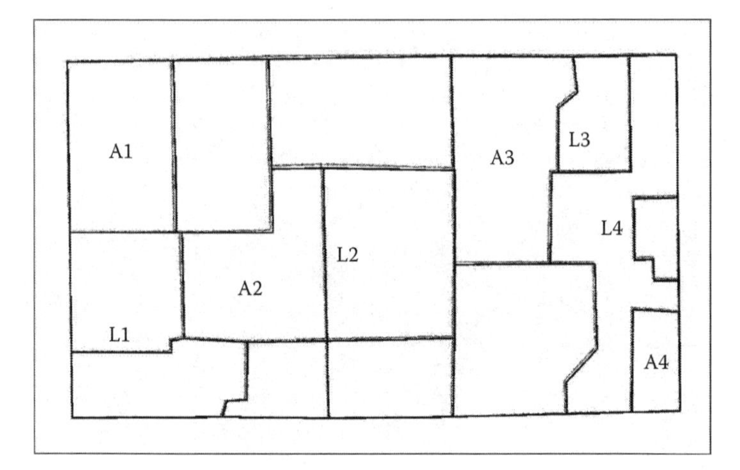

FIGURE 15.4 Farm boundary map and sample realization with $M = N = 3$ allowing maximum error of ± 60 m in row and column co-ordinates.

TABLE 15.1
Expected line lengths and standard deviations.

Line	Expected true length (m)	Standard deviation (m)
L1	1392·2	12·7
L2	1838·8	12·9
L3	1334·9	15·3
L4	1836·6	13·8
L5	6775·7	32·1

TABLE 15.2
Expected areas and standard deviations.

Area	Expected true area (ha)	Standard deviation (ha)
A1	239·2	3·1
A2	219·9	2·2
A3	259·0	3·0
A4	55·5	1·0

5 Simulation of maps

Generating realizations from (15.2) is important in developing an understanding of the model and for approximating the **distribution** of quantities such as lengths, perimeters, areas and the results of set operations such as overlaying. We briefly discuss the mechanics of doing this below. However, note that using this model, simulation is **not** required to calculate means and variances of lengths, perimeters and areas nor for computing the probability of class membership at a particular location. Uncertainty in set operations with maps, e.g., overlaying, also does not require simulation. At present simulation is required for assessing the uncertainty in more complex functions involving several uncertain maps.

5.1 Simulating maps

For a given *true map*, σ_α, σ_β, and specified basis functions, the simulation of maps is simply done by generating the iid (truncated) normal variables $\boldsymbol{\alpha}$ and $\boldsymbol{\beta}$ computing Q and using (15.3). In practice we require values of $g_0(\mathbf{u})$ on a finite number of values $\mathbf{u}$. If our map is defined by a set of arcs, i.e., each arc being a connected series of line segments, then we only need to compute $Q(\cdot)$ at the points defining the line segments for each realization of α and β. The function Q can be approximated by $\mathbf{Q}$ as in § 2.1 or by the method in the Appendix.

Raster or pixel data can be simulated by computing $\mathbf{P}(\bullet)$ on a rectangular grid of points $\{\mathbf{u}_{ij}$: $1 < i < m$, $1 < j < n\}$ and using the pixel centre closest to $\mathbf{P}(\mathbf{u}_{ij})$ to evaluate $g_0\ (\mathbf{u}_{ij})$. This typically requires much more computation than maps defined by series of arcs, however computation can be reduced by focusing on boundary pixels and their neighbours.

Several different maps can be simulated to study the effect of error on overlaying operations and other functions of the data.

5.2 Simulating distributions of quantities

Given a particular observed map, σ_α, σ_β, and specified basis functions, quantities such as (15.16) and (15.25) can be simulated simply by generating realizations of α and β. Thus their distributions can be approximated and moments compared with those given by exact or approximate formulae.

5.3 Example continued

The grey lines in Figure 15.4 depict a realization from the distortion model, assuming the black lines were the *true* lines.

To examine the distributions of (15.16) and (15.25) for the given observed map, 2000 simulations were conducted. Agreement between means and standard deviations calculated analytically with those from the simulation was good. Histograms of the areas and lengths suggested that the distributions were reasonably symmetric about their means and hence that use of the standard deviation to describe uncertainty was reasonable.

6 Parameter values for the model

Up until this point we have assumed that σ_α, σ_β, and M, N have been known. Of course in practise they are not, so we consider some methods for obtaining values for these parameters.

Firstly, we could supply values for these parameters in a subjective way using (15.13). To do this we might fix M, N at values so that (i) computation is not too intensive, (ii) the sum in (15.12) of terms from M, N to infinity is appropriately small for our purposes. Calculations with different values of M, N show that the variances in (15.9) settle down to two decimal place accuracy when M and N are both greater than or equal to 5. Using the relation (15.13) we can choose σ_α, and σ_β, according to our subjective beliefs or previous knowledge as to how much distortion we might expect in the map, e.g., for q=3 and a maximum 5 per cent distortion in the x direction we would choose $\sigma_\alpha \approx 0{\cdot}066$. Another possibility is the use of map scales and known digitizer errors to infer values for σ_α, and σ_β.

Secondly, with ground truthed data, i.e., measurements of true co-ordinate values, we can use likelihood methods or properties of our model to estimate σ_α, σ_β. Perhaps the simplest (but inefficient) method is to equate observed and expected values as follows:

Given a set of observed map points $\mathbf{u}_i$, i=1, …, P for which ground truth co-ordinates $\mathbf{v}i$ are available, define $\mathbf{x}_T$ to be the $p \times 1$ vector of first co-ordinates of $\mathbf{v}_i$ and $\mathbf{x}_0$ the first co-ordinates of $\mathbf{u}_i$. Then according to our model (ignoring measurement error in $\mathbf{x}_T$) $\mathbf{u}_i = \mathbf{Q}(\mathbf{v}_i)$ and approximating $\mathbf{Q}$ by $\hat{\mathbf{Q}}$ as in § 2.1.

$$E\{\mathbf{x}_0\} = \mathbf{x}_T, \quad \mathrm{V}\,\{\mathbf{x}_0\} = \text{ trace } (c\mathbf{E}\mathbf{E}^T)\sigma_\alpha^2\,\mathrm{t} \tag{15.29}$$

where the ith **row** of $\mathbf{E}$ is defined by the basis functions in (15.5) evaluated at $\mathbf{u}_i$. From (15.29) we can see that:

$$\begin{aligned} E\{(\mathbf{x}_0 - \mathbf{x}_T)^T(\mathbf{x}_0 - \mathbf{x}_T)\} &= \text{ trace } (c\mathbf{E}\mathbf{E}^T)\sigma_\alpha^2 \\ &= \text{trace } (c\mathbf{E}^T\mathbf{E})\sigma_\alpha^2 \end{aligned}$$

so we could estimate σ_α^2 by:

$$\hat{\sigma}_\alpha^2 = (\mathbf{x}_0 - \mathbf{x}_T)^T(\mathbf{x}_0 - \mathbf{x}_T)/\text{ trace } (c\mathbf{E}\mathbf{E}^T)$$

Similar expressions can also be derived for estimating . To get reasonable results P will need to be large since the observations are correlated. Of course we still have the problem of determining M and N. Using likelihood methods we can use a criterion such as BIC (Schwarz 1978) to obtain values of **M** and **N**. We can also predict $\boldsymbol{\alpha}$, $\boldsymbol{\beta}$ and consequently the 'true' map given the observed one. These and other ideas will be reported elsewhere.

7 Extensions and further work

The above is a first attempt at specifying a coherent probabilistic model for positional uncertainty in GIS and studying its implications. Considering the specification of the model in Section 2 suggests a number of avenues for further research. In equations (15.5) and (15.6), we could consider distributions for $\boldsymbol{\alpha}$ and $\boldsymbol{\beta}$ other than the normal distribution as well as correlations between $\boldsymbol{\alpha}$ and $\boldsymbol{\beta}$. We could also allow for biased distortions, i.e., systematic errors, by including non random terms as functions of $\mathbf{u}$ in (15.5) and (15.6). In this case, the expected value of the observed location is not equal to the true location.

Another important issue is the choice of basis functions. The choice in this paper is motivated by Fourier series. Other choices such as orthogonal polynomials are possible. The multi-resolution analysis of Mallat (1989) suggests that wavelets could be a useful class of basis functions to explore. These functions are attractive because of their local nature and would most likely allow for a more local correlation structure for the distortions. Work is proceeding on this.

The boundary of the map is problematical. With the boundary constraints mentioned in Section 2, realizations of the model are always (square) maps. The effect and appropriateness of these constraints need to be explored further. However, it should be noted that by placing a blank buffer zone around a map, any edge effects can be effectively minimized. It is possible to abandon the constraints on the edges totally and still derive results similar to the ones in this paper. For example, we could choose polynomial basis functions. However, there will be problems near the edges. Realizations of the model will have blank regions near the edges as well as 'mapped' areas beyond the edges of the map.

Model fitting has only been considered briefly in this paper. Likelihood methods for model fitting have been defined and implemented and we hope to report on these at a later date. These methods require the true and observed co-ordinates of a sample of ground locations. There are interesting questions here concerning how to produce these pairs of points when features in a map are difficult to locate on the ground, e.g., matching the boundary of a lake in an aerial photograph with its representation on a map.

Given appropriate model fitting strategies, tests using likelihoods, as well as examination of model residuals, can guide in selecting appropriate basis functions (both in form and number) and in testing boundary constraints.

Relaxing the requirement of invertibility of P opens up the possibility of models in which 'true' map features are lost and do not appear in the observed map.

Finally, note that positional uncertainty implies attribute errors near edges. Additional attribute errors could be introduced by including models for the probability of an observed attribute given the true attribute. These models could also allow for spatial dependence, i.e., the probability of a polygon attribute depends on the neighbouring attributes, see for example the models in Darroch *et al.* (1980).

8 Summary and conclusion

In this paper we have developed a model for positional uncertainty in GIS which can be applied for vector, raster and point data. The model leads to unified procedures for computing attribute probabilities at a given location as well as defining formulae for the variance in quantities such as length, perimeter and area. The model also allows for efficient simulation of maps.

The methods are computationally feasible. They were implemented by adding functions to the GRASS (1993) GIS and through stand alone code. Although not optimized for speed, the algorithms used to produce the examples in this paper required cpu time of the order of **minutes** on a Sparc 10 workstation.

More research is needed for the selection of appropriate basis functions and boundary constraints. The effects of relaxing the assumptions of independence of row and column distortions and invertibility of the distortion mapping P could also be fruitful areas to explore. Extensions to three-dimensional data could also be made.

The model is potentially a useful and flexible tool for assessing, representing and transmitting positional uncertainty in GIS. The model needs to be assessed and developed further through application to a wide variety of examples.

Acknowledgments

I would like to thank the three reviewers for their most helpful comments.

References

Altman, D., 1994, Fuzzy set theoretic approaches for handling imprecision in spatial analysis. *International Journal of Geographical Information Systems*, 8, 27–289.

Amit, Y., Grenander, U., and Piccioni, M., 1991, Structural image restoration through deformable templates. *Journal of the American Statistical Association*, 86, 376–387.

Darroch, J. N., Lauritzen, S. L., and Speed, T. P., 1980, Markov Fields and Log-linear interaction models for contingency tables. *Annals of statistics*, 8, 522–539.

Goodchild, M., and Gopal, S., 1989, *Accuracy of Spatial Data Bases* (New York: Taylor & Francis).

GRASS (1993), Geographical Resources Analysis Support System. US Army Construction Engineering Research Lab. Available by anonymous ftp from moon.cecer.army.mil.

Haining, R. P., and Arbia, G., 1993, Error propagation through map operations. *Technometrics*, 35, 293–305.

Heuvelink, G. B. M., Burrough, P. A., and Stein, A., 1989, Propagation of errors in spatial modelling with GIS. *International Journal of Geographical Information Systems*, 3, 303– 322.

Heuvelink, G. B. M., and Burrough, P. A., 1993, Error propagation in cartographic modelling using Boolean Logic and continuous classification. *International Journal of Geographical Information Systems*, 7, 231–246.

Johnson, N. L., and Kotz, S., 1972, *Distributions in Statistics: Continuous multivariate distributions* (New York: John Wiley and Sons).

Keefer, B. J., Smith, J. L., and Gregoire, T. G., 1991, Modelling and evaluating the effects of stream mode digitizing errors on map variables. *Photogrammetric Engineering and Remote Sensing*, 57, 957–963.

Lang, S., 1971, *Analysis I.* (Reading, Massachusetts: Addison-Wesley).

Mallat, S. L., 1989, A theory of multiresolution signal decomposition: the wavelet representation. *I.E.E.E. Transactions of Pattern Analysis and Machine Intelligence*, 11, 674–693.

Maguire, D. J., Goodchild, M. F., and Rhind, D. W., 1991, *Geographical Information Systems, Volume 1: Principles* (Essex: Longman).

Muirhead, R. J., 1982, *Aspects of Multivariate Statistical Theory*, (New York: Wiley).

Schwarz, G., 1978, Estimating the dimension of a model. *The Annals of Statistics*, 6, 461–464.
Stolk, R., and Ettershank, G., 1987, Calculating the area of an irregular shape. *Byte*, February, 135–136.
Wesseling, C. G., and Heuvelink, G. B. M., 1993, ADAM an error propagation tool for geographical information systems. User Manual, Department of Physical Geography, University of Utrecht.

Appendix 1

Positiveness of the Jacobian

We want to show that $|\mathbf{J}_{\alpha\beta}|$ in Section 2.1 is positive given that $(\boldsymbol{\alpha}^T \boldsymbol{\alpha}) \leq k^2 \sigma_\alpha^2$ and $(\boldsymbol{\beta}^T \boldsymbol{\beta}) \leq k^2\sigma_\beta^2$. By the Cauchy Schwartz inequality we have:

$$\left.\begin{aligned} \left|\boldsymbol{\alpha}^T \mathbf{e}_x\right| &\leq (\boldsymbol{\alpha}^T \boldsymbol{\alpha})^{1/2} (\mathbf{e}_x^T \mathbf{e}_x)^{1/2} \\ &\leq k\sigma_\alpha (1\cdot 195)^{1/2} \end{aligned}\right\} \tag{15.A.1}$$

since it can be shown that for $M = N \leq 9$, over S $\max(\mathbf{e}_x(\mathbf{u})^T \mathbf{e}_x(\mathbf{u})) = 1\cdot 195$ occurs at the four corners of the unit square.

Similarly:

$$\left|\boldsymbol{\beta}^T \mathbf{f}_y\right| \leq k\sigma_\beta (1\cdot 195)^{1/2} \tag{15.A.2}$$

Also, it can be shown that for $M = N \leq 9$ $\max \mathbf{f}_x^T \mathbf{f}_x = 0\cdot 225$ occurs at the centre $(0\cdot 5,\ 0\cdot 5)$ of the unit square. Hence:

$$\begin{aligned} \left|\boldsymbol{\beta}^T \mathbf{f}_x\right| &\leq (\boldsymbol{\beta}^T \boldsymbol{\beta})^{1/2} (\mathbf{f}_x^T \mathbf{f}_x)^{1/2} \\ &\leq \sigma_\alpha (0\cdot 225)^{1/2} \end{aligned} \tag{15.A.3}$$

and:

$$|\boldsymbol{\alpha}^T \mathbf{e}_y| \leq k\sigma_\alpha (0\cdot 225)^{1/2} \tag{15.A.4}$$

Given the above, it is not difficult to establish the following result.

Lemma 15.A.1 For $M, N \leq 9$ and k, s_α, s_β satisfying:

$$\begin{gathered} (1 - k\sigma_\alpha s_1)(1 - k\sigma_\beta s_1) - k^2 \sigma_\alpha \sigma_\beta s_2 > 0 \\ 1 - k\sigma_\alpha s_1 > 0 \\ 1 - k\sigma_\beta s_1 > 0 \end{gathered} \tag{15.A.5}$$

where $s_1 = (1\cdot 195)^{1/2}$, $s_2 = 0\cdot 225^{1/2}$ it follows that $\left|\mathbf{J}_{\alpha\beta}\right| > 0$ on S.

Proof Find lower and upper bounds for the first and second product terms in (15.10) respectively.

The proof of Lemma 15.A.1, and simulations, suggest that for a given value of k, values of s_α and s_β larger than those suggested by the Lemma will also have $|\mathbf{J}_{\alpha\beta}| > 0$, i.e., the result is sufficient but not necessary.

The inequalities in Lemma 15.A.1 are easily checked by simply evaluating the expressions for given k, s_α and s_β.

Appendix 2

Truncated normal distribution

Some of the results used in this paper concerning truncated normal distributions are given below. Suppose $\boldsymbol{\alpha}$ and $\boldsymbol{\beta}$ are independent $N(0, \sigma_\alpha^2 I)$ and $N(0, \sigma_\beta^2 I)$ respectively. Define the truncation sets $H_\alpha = \{\boldsymbol{\alpha} \in R^p: \boldsymbol{\alpha}^T\boldsymbol{\alpha} \le k^2\sigma_\alpha^2\}$ and $H_\beta = \{\boldsymbol{\beta} \in R^p: \boldsymbol{\beta}^T\boldsymbol{\beta} \le k^2\sigma_\beta^2\}$.

Lemma 15.A.2 Suppose the normal distribution for $\boldsymbol{\alpha}$ and $\boldsymbol{\beta}$ is truncated to $H_\alpha \times H_\beta$, then for this truncated distribution:

1.
$$E\{\boldsymbol{\alpha}\} = E\{\boldsymbol{\beta}\} = 0 \quad (15.A.6)$$

2.
$$Var\left\{\begin{bmatrix}\boldsymbol{\alpha}\\ \boldsymbol{\beta}\end{bmatrix}\right\} = c\begin{bmatrix}\sigma_\alpha^2 I & 0\\ 0 & \sigma_\beta^2 I\end{bmatrix} \quad (15.A.7)$$

where $c = \mathbf{Pr}\{\chi_{p+2}^2 \mathrm{k}^2\} / \mathbf{Pr}\{\chi_p^2 k^2\}$.

3. For vectors $\mathbf{p}, \mathbf{q}, \mathbf{r}, \mathbf{s} \in R^p$

(i)
$$\mathrm{Cov}(\mathbf{p}^T\boldsymbol{\alpha}, \mathbf{q}^T\boldsymbol{\alpha}) = \mathrm{c}(\mathbf{p}^T\ \mathbf{q})\sigma_\alpha^2$$

(ii)
$$E\{(\mathbf{p}^T\boldsymbol{\alpha}\mathbf{q}^T\boldsymbol{\alpha} - c\mathbf{p}^T\mathbf{q}\sigma_\alpha^2)(\mathbf{r}^T\boldsymbol{\alpha}\mathbf{s}^T\boldsymbol{\alpha} - c\mathbf{r}^T\mathbf{s}\sigma_\alpha^T)\} = b\sigma_\alpha^4\{(\mathbf{p}^T\mathbf{r})(\mathbf{q}^T\mathbf{s}) + (\mathbf{p}^T\mathbf{s})(\mathbf{q}^T\mathbf{r})\} + (b - c^2)\sigma_\alpha^4(\mathbf{p}^T\mathbf{q})(\mathbf{r}^T\mathbf{s}) \quad (15.A.8)$$

where $b = \Pr\{\chi_{p+4}^2 \le k^2\}/\Pr\{\chi_p^2 \le k^2)$ and c is defined as in (15.A.7).

4. The distribution of $y = \mathbf{e}^T\boldsymbol{\alpha}$ for any vector $\mathbf{e}$ is:

$$g(y) = \phi(y, 0, (\mathbf{e}^T\mathbf{e})\sigma_\alpha^2)\gamma(y, k, p)1_{\bar{H}}(y) \quad (15.A.9)$$

where $\phi(\cdot, 0, \sigma^2)$ is the univariate normal density function with mean 0 and variance σ^2:

$$\gamma(y,k,p) = \mathbf{Pr}\{\chi^2_{p-1} \leq k^2 - [y^2/(\mathbf{e}^T\mathbf{e}\sigma^2_\alpha)]\}/\Pr\{\chi^2_p \leq k^2\}$$

and:

$$1_{\bar{H}}(y) = \begin{cases} 1, & |y|k(\mathbf{e}^T\mathbf{e})\sigma_\alpha^{1/2} \\ 0, & \text{otherwise} \end{cases}$$

Proof

1. The fact that the defining integrals are odd functions proves that expected values and covariances are zero.
2. Equation (15.A.7) follows from Johnson and Kotz (1972, p. 72).
3. The first result is clear from (15.A.7). The second result follows on rewriting the expectation in (15.A.8) as:

$$E\{(\mathbf{p}^T\boldsymbol{\alpha}\mathbf{q}^T\boldsymbol{\alpha}\mathbf{r}^T\boldsymbol{\alpha}s^T\boldsymbol{\alpha}\} - c^2(\mathbf{p}^T\mathbf{q})(\mathbf{r}^T\mathbf{s})\sigma^4_\alpha \tag{15.A.10}$$

The first term in (15.A.10) is:

$$\sum_{ijkl} p_i q_j r_k s_l E\{\alpha_i \alpha_j \alpha_k \alpha_l\} \tag{15.A.11}$$

which by the result in Johnson and Kotz (1972, p. 72) and a standard result concerning 4th order moments of the normal distribution reduces to:

$$\sum_{ijkl} p_i q_j r_k s_l b\sigma^4_\alpha \{I_{ik}I_{jl} + I_{il}I_{jk} + I_{ij}I_{kl}\} \tag{15.A.12}$$

where for example I_{ik} is the (i, k) th element of the identity matrix. Simplifying (15.A.12) and substituting into (15.A.10) gives the result.

4. Given $\mathbf{e}$, define orthonormal rows $\mathbf{E}$ such that $\mathbf{Ee} = 0$ and $\mathbf{EE}^T = 1$. Define:

$$R = \begin{bmatrix} \mathbf{e}^T \\ \mathbf{E} \end{bmatrix} = R^{-1}[\mathbf{e}/\mathbf{e}^T\mathbf{e} \vdots \mathbf{E}^T] \tag{15.A.13}$$

Let $\mathbf{y} = \mathbf{R}\boldsymbol{\alpha}$. Now the density of $\boldsymbol{\alpha}$ is $\mathbf{f}(\boldsymbol{\alpha}) = (\exp\{-1/2(\boldsymbol{\alpha}^T\boldsymbol{\alpha}/\boldsymbol{\sigma}^2_\alpha)\}1_H(\boldsymbol{\alpha}))/D$, where:

$$1_H(\alpha) = \begin{cases} 1, & \boldsymbol{\alpha}^T\boldsymbol{\alpha} \leq k^2\sigma^2_\alpha \\ 0, & \text{otherwise} \end{cases} \tag{15.A.14}$$

and:

$$D = \Pr(\chi_p^2 \le k^2)(\pi)^{p/2}\sigma_\alpha^p \qquad (15.A.15)$$

Using the standard formula for the density of transformed variables (Muirhead 1982, p. 50) the density of **y** is $f(\mathbf{R}^{-1}\,\mathbf{y})\,|\mathbf{R}^{-1}|$. Now computing $\mathbf{RR}^T$ it is easy to see that $|\mathbf{R}^{-1}| = (\mathbf{e}^T\,\mathbf{e})^{-1/2}$.

Hence the density of **y** is:

$$g(y) = \left(\exp\left\{ 1/2_y^T \begin{bmatrix} (\mathbf{e}^T\mathbf{e}) & 0 \\ 0 & I \end{bmatrix}^1 \mathbf{y}\sigma_\alpha^{-2} \right\} 1_H(\mathbf{R}^{-1}\mathbf{y}) \right) \Big/ (D(\mathbf{e}^T\mathbf{e})^{1/2})\mathbf{R}^{-1}\mathbf{y} \qquad (15.A.16)$$

Now $g(\mathbf{y})$ is defined on $R(H_\alpha)$, the range of H_α under R, and writing R as:

$$R = \begin{bmatrix} (\mathbf{e}^T\mathbf{e})^{1/2} & 0 \\ 0 & I \end{bmatrix} \begin{bmatrix} \mathbf{e}/(\mathbf{e}^T\mathbf{e})^{1/2} \\ \mathbf{E} \end{bmatrix} \qquad (15.A.17)$$

shows that $R(H_\alpha)$ is the original Hyper-sphere H_α rotated in space and then having the y_1 axis rescaled. In other words, $g(\mathbf{y})$ is defined on $\{\mathbf{y} = (\mathbf{y}_1\mathbf{y}_2)^T : (y_1^2/e_1^T e_1) + \mathbf{y}_2^T\mathbf{y}_2 \le k^2\,\sigma^2{}_\alpha\}$. Integrating out y_j for $j \ne 1$ gives the density of $y_1 = \mathbf{e}^T\alpha$ as:

$$g(y) = \phi(y, 0, (\mathbf{e}^T\mathbf{e})\sigma_\alpha^2)\gamma(y_1, k, p)1_{\bar{H}}(y_1) \qquad (15.A.18)$$

as stated in (15.A.9).

Remark Computing probabilities from (15.A.9) could be done directly using appropriate numerical techniques, however empirical work shows that a good approximation to $g(y)$ is:

$$g(y) = \phi(y, 0, d(k,p)(\mathbf{e}^T\mathbf{e})\sigma_\alpha^2)\gamma(y, k, p)1_{\bar{H}}(y) \qquad (15.A.19)$$

where:

$$d(k,p) = \{\mathbf{Pr}(\chi_p^2 \le k^2) / \mathbf{Pr}(\chi_{p-1}^2 \le k^2)\}$$

This result is obtained by matching the modes of the left- and right-hand sides of (15.A.19). Strictly speaking we should include a normalization factor in the right-hand side of (15.A.19) however for all practical purposes it is usually indistinguishable from unity.

Tables of values of the quantities $b = b(k, p)$, $c = c(k, p)$ and $d(k, p)$ in *Lemma 15.A.1* are given for k^2 equal to the 95th and 99th percentile of the chi squared distribution on p degrees of freedom. In the tables the number of basis functions is $p = M^*(M + 1)$ since $M = N$.

TABLE 15.A.1

$M = N$	p	k	k^2	b	c	d
1	2	2·4477	5·9915	0·60616	0·84233	0·96386
2	6	3·5485	12·5916	0·79220	0·91927	0·97689
3	12	4·5854	21·0260	0·86578	0·94635	0·98252
4	20	5·6045	31·4104	0·90284	0·95998	0·98586
5	30	6·6161	43·7730	0·92456	0·96814	0·98810
6	42	7·6239	58·1240	0·93863	0·97356	0·98972
7	56	8·6295	74·4683	0·94841	0·97741	0·99095
8	72	9·6337	92·8083	0·95557	0·98029	0·99191
9	90	10·6370	113·1453	0·96102	0·98252	0·99268

k^2 = 95th Percentile

TABLE 15.A.2

$M = N$	p	k	k^2	b	c	d
1	2	3·0349	9·2103	0·84637	0·95348	0·99239
2	6	4·1002	16·8119	0·93068	0·97765	0·99485
3	12	5·1202	26·2169	0·95853	0·98557	0·99600
4	20	6·1291	37·5662	0·97133	0·98941	0·99671
5	30	7·1339	50·8922	0·97840	0·99166	0·99721
6	42	8·1367	66·2062	0·98280	0·99313	0·99757
7	56	9·1386	83·5134	0·98577	0·99417	0·99785
8	72	10·1398	102·8163	0·98789	0·99493	0·99807
9	90	11·1408	124·1163	0·98948	0·99552	0·99825

k^2 = 99th Percentile

Appendix 3

Computation of the inverse mapping $\mathbf{Q} = \mathbf{P}^{-1}$

Since our model has been developed for $\mathbf{P}$ in (2), to evaluate the probabilities in Section 3 we need to evaluate $\mathbf{Q} = \mathbf{P}^{-1}$, at least at a given set of points.

The problem can be posed as follows. Given $\mathbf{v} = \mathbf{P}(\mathbf{u})$ find $\mathbf{u}$, or solve the equation $\mathbf{v} - \mathbf{P}(\mathbf{u}) = 0$ for $\mathbf{u}$. Newton's Method can be applied to obtain a sequence of approximations to $\mathbf{u}$ via:

$$\mathbf{u}_{n+1} = \mathbf{u}_n + \left(\frac{\partial \mathbf{P}}{\partial \mathbf{u}_n}\right)^{-1} (\mathbf{v} - \mathbf{P}(\mathbf{u}_n)) \tag{15.A.20}$$

Evaluating (15.A.20) gives a one step approximation ($\mathbf{u}_1 = \mathbf{u}_1$, $\mathbf{u}_0 = \mathbf{v}$) as:

$$\begin{aligned} \mathbf{u} &\doteq \mathbf{v} - (I + \mathbf{A}_0)^{-1} \begin{bmatrix} \mathbf{e}(\mathbf{v})^T \boldsymbol{\alpha} \\ \mathbf{f}(\mathbf{v})^T \boldsymbol{\beta} \end{bmatrix} \\ &= \mathbf{Q}(v) \end{aligned} \tag{15.A.21}$$

Here $\mathbf{A}_0 = \mathbf{A}(\mathbf{v})$, $\mathbf{v} = (x, y)^T$ and:

$$\mathbf{A}(\mathbf{v}) = \begin{bmatrix} \mathbf{e}_\mathbf{x}(\mathbf{v})^T \alpha & \mathbf{f}_x(\mathbf{v})^T \beta \\ \mathbf{e}_\mathbf{y}(\mathbf{v})^T \alpha & \mathbf{f}_y(\mathbf{v})^T \beta \end{bmatrix}$$

with $\mathbf{e}_x = \partial/\partial x\mathbf{e}$ and $\mathbf{e}_y$, $\mathbf{f}_x$, $\mathbf{f}_y$ defined similarly. If the matrix A is sufficiently small then (15.A.21) can be approximated by:

$$\mathbf{u} \doteq \mathbf{v} - \begin{bmatrix} \mathbf{e}(\mathbf{v})^T \alpha & \mathbf{e}(\mathbf{v})^T \alpha \\ \mathbf{f}(\mathbf{v})^T \beta & \mathbf{f}(\mathbf{v})^T \beta \end{bmatrix} + \mathbf{A} \begin{bmatrix} \mathbf{e}(\mathbf{v})^T \alpha \\ \mathbf{f}(\mathbf{v})^T \beta \end{bmatrix} \tag{15.A.22}$$

Ignoring the last term of (15.A.22) shows that in this case $\mathbf{Q}$ has essentially the form:

$$\mathbf{Q}(\mathbf{v}) \doteq 2\mathbf{v} - P(\mathbf{v}) = \mathbf{v} - \delta(\mathbf{v}).$$

The above argument requires $\mathbf{A}^2$ to be sufficiently small (i.e., $(\mathbf{I} + \mathbf{A})^{-1} \approx \mathbf{I} - \mathbf{A}$), as well as the second term in (15.A.22). Some intuitive justification for this is given by noting that the elements of vec{$\mathbf{A}$}=$(a_{11}, a_{12}, a_{21}, a_{22})T$ have mean 0 and covariance matrix Σ where:

$$\Sigma = c(k, p) \begin{bmatrix} \mathbf{G} & \mathbf{0} \\ 0 & \mathbf{H} \end{bmatrix} \tag{15.A.23}$$

and:

$$\mathbf{G} = \sigma_\alpha^2 \begin{bmatrix} \mathbf{e}_x^T \mathbf{e}_x & \mathbf{e}_x^T \mathbf{e}_y \\ \mathbf{e}_y^T \mathbf{e}_x & \mathbf{e}_y^T \mathbf{e}_y \end{bmatrix} \quad \mathbf{H} = \sigma_\beta^2 \begin{bmatrix} \mathbf{f}_x^T \mathbf{f}_x & \mathbf{f}_x^T \mathbf{f}_y \\ \mathbf{f}_y^T \mathbf{f}_x & \mathbf{f}_y^T \mathbf{f}_y \end{bmatrix} \tag{15.A.24}$$

It can be shown by calculation that $\mathbf{e}_x^T \mathbf{e}_x$ and $\mathbf{f}_y^T \mathbf{f}_y$ have a maximum at (0, 0), (0, 1), (1, 0) and (1, 1) and the maximum value is 1·195, for $M, N \leq 9$. Similarly $\mathbf{e}_y^T \mathbf{e}_y$ and $\mathbf{f}_x^T \mathbf{f}_x$ have a maximum value of 0·225 at (0·5, 0·5) for $M, N \leq 9$. With $\sigma_\alpha = \sigma_\beta = 0 \cdot 1$ it is easy to see that the elements of **G** and **H** are small and thus the elements of $\mathbf{A}^2$ will be small.

Similarly, $(\mathbf{e}^T\boldsymbol{\alpha}, \mathbf{f}^T\boldsymbol{\beta})$ has mean zero and variance $\Sigma(\mathbf{v})$ as defined in § 3, equation (15.15) with d replaced by c. The maximum value of $\mathbf{e}^T\mathbf{e}$ and $\mathbf{f}^T\mathbf{f}$ is $(0 \cdot 24)^2$ (§ 2.2, equation (15.12)) so that the last term of (15.A.11) will be the product of small terms and hence will be small.

Simulations show that on the unit square the approximation $\mathbf{Q}(\mathbf{v}) = 2\mathbf{v} - \mathbf{P}(\mathbf{v})$ differs in the 4th decimal place from the actual inverse when σ_α and $\sigma_\beta < 0 \cdot 1$.

Assessing, Representing, and Transmitting Uncertainty in GIS: Ten Years On

Harri T. Kiiveri

Background

The motivation for the work presented in this paper came from the Land Monitor project (Caccetta et al., 2000, see also http://www.landmonitor.wa.gov.au/), a large environmental monitoring project concerned with mapping and monitoring dryland salinity over the entire southwestern area of Western Australia. This project involved the use of a time series of mosaiced satellite images, digital elevation models, and expert opinion, both in the form of ground truthing and in knowledge of the process of salinisation. A ground area of 24 million hectares was mapped over a series of times.

In coming to terms with this project, it was clear that there were many sources of error and uncertainty. For example, there were a lot of issues in accurately registering and rectifying a large number of images — classification uncertainty, the production of digital elevation models and suitable variables constructed from them, and making use of expert knowledge (rules), which in themselves were uncertain.

The approach of producing classified images at each date and simply comparing them to identify changes was quickly abandoned due to the difficulty of accurately classifying saline ground using one date alone. Clearly we needed to enforce consistency over time, and also use that consistency to help distinguish the classes of interest. We decided to use a spatial version of Bayesian networks (Kiiveri et al., 1999) to integrate all this data. The inputs to this network were sets of probability surfaces that represented the uncertainty in the classifications at each date, and a spatial variable derived from the digital elevation model. The question then arose as to how to include positional uncertainty in the analysis.

The generic topic of the paper was identified against the background of the above work. For Bayesian networks, uncertainty propagation was relatively straightforward, however the main issucs concerned the uncertainties in the input maps. Using the results in the paper, it was shown that ignoring positional uncertainty in the input maps had a negligible effect on the resulting output maps. I think this was a testament to the efforts of the group in getting the registration, rectification, and mosaicing right. On a slightly different tack, another motivation I had at this time for this paper was to try to get some results for uncertainty propagation in basic GIS operations without resorting to simulation.

I originally tried to publish this paper in a mainstream statistical journal, statistics being my background, but received a response of the form "it's a serious attempt to solve the uncertainty propagation problem, however it's unlikely to interest many of our readers" — a comment I shall return to later in this chapter. I think the paper eventually found its correct home.

Developments following publication

At this point I should confess that I have not actively worked in this area since 2000. I have spent some time attempting to catch up with recent developments, however clearly not enough. My comments in this section should be viewed with this in mind.

I am aware of some further work on the topic of this paper (Leung and Yan, 1998; Arbia et al., 1998; Shi and Liu, 2000; Leung et al. 2004a, 2004b, 2004c, 2004d). However, while there are references to my paper in a number of articles, (an interesting one being Zhang and Stuart, 2001), I could find no evidence of anyone actually using the model on real data. I can only speculate as to the reasons for this. Some possible explanations are:

1. It is an issue about everyone working on their own research "turf";
2. The presentation was too technical and that has scared some potential users away;
3. Too much effort is required to implement the methods from scratch;
4. It is just not a good model.

I will deal with these possibilities one at a time. Concerning (i), I would like to make it clear that I do not regard this area as my own turf. While it is human nature to think about problems in our own way, an occasional visit to foreign lands can be useful.

As for (ii), the presentation was originally intended for a statistical audience. I attempted to simplify things, however a certain level of technical argument is required to deal with uncertainty adequately. Later publications on this topic also bear this out.

Concerning (iii), perhaps access to software implementing the methods would have helped. In fact, GRASS functions were implemented to enable use of the model, as well as some stand-alone software to fit the model given observed "distortions" in maps.

The conclusion (iv) should be based on shortcomings observed in applying the model to a number of data sets, and an inability to adjust the model to fit real-life situations. I am not currently aware of any work of this nature.

In regards to (ii) and (iii) above, I would like to point out a couple of easy-to-use applications of the results in the paper. First, as an approximation, positional uncertainty could be incorporated into raster maps by applying a Gaussian moving average filter with extents determined by the likely size of the positional errors. This could be applied to both hard and so-called *fuzzy* maps. Second, the model in the paper makes for easy simulation of "distortions" in the location of points, lines, and curves, which preserve topological relationships.

An interesting extension would be to apply the ideas to maps with continuous attributes such as digital elevation models. The ideas in the paper might also be useful for generating "distorted" digital elevation models efficiently.

Finally, I found no evidence of any papers pursuing suggestions for further research in the paper topics, which I still consider potentially very fruitful.

Some observations

On reading some of the subsequent work on uncertainty propagation in GIS, what strikes me is the popularity of Monte Carlo simulation. This is understandable as some of the problems are very difficult to get a handle on analytically. Monte Carlo methods can also produce deceptively nice looking maps. However, the problem with this method is that it can simply push the uncertainty to another less obvious place, namely the models and assumptions used in simulating input maps. Should we simulate from models simply because they are easy to implement or convenient in some manner (for example, mathematically nice)? Should we assume independence when the honest answer would be that we don't know what the dependence is? Personally, I think not. I think there is a fair chance that our conclusions will be plain wrong. If the problem we are working on really matters, then I think we should be very careful with our models and assumptions. I'm reminded of a story of an airport built on an artificial island. After an uncertainty analysis, engineers estimated the island would sink somewhere between 19 and 26 feet. The planners designed for the island sinking 19 feet. Apparently the island has sunk 28 feet to date and is still sinking at the rate of 1 foot per year!

I recently had the good fortune to attend a workshop on uncertainty and risk analysis, which brought home to me how times have changed. In the early days of the Land Monitor project, we tried many times to interest land managers and potential collaborators in the idea of uncertainty and uncertainty propagation in spatial data. However, our efforts fell on deaf ears. Apparently the time for these ideas had not yet come. At this workshop there was such an emphasis on uncertainty that I almost felt the pendulum had swung too far the other way. Monte Carlo simulation methods and Bayesian-Markov chain Monte Carlo methods (Gilks et al., 1995) attracted a fair amount of criticism due to their copious and often unwarranted assumptions concerning distributional forms and (conditional) independence. It was also clear that serious efforts have been made to get around the need to make these assumptions. However, the cost for making fewer assumptions, was the need to accept a final answer in the form of bounds on the quantity(s) of interest. I think such an approach could be very useful for GIS scientists. For those interested, I would recommend Ferson (2002), Berleant et al., (2003) and a visit to http://www.ramas.com/resnew.htm. For those who care to delve into this, controversy may not be far from your door.

Finally, returning to my journal editor's remark about uncertainty in GIS being of little interest to mainstream statisticians, I think that this is probably still a fair comment. However, I can see substantial pockets of interest beginning to emerge. I recently reviewed a paper on a Bayesian model for positional error in maps that discussed the use of Markov chain Monte Carlo methods in this context. I hope that this will generate discussion between GIS scientists and statisticians. I think that statistical work on hierarchical Bayesian methods (Berliner, 1999; Berliner et al., 2000) is clearly of relevance to GIS scientists. For further information see http://www.stat.ohio-state.edu/~sses/research.html.

The harder uncertainty problems would benefit from collaboration between GIS scientists and statisticians. For example, some progress on handling model uncertainty can be made with Bayesian statistical methods (via Bayesian model averaging). This is a source of uncertainty that I suspect is being swept under the carpet at the present time. However, this is an inherently difficult problem that is often hampered by an inability to actually fully specify the class of alternative models.

Another difficult problem is how to model dependence among maps and among parameters. A possible way to deal with this issue could be the use of copulas (Clemen and Reilly, 1999), which allow joint distributions to be created from specifications of marginal distributions.

A desirable future

Concerning the topic of my paper, in an ideal future, I would like to see models and methods for propagating uncertainty, including the one discussed in my paper, subjected to rigorous testing with accurate ground truth data. There would be data-based answers to questions like:

- What do the error distributions in real life *actually* look like?
- Are there useful classes of distributions that appear regularly?
- What sort of spatial correlation models describe real-world errors?
- How much dependence/independence is really there?
- Is a particular model flexible enough to cover a lot of real situations?
- Are the approximations we are making good or not?

There would be lots of data sets with highly accurate ground truthing with people willing and able to do the necessary hack work. There would also be strong collaborations between GIS scientists and statisticians and new ways of thinking about things in both disciplines.

References

ARBIA, G., GRIFFITH, D., AND HAINING, R., 1998, Error propagation modelling in raster GIS: Overlay operations, *International Journal of Geographical Information Science*, 12(2), 145–167.

BERLEANT, D., XIE, L., AND ZHANG, J., 2003, Statool: A tool for Distribution Envelope Determination (DEnv), an interval-based algorithm for arithmetic on random variables, *Reliable Computing*, 9(2), 91–108.

Berliner, L.M., 1999, Hierarchical Bayesian modeling in the environmental sciences, *Allgemeines Statistische Archiv* (2000), 84, 141–153.

Berliner, L.M., Wikle, C.K., and Cressie, N., 2000, Long-lead prediction of Pacific SSTs via Bayesian dynamic modelling, *Journal of Climate*, 13, 3953–3968.

Caccetta, P.A., Allan, A., and Watson, I., 2000, The land monitor project. In *Proceedings of the Tenth Australasian Remote Sensing Conference*, Remote Sensing and Photogrammetry Association of Australasia, Adelaide.

Clemen, R. and Reilly, T., 1999, Correlations and copulas for decision and risk analysis, *Management Science*, 45, 208–224.

Ferson, S., 2002, *RAMAS Risk Calc 4.0 Software: Risk Assessment with Uncertain Numbers*, Lewis Publishers, New York.

Gilks, W.R., Richardson, S., and Spiegelhalter, D.J., 1995, *Markov Chain Monte Carlo in Practice*, Chapman & Hall/CRC Press, Boca Raton, FL.

Kiiveri, H.T., Caccetta, P.A., and Evans, F.H., 1999, Use of conditional probability networks for environmental monitoring, *International Journal of Remote Sensing*, 22(7), 1173–1190.

Leung, Y. and Yan, J., 1998, A locational error model for spatial features, *International Journal of Geographical Information Science*, 12(6), 607–620.

Leung, Y., Ma, J., and Goodchild, M., 2004a, A general framework for error analysis in measurement-based GIS Part 1: The basic measurement-error model and related concepts. *Journal of Geographical Systems* , 6(4), 325–354.

Leung, Y., Ma, J., and Goodchild, M., 2004b, A general framework for error analysis in measurement-based GIS Part 2: The algebra-based probability model for point-in-polygon analysis, *Journal of Geographical Systems*, 6(4), 355–379.

Leung, Y., Ma, J., and Goodchild, M., 2004c, A general framework for error analysis in measurement-based GIS Part 3: Error analysis in intersections and overlays, *Journal of Geographical Systems*, 6(4), 381–402.

Leung, Y., Ma, J., and Goodchild, M., 2004d, A general framework for error analysis in measurement-based GIS Part 4: Error analysis in length and area measurements, *Journal of Geographical Systems*, 6(4), 403–428.

Shi, W. and Liu, W., 2000, A stochastic process-based model for the positional error of line segments in GIS, *International Journal of Geographical Information Science*, 14(1), 51–66.

Zhang, J.X. and Stuart, N., 2001, Fuzzy methods for categorical mapping with image-based land cover data, *International Journal of Geographical Information Science*, 15(2), 175–195.

International Journal of Geographical Information Science,
1998, Vol. 12, No. 7, 699–714.

16 Loose-Coupling a Cellular Automaton Model and GIS: Long-Term Urban Growth Prediction for San Francisco and Washington/Baltimore

Keith C. Clarke and Leonard J. Gaydos

Abstract. Prior research developed a cellular automaton model, that was calibrated by using historical digital maps of urban areas and can be used to predict the future extent of an urban area. The model has now been applied to two rapidly growing, but remarkably different urban areas: the San Francisco Bay region in California and the Washington/Baltimore corridor in the Eastern United States. This paper presents the calibration and prediction results for both regions, reviews their data requirements, compares the differences in the initial configurations and control parameters for the model in the two settings, and discusses the role of GIS in the applications. The model has generated some long term predictions that appear useful for urban planning and are consistent with results from other models and observations of growth. Although the geographical information system (GIS) was only loosely coupled with the model, the model's provision of future urban patterns as data layers for GIS description and analysis is an important outcome of this type of calculation.

1 Introduction

The human geographical processes of urbanization and urban spread will apparently continue unabated into the twenty-first century. By the last decade of this century (1995), 21 world cities had total metropolitan area populations of over 6 million, led by Tokyo (30 300 000), New York (18 087 251), Seoul (15 850 000), Mexico City (14 100 000) and Moscow (13 150 000). Also by 1995, the number of people worldwide living in settlements of five thousand or more reached 51%, a majority of humankind and a dramatic increase from 29% in 1950. Gottmann (1961) coined the term *megalopolis* to describe the coalescence of metropolitan areas in the north-eastern United States. In the era of GIS, remote sensing and digital map products have recorded the birth and growth of similar megalopolises in California and in Mexico, South America, Europe, and Asia. New estimates of the world population in 2100 AD indicate an increase from the present population of 5.5 billion to 10 to 20 billion. We have termed the resultant super cities *gigalopolis*, the twenty-first century system of world cities containing billions of people centered on the world's major urban areas. Gibson's fictional view of the future urban United States, for example, talks only of the 'East Sprawl' and the 'West Sprawl' (Gibson 1984).

The magnitude of *gigalopolis* in population terms, however, understates the most critical permanent impact of increased urban-space consumption, which is usually at the expense of prime agricultural land essential for food production. Looked at spatially, each expanding metropolitan area will become both physically and virtually connected to other growing concentrations of people in the coming century through raw gain of territory as well as broader communication, transportation, and economic ties. Not since the dominance of agriculture within human affairs millennia ago has humankind's habitat changed so quickly and irreversibly. Simultaneously, recent trends show first that people are consuming more space per person within their urban environment, but also that the average household size is decreasing, at least in western cities. Urbanization and urban growth go hand-in-hand, and generate many other land transitions, with several varied land use types eventually converting to urban use. The spatial consequences of the urban transition deserve serious study by scientists and policy makers concerned with global change because they will impact humankind directly and profoundly. Vitousek (1994) called land use/land cover changes, including the urban transition, one of the few certainties of global change, because we are 'certain that they are going on, and certain that they are human-caused'.

This paper summarizes research conducted to describe, model, and predict future urban transitions. We constructed a model using a cellular automaton that simulates the urban growth process. We calibrated the model with historical data for two major metropolitan regions in the United States, and used it to produce one-hundred-year projections of their urban growth. GIS has been an indispensable tool in the model construction and calibration, and will play far more critical a role when the predictions are distributed and reproduced for other areas. The broader purpose of this work is to model and predict the spatial consequences of future urbanization, so that the impact of human-induced land transformations can be better understood.

Cellular modelling grew out of earlier environmental simulation work on wildfire behaviour (Clarke *et al.* 1995), that in itself was based on pioneering work by Michael Batty (Batty and Xie 1994). The role of cellular automata as potential powerful contributions to urban process modelling was demonstrated by Couclelis (1997) and Takeyama and Couclelis (1997). Influencing model choice was the need for a model that was scale independent, so that local, regional and continental scale processes could be described in a single context. Cellular automata are simple models for the simulation of complex systems (Waldrop 1992, Wolfram 1986). A cellular model assumes only an action space (usually a grid), a set of initial conditions, and a set of behaviour rules. Characteristic of such models of complexity is that behaviour is *emergent*, that is, it is generated by repetitive application of the rules beyond the initial conditions. Complex systems are also termed *self-organizing* and are remarkably suitable to computational simulation (Wolfram 1984). Simple cellular automata are characterized by phase transitions between behaviour types, so that a single model can result in stability, stochastic instability or chaos. As such they seem ideally suited to modelling the complexity of urban systems, which typically have many more unknowns than measurable variables. Cellular models are known in ecology as individual based models and this concept, that complex aggregate behaviour results from many interacting self-motivated agents, has great value for both urban modelling and for the data rich environment of GIS. This is especially appropriate in urban modelling, where the process of urban spread is entirely local in nature and aggregate effects, such as growth booms, are emergent.

Modelling cities with cellular automata is a new approach, and one that was virtually impossible without the data management capabilities of GIS and powerful workstation technology. The approach has distant roots in geography in the work of Hagerstrand (1965) and Tobler (1979). Links to prior urban modelling are less clear (Clarke, Gaydos and Hoppen, 1997), though Batty and Longley (1994), Makse *et al.* (1995) and White and Engelen (1992) have used essentially similar approaches.

2 The role of GIS in urban cellular modelling

Much has been written on the integration of GIS and modelling, especially for environmental issues (Wilson 1995, Goodchild *et al.* 1996, Wagner 1997). In a recent paper, Park and Wagner (1997) implemented a tight coupling of several cellular automaton models (including ours) within Idrisi using the Cellang CA language. In the context of our cellular model, GIS served at least three important roles, none of which could be called tight coupling. The first of these was as data integrator. In each of the initial applications, data were either already available as ARC/INFO coverages, or were captured by scanning and digitizing (Crawford-Tilly *et al.* 1996). Although the coverages existed, new map extents, projections, and grid resolutions were required, and the GIS was invaluable in ensuring co-registered input data layers for thc model. All further modelling and analysis depended on this essential first step, what Chrisman has called a 'universal requirement' for GIS (Chrisman 1997, p. 108). The input data layers were each raster grids, exported as image files and converted by specialized software. Thus the relation of our modelling to GIS is one of loose coupling, as classified and described by Anselin *et al.* (1993).

Secondly, GIS allowed the results to be visualized. This was the weakest component of the loose coupling. Stand-alone versions of software for the model were written to generate map displays, multiple display windows, and results (Clarke *et al.* 1996, Acevedo and Masuoka 1997). The role for the GIS in this instance was one of taking intermediate results, and facilitating their comparison with the original and other layers of information.

Thirdly, the predictions generated were reintroduced into the GIS data sets available for application, allowing planning decisions to be made with the data. An example is the ArcView window interface developed to display the three development scenarios for the Sterling Forest area in New York State (Kramer 1996). It is this third application of GIS that is by far the most powerful from the modelling point of view. Having 'what if' model projections available to perform the more traditional GIS operations and analyses such as buffering and overlay is a very powerful GIS capability. This function alone favours the use of loose coupling. In spite of efforts to build cellular modelling functions into GIS directly (Takeyama and Couclelis 1997; Park and Wagner 1997) and the suitability both of specific GIS packages (e.g. GRASS) and of control languages (e.g. MapObjects and Avenue in ArcView), it is likely that most numerical modelling, especially that requiring exhaustive or rigorous calibration, will need to parallel the GIS rather than work within the software. This approach is similar to the way that statistical functions and spatial analytical capabilities have been integrated with GIS. By broadening the definition of GIS, such as that of GIScience, the model coding can even be viewed as part of the science, if not part of the system (Goodchild 1992, Clarke 1997).

3 Data for the model

Data for the model's calibration and initial use in prediction came from a variety of sources. For the San Francisco Bay area, we assembled data from historical maps and air photos, analog and digital maps for different time periods, from data supplied by such agencies as local governments, and from satellite images (Kirtland *et al.* 1994). After success of the initial model application equivalent data, but to more specific and documented standards, was compiled for the Washington/Baltimore region (Crawford-Tilley *et al.* 1996). These data are available on-line as part of the Washington/Baltimore Regional Collaboratory (Clark *et al.* 1996).

Data assembly problems included inconsistent feature definitions over time, especially for urban areas and major roads, extensive manual generalization present in historical maps, and the need to integrate multiple image and map sources from different projections, datums, and coordinate systems. Although much of the delineation of urban versus non-urban use had to be interpreted, skilled image interpreters were available to make the interpretation to acceptable standards.

The model's input layers are fourfold: (1) A digital elevation model, for which the GIS was used to create a grid of slopes in percentage, (2) A layer showing the initial or seed configuration of urban areas, plus as many additional historical layers as possible, to calibrate the model, (3) As many historical transportation layers as possible, which the model reads and uses sequentially as their year of construction is reached, (4) A layer of excluded areas unlikely or impossible to urbanize, such as national parks, water bodies and protected wetlands. The latest version for the *gigalopolis* project, in which several additional land cover types such as agricultural and forest are included in the model's behaviour, requires a land use layer as well.

Since the growth rules in this model are defined primarily by physical factors, the San Francisco Bay area was an ideal first test site. Elevations range from sea level to 2500 m, with some clear topographic control over growth by the dichotomy between slopes and flats. The region also is diverse in its patterns of urbanization, reflecting its beginnings in small enclaves clustered around the inland waterway networks, the emergence of San Francisco as a transportation hub, and the more recent urbanization of the surrounding valleys closely reflecting the extended highway system. The model's temporal input comprised six raster image maps of urban extent for the years 1900, 1940, 1954, 1962, 1974, and 1990, with 1900 as the initial conditions or 'seed' year. An 1850 data set was too sparse for modelling purposes. Roads were available for 1900, 1920 and 1970. The San Francisco data were included in Clarke *et al.* (1997) and can be viewed on the World Wide Web at http://edcwww2.cr.usgs.gov/umap/umap.html.

Similarly, for Washington/Baltimore, urban and road layers were constructed for 1792, 1850, 1900, 1925, 1938, 1953, 1966, 1972, 1982, and 1992 (Figure 16.1 and Figure 16.2). Again due to the sparse and least reliable early data, 1900 was used as the initial conditions or seed layer. While the Washington/Baltimore Collaboratory digitized layers for railroads also, these were excluded to ensure reliability and compatibility between the two applications (Clark *et al.* 1996).

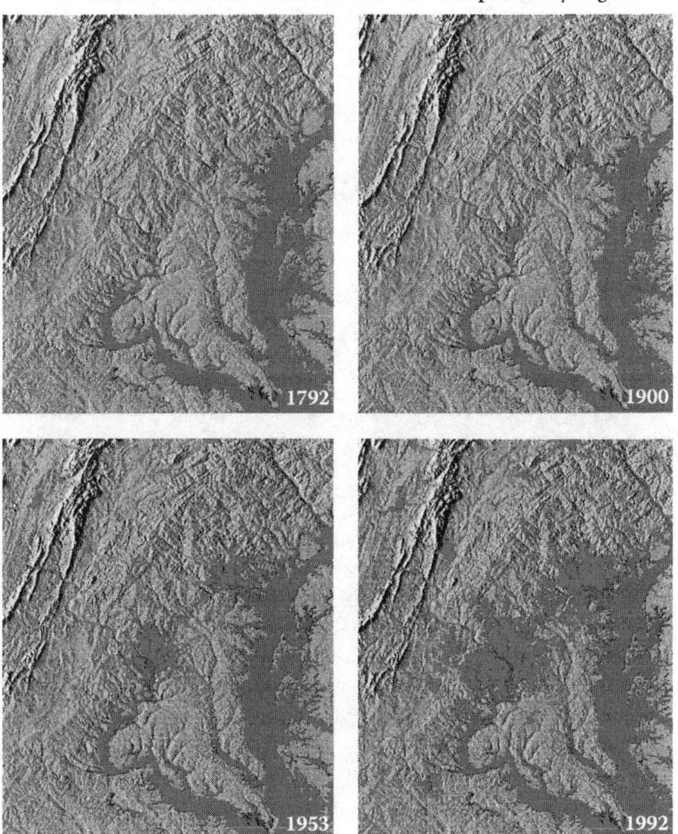

FIGURE 16.1 Historical urban development in the Washington/Baltimore area. (Figure originally published in colour.)

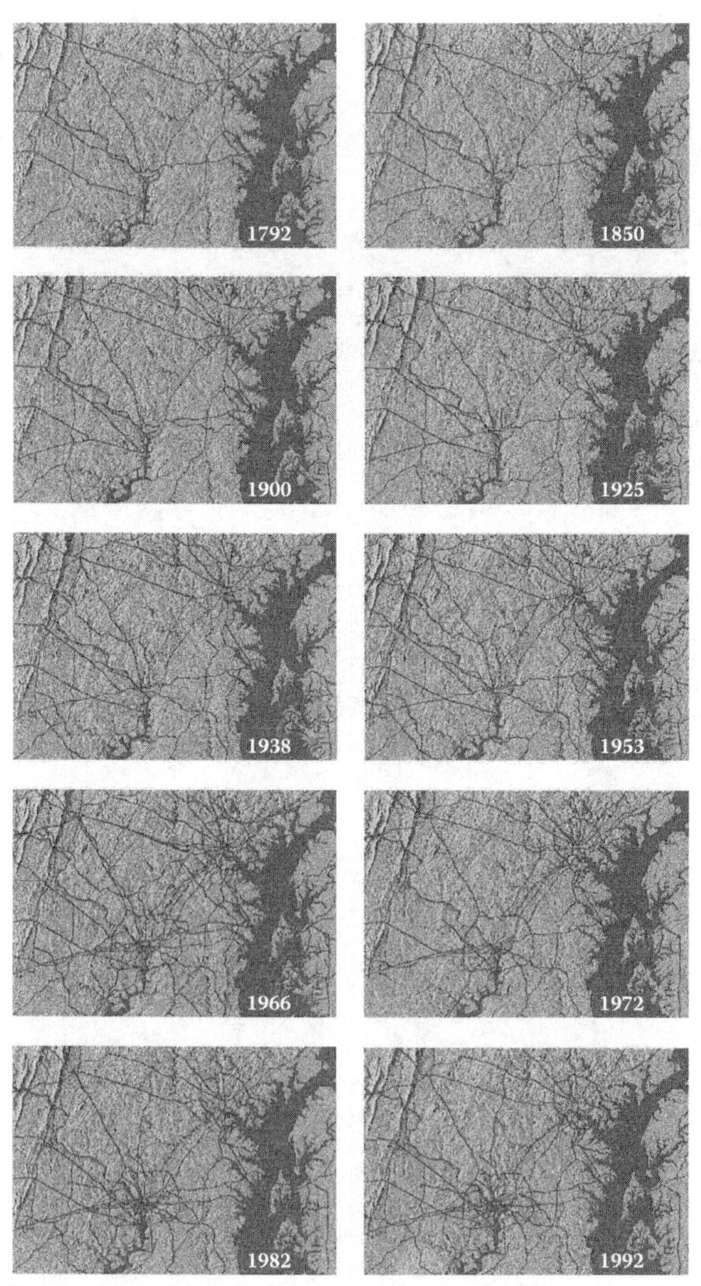

FIGURE 16.2 Historical growth of transportation in the Washington/Baltimore area. (Figure originally published in colour.)

4 Model description

The urban model is a scale-independent cellular automaton (CA), model with some variations from the traditional CA, and multiple behaviour types. The growth rules are uniform throughout a gridded representation of geographical space and are applied on a cell-by-cell basis. A single time span is one iteration of the CA, and all changes are applied synchronously at the end of each time period. The set of growth rules, and the initial condition (map of urban extent at a period in time) are integral to the data set being used because they are defined in terms of the physical nature of the location under study. In-site calibration then adapts the model to its local environment. One interpretation is that the urban area corresponds to an organism. The initial condition is the start 'seed' layer, growth occurs one cell at a time with each cell acting independently of all others, and patterns emerge during growth as the organism 'learns' more about its environment.

The model was implemented as a computer program written in the C language. The program operates as a set of nested loops: the outer control loop repeatedly executes each growth 'history', retaining cumulative statistical data, while the inner loop executes the growth rules for a single 'year'. The starting point for urban growth is an input layer of 'seed' cells, the urban extent for a particular year identified from historical maps, atlases, and other sources. The rules apply a cell at a time and the whole grid is updated as the 'annual' iterations complete. The modified array forms the basis for urban expansion in each succeeding year. Potential cells for urbanization are selected at random and the growth rules evaluate the properties of the cell and its neighbours (e.g., whether or not they are already urban, what their topographic slope is, how close they are to a road). The decision to urbanize is based on mechanistic growth rules as well as a set of weighted probabilities that encourage or inhibit growth. The model is described in detail in Clarke *et al.* (1997).

Five factors control the behaviour of the system. These are: a diffusion factor, which determines the overall outward dispersive nature of the distribution; a breed coefficient, which specifies how likely a newly generated detached settlement is to begin its own growth cycle; a spread coefficient, which controls how much diffusion expansion occurs from existing settlements; a slope resistance factor, which influences the likelihood of settlement extending up steeper slopes; and a road-gravity factor, which attracts new settlements toward and along roads. These values, which affect the acceptance level of randomly drawn numbers, are set by the user at the outset of every model run.

Four types of growth are possible in the model: spontaneous, diffusive, organic, and road influenced growth. Spontaneous growth occurs when a randomly chosen cell falls in a suitable location for urbanization at the boundary of an existing settlement, simulating the fragmenting influence urban areas have on their surroundings. Diffusive growth permits the urbanization of cells which are flat enough to be desirable locations for development, even if not near an established urban area. Organic growth spreads outward from existing urban cores, representing the tendency of all urban areas to expand. Road influenced growth encourages urbanized cells to develop along the transportation network, reflecting increased accessibility.

A second hierarchy of growth rules, termed self-modification, is prompted by an unusually high or low growth rate above or below a threshold. The growth rate is computed by comparing the number of new pixels urbanized in any time period to the total existing urban area. The limits of 'critical high' and 'critical low' begin an increase or decrease in three of the growth-control parameters. The increase in the parameters is by a multiplier greater than one, 'boom', imitating the tendency of an expanding system to grow ever more rapidly, while the decrease is by a multiplier less than one, 'bust', causing growth to taper off as it does in a depressed or saturated system. However, to prevent uncontrolled exponential growth as the system increases in overall size, the multiplier applied to the factors is slightly decreased or lagged in every subsequent growth year.

Other effects of self-modification are an increase in the road-gravity factor as the road networks enlarges, prompting a wider band of urbanization around the roads, and a decrease in the slope resistance factor as the percentage of land available for development decreases, permitting expansion onto steeper slopes. Under self-modification, the parameter values increase most rapidly in the beginning of the growth cycle when many cells are available for urbanization, and decrease as urban density increases in the region and expansion levels off. Without self-modification the model produces linear or exponential growth as long as new land remains available; self-modification generates the typical *S*-curve growth rate of urban expansion observed within a region.

5 Model calibration

Calibration of the original model was described in Clarke *et al.* (1996). Since then, a new procedure has been developed and tested for the two study areas. After assembly of the various data sets and their conversion to the input format for the model, the calibration problem may be stated: given a starting image of urban extent, which set of initial control parameters leads to a model run which best fits the observed historical data? Pursuit of this question implies that 'best' can be quantified, and used to test statistically the observed against the expected. These parameters are then used for prediction.

Calibration then proceeded with a UNIX script written in the PERL shell language. The script generated a control file for the program, executed the program so that it iterated over four Monte Carlo runs, wrote log files of its outputs, then executed a stand-alone program that computed the four measures, and wrote these into a master log file. The first of two PERL scripts performed coarse iterations over the control parameters, with increments of 20 units at a time and 25 for Washington/Baltimore. Even so, this procedure involved iterating over the 0–100 range of the diffusion coefficient, the breed coefficient, the spread coefficient, the slope-resistance and the road-gravity control parameters. For San Francisco, six, six, six, five and seven combinations were tested for the parameters respectively, starting in 1900 and terminating at the 1990 data set. This gave *r*-squared values computed for 1940, 1954, 1962, 1974, and 1990, for only five 'observations' although each *r*-squared was a composite of four measures and an average over four model runs. The 7560 combinations executed in about 252 hours of CPU time on a Silicon Graphics Indigo 2

Impact 10000 workstation. The longer runs for the much larger Washington/Baltimore data set mandated fewer combinations, on the order of 3000. This broad coverage phase of the calibration gave those combinations of parameter setting that tended to produce the best 'projections' of the present day. We chose four ways to test statistically the degree of historical fit, and are investigating twelve measures in our latest work. The four initial tests are correlation coefficients of predicted outcomes from the model with values computed from the historical map layers. The specific measures were: (1) the r-squared fit between the actual and predicted number of urban pixels; (2) the r-squared fit between the actual and the predicted number of edges in the images (i.e. those pixels that have contact between urban and non-urban on any side, so that a single isolated urban pixel counts as 4 edges); (3) the r-squared fit between the actual and the predicted number of separate clusters in the urban distribution, computed by eroding cluster edges with an image processing routine until all separate blobs collapse onto just one pixel, then counting these pixels; and (4) a modified Lee-Sallee shape index, computed by combining the actual and the predicted distributions as binary urban/non-urban layers, and computing the ratio of the intersection over the union. For perfect correspondence the value is 1.0. Practically, values of about 0.3 were obtained. The four measures were computed as averages of multiple runs. We used 4, 10, and 100 runs in the tests, although more rigorous multi-start tests were conducted in the initial calibration (Clarke *et al.* 1996). We devised a single composite measure, the first three summed and multiplied by the fourth, to rank overall outcomes.

The initial test was then repeated for unit increments of the control parameters above and below the 'winning' values for the coarse calibration. The additional step or fine calibration involved another $6 \times 5 \times 5 \times 5$ combinations of four iterations each for 3000 combinations and another 100 hours of CPU time. Again, for the larger data set only about 1800 combinations were computed. A final set of parameters was then used for a 100-run iteration, saving the terminating values of the control parameters (changed by the self-modification) as input for a 100-iteration prediction run into the future, stopping at the year 2100.

One obvious problem with this approach is the processing time. Therefore, the Washington/Baltimore data set was used in an experiment to test sensitivity of the calibration process to scale dependence. In the original calibration, at the full resolution and extent of the data set, the mapped area was 486 by 720 pixels with a resolution of 210 m. The entire two step calibration process described above was then repeated at resolutions of 410 m, 820 m, and 1640 m. New data sets were created by direct sub-sampling, i.e. taking every pixel in each case. To prevent the drop out problem or loss of connectivity in the roads map, the roads were thickened prior to re-sampling. This did indeed cut the processing time. The 1640 m data set was fully calibrated in only 6 hours of CPU time, a 42 fold speed-up.

6 Results

The self-modification internal to the cellular model enhances phase changes. In geographical terms, this means that different periods of time should be dominated by different growth behaviour, and by increasing spatial adaptation to the local

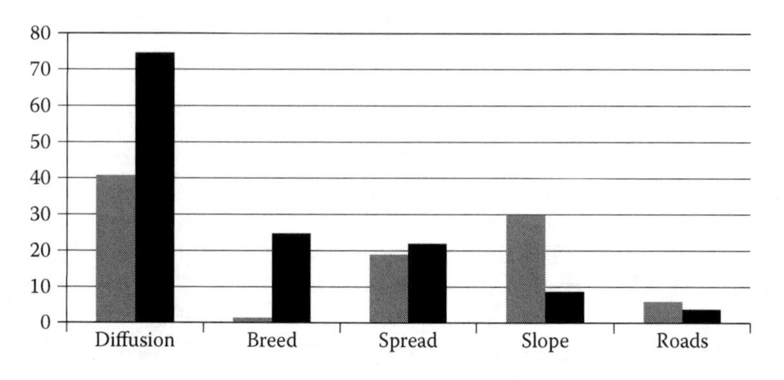

FIGURE 16.3 Calibration results: initial values for the model control parameters.

environmental conditions. As a result, it was expected that the impact of slope would be more pronounced in San Francisco at first, followed by a 'spilling out' of growth into the Central Valley. On the other hand, topography is clearly less of a constraint to growth in the Washington/Baltimore corridor, so that little difference in behaviour would be expected to result from the slope factor. There were indeed significant differences in slope behaviour. For example, for the population (number of urban pixels) correlation, the San Francisco fit was high but varied, and increased as the slope resistance increased. On the contrary, in Washington/Baltimore, there was very little variance in response, and fit declined with the slope resistance (Figure 16.3). This implied that slope explains far more of the urbanization in San Francisco than in Washington and Baltimore. While a rather obvious point, it confirms the basic soundness of the model.

Phase changes are obvious from the coarse calibration data. In each case, the five control parameter initial settings were compared to the scores. Some remarkable behaviour was evident within the full range of model applications. For example, for the population r-squared (actual vs. predicted urban extent for each observation period), the mean score decreased as the diffusion coefficient increased, but the lowest setting had four iterations that scored higher than all others. A phase transition was clearly precipitated by changing the spread parameter. Extremely high correlation (in excess of 0.99 r-squared) dropped markedly between spread settings of 1 and 20, but then increased linearly for the remainder of the values. In other words, at a setting between 1 and 20, some interaction was triggered that precipitates a behaviour change. Spatially, for example, this may be the breaking out of the initial constrained urban areas and valleys. The effect is evident, though less clearly, in the Washington/Baltimore data also. Phase transitions were evident in all four of the test measures.

If the success measures can be characterized, then the four are increasingly spatial in nature. The population value is, as expected, the easiest to simulate with the model. Most fits were in the 0.99 range, partly due to the low number of data years. The number of clusters was an indication of how many separate spreading centres or independent communities were present. This value proved a far better discriminator in San Francisco than in Washington/Baltimore. The number of edges was the total number of pixel contact edges between urban and non-urban within the grids. As such, it has already been shown to be influenced strongly by the map source. Remotely sensed data are more speckled than map sources, and thus have more edges. Finally, the Lee Sallee measure is a modified shape index, defined as a ratio of the AND to the OR of the actual and predicted urban images as binary layers. This measure penalizes spatial mismatch twice; for example if a good spatial shape match was displaced slightly, the intersection would be smaller but the union would be larger by twice the displacement area. Values rarely approached even a 30% match, and the ability to discriminate spatial match with this measure was quite good. Again, the spread-parameter phase change was evident, both in San Francisco and Washington/Baltimore. In a few cases, evidence pointed to a local maximum, for example in the breed coefficient for both data sets.

The second round of calibration selected the best outcome from the coarse calibration and iterated single parameter-value increments above and below. Not all values were permuted, since some were already clearly at maximum. Most striking from the fine tuning results was the scaling nature of the model. Phase transitions, trends, maxima and even oscillations (e.g., breed in Washington/Baltimore for the Lee-Sallee measure) were evident. While in each case a single value was chosen as maximum, in fact many individual runs can attain far higher success, in a specific measure and overall. Nevertheless, stability across many runs was considered important for Monte Carlo modelling. We made 100 permutations of the final predictive runs. This was done by starting the model at the present time, but using the terminating values (because of the self-modification) of the control parameters.

Predictions made for over 100 Monte Carlo runs allow us to map probabilities. We decided to divide the probabilities into three categories. Stopping the predictions at different future dates created multiple frame displays (Figure 16.4 and Figure 16.5). The images of the future urban extent are most convincing when animated with the historical data. In a first attempt to animate the San Francisco data, it was difficult to tell where the observed data ended and the simulations began.

Finally, the results of the multiscale calibration were very encouraging. Table 16.1 shows the results of the 'best' combinations for the Washington/Baltimore data set arrived at by successive halving of the resolution. As can be seen, the only major differences in the degree of success came after the change from 820 to 1640 m. We attribute this change to the severe distortion of the roads layer when roads must be over one and a half kilometers wide, and occasionally disconnect. Slope also suffers from the resolution change, with maximum and average slopes diminishing quickly as the scale is made smaller. From Table 16.1, is was clear that the diffusion factor seems most scale sensitive, and that the breed factor seems almost a self-similar fractal, with complete scale invariance. The obvious conclusion from the multi-scale calibration is that a nested or hierarchical approach to calibration would be optimal, first using coarse data to investigate the scaling nature of each parameter in a different city setting, then scaling up once the best data ranges are found. As a result, far fewer combinations need be run at the finest scales, and CPU times can be significantly reduced, with no loss of calibration rigour.

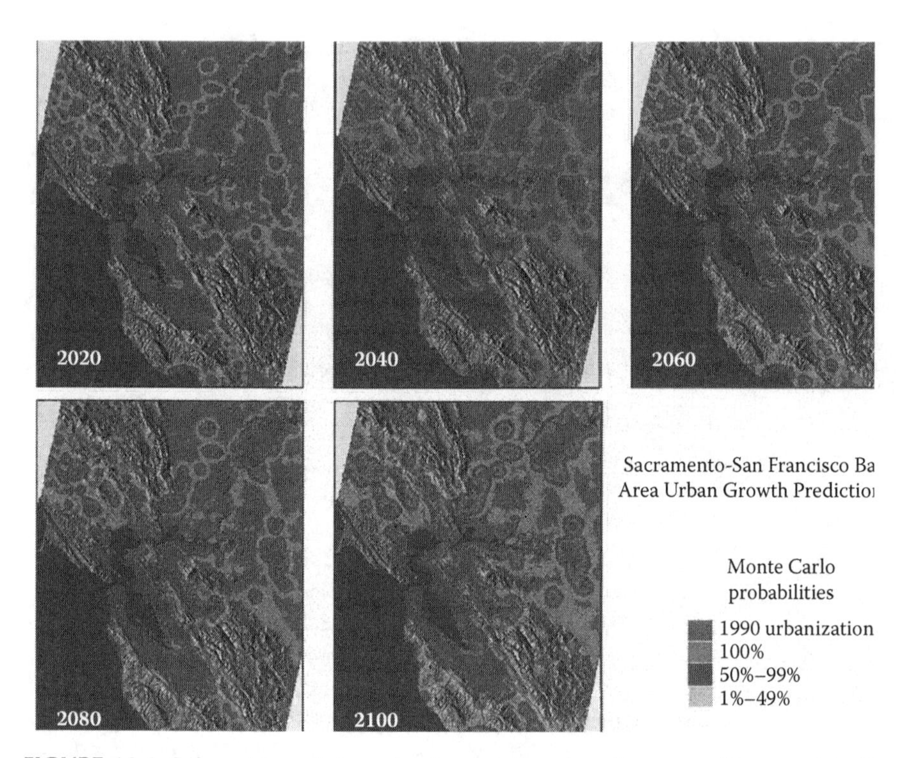

FIGURE 16.4 CA model predictions for the San Francisco Bay area. (Figure originally published in colour.)

7 Predictions

Figures 16.4 and 16.5 show model predictions of future urbanization patterns for the two applications areas. Obviously, the short-term predictions are more reliable than the long term. The Monte Carlo probabilities have been reduced to three classes. Shaded pink are those areas that the model predicts as 100% certain of growth. Some of these areas are the existing urbanized areas, but as can be seen from the figures, this area expands over time. Most conservatively, these would be judged as potential growth zones. Because ordinary outward expansion and infilling of existing settlements are the most predictable types of growth, the organic spread of the settlements has the highest degree of confidence. Shaded green are areas with less than 100%, but greater than 50% or even chance of growth. The greater-than-even-odds criterion may be a good compromise for spatial forecasting. Taking in this class of prediction much of the highway-based growth is captured. Finally, a high-risk projection category of greater than zero but less-than-even-odds is plotted in light green. This zone combines more extended and fragmented growth of transportation, but also includes many entirely new urban centers. When a new spreading centre forms in repeated model runs, it could be identified as a 'city waiting to happen', a site so potentially ripe for growth that it is merely a matter of time before urbanization

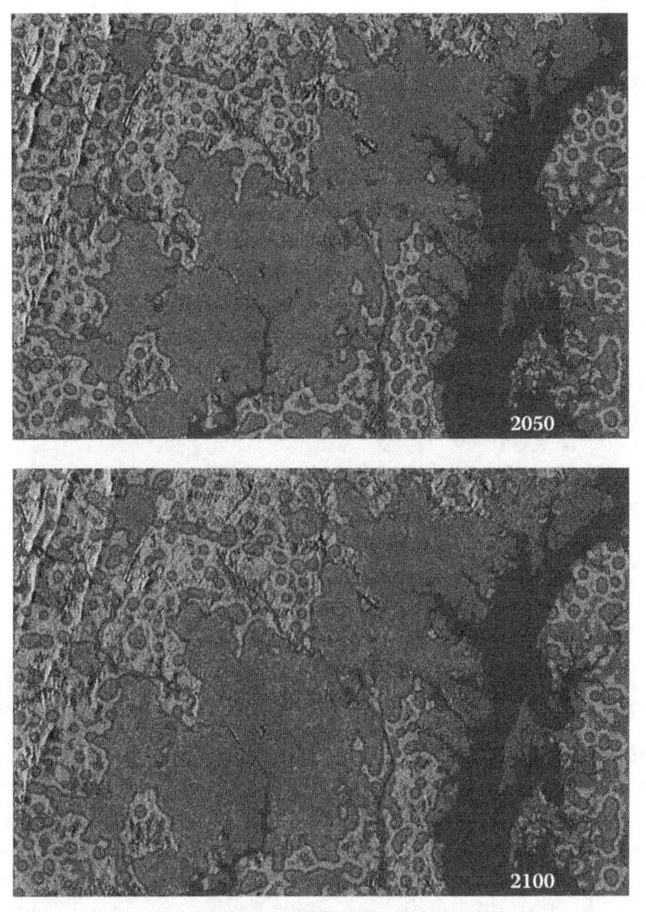

FIGURE 16.5 CA model predictions for the Washington/Baltimore area. (Figure originally published in colour.)

arrives. Examples of such settlements are: Vacaville and Morgan Hill in California, and Chantilly and Manassas in Virginia.

The utility of these predictive layers clearly lies in their being placed in the public realm. The San Francisco Bay area projections have been posted on the World Wide Web for some time. The Washington D.C./Baltimore projections are available at http://landcover.usgs.gov/urban.html. They will be most useful as data layers in a GIS data format, for integration with existing information.

TABLE 16.1
Best overall calibrations by resolution: Washington/Baltimore.

Resolution	210 m (486 × 720)	420 m (243 × 360)	840 m (121 × 180)	1680 m (60 × 100)
Composite score 1	0.900014	0.950593	0.993430	0.841441
Composite score 2	3.223740	3.237434	3.226773	3.213986
r-squared urban	0.971070	0.969425	0.960926	0.967818
r-squared edges	0.965636	0.972149	0.972061	0.985269
r-squared clusters	0.978280	0.969293	0.949094	0.973370
Shape match	0.308754	0.326567	0.344692	0.287529
Diffusion	75	10	50	75
Breed	25	2	2	3
Spread	22	26	20	22
Slope	9	1	5	5
Roads	4	25	3	9

8 Conclusions

Work on the model so far has established that the approach can produce useful results. While the calibration phase is slow and highly dependent on the size of the input data set and on the quality and quantity of historical data, historical map and image data as the basis for future settlement predictions seems suitable and the multi-scale calibration process could speed the process. Monte Carlo methods, coupled with the variance estimates they generate, allow clear confidence limits to be placed on projected future spatial patterns. As sensitive and potentially controversial as growth predictions are from the political and societal viewpoint, the rigour of the calibration, the repeatability of the experiment and the utility of a prediction-probability layer for further GIS work justify the outcomes of the modelling. Only the real future, as it slowly unfolds, can verify our model. But then, is the purpose of modelling to actually predict, or to help imagine, test, and choose between possible futures? If so, then this model and approach can indeed be useful for scenario planning (Kramer 1996).

While the role of the model as both a consumer and provider of GIS data has been evident, the functional coupling of the model with any particular GIS has been loose, at best. On the other hand, we see such models as valuable enriching sources of GIS data layers, and layers that have real value for planning and GIS application. For example, what location decisions pending today might change in the light of predicted future patterns? Could better decisions on siting waste disposal be made given predictions of urban spread? Would zoning or land-preservation policies change if their future consequences were better known, or even better understood? Could cities test and evaluate alternative growth-control strategies by spatial modeling? Experience with existing models has shown this to be the case.

Future work with the model will involve extending the land transition to a broader set of land use and land-cover changes. We have set the Anderson Level I classes as a first goal for the new work (Anderson *et al.* 1976). In addition, the model needs to be ported to and repeatedly applied to new study areas and at different map scales. We intend to further examine the impact of scale and locale of application of the model on the initial conditions, on the best start parameters, and on the process of self-modification. Our plans call for testing the model at about 1 km resolution for the entire lower 48 United States for the full Anderson Level I classification, and for applications in New York, Chicago, Philadelphia, and Portland. Finally, can the model be used to predict the extent and impact of *gigalopolis*? Can the model be coupled with other models of the global environment to provide multiple assessments of the impact of global change? If these questions are answerable in the affirmative, the cellular model could become a valuable tool for anticipating and effectively addressing some aspects of our children's, and our childrens children's, urban future.

Acknowledgments

The Human-Induced Land Transformations (HILT) project was supported by the United States Geological Survey (USGS) under a NASA Joint Research Initiative (JRI NCC2-5091). The Gigalopolis project is supported by the USGS National Mapping Division's EROS Data Center under grant 1434-CR-96-SA-01235 (Department of the Interior). Support is gratefully acknowledged.

William Acevedo, Michael Figueroa and Stacy Hoppen provided programming and data analysis support. Data were compiled and provided by USGS under the HILT and Temporal Urban Mapping projects (see http://edcwww2.cr.usgs.gov/umap/umap.html). Jeannette Candau and John Lin prepared the graphics. Additional support for data for Washington/Baltimore came from the Washington/Baltimore Regional Collaboratory courtesy of Dr. Tim Foresman of the University of Maryland-Baltimore County (see http://research.umbc.edu/~tbenja.bwhp/main.html). Dr. Richard J. Pike and four anonymous reviewers gave valuable comments on the manuscript.

References

Acevedo, W., and Masuoka, P., 1997, Time-series animation techniques for visualizing urban growth. *Computers & Geosciences*, **23**, 423–435.

Anderson, J. R., Hardy, E. E., Roach, J. T., and Witmer, R. E., 1976, *A Land Use and Land Cover Classification System for use with Remote Sensor Data*, U.S. Geological Survey Professional Paper 964 (Washington, DC: US Government Printing Office).

Anselin, L., Dodson, R. F., and Hudak, S., 1993, Linking GIS and spatial data analysis in practice. *Geographical Systems*, **1**, 3–23.

Batty, M., and Longley, P., 1994, *Fractal Cities* (London: Academic Press).

Batty, M., and Xie, Y., 1994, Modelling inside GIS: Part 2. Selecting and calibrating urban models using Arc/Info. *International Journal of Geographical Information Systems*, **8**, 429–450.

Chrisman, N., 1997, *Exploring Geographic Information Systems* (New York: Wiley).

Clark, S., Starr, J., Foresman, T. W., Prince, W., and Acevedo, W., 1996, Development of the temporal transportation database for the analysis of urban development in the Baltimore-Washington Region. In *Proceedings, ASPRS/ACSM Annual Convention and Exhibition, Baltimore, MD, 22–24 April.*

Clarke, K. C., 1997, *Getting Started With Geographic Information Systems* (Upper Saddle River, NJ: Prentice Hall).

Clarke, K. C., Gaydos, L. J., and Hoppen, S., 1997, A self-modifying cellular automaton model of historical urbanization in the San Francisco Bay area. *Environment and Planning B,* **24,** 247–61.

Clarke, K. C., Hoppen, S., and Gaydos, L. J., 1996, Methods and techniques for rigorous Calibration of a cellular automaton model of urban growth. In *Proceedings, Third International Conference/Workshop on Integrating GIS and Environmental Modeling CD-ROM, Santa Fe, NM, 21–26 January, 1996* (Santa Barbara, CA: National Center for Geographic Information and Analysis). http://www.ncgia.ucsb.edu/conf/SANTA_FE_CD-ROM/main.html.

Clarke, K. C., Riggan, P., and Brass, J. A., 1995, A cellular automaton model of wildfire propagation and extinction. *Photogrammetric Engineering and Remote Sensing*, **60,** 1355–1367.

Couclelis, H., 1997, From cellular automata to urban models: New principles for model development and implementation. *Environment and Planning B-Planning & Design*, **24,** 165–174.

Crawford-Tilley, J. S., Acevedo, W., Foresman, T., and Prince, W., 1996, Developing a temporal database of urban development for the Baltimore/Washington region. In *Proceedings, ACSM/ASPRS Annual Meeting, Washington DC*, **3,** 101–110. http://edcwww2.cr.usgs. gov/umap/pubs/asprs_jt.html.

Gibson, W., 1984, *Neuromancer* (New York: Ace Books).

Goodchild, M. F., 1992, Geographical information science. *International Journal of Geographical Information Systems*, **6,** 31–46.

Goodchild, M. F., Steyaert, L. T., and Parks, B. O., editors, 1996, *GIS and Environmental* Modeling (Fort Collins: GIS World).

Gottmann, J., 1961, *Megalopolis: the urbanized northeastern seaboard of the United States* (New York: Twentieth Century Fund).

Hagerstrand, T., 1965, A Monte-Carlo approach to diffusion. *European Journal of Sociology*, **VI**, 43–67.

Kirtland, D., Gaydos, L. J., Clarke, K. C., DeCola, L., Acevedo, W., and Bell, C., 1994, An analysis of human-induced land transformations in the San Francisco Bay/Sacramento Area. *World Resources Review*, **6,** 206–217.

Kramer, J., 1996, Integration of a GIS with a Local Scale Self-Modifying Cellular Automaton Urban Growth Model in Southeastern Orange County, New York. MA Thesis. Hunter College CUNY. http://everest.hunter.cuny.edu/~jkramer/paper.html.

Makse, H. A., Havlin, S., and Stanley, H. E., 1995, Modelling urban growth patterns. *Nature*, **377,** 608–612.

Park, S., and Wagner, D. F., 1997, Incorporating Cellular Automata simulators as analytical engines in GIS, *Transactions in GIS,* **2,** 213–231.

Takeyama, M., and Couclelis, H., 1997, Map dynamics: integrating cellular automata and GIS through geoalgebra. *International Journal of Geographical Information Science*, **11,** 73–91.

Tobler, W. R., 1979, Cellular geography. In *Philosophy in Geography*, edited by S. Gale and G. Olssen, (Dordrecht: Reidel), pp. 379–386.

Vitousek, P. M., 1994, Beyond global warming: Ecology and global change. *Ecology*, **75,** 1861–1876.

Wagner, D., 1997, Cellular automata and geographic information systems. *Environment and Planning B*, **24,** 219–234.

Waldrop, M. M., 1992, *Complexity: The Emerging Science at the Edge of Order and Chaos* (New York: Simon and Schuster).

White, R., and Engelen, G., 1992, Cellular automata and fractal urban form: a cellular modelling approach to the evolution of urban land use patterns, Working Paper no. 9264, Research Institute for Knowledge Systems (RIKS), Maastricht, The Netherlands.

Wilson, J., editor, 1995, Special issue on the integration of GIS and environmental modelling. *International Journal of Geographical Information Systems*, **9**.

Wolfram, S., 1984, Cellular automata as models of complexity. *Nature*, **311,** 419–424.

Wolfram, S., 1986, *Theory and Applications of Cellular Automata* (Singapore: World Scientific).

A Decade of SLEUTHing: Lessons Learned from Applications of a Cellular Automaton Land Use Change Model

Keith C. Clarke, Nicholas Gazulis, Charles K. Dietzel, and Noah C. Goldstein

History of the SLEUTH urban growth model

SLEUTH is a model for simulating urban growth and other land use changes that are caused by urbanization over time. The name is a moniker derived from its six input layers including slope, land use, exclusion, urban extent, transportation, and hillshade, and was first created by Keith Clarke while visiting the United States Geological Survey (USGS) as a research scientist in 1992. The model's cellular automaton (CA) approach was a direct descendent of prior work on modeling wildfire, which involved extensive C-language coding of CA and anticipated many of the subsequent issues for SLEUTH. While a NASA-American Society for Engineering Education fellow at the NASA Ames Research Center in 1991, Clarke had a series of discussions on land use change and urbanization in the San Francisco Bay Area with USGS geographer Len Gaydos, who at the time was involved in various land use mapping and analysis tasks. After 1992, with the development of the first operational version of the SLEUTH model (then termed UGM for urban growth model), Gaydos pursued an internal USGS research initiative that funded Clarke's further developmental work and established the urban dynamics research program at USGS. This funding allowed for further experimentation with the model, ultimately resulting in the addition of a land use change coupling available under project Gigalopolis (Clarke, 1997; Clarke et al., 1997). With further funding from USGS and the EPA (Environmental Protection Agency), the team transitioned the model code to a flat memory model and included calls to the message passing interface, suitable for a Cray supercomputer. Key programming work was conducted at the EPA by Tommy Cathey, and at USGS's Rocky Mountain Mapping Center by Mark Feller, resulting in the release of version 3.0 of the computer code.

The purpose of the USGS's urban dynamics program was to model the growth of major American cities as a vehicle for raising public awareness about the consequences of rapid urbanization. This came at a time when the Internet bubble was at its peak and urban sprawl continued unchecked across the American landscape. This public role was originally the consequence of a press conference at the 1994 Association of American Geographers meeting in San Francisco (Clarke et al., 1996). Unexpectedly, the animated urban growth sequences created in the modeling work became a television success. Later work in Washington, D.C. and Maryland redoubled this public success, as the video sequences were used in various smart growth initiatives at the federal political level. The original inspiration for using historical data in the calibrations was the awareness of the value of the USGS historical map holdings in their Reston map library, where a single paper copy of every map is kept, representing a remarkable historical record of urban growth. While the incompatibility of paper map products and remote sensing was a challenge, eventually many and varied sources were used for SLEUTH data creation, including historical road maps (for highways) and the CORONA declassified spy satellite data.

A decade of SLEUTH applications

With such a substantial number of model applications, it is difficult to distinguish the individual contributions of each application, as they have all built upon prior work, experiences, and lessons learned. What follows is a set of selected applications that stand out as essential works in the history and development of the SLEUTH model.

The application of SLEUTH to the San Francisco Bay area by Clarke et al. (1997) was the first major application of the model to a metropolitan region. In this research, the historical urban extent was determined from cartographic and remotely sensed sources from 1850 to 1990. Based on historical urban extent, along with other input layers for slope and transportation, forecasts of urban growth were created. In addition to forecasts of urban extent, animations of spatial growth patterns were produced and spatial growth statistics were calculated. This foundational publication documents the details of the SLEUTH model, describing the necessary data layers, the five coefficients and what they control, and the four types of urban growth. In addition to the basics of the model, the concept of self-modification was introduced, and a detailed explanation of how the model controls itself and its parameter values was presented. While the method of calibration used in this work was primitive compared to current methods (Goldstein 2005), this research introduced the complete SLEUTH model and led to other applications.

The 1998 paper in the *International Journal of Geographic Information Systems* (Clarke and Gaydos, 1998) detailed a further application to Washington–Baltimore, and allowed comparison of results between the cities. The paper introduced the model to a broader GIScience audience, attracting several other applications. Most important, the paper introduced the model coupling debate, which was a significant feature of the 1996 GIS/Environmental Modeling Conference in Santa Fe, New

Mexico. The support for loose coupling has since been echoed and contradicted, and the debate remains open to this day.

While early applications of SLEUTH focused solely on the modeling of urban growth, the addition of the Deltatron land use change coupling increased the model's ability to represent multiattribute landscape change over time. Candau and Clarke (2000) document this submodel and its use for modeling land use change in the EPA's MAIA (Mid-Atlantic Integrated Assessment) region. Much like Clarke et al. (1997), the paper is the documentation of how the land use change coupling functions. In modeling the eight-state MAIA region, Candau and Clarke were able to compile land use data classified at Anderson Level I for 1975 and 1992, produce a map of predicted land use in the year 2050, and introduce the concept of the uncertainty map, whereby the number of times a particular land use class is predicted at a given location during Monte Carlo simulation as a function of that prediction holds true. These three accomplishments were critical in allowing SLEUTH to evolve from an urban growth model to a more dynamic model in its simulation abilities — an important breakthrough that incorporates the feedbacks between urban growth and land use change that are seen in real-world systems.

In any application of SLEUTH, one of the most time-consuming processes is the calibration. While briefly discussed in Clarke et al. (1997) and in Clarke and Gaydos (1998), Silva and Clarke (2002) present a more refined focus on the calibration of SLEUTH during the application of the model to Lisbon and Porto, Portugal. The paper presents four key findings from the application: (1) SLEUTH is a universally portable model that can be applied not only to North American cities, but to European and other international cities as well; (2) increasing the spatial resolution of the input datasets makes the model more sensitive to local conditions; (3) using a multistage "brute force" calibration method can better refine the model parameters to find those that best replicate the historical growth patterns of an urban system; and (4) the parameters derived from model calibration can be compared among different systems, and the interpretation can provide the foundation for understanding the urban growth processes unique to each urban system. The work of Silva and Clarke in documenting the calibration process of SLEUTH in their application to Lisbon and Porto provided a basis for the work of others applying SLEUTH (Jantz et al., 2003; Yang and Lo, 2003; Dietzel and Clarke, 2004a) and made those applications more robust through a better understanding of the calibration process.

Most recently, an important advance has been made in the use of SLEUTH wherein users have successfully coupled SLEUTH output with other spatiotemporal models to provide greater insight into problems dealing with future urbanization. Some of these applications involved the coupling of SLEUTH outputs with social modeling efforts, while others rest more in the domain of physically based modeling. Claggett et al. (2004) were successful in coupling SLEUTH with the Western Futures Model (Theobald 2001). By doing so, they demonstrated the ability of SLEUTH to move beyond simply providing a spatiotemporal picture of urban growth, by actually categorizing the growth into different classes of "development pressure" based on forecasted population growth. Working almost in parallel, Leão et al. (2004) coupled SLEUTH outputs with a multicriteria evaluation of landfill suitability (Siddiqui et al., 1996) to determine zones around Porto Alegre city (Brazil), where future land would not be urbanized, yet was suitable for landfills. Arthur (2001) coupled SLEUTH to an urban runoff model in Chester County, Pennsylvania. Syphard et al. (2005) examined the consequences of urban development on wildfire regime and vegetation succession in Southern California's Santa Monica Mountains. Cogan et al. (2001) compared using SLEUTH outputs with the California Urban Futures Model (Landis 1994) of urban development to assess stresses on biodiversity. This research demonstrates that SLEUTH can be successfully coupled with other models displaying the potential of the model to be incorporated into a wide array of applications ranging from urban development to environmental assessment and beyond.

Through the Urban Change Integrated Modeling Environment (an NSF–funded research project), the value of using scenarios as a presentation of SLEUTH results, became evident. The application of SLEUTH to Santa Barbara, California, reported by Herold et al. (2003) was part of a broader study that sought to increase local residents' awareness of smart growth principles and planning options through modeling. Both SLEUTH and SCOPE (the South Coast Outlook and Participation Experience) were used to create a set of scenarios that could be used to experiment with alternative futures. SCOPE is a systems dynamics model in the Forrester tradition, coded in the STELLA modeling language, and includes various social, economic, and demographic variables (Onsted 2002). SLEUTH allows policy and plans to be incorporated through new transportation layers and through variations in the excluded layer. The UCSB (University of California at Santa Barbara) Urban Change Integrated Modeling Environment was the result (UCIME, 2001). Choosing scenarios and using models including SLEUTH, led to further research on the nature of scenario planning (Xiang and Clarke 2003) and on simplicity in modeling (Clarke 2005). More recently, the links among parameters, model behavior, and scenario generation have been the subject of further investigation (Dietzel and Clarke 2004b).

SLEUTH has been used for exploratory visualization as well. Acevedo and Masuoka (1997) presented the general methodologies used to create 2-D and 3-D animations of the Baltimore-Washington DC region. Candau (2000) presented the possible ways of visualizing the uncertainty of the location of urban growth in a simulated landscape. Aerts et al. (2003) continued this thread by experimenting with subjects concerning their understanding of the uncertainty of the forecasted urban growth in a section of Santa Barbara using two different techniques. While the myriad of visualization techniques has expanded since SLEUTH's introduction, the understanding and interpretation of simulation forecasts is still a nascent research topic — especially given stakeholders' access to modern visualization techniques.

Current and future research on and with SLEUTH

Current research using SLEUTH is dual faceted. While some researchers are using the model as an applied urban modeling tool (Jantz et al., 2003; Yang and Lo, 2003; Claggett et al., 2004; Leão et al., 2004), others are still investigating simple properties of the model and making major refinements (Dietzel and Clarke, 2004a; Goldstein,

2005). While the list of applications continues to grow, research into the model has been mainly focused on its calibration and how to improve the process.

While the work of Silva and Clarke (2002) was an important step forward in our understanding of the calibration process, it illustrated not only that considerable computational power was necessary to calibrate SLEUTH, but that depending on how a user interpreted the goodness of fit measures, a variety of parameter sets could lead to the "best fit." Goldstein's genetic algorithm (Goldstein, 2005) has tackled the problem of the burdensome amount of time necessary to calibrate SLEUTH using the brute force method. While the algorithm still requires a fast CPU on a computer, a calibration that previously took several days now takes minutes or a few hours — an order of magnitude decrease in the time required to execute model calibration. Through their work with self-organizing maps, Dietzel and Clarke (2004b) have determined the optimal metrics to use in making a more robust calibration of the model to historical data. When coupled together, these two research papers completely overhaul the previous calibration methods for SLEUTH, and mark a new era in model efficiency and calibration robustness.

There are still some very basic ideas behind the model that are yet to be fully explored. In the greater modeling context, one challenge central to the understanding of SLEUTH forecasting, and to a lesser extent back-casting, is the amount of information produced by the sometimes diverse Monte Carlo simulation runs. Goldstein (2004) has begun to explore this issue, and much is being learned about the spatial differences between Monte Carlo results and how to quantify their differences. Averaging over Monte Carlo runs hides the details of individual emergent, atypical results. In order to tap into the lost information present in the Monte Carlo realizations, Goldstein has devised the *area weighted summation metric*, which can lend insight into the possibly chaotic Monte Carlo runs, and has used the idea of "momentum" to explore the difference in spatial spread of each temporal simulation.

Other concepts that will be examined in the near future will include practical, computational, and analytical issues of SLEUTH. For example, the number of Monte Carlo iterations needs elucidation, so as to reduce computational load while providing enough variety in the results. A multidimensional sensitivity testing of SLEUTH is needed. This would include sensitivity to the data layers, the calibration coefficients, and the many parameters preset inside the model, such as *critical slope* and the self-modification control parameters. For more information about SLEUTH and to download the model, visit the online documentation and data repository at http://www.ncgia.ucsb.edu/projects/gig/.

An idea necessary for SLEUTH development, as well as for the greater spatiotemporal modeling community, is that of ontology. There are many theoretical questions that have applied value. This includes the formalization of what constitutes an urban pixel, a landscape type, and more challengingly, a time step. When SLEUTH input data comes from different sources, such as both maps and imagery, does it have the same meaning? The significance of spatial and temporal scale is necessary in this exploration as well. Currently, model applications give little or no heed to these issues. These ontological questions are even more pertinent to model coupling exercises, since each model has its own history and rests in a different intellectual domain, let alone spatial and temporal scale and computing environment.

Lessons learned

With SLEUTH's first decade now behind us, we note that many of the restrictions faced in the early days on use of the model have now been overcome. The computational problems of calibration and forecasting have yielded to better methods and more computing power. The type of data that SLEUTH requires is now also far more ubiquitous, reliable, and consistent. One of the original goals of the SLEUTH work was to scale the model upward toward continental and global scales. We contend that the global impact of urbanization is just as much a factor in global change as many other physical factors, such as methane production and climate; indeed, these factors are highly interconnected.

We present summaries of known SLEUTH applications and parameter values in Table 16.2 and Table 16.3, and the maps of SLEUTH parameters in Figure 16.6 and Figure 16.7 are a first effort at suggesting that SLEUTH is both regionally and universally applicable. Of interest is the spatial variation in the parameters and their links to other geographical, historical, and cultural characteristics. Filling in the gaps, or applying SLEUTH globally, could lead to some useful research about these macrogeographical issues.

TABLE 16.2
Known SLEUTH applications.

Geography	Research group/affiliation	Application	Reference
Albuquerque, NM	USGS/GD/RMMC	Urban Change	Hester, 1999; Hester and Feller, 2002
Alexandria, Egypt	Newcastle University	Urban Change	Azaz 2004
Atlanta, GA	Florida State University, Tallahassee	Urban Change	Yang and Lo, 2003; Yang, 2004
Austin, TX	USGS/GD/RMMC	Urban & Land use Change	USGS/RMMC, 2002
Chester County, PA	Penn State Meteorology and Atmospheric Science	Coupled Modeling	Arthur et al., 2000; Arthur, 2001
Chicago, IL	USGS Urban Dynamics	Urban Change	Xian et al., 2000
Colorado Front Range	USGS/GD/RMMC	Urban & Land use Change	USGS/RMMC, 2002
Detroit, MI	USGS Eros Data Center	Urban Change	Richards, 2003
Houston, TX	Texas A&M University	Urban Change	Oguz et al., 2004
Lisbon, Portugal	University of Massachusetts UCSB Geography	Urban Change	Silva, 2001; Silva and Clarke, 2002
Mexico City, Mexico	UCSB Bren School	Urban Change	UCIME, 2001
Monterey Bay, CA	UC Santa Cruz Environmental Studies UCSB Geography	Biodiversity loss/model integration	Cogan et al., 2001
Netherlands	Berlage Institute	Urban Change	Tack, 2000

TABLE 16.2 (continued)
Known SLEUTH applications.

Geography	Research group/affiliation	Application	Reference
New York, NY	Montclair State University	New York Climate and Health Project	Oliveri, 2003; Solecki and Oliveri, 2004
New York, NY	USGS/GD/RMMC	Urban Change	USGS/RMMC, 2002
Oahu, HI	University of Hawaii at Manoa	Urban Change	James, 2004
Phoenix, AZ	Arizona State University, School of Life Sciences	Urban Change	Breling-Wolf and Wu, 2004
Porto, Portugal	University of Massachusetts	Urban Change	Silva, 2001; Silva and Clarke, 2002
Porto Alegre, Brazil	The University of Melbourne, Department of Geomatics	Coupled Modelling	Leão et al., 2001, 2004
San Antonio, TX	USGS/GD/RMMC	Urban & Land use Change	USGS/RMMC, 2005
San Francisco, CA	UCSB Geography USGS UrbanDynamics	Urban Change	Clarke et al., 1997
San Joaquin Valley, CA	UCSB Geography	Urban Change/Calibration testing	Dietzel and Clarke, 2004a; Dietzel et al., 2005
Santa Barbara, CA	UCSB Geography	Urban Change/ Coupled modelling	Candau and Clarke, 2000 Goldstein et al., 2000, 2004 Herold et al., 2002, 2003
Santa Monica Mountains, CA	San Diego State University	Vegetation Succession	Syphard et al., 2005
Seattle, WA	USGS/GD/RMMC	Urban Change	USGS/RMMC, 2001
Sioux Falls, SD	UCSB Geography	Development of GA	Goldstein, 2004
Sydney, Australia	The University of Southern Queensland	Urban Change	Liu and Phinn, 2005
Tampa, FL	USGS/GD/RMMC	Urban Change	USGS/RMMC, 2001
Tijuana, Mexico	Université Paul Valéry	Urban Change	Le Page, 2000
Washington, D.C./Baltimore	University of Maryland, Geography	Urban Change	Jantz et al., 2003
Washington, D.C./Baltimore	USGS/GD/RMMC/UCSB Geography	Urban Change/Change Visualization	Acevedo, 1997; Clarke et al., 1997
Yaounde, Cameroon	University of Melbourne, School of Anthropology	Urban Change	Sietchiping, 2004

TABLE 16.3
Known SLEUTH parameter values gathered from the applications and references in Table 16.2.

Location	Parameter value				
	Diffusion	Breed	Spread	Slope	RG
	US Applications				
Atlanta, GA	55	8	25	53	100
Austin, TX	47	12	47	1	59
Colorado, Front Range	11	35	41	1	91
Houston, TX	1	3	100	22	17
New York, NY	100	38	41	1	42
New York, NY	21	41	9	90	45
San Joaquin Valley, CA	2	2	83	10	4
Santa Barbara, CA	40	41	100	1	23
Santa Monica Mountains, CA	31	100	100	1	33
Seattle, WA	87	60	45	27	54
Sioux Falls, SD	1	1	12	34	29
Tampa, FL	90	95	45	50	50
Washington, D.C./Baltimore	52	45	26	4	19
Washington, D.C./Baltimore	55	50	26	6	18
Oahu, HI	5	96	12	1	50
	International Applications				
Lisbon, Portugal	19	70	62	38	43
Mexico City, Mexico	24	100	100	1	55
Netherlands	2	80	5	4	5
Porto, Portugal	25	25	51	100	75
Tijuana, Mexico	3	8	70	42	22
Yaounde, Cameroon	10	12	25	42	20

With a new generation of remote sensing instruments capable of timely land cover mapping globally at medium resolution, and with the release of data, such as the Shuttle Radar Topographic Mapping Mission's global digital elevation model, a global application at the scale of 1km, or even nested continental and regional applications at finer scale may be possible in the near future. Given SLEUTH's ability to raise awareness of urbanization issues locally and regionally, we can only imagine what impact global scenarios might have. Should this work become possible, the authors would welcome its contribution in the tradition of the first decade of SLEUTHing.

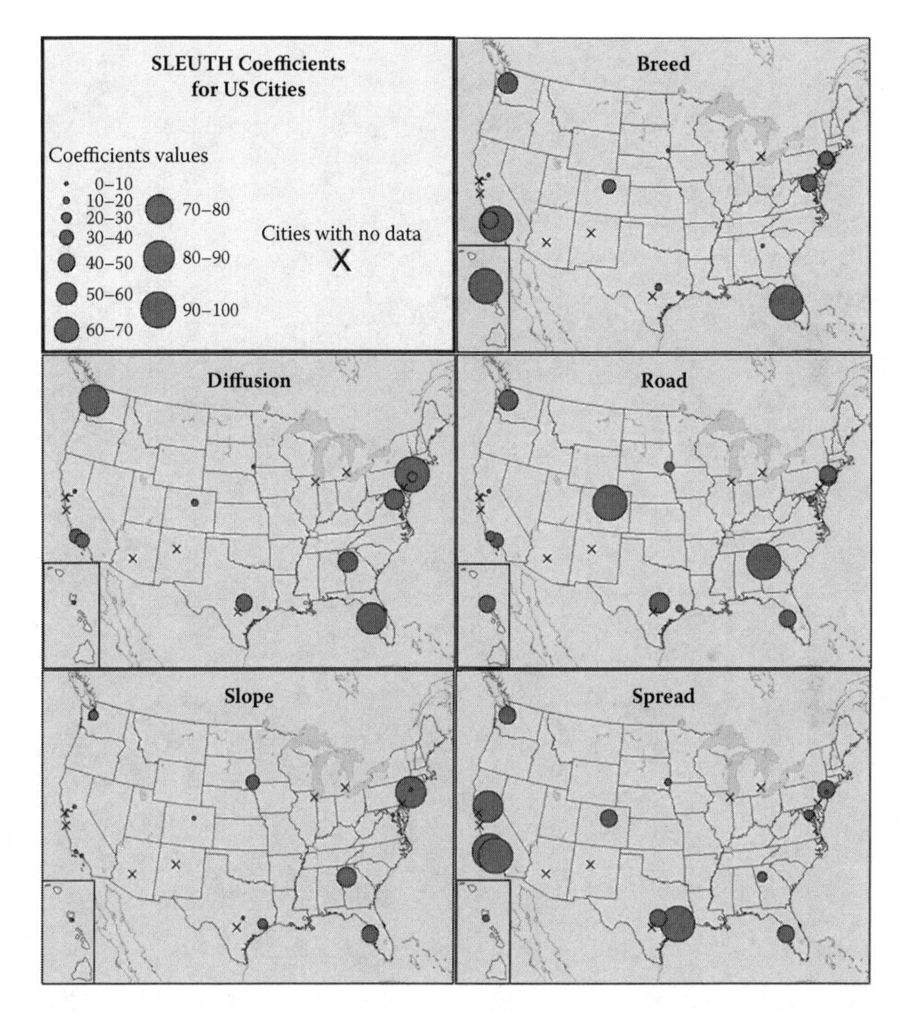

FIGURE 16.6 Maps of SLEUTH parameter values for known cities within the United States.

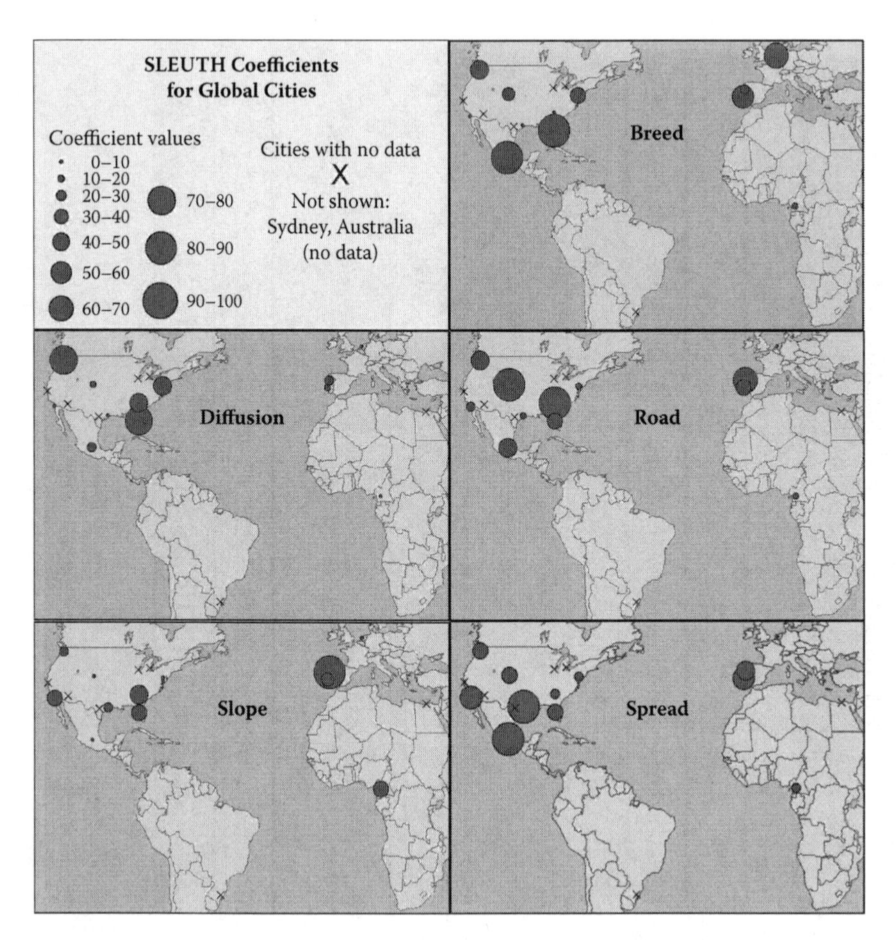

FIGURE 16.7 Global maps of SLEUTH parameter values for known cities.

References

Acevedo, W. and Masuoka, P., 1997, Time-series animation techniques for visualizing urban growth, *Computers and Geosciences*, **23**, 423–435.

Aerts, J.C.J.H., Clarke, K.C., and Keuper, A.D., 2003, Testing popular visualization techniques for representing model uncertainty, *Cartography and Geographic Information Science*, **30**, 249–261.

Arthur, S.T., 2001, A satellite based scheme for predicting the effects of land cover change on local microclimate and surface hydrology. Ph.D. dissertation, Department of Meteorology, Pennsylvania State University.

Arthur, S.T., Carlson, T.N., and Ripley, D.A.J., 2000, Land use dynamics of Chester County, Pennsylvania, from a satellite remote sensing perspective, *Geocarto International*, **15**, 25–35.

Azaz, L.K.A., 2004, Monitoring and modelling urban growth in Alexandria, Egypt using satellite images and geographic information systems. Ph.D. dissertation, School of Architecture, Planning and Landscape, Newcastle University.

Breling-Wolff, S. and Wu, J., 2004, Modeling urban landscape dynamics: A case study in Phoenix, *Urban Ecosystems*, **7**, 215–240.

Candau, J., 2000, Visualizing modeled land cover change and related uncertainty. In *Proceedings of the First International Conference on Geographic Information Science (GIScience 2000)*, Savannah, GA.

Candau, J. and Clarke, K.C., 2000, Probabilistic land cover modeling using deltatrons. In *Proceedings of the 38th Annual Conference of the Urban Regional Information Systems Association*, NCGIA, Orlando, FL.

Claggett, P., Jantz, C.A., Goetz, S.J., and Bisland, C., 2004, Assessing development pressure in the Chesapeake Bay watershed: An evaluation of two land-use change models, *Environmental Monitoring and Assessment*, **94**, 129–146.

Clarke, K.C., 1997, Land use modeling with deltatrons. In *Proceedings of The Land Use Modeling Conference*, NCGIA, Sioux Falls, SD. http://www.ncgia.ucsb.edu/conf/landuse97/, accessed 18 December 2005.

Clarke, K.C., 2005, The limits of simplicity: Toward geocomputational honesty in urban modeling. In *GeoDynamics*, Atkinson, P., Foody, G., Darby, S., and Wu, F., Eds., CRC Press, Boca Raton, FL, 215–232.

Clarke, K.C. and Gaydos, L., 1998, Loose-coupling a cellular automaton model and GIS: Long-term urban growth prediction for San Francisco and Washington/Baltimore, *International Journal of Geographic Information Science*, **12**, 699–714.

Clarke, K.C., Gaydos, L., Acevedo, W., and Hoppen, S., 1996, Communicating land cover model outputs to the general public. In *Proceedings of the Third International Conference/Workshop on Integrating Geographic Information Systems and Environmental Modeling*, Santa Fe, NM.

Clarke, K.C., Hoppen, S., and Gaydos, L., 1997, A self-modifying cellular automaton model of historical urbanization in the San Francisco Bay area, *Environment and Planning B*, 24, 247–261.

Cogan, C.B., Davis, F.W., and Clarke, K.C., 2001, *Application of urban growth models and wildlife habitat models to assess biodiversity losses.* University of California-Santa Barbara Institute for Computational Earth System Science. U.S. Department of the Interior, U.S. Geological Survey, Biological Resources Division, Gap Analysis Program, Santa Barbara, CA.

Dietzel, C.K. and Clarke, K.C., 2004a, Spatial differences in multi-resolution urban automata modeling, *Transactions in GIS*, 8, 479–492.

Dietzel, C.K. and Clarke, K.C., 2004b, Determination of optimal calibration metrics through the use of self-organizing maps. Institute for Environmental Studies, Amsterdam. Integrated Assessment of the Land System Workshop.

Dietzel, C., Herold, M., Hemphill, J.J., and Clarke, K.C., 2005, Spatio-temporal dynamics in California's Central Valley: Empirical links to urban theory, *International Journal of Geographic Information Science*, **19**, 175–195.

Goldstein, N.C., 2004, A methodology for tapping into the spatiotemporal diversity of information in simulation models of spatial spread. In *Proceedings of the Third International Conference on Geographic Information Science* (GIScience 2004), College Park, MD.

Goldstein, N.C., 2005, Brains vs. brawn — Comparative strategies for the calibration of a cellular automata–based urban growth model. In *GeoDynamics*, Atkinson, P., Foody, G., Darby, S., and Wu, F., Eds., CRC Press, Boca Raton, FL, 249–272.

Goldstein, N.C., Candau, J.T., and Clarke, K.C., 2004. Approaches to simulating the "March of Bricks and Mortar." *Computers, Environment and Urban Systems,* 28, 125–147.

Goldstein, N.C., Candau, J., and Moritz, M., 2000, Burning Santa Barbara at both ends: A study of fire history and urban growth predictions. In *Proceedings of the 4th International Conference on Integrating GIS and Environmental Modeling (GIS/EM4)*, Banff, Alberta, Canada. http://www.colorado.edu/research/cires/banff/pubpapers/60/, accessed 18 December 2005.

Herold, M., Goldstein, N.C., and Clarke, K.C., 2003, The spatio-temporal form of urban growth: Measurement, analysis and modeling, *Remote Sensing of Environment*, **86**, 286–302.

Herold, M., Goldstein N., Menz, G., and Clarke, K.C., 2002, Remote sensing based analysis of urban dynamics in the Santa Barbara region using the SLEUTH urban growth model and spatial metrics. In *Proceedings of the 3rd Symposium on Remote Sensing of Urban Areas*, Istanbul, Turkey.

Hester, D.J., 1999, Modeling Albuquerque's urban growth (case study: Isleta, New Mexico, 1:24,000-scale quadrangle). In *Proceedings of the Third Annual Workshop on the Middle Rio Grande Basin Study*, Albuquerque, New Mexico. U.S. Geological Survey Open-File Report 99-203, U.S. Government Printing Office, Washington, D.C., 8–12.

Hester, D.J. and Feller, M.R., 2002. Landscape change modeling: Ground-water resources of the Middle Rio Grande Basin, New Mexico. In *U.S. Geological Survey Circular 1222*, Bartolino, J.R. and Cole, J.C., Eds., U.S. Government Printing Office, Washington, D.C., 20–21.

Hill, L.L., Crosier, S.J., Smith, T.R., and Goodchild, M.F., 2001, A content standard for computational models, http://www.dlib.org/dlib/june01/hill/06hill.html, *D-Lib Magazine* **7**.

James, R., 2004, *Predicting the spatial pattern of urban growth in Honolulu County using the cellular automata SLEUTH urban growth model*. Masters thesis, Department of Geography, University of Hawaii at Manoa.

Jantz, C.A., Goetz, S.J., and Shelley, M.K., 2003, Using the SLEUTH urban growth model to simulate the impacts of future policy scenarios on urban land use in the Baltimore/ Washington metropolitan area, *Environment and Planning B*, **31**, 251–271.

Landis, J.D., 1994, The California Urban Futures Model — a new generation of metropolitan simulation models, *Environment and Planning B*, **21**, 399–420.

Leão, S., Bishop, I., and Evans, D., 2001, Assessing the demand of solid waste disposal in urban region by urban dynamics modelling in a GIS environment, *Resources Conservation & Recycling*, **33**, 289–313.

Leão, S., Bishop, I., and Evans, D., 2004, Spatial-temporal model for demand allocation of waste landfills in growing urban regions, *Computers Environment and Urban Systems*, **28**, 353–385.

Le Page, M., 2000, *Expansion urbaine à la frontière du 1er monde: Analyse et modélisation de la croissance spatiale de Tijuana, Mexique*. Ph.D. dissertation, Université Paul Valéry.

Liu, Y. and Phinn, S.R., Modelling the driving forces of Sydney's urban development (1971–1996) in a cellular environment, *Applied GIS*, 1, 3, 27.1–27.18.

Oguz, H., Klein, A., and Srinivasan, R., 2004, Modeling urban growth and landuse and landcover change in the Houston metropolitan area from 2002 to 2030. In *Proceedings of the ASPRS 2004 Fall Conference*, CD ROM, Kansas City, MO.

Oliveri, C., 2003, Land Use Change Assessment (LUCA) Group. http://www.csam.montclair.edu/luca/, accessed 18 December 2005.

Onsted, J.A., 2002, *SCOPE:* A modification and application of the Forrester Model to the south coast of Santa Barbara County. Master's thesis, Department of Geography, University of California–Santa Barbara.

RICHARDS, L., 2003, Detroit River Corridor Preliminary Assessment of Land Use Change. USGS Web publication, http://landcover.usgs.gov/urban/detroit/intro.asp, accessed 18 December 2005.

SIDDIQUI, M.Z., EVERETT, J.W., AND VIEUX, B.E., 1996, Landfill siting using geographic information systems: A demonstration, *Journal of Urban Planning and Development*, **122**, 515–523.

SIETCHIPING, R., 2004, Geographic information systems and cellular automata-based model of informal settlement growth. Ph.D. dissertation, School of Anthropology, Geography and Environmental Studies, University of Melbourne.

SILVA, E.A., 2001, The DNA of our regions. Artificial Intelligence in Regional Planning, In *Proceedings of the World Planning Schools Congress: Planning for Cities in the 21st Century*, Shanghai, China.

SILVA, E.A. AND CLARKE, K.C., 2002, Calibration of the SLEUTH urban growth model for Lisbon and Porto, Portugal, *Computers, Environment and Urban Systems*, **26**, 525–552.

SOLECKI, W.D. AND OLIVERI, C., 2004, Downscaling climate change scenarios in an urban land use change model, *Journal of Environmental Management*, **72**, 105–115.

SYPHARD, A.D., CLARKE, K.C., AND FRANKLIN, J., 2005, Using a cellular automaton model to forecast the effects of alternate scenarios of urban growth on habitat pattern in southern California, *Ecological Complexity*, **2**, 185–203.

TACK, F., 2000, Emulating the future. *ArchiNed News*, 31 July 2000, http://www.classic.archined.nl/news/0007/bia_eng.html, assessed 18 December 2005.

THEOBALD, D., 2001, Land-use dynamics beyond the urban fringe, *Geographical Review*, **91**, 544–564.

UCIME (Urban Change — Integrated Modeling Environment), 2001, Urban change — Integrated modeling environment. http://www.geog.ucsb.edu/~kclarke/ucime/index.html, accessed 18 December 2005.

USGS (United States Geological Survey), 2001, Tampa Bay integrated science pilot study. Open-File Report 01-398 Available online at: http://gulfsci.usgs.gov/tampabay/reports/crane1/pdf_cran.html, accessed 18 December 2005.

USGS/RMMC, 2005, Assessment of impacts and stresses on the Edwards-Trinity Springs Ecosystem in Hays County, Texas, http://rockyitr.cr.usgs.gov/rmgsc/pdf/etcrisp05.pdf, accessed 18 April 2006.

USGS/RMMC (United States Geological Survey), 2002, (National Mapping Division) NMD's Urban Dynamics Research Program. USGS, U.S. Department of the Interior. http://rockyweb.cr.usgs.gov/html/growth/dynamics.html, accessed 18 December 2005.

USGS/RMMC, 2001, Seattle Area Natural Hazards Project, http://rmmcweb.cr.usgs.gov/html/hazards, accessed 18 April 2006.

XIAN, G., ACEVEDO, W., AND NELSON, J., 2000, Urban development in the Chicago area — A dynamic model study. In *Proceedings of the 4th International Conference on Integrating GIS and Environmental Modeling (GIS/EM4)*,. Banff, Alberta, Canada. http://www.colorado.edu/research/cires/banff/pubpapers/46/, accessed 18 December 2005.

XIANG, W.-N. AND CLARKE, K.C., 2003, The use of scenarios in land use planning, *Environment and Planning B*, **30**, 885–909.

YANG, X., 2004, Predicting urban growth with remote sensing and dynamic spatial modelling. In *Proceedings of the XXth ISPRS Congress*, Istanbul, Turkey.

YANG, X. AND LO, C.P., 2003, Modelling urban growth and landscape change in the Atlanta metropolitan area, *International Journal of Geographical Information Science*, **17**, 463–488.

International Journal of Geographical Information Science,
1998, Vol. 12, No. 4, 299–314.

17 Overcoming the Semantic and Other Barriers to GIS Interoperability

Yaser Bishr

Abstract. An increasing number of applications need to work on data that are scattered over several independent geographical information systems. This has led to extensive research efforts in the field of interoperable information systems. The term interoperability is used in a number of different contexts. In this paper a concise explanation of the meaning of interoperability in the context of geographic information systems (GIS) is given. Current efforts in this realm are presented. Problems and challenges of the next interoperable GIS generation are identified. A proposal to meet these challenges is presented.

1 Interoperability: A worldwide awareness

Through the 1970s and the early 1980s, most GIS applications were considered islands of information. They were self-contained independent systems where spatial data were digitally captured, stored, analysed, and displayed. Data were rarely acquired from digital sources because of the closed nature of the application's file formats which is true except for satellite imagery. The advances in information technology and the growing demands from GIS users to overcome the bottleneck and costs of data capture, led users to share data by transferring them from one island to the other. Such a transfer was accomplished using either special purpose translators or a neutral format that was understandable to the source and target systems. In either case, the transfer was characterized by the fact that it was batch-oriented where an entire data set was converted and transferred on the file level. The transfer was eventually done via physical means, e.g., magnetic tapes, or more recently, electronically, e.g., the Internet.

In more recent years, users began to realize the inefficient redundant data provided by the batch-oriented approach. With the rapid development in information system and distributed database paradigms, GIS users realized the need for interoperable geographical information systems, IGIS. IGIS provide the means by which spatially distributed GIS can be connected together in a web in order to transparently exchange data and some remote access to GIS services.

Interoperability is the ability of a system, or components of a system, to provide information portability and inter-application cooperative process control. Two geographical databases X and Y can interoperate if X can send requests for services R to Y on a mutual understanding of R by X and Y, and Y can return responses S to X based on a mutual understanding of S as responses to R by X and Y (Bordie 1992).

Interoperability could be characterized as a form of a system's intelligence that enhances the cooperation between component information systems. The intelligence is required to provide services, find resources, cooperate and carry out complex functions across component information systems without the need to know in advance precisely what resources are available, or how to acquire them.

Despite the clear definition of interoperability, some software vendors and research groups are still not clear of what is meant by interoperability. The question is: when to call an information system interoperable? The following section is an attempt to answer this question.

2 Levels of interoperability

There are six levels of interoperability between two or more spatially distributed independent GIS, Figure 17.1. Network protocols lie at the lowest level of interoperability. At this level users can usually communicate without any direct service from the remote GIS, they merely download flat files. A good example of this level of communication is TELNET (Krol 1994), where users log into a remote system, called the host, without any knowledge of the network protocol it supports. The major problem of this method is that users have to have knowledge of the operating system that runs at the remote machine in order to interact with it.

The second level provides a higher level network protocol where users can connect to a host and interact seamlessly regardless of its operating system. A good example of this level of interoperability is the FTP protocol where users connect to

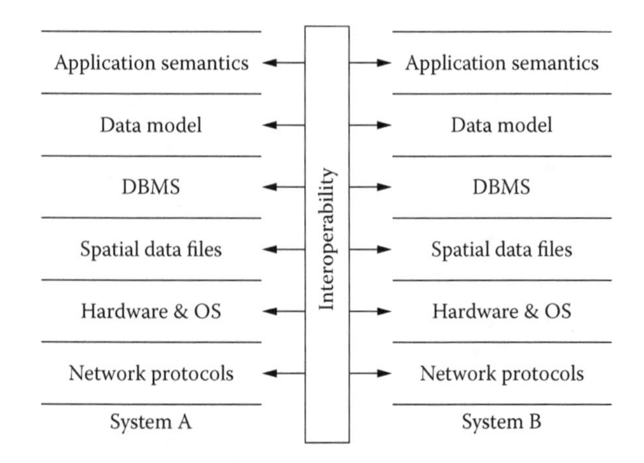

FIGURE 17.1 Levels of interoperability.

the host and interact using the FTP's proprietary commands. At this level users can merely transfer data file between systems. Current advances in computer networking provide transparent interoperability between different protocols (Tanenbaum 1989). The major disadvantage of providing communication at this level is that users have to have prior knowledge of the formats of the transferred files, and hence have the proper converters.

The third level of interoperability, the spatial data files, solves the above disadvantage. Users can download data files in a standard format and the system automatically identifies, and consequently converts, the format to that of the user. DeltaX is a good example of this level of interoperability (Allam 1994, Otoo 1994). The main disadvantage is that, users have to have prior knowledge of what data exist at the source, and have to search its contents using its user interface and query language.

Currently there are several GIS software that provide communication at the fourth level of interoperability, e.g., Intergraph's new technology Jupiter, ODBC from Microsoft (Geiger 1995). Moreover, advances in client/server technology (Boar 1993, Gorman 1994), allow users to establish a connection between their own GIS and other remote GIS. Once established they can query the remote systems using their own, local, query language. It is also possible to analyse and display the remote data sets. However, the main disadvantage is that users have to acquire knowledge on the data models and semantics of the underlying remote database.

At the fifth level users are provided with a single 'virtual' global data model, that is an abstraction of all the underlying remote databases. Queries are sent to the global model that in turn maps between the user's and the remote data models. Although users can query remote databases seamlessly at this level, a proper knowledge on their semantics is still required. Two GIS with two different thematic views can have the same data model, but they vary in their assumption and semantics. The intention of the sixth level is to provide seamless communication between remote GIS without having prior knowledge of their underlying semantics.

Interoperability could occur at any of the above six levels. Each level corresponds to a wide field of technology. Interoperability could be improved or its scope increased by advances in the underlying technology. Some of these advances can be expressed in terms of various forms of transparency in that differences are hidden and a single homogeneous view is provided. There is no known GIS that provides interoperability at the data model and application semantics levels, i.e., the fifth and the sixth levels. However there are several research activities in the field of federated database systems (Bererra 1993, Brackett 1994, Brenner 1993). We anticipate that the future generations of IGIS will provide transparent communications at data model and application semantics levels.

The intention of the subsequent sections is to provide a detailed breakdown of interoperability problems at the fifth and the sixth levels, as shown in Figure 17.1, as well as to propose a mechanism for resolving these problems. The problem in hand can be described as follows: geographical objects stored in independent remote databases can vary in their geometric syntactic representation, in their class hierarchies, or in their semantics, even though they may refer to the same Real World feature. A user might need to search for an object X that is presented as Y in a remote database. Both X and Y refer to the same Real World feature. Usually objects are presented as a 3-tuple {thematic attributes, geometric attribute, ID} or {T, G, ID}. Objects could be represented using 4-tuple, if the temporal dimension is considered. Sharing temporal databases is another complex issue that is outside the scope of this paper. Then X is presented as {Ti, Gi, IDi} in the user's database, and as {Tj, Gj, IDj} in the remote database. The user might search the remote database based on thematic attributes, geometric descriptors or ID. But the three are represented differently in the two databases. The only solution is to provide the user with metadata that describes Tj and Gj before searching the remote database. This solution could be acceptable if the user searches only one database. But what is the solution if he/she has to search several databases?

A viable solution is to base the search on what X and Y actually represent in the Real World, i.e., semantics, not on the way they are presented in the database.

3 Abstraction levels of real world objects

To formalize the definition of semantics, several levels of abstraction, that are implicit when modelling Real World facts to the GIS domain, are discussed, as depicted in Figure 17.2. The figure is an improvement of the one presented by Kottman in the OpenGIS guide (Buehler *et al.* 1996). The first three levels, from the Real World to the Dimensional World, deal with the abstraction of Real World facts, and are usually not modelled in GIS software. We mean by a fact, an event known to have happened or something known to have existed. A fact, then, is associated with a phenomenon or a geographical feature in the Real World. Soil scientists observe the process of soil erosion in a watershed, while biologists observe natural habitats of the same area. Two geographical information communities frequently give the same names to different phenomena, or different names to the same phenomenon. The result is semantic heterogeneity that can be a significant barrier to interoperability.

Semantics in this paper is defined as the relationship among the computer representations and the corresponding Real World feature within a certain context. Furthermore, semantics in linguistics is defined as the relationships between words and the things to which these words refer (ter Bekke 1992), which is similar to our definition. It is essential to distinguish between semantics, as defined above, and computer model's semantics as defined in computer science. In the latter, semantics is the mathematical interpretation of the formal language expressions of the database (Leeuwen 1990).

The Discipline Perception World, Figure 17.2, represents the domain where the first three abstraction levels dwell. A discipline is a state of order where a branch of human knowledge exists and a set of rules and methods could be applied. All disciplines that deal with geospatial information, share at least one thing; the earth. The rules and methods could not be applied in the chaotic, complex Real World. Usually people in a particular discipline simplify this world and build a mental model of the relevant facts. The Discipline Perception World represents the mental model of a group of people in a particular discipline. There are three levels of abstraction in this world. In the Cognitive Content World facts and their relationships are abstracted as categories. In the Conceptual World a mental association is performed in order to assign the actual facts to their corresponding categories. People

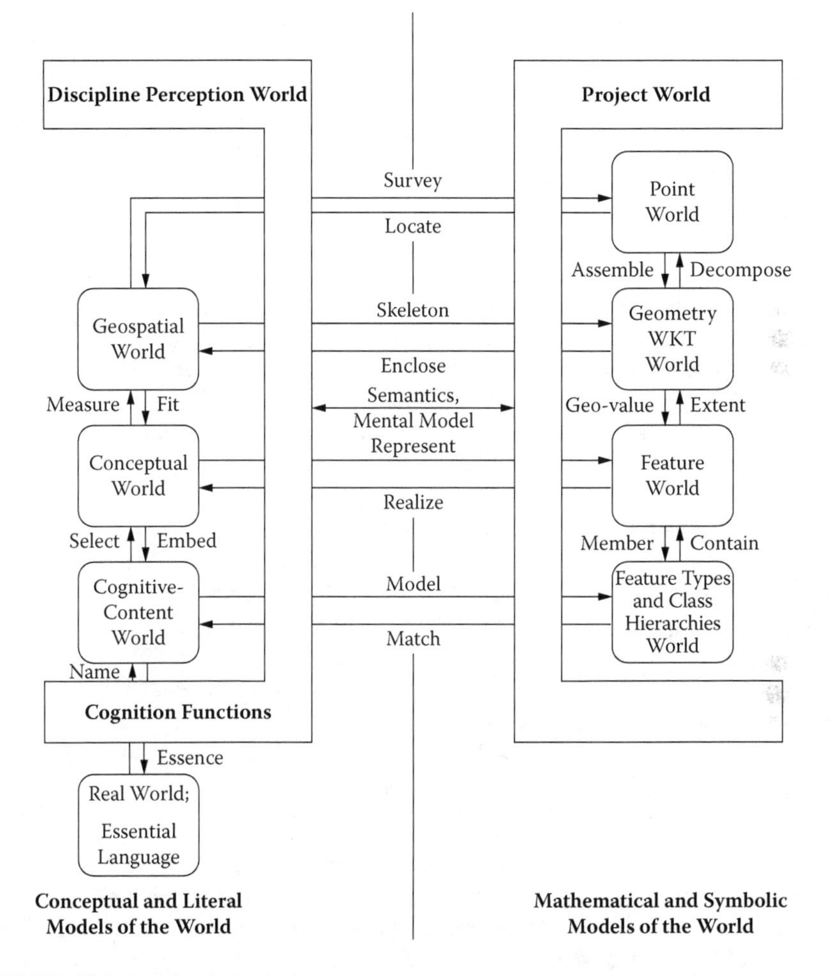

FIGURE 17.2 Levels of abstraction.

in the Geospatial World tend to build a discrete geometric mental model of the facts in order to be able to apply the set of rules and methods of the underlying discipline for the purpose of analysis.

GIS, naturally, is developed such that it can represent Real World objects in computers. The Project World in our definition encompasses all the basic assumptions made by a GI community in order to represent Real World objects in the GIS domain. The representation was made as close as possible to the mental model of human beings. There are four levels of abstractions that occur within the Project World in order to achieve this objective. The levels correspond to the three abstraction levels in the Discipline Perception World, as shown by the arrows in Figure 17.2. The Project World is different from the term World View used by some authors (Coad *et al.* 1990). The latter implies the mental model, of users in a discipline, of the Real World facts and hence is equivalent to the Discipline Perception World.

The four levels of abstraction in the Project World deal with the syntactic and schematic representation of geographical objects in the GIS. The relationship between the Discipline Perception World and the Project World is implicitly maintained in the user's mental model. This is acceptable in a single GIS environment. However, when data are transferred between two different disciplines, the relationship ceases to exist. The receiver of the data set is not informed about the Discipline Perception World of the provider.

The key to enabling interoperability at the semantic level is to ensure that this relationship is maintained during data transfer. There are two possible approaches to solving this problem:

1. Provide the data receiver with information on the Discipline Perception World of the provider.
2. Provide a mechanism that allows automatic transformation from one Discipline Perception World to the other.

The first solution will leave the user with two difficult tasks. First understand and interpret the Discipline Perception World of the provider and then transform the retrieved data set from that world to his own. The second approach to this problem is twofold: firstly, the formalization of the notion of semantics in the geospatial information community. Secondly to develop a mechanism that allows capturing semantics from the Discipline Perception World and representing them in the Project World, that is the focus of the next sections.

4 Heterogeneity in a multi-GIS environment

Data are collections of facts from which conclusions can be drawn. Facts are communicated using a language, e.g., English. Language is then the means by which that data are communicated. Therefore, to understand the reasons for the complexity of the problem of data sharing, it is useful to consider the close analogy between language and data.

In an application of language, three chief factors may be distinguished: the speaker, the expression uttered, and the designatum of the expression, i.e., to which the speaker intends to refer by the expression. Accordingly, an investigation of a language belongs to pragmatics if explicit reference to a speaker is made; it belongs to semantics of designata are referred to; it belongs to syntax if only expressions and their relationships are dealt with. Syntax is then the grammar that defines how the language is used. The whole science of language, consisting of the three parts mentioned, is called semiotics (Carnap 1964).

The above definition leads to the following assertions:

1. An example of pragmatic investigation is the study of the way people categorize, or classify, and relate Real World facts. The outcome of this process is a schemata of the underlying facts. Naturally disciplines have different schemes of the Real World facts, that result in schematic heterogeneity.

2. Expressions always refer to facts. A group of people may use different expressions to refer to the same Real World fact, or they may use the same expression to refer to different facts, which result in semantic heterogeneity.
3. Syntactical investigation deals with formalizing the grammar of schemata and semantics expressions, without any reference to what they actually mean. There could be different grammar that results in syntactic heterogeneity.

In the GI community people recognize Real World facts and give them names, categorize them and create a mental model, i.e., schemata, and then present them in a database using its underlying syntax. Therefore, the above assertions of types of heterogeneity could be redefined in the context of database as:

1. Semantic heterogeneity: a Real World fact may have more than one description in the underlying databases to comply with various disciplines, giving as a consequence semantic heterogeneity. For example a road network for pavement management has different semantic descriptions from transportation data maintained in a GIS database designed for small scale topographic mapping applications.
2. Schematic heterogeneity: objects in one database are considered as properties in another, or object classes can have different aggregation or generalization hierarchies, although they might describe the same Real World facts.
3. Syntactic heterogeneity: each database may be implemented in a different DBMS of different paradigms, such as relational or object oriented models. Syntactic heterogeneity is also related to the geometric representation of geographic objects, e.g., raster and vector representations.

Interoperable GIS means that systems should be independent and yet can transparently communicate at a high level of semantics (Bishr *et al.* 1996). Due to the fact that geographical information systems were developed independently, several heterogeneity problems arise. Heterogeneity occurs in three facets: syntactic, schematic, and semantic. A clear definition of these aspects will help to key out the sources of problems that have to be tackled. In the subsequent sections, these facets are discussed. The discussion is based on the levels of abstraction shown in Figure 17.2. Syntactic, schematic, and semantic heterogencity have their origin from the multiplicity of the Discipline Perception World and the Project World.

4.1 Semantic heterogeneity

Semantic heterogeneity is usually the source of most data sharing problems. It occurs within the Discipline Perception World due to the variation of the mental models of the different disciplines. In the Cognitive Content World we are concerned with facts that are abstracted from the Real World. Individuals from the same discipline are likely to share a common interest in the same Real World facts. For example, a pavement management group is interested in road networks, their number of lanes, their intersections, and their traffic flow. This is different from the way the marketing group recognizes road networks. They observe houses and streets as addresses of clients. In the pavement management and the marketing group streets serve different purposes. In this case there is no common base of definitions of the underlying facts between the two disciplines. We call this type of semantic heterogeneity cognitive heterogeneity.

The second type of semantic heterogeneity that occurs in the Cognitive Content World is the naming heterogeneity. Semantically alike entities in the Cognitive Content World, i.e., refer to the same Real World fact, might be named differently. For instance, watercourse and river might be two names describing the same thing.

The two types of semantic heterogeneity persist when the Cognitive Content World is abstracted into the Feature Type and Class Hierarchy World through the method model. Naming heterogeneity could be corrected through simple mapping, using a thesaurus. The cognitive heterogeneity is, however, more difficult to solve, especially when there is no minimum set of common definitions. The difficulty stems from the fact that it is difficult to capture the role and the set of definitions of Real World facts from the mental model and represent them in the database. A possible solution would be to capture the roles of Real World objects, within a certain context, as a set of rules and constraints into the database and then devise a mechanism to map between different roles using a rule base as will be described in Section 5.1.

4.2 Schematic heterogeneity

The schemata of the mental model, i.e., the classification and hierarchical structure of the Real World categories could vary within or across disciplines. The schemata of the mental model is presented as a conceptual model during the database design phase. Schematic heterogeneity can be classified as follows:

1. Entity versus Entity
   (*a*) 1:M and N:M relationships
   (*b*) missing attributes
   (*c*) missing but implicit attributes, i.e., an attribute that can be calculated from other ones
   (*d*) entity constraints
   (*e*) behaviours
2. Attribute versus Attribute
   (*a*) 1:M and N:M relationships
   (*b*) behaviours
   (*c*) default values
   (*d*) domain
3. Entity versus Attribute, i.e., an instance of a class in one database can be an attribute in another
4. Different representation for equivalent data
   (*a*) different units
   (*b*) different quality parameters

A possible solution to this problem is to provide a unified schema of the underlying lying heterogeneous schemes. This is known as schema integration. The schema integration process can be thought of as deriving a single schema through a sequence of simple functions, each of which addresses a type of schematic discrepancy (Kim 1995). Schema integration and related issues have been discussed in the literature (Batini *et al.* 1986, Kaul *et al.* 1990, Tari 1992). However, to the best of our knowledge, there exist no general framework for the comprehensive enumeration and systematic classification of resolution techniques for schematic conflicts.

4.3 *Syntactic heterogeneity*

There are two types of syntactic heterogeneity, one is related to the logical data model and its underlying DBMS, e.g., relational and object oriented, and the other is related to the representation of the spatial objects in the database.

The implementation of a conceptual model in a DBMS is the process of translating a conceptual model into a logical one. A conceptual model could be implemented into an object oriented or relational logical model. The logical model is dependent on the DBMS. The paradigms of such models are different. A bi-directional mapping between OO and relational models, or between similar paradigms, is required in order to resolve such a problem. Approaches of schema integration are explained elsewhere (Tari 1992, Yan *et al.* 1992).

The representation of facts as entities or as a spatio-temporal continuum in the database domain is syntactically different. There are two principal structures for representing spatial objects and linking their thematic and geometric data (more details found in Molenaar, 1992):

1. Raster data structure: a collection of points or cells distributed in a regular grid. Each cell in a grid is assigned a thematic value that refer to an area segment in the real world. Raster can be either single value, or multi-valued. In the former raster have only one attribute representing one particular thematic aspect of the terrain. In multi-valued raster, each raster has several attributes of the same terrain segment. The topology of the raster structure is based on the adjacency of the raster point or cells.
2. Object data structure: the link between the thematic data and the geometric data is made through an object identifier. The geometry of objects can be a collection of point, lines, or area (in case of planar graph geometry).

A proper solution to this problem would be to provide a common syntax for spatial objects representation to all GIS. OGIS (Buehler *et al.* 1996), and the formal data structure, FDS, (Molenaar 1994), provide a formal way for syntactic representation of the Real World facts.

5 Resolving heterogeneity of spatial databases

In section 4, three types of heterogeneity were shown, semantic, schematic and syntactic. A conceptual model in the project world is a reflection of the discipline perception world of the underlying discipline. Resolving the schematic or the syntactic heterogeneity could not be achieved without knowledge of the mental models of the underlying disciplines. In other words, resolving the semantic heterogeneity is the key to resolving schematic and syntactic heterogeneity.

We argue that a successful mapping between two heterogeneous schemes, and consequently application semantic interoperability, requires semantic rules that were extracted from the discipline perception world. In this section we introduce a salient formal model for capturing Real World semantics into database. The model is considered an extension of the Formal Data Structure (FDS) proposed by Molenaar.

5.1 Capturing semantics

In Section 4.1 we distinguished two types of semantic heterogeneity, cognitive and naming. The former occurs when the definitions of the facts that are shared between two disciplines are different. The latter type of heterogeneity occurs when two Real World facts have the same name, or vice versa. Intuitively, Real World facts acquire meaning from the way they are used. For example, the meaning of a road network lies in the fact that cars use it to drive from point A to B. The meaning of a house lies in the fact that John lives in it with his family. In this case road object cannot be confused with a runway or a house with a barn.

Such an understanding of semantics derived from the use contrasts with the traditional way of defining meaning by geometry and attributes, as shown in Figure 17.3. An object that belongs to a class has geometric, thematic descriptors, and an object identifier that uniquely identifies it in the database. For example, a house could be defined by its rectangular shape, number of windows, owner, and so forth. Still this information is not enough to distinguish a house from a barn.

In general, it is extremely difficult and often impossible to find a complete set of characteristic attributes for a Real World object (Kuhn 1995). The major problem of defining semantics by attributes, is to identify which attributes to choose. For any chosen set of objects, it is usually possible to find an instance of the category that does not have all of them, or an object having them, but not falling in a category. Moreover, different people consider different features to be relevant, leading to a lack of agreement on formalized meaning. It is clear that this approach cannot fully capture how we deal with meaning in normal human communication.

Two fundamental problems with this approach were identified: meaning derived from names and meaning implied in the use of an object. Recognizing objects based

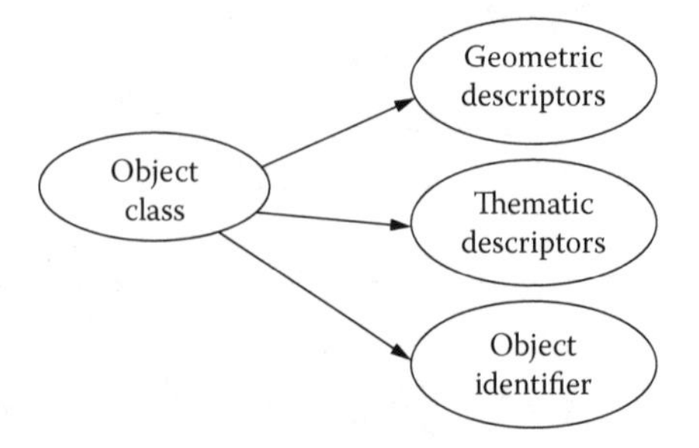

FIGURE 17.3 Classical representation of geographical objects.

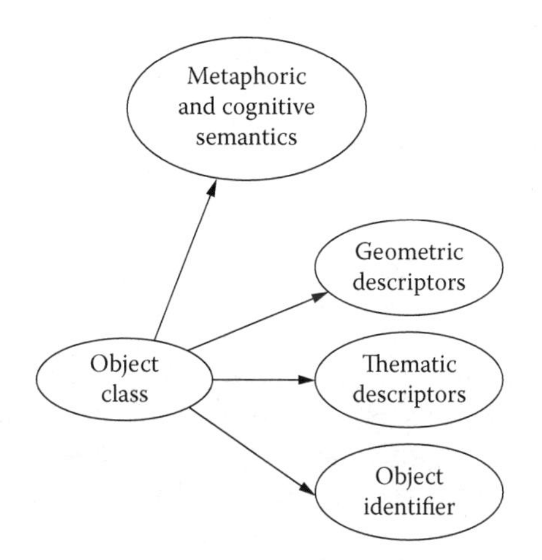

FIGURE 17.4 Adding semantics to the classical representation of geographical objects.

on their use, i.e., based on their function in the cognitive world, has the potential to overcome this problem.

Figure 17.4 shows an extension of the classical representation of geographical objects. Objects will have geometric, thematic, and semantic descriptors. The semantic descriptors differ from the other ones in such a way that the former are fixed within an object class, and hence all instances of that class will have the same semantic descriptors. In this case an object can semantically belong to a class, although it has different geometric and/or thematic descriptors. This requirement contrasts the FDS requirement, that is, each object belongs to one and only one class. Classes are exhaustive and disjoint with respect to the mapped objects. This statement is true under the assumption of single context. In a multi-database environment the geometric and thematic constraints of object membership to a class are relaxed and semantic constraints are introduced.

5.1.1 Representing naming semantics

The naming semantics is resolved by developing a thesaurus that has all alternative names of a particular class, or descriptors of that class. This provides a direct mapping between heterogeneous lexicons used by different disciplines.

5.1.2 Representing cognitive semantics

The way by which experts of a particular discipline define Real World facts is heuristic. Heuristics are common sense knowledge, or rules of thumb, that originate from the expert's past experience (Avelino *et al.* 1993). There are three categories of heuristic knowledge: associational, motor skills, and theoretical. We argue that it is important to identify under which category the knowledge applied to map objects from Cognitive Content World to the Real World falls. Once this is achieved, a proper methodology for capturing such knowledge can be identified.

1. Associational knowledge: this type of knowledge is mostly acquired through observation. It is typically in the form of rules (IF–THEN relationships). Expert systems can easily represent this type of knowledge.
2. Motor Skills: it is physical rather than cognitive. Humans learn these skills by repeatedly performing them. Neural Networks can emulate this type of knowledge.
3. Theoretical Deep Knowledge: finding a solution to a technical problem often requires going beyond our basic understanding of the domain. We must apply creative ingenuity that is based on our theoretical knowledge of the domain. This type of knowledge allows experts to solve problems that have not been seen before and, therefore, are not associational in nature. Model-based reasoning systems are a notable attempt to encapsulate this deep knowledge and reason with it.

With a closer look to the original problem, that is, associating Real World facts to objects in the Discipline Perception World, we can easily discern that cognitive knowledge is similar to associational knowledge. Experts apply a set of rules by which they can map objects between Real World and Discipline Perception World. These are heuristic rules that were originated from an expert's past experiences. Hence, representing such knowledge as a set of rules is a viable solution. Data, geometric and thematic, and knowledge can be encapsulated into an abstract object type with sufficient discipline knowledge to be able to interpret the appropriate meaning of an object. This is known as data/knowledge packets that allow both knowledge and data to be represented in a unified model (Doyle and Kerschberg 1991). Data/knowledge packets are formed using triples of the form <Operation, Object, Meaning> where: operations are functions, procedures, or constraints; objects are database entities and attributes; and meaning is the corresponding abstracted Real World fact. A data/ knowledge packet could look like this:

{
ObjectClass Y
Properties (A_1, A_2, …, A_n)
IF X is human and Y is building and X Lives in Y Then Y is a house
IF X is animal and Y is building and X Lives in Y Then Y is a barn
}
Y is an object class with a set of properties A, The two rules are operations; Y-is-a-house and Y-is-a-barn are the assertions of the meaning of Y.

Now consider the following two rules

IF X is human and Y is building and X Lives in Y Then Y is a house
IF X is human and Y is building and X Lives in Y Then Y is a hotel

The two assertions are different although the two rules are similar. We can overcome this problem by indicating the discipline where these rules are applied, e.g., marketing and hotel booking applications. We will refer to the discipline where a set of rules is applied as context. Hence, a context in our definition refers to the assumptions underlying the way in which a group of people in a discipline interpret or represent data. In other words when we are in the Discipline Perception World then we are in a discipline, but when we abstract it to the Project World we are in a context.

The two rules then become

IF X is human and Y is building and X Lives in Y and Context is marketing Then Y is a house
IF X is human and Y is building and X Lives in Y and Context is hotel-booking Then Y is a hotel

Our proposal to encapsulate knowledge with the classical object definition lends itself to an active field of research known as Deductive Database. Deductive Database stemmed from the desire to combine the traditional DBMS with the power logic of programming technologies. While the term deductive database has no precise definition, it is typically associated with a system having at least the following properties (more details can be found in Kim, 1995):

1. The capability to encapsulate knowledge and object definition.
2. The capability to express queries by means of logical rules.
3. Support for recursive query processing.
4. Support of usual database amenities.

In this section, we presented a model for capturing the naming and the cognitive semantics and representing them as an extension of the classical object representation. Naming semantics were presented as a thesaurus of different names that can be assigned to a certain Real World fact. Cognitive semantics were captured as a set of rules which is encapsulated with object definition as data/knowledge packets. In the next section we will introduce a mechanism by which we resolve the semantic, schematic, and syntactic heterogeneity between two or more GIS. The concept is based on providing a Proxy Context which is a mediator for sharing data between classes that exist in two or more GIS.

6 Handing semantics in multi-databases

There are two approaches for geographical object representation in GIS: (1) field approach where the Earth's surface is presented as a spatio-temporal continuum; (2) object-structure planar graph approach where terrain features are defined by their geometry, shape and position (in addition to thematic descriptors). Laurini (1994) listed some discrepancies that could exist in different Project Worlds: spatial representation, spatial scale, projection and coordinated systems, and spatio-temporal difference are some examples. Classification of the types of geometric information can then be structured according to the geometric discrepancies that could exist between heterogeneous GIS. There are three aspects of geometry: topology, size and shape, and position and orientation; three levels of topology: connectivity of geometric elements, connectivity of objects, and connectivity of geometric elements to objects. Geometric and thematic descriptors of geographical objects are prone to change in a database during the lifetime of an object while their semantics are more stable.

In this section we introduce the notion of Proxy Context. A proxy context has proxy class hierarchies, that are formed by a set of proxy classes. A proxy class is a substitute for two semantically similar classes, in two independent contexts, where their instances are to be shared. In our definition we distinguish between Classes and Proxy Classes. The former exist in a single context, while the latter exist in a proxy context that is the summation of the contexts that need to share data. Objects' membership to proxy classes are based only on their semantics, not geometry or thematic attributes.

Classes

In a single database we adopt the FDS for a Single Valued Vector Map, SVVM, presented by Molenaar. The only difference is the addition of the semantic component. We assume that the geometry and the class hierarchies of the underlying databases comply to FDS specifications. We mention here only the object class membership, as it is relevant to our discussion. A detailed description of the geometric and thematic components of the FDS is mentioned elsewhere (Molenaar 1993, 1994, 1995, Molenaar and Richardson 1994, Kufoniyi 1995).

Let C_j be a class that has a list of attributes A_n

$$\text{LIST }(C_j) = \{A_1 ,\ldots,A_r ,\ldots,A_n\}$$

If an object O_i passes a test formulated in a decision function for a class C_j then it will be a member of that class and that will be expressed by the membership function:

$$M[O_i,C_j] = 1 \text{ if } O_i \text{ is a member of } C_j$$

$$= 0 \text{ otherwise}$$

The attribute structure of objects is determined by the class to which they belong, so that each object has a list containing one value for every attribute of its class. An object inherits the class attribute structure.

Hence from 1 and 2

$$\text{VLIST }(O_i) = \{a_{1j} ,\ldots,a_{ri} ,\ldots,a_{ni}\}$$

$$\text{with } a_{ri} = A_r\ [O_i] \text{ is value of } A_r \text{ for object } O_i$$

$$\text{and } A_r \in \text{ (LIST }(C_j)$$

$$\text{and } a_{ri} \in \text{ (DOMAIN }(A_r)$$

Proxy context

The proxy context is a mediator between two or more contexts. The semantics of a proxy context are similar to the contexts that are sharing data. It is composed of proxy class hierarchies. Proxy classes have the following characteristics:

1. The semantic similarity of two classes, in two different contexts, can be realized in a proxy context.
2. Proxy classes are media by which objects are transferred from one database to the other.
3. A proxy class must be semantically similar to two or more classes.
4. An object that belongs to a class, is a member of a proxy class if both the class and the proxy class are semantically similar.
5. Membership of an object to a proxy class is determined based on semantics, not geometry nor attribute.
6. Instance of a proxy class could be complex objects that are aggregated from other objects scattered in several databases.
7. Properties of a proxy class are not by default inheritable to its member objects. Instead an inheritance function is applied in order to select inheritable objects.

A careful analysis of point 5, reveals that a complex object could be formed from other objects with different geometric primitives. For instance, a class watershed might consist of point soil, linear river information, and areal land cover information. For this reason the geometric component of the Data Model for Multi-valued Vector Map, DMMVM for object representation is adopted in proxy contexts. DMMVM allows the geometric primitives arcs and nodes to have many-to-many links with the terrain objects so that objects from one or more layers can be represented in a single structure (Hoop *et al.* 1993). This will provide more flexible handling of spatial objects in the Proxy Context especially when objects with similar semantics and different geometric descriptors are retrieved from several databases for further aggregation or generalization in the requester's database.

In the following formalism we intend to explain the idea of mapping an instance which belongs to a class in a context, to a semantically proxy class, in a proxy context.

$$\text{Let } CON_i \text{ and } CON_j \text{ be two distinct contexts}$$

$$\text{and } LIST\ (C_k) = \{A_1,\ldots,A_r,\ldots,A_m\}$$

$$\text{and } C_k \text{ is a class in } CON_i$$

$$\text{then } VLIST\ (O_n) = \{a_{1i},\ldots,a_{ri},\ldots,a_{mi}\} \text{ in } CON_i$$

$$\text{Let } LIST\ (C_p) = \{A_1,\ldots,A_s,\ldots,A_q\}$$

$$\text{and } C_p \text{ is a class in } CON_j$$

If an object O_z passes a test formulated in a decision function for a class C_p then it will be a member of that class and that will be expressed by the membership function:

$$M[O_z,C_p] = 1 \text{ if } O_z \text{ is a member of } C_p$$

$$= 0 \text{ otherwise}$$

then VLIST $(O_z) = \{a_{1z},\ldots,a_{sz},\ldots,a_{qz}\}$ in CON_j

We will use the operator (to indicate semantic similarity

Let PCON be a proxy context that is created such that

$$PCON \subseteq (CON_i \text{ and } PCON \subseteq (CONj$$

Let PC_f be a proxy class in PCON

and

$$LIST\ (PC_f) = \{A_1,\ldots,A_t,\ldots,A_y\} \text{ in PCON}$$

and

PC_f is semantically similar to C_k and C_p

then

$$PC_f\ (C_k \text{ and } PC_f\ (C_p \text{ in PCON}$$

An object O_z passes a test formulated in a decision function for a class PC_f then it will be a member of that class.

$$M[O_z,PC_f] = 1 \text{ if } O_z\ (PC_f$$
$$\text{then } O_z \in\ (PC_f$$

$$= 0 \text{ otherwise}$$

then O_z $(PC_f$

The function is only based on the semantics. The attribute structure of the proxy class is not inherited directly from the proxy class it belongs to. Instead, a set of functions is applied that examines the context of the original object requester and determines which attribute to inherit. The values of the attributes are either mapped from the original object or either given default or null value. A detailed discussion on proxy contexts and their development can be found in Bishr (1997).

7 Conclusions and future directions

Six levels of interoperability are presented in this paper. It is concluded that the GI community is still far from achieving interoperability at the data model and the application semantics levels. A structured analysis of the levels of abstractions revealed that interoperability at these two levels can be achieved by capturing semantics from the mental model in the discipline perception world and using them as tools to resolve the schematic heterogeneity. In other words semantics are used to resolve schematic heterogeneity.

A mechanism for capturing and handling semantics was introduced in this paper. Semantics were introduced as sets of rules and constraints attached to object class definitions. The concept of proxy context was briefly introduced. The proxy context acts as a mediator between two heterogeneous data models.

This research runs under the ongoing project between OEEPE and ITC, of the Netherlands, entitled 'Open System Perspective to Federate Heterogeneous Spatial Databases'. Currently the research team is developing a prototype that implements the developed concepts. The prototype forms the basis for designing a case tool that allows users to access geographical information that resides in remote databases.

References

Alaam, M., 1994, Management perspective of an infrastructure for GIS interoperability — the data-x project. In *Proceedings of ISPRS Commission II Symposium on System for Data Processing Analysis and Presentation, Ottawa, Canada*, edited by M. Alaam and G. Plunkett (Ottawa: ISPRS), **30,** pp. 347–351.

Avelino, J. G., and Douglas, D. D., 1993, *The Engineering of Knowledge-Based Systems, Theory and Practice* (Englewood Cliffs, NJ, Prentice Hall).

Batini, C., Lenzerini, M., and Navathe, S. B., 1986, A comparative analysis of methodologies for database schema integration. *ACM Computing Surveys*, **18,** 323–364.

Bererra, R., 1993, Architecture for a federated GIS. Research Report, Intergraph Corporation, Mapping Science Division.

Bishr, Y., 1997, Semantic aspects of interoperable GIS. Ph.D. thesis (Enschede, NL: ITC).

Bishr, Y., Molenaar, M., and Radwan, M., 1996, Spatial heterogeneity of federated GIS in a client/server architecture. In *Proceedings of the GIS'96 Conference* (Boardwalk Drive, Colorado: GIS World, Inc.), CD ROM.

Boar, H. B., 1993, *Implementing Client/Server Computing, A Strategic Perspective* (Maidenhead, UKL McGraw-Hill).

Brackett, H. M., 1994, *Data Sharing Using a Common Data Architecture* (Chichester: John Wiley).

Brenner, J., 1993, *Distributed Application Services, Open Framework, Systems Architecture* (Englewood Cliffs, NJ, Prentice Hall).

Brodie, M. L., 1992, The promise of distributed computing and the challenge of legacy Information systems. In *Proceedings of the IFIP WG2´6 Database Semantics Conference on Interoperable Database Systems* (DS-5), *Lorne, Victoria, Australia* (Amsterdam: North Holland), pp. 1–31.

Buehler, K., 1996, *The Open Geodata Interoperability Specification Part II: Abstract Specification, Draft version.* (Wayland, USA: Open GIS Consortium).

Carnap R., 1964, *Meaning and Necessity, a Study in Semantics and Model Logic* (Chicago: The University Press).

Coad, P., and Yourdon, E., 1990, *Object-oriented Analysis*, 2nd edition (Englewood Cliffs, NJ, Prentice Hall).

Date, C. J., 1995, *An Introduction to Database System (The Systems Programming)*, 6th edition, (Reading, MA: Addison-Wesley).

Doyle, W., and Kerschberg, L., 1991, Data/knowledge packets as a means of supporting semantic heterogeneity in multidatabase systems. *SIGMOD Record*, **20,** 69–73.

Geiger, K., 1995, *Inside ODBC* (Redmond: Microsoft Press).

Getta, J., 1992, Translation of extended entity-relationship database model into object oriented database model. In *Proceedings of the IFIP WG2´6 Database Semantics Conference on Interoperable Database Systems* (DS-5), *Lorne, Victoria, Australia* (Amsterdam: North Holland), pp. 87–100.

Gorman, M., 1994, *Enterprise Database in a Client/server Environment* (New York: John Wiley & Sons).

Hoop, S. de, Oosterom, P., and Molenaar, M., 1993, Topological querying of multiple map layers. In *Proceedings of the Conference On Spatial Information Theory, COSIT '93* edited by A. Frank, Lecture Notes in Computer Science (Berlin: Springer), pp. 139–157.

Kaul, M., Drosten, K., and Neuhold, E. J., 1990, ViewSystems: integrating heterogeneous information bases by object oriented views. In *Proceeding of the Sixth Data Engineering Conference,* ICDE (Los Almitos: IEEE Computer Society), pp. 2–10.

Kim, W., 1995, *Modern Database Systems, The Object Model, Interoperability, and Beyond*, (Reading, MA: Addison-Wesley).

Krol, L., 1994, *The Whole Internet, Users Guide & Catalogue*, 2nd edition (Sebastopol, CA: O'Reilly & Associates).

Kufoniyi, O., 1995, Spatial Coincidence Modeling, Automated Database Updating and Data Consistency in Vector GIS. Ph.D. thesis, ITC Enschede Netherlands.

Laurini, R., 1994, Sharing geographic information in distributed databases. In *Proceedings of Conference on Urban and Regional Information System Association* (Washington, DC: Urban and Regional Information Systems Association), pp. 441–454.

Leeuwen, J., 1990, *Formal Models and Semantics. Handbook of Theoretical Computer Science,* XIV (Cambridge, USA: MIT Press).

Molenaar, M., 1993, Object hierarchies and uncertainty in GIS or why is standardization so difficult? *Journal of Geoinformation Systems,* **6,** 22–28.

Molenaar, M., 1994, A syntax for the representation of fuzzy spatial objects. In *Proceedings of Conference on Spatial data Modeling and Query Languages for 2D and 3D Applications,* edited by M. Molenaar and S. de Hoop (Delft, NL: Geodesy-New Series), pp. 155–169.

Molenaar, M., 1995, An introduction into the theory of topologic and hierarchical object modeling. In *Geo-Information Systems. Lecture Notes.* (Department of Land-Surveying & Remote Sensing, Wageningen Agricultural University, The Netherlands).

Molenaar, M., and Richardson, D. E., 1994, Object hierarchies for linking aggregation levels in GIS. In *Proceedings of ISPRS Comm.IV, Archives of Photogrammetry and Remote Sensing* (Athens, USA: ISPRS), **30,** pp. 610–617.

Otoo, J. E., and Mamhikoff, A., 1994, Delta-X federated spatial information management system. In *Proceedings of ISPRS Commission II Symposium on System for Data Processing Analysis and Presentation,* Ottawa, Canada, edited by Mosaad Alaam, and Gordon Plunkett (Ottawa: ISPRS), **30,** pp. 501–514.

Tanenbaum, A. S., 1996, *Computer Networks,* 3rd edition (Englewood Cliffs, NJ, Prentice Hall).

Tari, Z., 1992, Interoperability between database models. In *Proceedings of the IFIP WG26 Database Semantics Conference on Interoperable Database Systems* (DS-5), *Lorne, Victoria, Australia* (Amsterdam: North Holland), pp. 101–118.

ter Bekke, J. H., 1991, Semantic Data Modeling in Relational Environments. Ph.D. thesis, University of Delft.

Werner, K., 1995, *Semantics of Geographic Information.* GeoInfo Vol. 7. (Vienna: Department of Geoinformation Technical University Vienna).

Yan, L., and Link, T., 1992, Translating relational schema with constraints in to OODB schema. In *Proceedings of the IFIP WG2·6 Database Semantic Conference on Interoperable Database Systems* (DS-5), *Lorne, Victoria, Australia,* (Amsterdam: North Holland), pp. 69–86.

Overcoming the Semantic and Other Barriers to GIS Interoperability: Seven Years On

Yaser Bishr

Background

In 1997 Sir Tim Berners Lee coined the term *"Semantic Web"*. The Semantic Web is defined as "… not a separate Web but an extension of the current one, in which information is given well-defined meaning, better enabling computers and people to work in cooperation." (Berners-Lee et al., 2001, 35).

"Overcoming the Semantic and Other Barriers to GIS Interoperability" was written in 1998 and included work that was mainly conducted between 1994 and end of 1997. In early 1994 the term "Semantic Web" simply did not exist, let alone research on geospatial semantics. Currently, the Semantic Web is becoming a reality. Among other activities, the W3C (World Wide Web Consortium) published the Web Ontology Language (OWL) specifications (Miller and Hendler, 2005) and the Semantic Web Query Language (SPARQL) (Prud'hommeaux and Seaborne, 2005). The geospatial research community is currently producing a substantial body of work on the subject (Cohn and Hazariki, 2001; Bateman and Farrar, 2004; Niles and Pease, 2001; Casati and Varzi, 1994; Gärdenfors, 2000). With the exception of a few researchers (Grenon and Smith, 2004; Miller and Hendler, 2005; Smith and Varzi, 1999), the geospatial semantics research community at large has yet to define a framework for the Geospatial Semantic Web. The geospatial community has long recognized the need for a standard to enable the exchange of geospatial content. The result was a set of ISO specifications known as the 19100 series (Kresse and Fadaie, 2004). Extending this same principle to the domain of geospatial semantics, I argue that there is an immediate need for a Geospatial Ontology Framework (GOF) that will be the catalyst for advancing the Geospatial Semantic Web.

Ontology is the discipline that seeks to establish the types and structures of objects, properties, events, processes, and relations in all domains of reality. In knowledge representation in general, and the Semantic Web in particular, ontology refers to formal specifications of the systems of concepts employed by different groups of users. The W3C has developed the OWL to represent knowledge on the Web (Miller and Hendler, 2005).

The goal of a GOF is to set out consistent and well-specified general geospatial ontology, free from contradiction, and from which follows a set of generic properties that necessarily hold over the entities covered by specific specialized domains. This general ontology will support problem solving and inference in a wide range of applications. Given the increasing need for accounts of space and a similarly growing awareness of the potential application of ontological methods, it is worth mentioning here that the development of a GOF is similar in several respects to some other current ongoing efforts or proposals.

It is important to note that a GOF assumes a crisp world, no vague semantics, and no fuzzy axioms. We do not have any technical or scientific reason to opt out of a non-crisp world model. Our only rational is for practical reasons, because current Semantic Web technology, more specifically OWL, has Description Logic as its model theoretical semantics.

Developing ontologies is not similar to developing DB schema

There are now many proposals for describing space, spatial relationships, and relations among entities and their locations that have been developed within broad 'ontological' frameworks. We have investigated the expressiveness of a number of publicly available ontologies (Islam et al., 2005). Similar to database schema design, those ontologies do not go beyond simple classifications and property associations. It is important to note here that developing ontologies must be executed with full knowledge of the kinds of inferences that applications may require.

High-level description of GOF

The Geospatial Ontology Framework is a multi-layered ontology as shown in Figure 17.5. The underlying theories that form the core of those layers are defined in the subsequent following sections. The layers of GOF include:

1. *Geospatial upper-level ontology:* describes very general concepts such as space, time, mereotopology, and frame of reference, which are independent of a particular problem or domain.
2. *Core geospatial knowledge:* describes essential inferences and relationships derived from the upper-level ontology. This layer of ontology will make inferred knowledge explicit and hence enable the construction of richer domain ontologies.
3. *Domain ontologies:* describe concepts depending on both on a particular domain and task, which are often specializations of *both* the related ontologies.

Identity Criteria

Identity criteria (IC) determine sufficient conditions for determining the identity of concepts defined in an ontology (Renz and Nebel, 1999; Renz, 1999; Bennett et al., 2002; Gangemi et al., 2002; Masolo et al., 2003). We believe this to be the most

Domain Ontology	Domain Ontology	(Road crosses River) and (Road is above River) then (Road is Bridge). (Road Below River) then (Road is Tunnel).
Core Geospatial Ontology		X Adjacent to Y ≡ Y Adjacent to X. X Above Y → Y Below X. X Entered Y → (X Inside Y) and (Y = Polygon) and (Y Contains X)
Upper level Geospatial Ontology		RCC8, Geometry, Location

FIGURE 17.5 Layers of GOF.

important criteria to create classification hierarchies in an ontology. From an ontological perspective, IC is used to:

- *classify* an entity as an instance of a class C
- *individuate* an entity as a countably distinct instance of C
- *identify* two entities at a given time (synchronic identity)
- *re-identify* an instance of C across time (persistence, or *diachronic identity*)

One important issue to note here is the relationship between identity criteria and location. Current GIS and even modern spatial ontologies adopt the premise that an object must have some location. That location is, in general, arbitrary, but where an object can be is constrained by its physical constitution.

The above thesis flies against the realities of many geospatial applications. For example, image interpretation experts determine the identity of objects by analyzing their relative location. Thus, looking at a tilted circular building in a high-resolution satellite image of northern Italy would lead to the interpretation that it could only be the Leaning Tower of Pisa. On the other hand, looking at a tilted building in Paris would mean that it could not be the Pisa Tower.

Following the above example, it then becomes important to always ask which spatial relationships can be used to define distinct ontological entities. Physical entities, for example, are typically distinguished from abstract entities precisely by virtue of their necessary location in time and space. In general, we want to characterize precisely in what ways physical objects (or events) can be said to be located at particular locations in space. This raises questions about how the objects concerned are to be identified and how the locations are to be identified.

Spatial Reference System

There are two distinct schools in representing location; absolutist and relativist position. Some researchers argue that for the purposes of linguistic semantics, it is the relativist view that is compelling; another group favors the absolutist position ignoring the argument that there is no such thing as an absolute reference system. This is because there is no single point from which the entire environment can be considered. Therefore, GOF adopts the relativist view where location must always have a frame of reference, be it egocentric or exocentric.

Mereotopology

Mereology is the formal theory of part-relations: of the relations of part to whole and the relations of part to part within a whole. Mereotopology is the combined logical account of mereology and topology. Mereotopology is the theory of boundaries, contact, and separation, built upon a mereological foundation. In the last few years, several researchers have shown that mereotopology is more than a trivial formal variant of point-set topology. The central aims of point-set and set-theoretic topology are to investigate and prove results concerning the properties of entire classes of spaces. On the other hand, mereotopology aims are to find perspicuous ways to represent and reason about the topological properties and relations of spatial entities that exist in space.

Boundaries

A GOF distinguishes between *bona-fide* and *fiat* boundaries. *Bona-fide* boundaries are boundaries *in the things themselves*. Bona-fide boundaries exist independently of human cognitive acts. They are a matter of qualitative differentiations or discontinuities of the underlying reality. Examples are surfaces of extended objects like cars, walls, and parking lots. *Fiat* boundaries exist only by virtue of different sorts of demarcation effected cognitively by human beings. Such boundaries may lie skew to boundaries of a bona-fide sort, as in the case of the boundaries of a parking spot in the center of a parking lot. They may also, however, as in the case of a parking spot at the outer wall of the parking lot, involve a combination of fiat and bona-fide boundaries. The classification of boundaries generalizes to a classification of objects. Bona-fide objects have a single topologically closed bona-fide boundary. Fiat objects have fiat boundary parts.

Time

Events and relations are naturally anchored in time. For this reason, temporally grounded events and relations are the very foundation from which we reason about how the world changes. Moreover, because feature properties and their relationships change over time, a database of assertions about features will be incomplete or incorrect if it does not capture how these properties are temporally updated. An ontology that has temporal dimension must account for: time stamping of events, ordering events with respect to one another, reasoning with contextually underspecified temporal expressions (e.g., temporal functions such as last week and two weeks before), and reasoning about the persistence of events in space and time.

Shape and Size

In a purely qualitative ontology, it is difficult to describe shape. In such an environment, very limited statements can be made about the shape of a region. Possible statements include whether the feature has holes, is hollow, and whether it is one piece or not. One approach that has not been explored by the research community to qualitatively describe shapes is to use analogical reasoning and similarities. For example, one can say that the shape of this feature is analogous to a crown, and then one can create assertions about crown-shaped objects.

Conclusions

In this commentary I have tried to establish the need for a general framework for geospatial ontologies. I have also briefly outlined the fundamental theoretical foundation for that Framework. No doubt more coordinated work is urgently needed by all stakeholders: users, software vendors, data producers, and the research community. I postulate that the lack of GOF will surely slow our progress and will limit our ability to innovate in the research and development of Geospatial Semantic Web technologies. The GOF must be developed by knowledgeable engineers and researchers who are familiar with theories, tools, and methods to exploit (e.g., inference services) ontologies. It is important to stress that constructing ontology is more than creating a database schema or a classification hierarchy.

Acknowledgement

The section's theoretical foundation of GOF was partly developed under Contract: W9132V-04-0020, from the National Geospatial Intelligence Agency (NGA). The author would like to thank the NGA for its support. The results of this work, however, do not reflect the adopted views of the NGA.

References

Bateman, J. and Farrar, S., 2004, Spatial ontology baseline. SFB/TR8 internal report I1-[OntoSpace]: D2, Collaborative Research Center for Spatial Cognition, University of Bremen, Germany.

Bennett, B., Wolter, F., and Zakharyaschev, M., 2002, Multi-dimensional modal logic as a framework for spatio-temporal reasoning, *Applied Intelligence*, 17(3), pp. 239–251.

Berners-Lee, T., Hendler, J., and Lassila, O., 2001, The semantic web, *Scientific American,* 284(5), pp. 34–43.

Casati, R. and Varzi, A.C., 1994, *Holes and other superficialities* (Cambridge: MIT Press).

Cohn, A.G. and Hazariki, S.M., 2001, Qualitative spatial representation and reasoning: An overview, *Fundamenta Informaticae*, 43, pp. 2–32.

Gangemi, A., Guarino, M., Masolo, C., Oltramari, A., and Schneider, L., 2002. Sweetening ontologies with DOLCE. In *Proceedings of the 13th International Conference on Knowledge Engineering and Knowledge Management (EKAW02)*, Lecture Notes in Computer Science 2473, (Berlin: Springer-Verlag), 166–178.

Gärdenfors, P., 2000, *Conceptual spaces: the geometry of thought*, (Cambridge: MIT Press).

Grenon, P. and Smith. B., 2004, SNAP and SPAN: Prolegomenon to geodynamic ontology, *Spatial Cognition and Computation*, 4, pp. 69–104.

Islam, A.S., Bermudez, L., Beran, B., Fellah, S., and Piaseki, M., 2005, List of OWL ontologies based on Norms, http://loki.cae.drexel.edu/~wbs/ontology/list.htm, accessed 3 November 2005.

Kresse, W. and Fadaie, K., 2004, *ISO Standards for Geographic Information*, (Berlin: Springer-Verlag), Berlin.

Masolo, C., Borgo, S., Gangemi, A., Guarino, N., and Oltramari, A., 2003, Ontologies library (final), WonderWeb Deliverable D18, ISTC-CNR, Padova, Italy, December 2003.

Miller, E. and Hendler, J., 2005, Web Ontology Language (OWL), http://www.w3.org/2004/OWL/, accessed 3 November 2005.

Niles, I. and Pease, A., 2001, Origins of the standard upper merged ontology: A proposal for the IEEE standard upper ontology. In *Working Notes of the IJCAI-2001 Workshop on the IEEE Standard Upper Ontology*, Seattle, WA, pp. 37–42.

Prud'hommeaux, E., and Seaborne, A., 2005, SPARQL Query language for RDF; W3C working draft, http://www.w3.org/TR/rdf-sparql-query/, accessed 3 November 2005.

Renz, J., 1999. Maximal tractable fragments of the region connection calculus: A complete analysis. In *Proceedings of the 16th International Joint Conference on Artificial Intelligence (IJCAI-99)*, T. Dean, T., (Ed.), (San Francisco: Morgan Kaufman), pp. 448–454.

Renz, J. and Nebel, B., 1999. On the complexity of qualitative spatial reasoning: a maximal tractable fragment of the region connection calculus, *Artificial Intelligence*, 108(1–2), pp. 69–123.

Smith, B. and Grenon, P., 2004, The cornucopia of formal-ontological relations, *Dialectica*, 58, pp. 279–296.

Smith, B. and Varzi, A.C., 1999, The niche, *Nous*, 33(2), pp. 214–238.

International Journal of Geographical Information Science, 1999, Vol. 13, No. 4, 355–374.

18 Interactive Maps for Visual Data Exploration*

Gennady L. Andrienko and Natalia V. Andrienko

Abstract

Descartes (formerly called IRIS) is a software system designed to support visual exploration of *spatially referenced data*, e.g. demographic, economical, or cultural information about geographical objects or locations such as countries, districts, or cities. Descartes offers two integrated services: **automated presentation** of data on maps, and facilities to **interactively manipulate** these maps. Automated mapping is enabled by incorporating generic knowledge on map design into the system. Descartes selects suitable presentation methods according to characteristics of the variables to be analysed and relationships among those variables — if more than one were selected simultaneously. The cartographic knowledge of Descartes allows non-cartographers to receive proper presentations of their data, and the automation of map construction helps the users to save valuable time that can better be used for data analysis and problem-solving.

Exploratory data analysis requires highly interactive, dynamic data displays. We strive to develop various interactive techniques for map manipulation that could enhance the expressiveness of maps and thus promote data exploration. We are convinced that a technique can be made especially productive if it is directed towards a particular presentation method: it can utilise peculiarities of this method and support those analytical operations that best fit to the method.

1 Introduction

The notion of exploratory data analysis developed in statistics has over the past decade spread to cartography, see DiBiase (1990) for one of the first discussions of the potential of exploratory data analysis (EDA) for cartography. Cartographers have recognised the demand in a new software allowing specialists in various disciplines (i.e., not professional map designers) to generate thematic maps and use them as tools facilitating 'visual thinking' about spatially referenced data (Kraak 1998).

Our system Descartes has been designed as such an environment for exploratory analysis of spatially referenced data. In this role it offers the user two complementary instruments: *automated, knowledge-based cartographic visualisation* and *interactive*

* Web supplement is located at: //www.ais.fraunhofer.de/and/IcaVisApplet/.

manipulation of the generated maps and supplementary data displays. The user interface of the system is implemented in Java 1.0.2, and therefore expected to run within any Java-enabling WWW browser. The interface part communicates with the server, which performs data management and visualisation design. The server runs under UNIX. Descartes will not work on a computer connected to the Internet through a firewall that disables communication between the client and the server.

The present paper is mostly devoted to the interactive techniques enabling the use of maps as aids in 'visual thinking'. The next section contains a survey of the literature devoted to interactive, dynamically changing data displays. The principles of map design in Descartes are briefly described in §3. A more detailed consideration can be found in (Andrienko and Andrienko 1997, 1998a).

Our research is map-centred: we strive to advance traditional data presentation principles from thematic cartography by adding interactivity and dynamics. From here on, we use the words 'dynamics' and 'dynamic' in the sense that a display changes in real time in response to a user's actions rather than to denote animated presentations of time-series data. The primary goal of this paper is to introduce the variety of techniques for map manipulation that we have developed thus far. As there is no solid theory yet about how to create and validate such techniques, in the course of our work we roughly adhered to the following strategy:

1. Address each cartographic presentation method individually.
2. Try to compensate for potential weaknesses of the target presentation method.
3. Try to enhance the advantages of the target presentation method with regard to the analysis task it is best suited for.
4. Exploit and affect specific visual properties of the target method.

Following the brief presentation of the interface for map viewing and manipulation in §4, the interactive manipulation techniques of Descartes are described in §5. Section 6 contains some concluding remarks concerning the introduced techniques.

2 Interactive manipulation of graphic displays in exploratory data analysis (literature survey)

A graphical presentation intended for use in exploratory data analysis should not be static. Recognising this, Bertin (1967) even proposes some methods of building transformable 'paper' presentations. It is therefore quite natural that researchers in statistical graphics, where the concept of the exploratory data analysis emerged (Tukey 1977), rather early started to exploit the potential of computers for creating interactive, transformable, and later dynamic (i.e., with real time motion) data displays. An overview of interactive techniques proposed in statistics can be seen in Cleveland and McGill (1988).

Buja *et al.* (1996) suggest a taxonomy of tools for manipulating graphical displays of data. They classify the tools into three fundamental categories: *focussing* individual views, *linking* multiple views, and *arranging* many views. These categories approximately correspond to three principal tasks the authors distinguish in data exploration: finding Gestalt, posing queries, and making comparisons.

Focussing techniques include selecting subsets and variables (projections) for viewing and various manipulations of the layout of information on the screen: choosing an aspect ratio, zooming and panning, 3-D rotations, etc. Tools of this kind are present in all contemporary systems for graphical data analysis.

As stated in Buja *et al.* (1991), 'a consequence of focussing is that each view will only convey partial information about the data'. This can be compensated for by displaying multiple views that 'need to be linked so that the information contained in individual views can be integrated into a coherent image of the data as a whole'. The method of linking depends on whether the views are displayed in sequence over time or in parallel, simultaneously. In the first case, linking is provided by smooth animation. The most popular method for linking parallel views is identical marking of corresponding parts of multiple displays, e.g., with the same colour or some other form of highlighting. Usually highlighting is applied to objects interactively selected by the user in one of the displays. This method is a generalisation of the 'scatterplot brushing' technique first implemented by Newton (1978) and later elaborated in several directions: colour encoding of distance from the moveable focus point (McDonald 1982), simultaneous painting of several subsets using multiple colours and mixing of colours (Carr *et al.* 1986), various ways of selection and combination of selection operations (a synopsis is given in Wills 1996), and application to various types of data displays such as time-series plots, tables, raster images (Buja *et al* 1991), bar charts (Hurley 1993, Wills 1995, Dawkes *et al.* 1996, Hummel 1996, Unwin *et al.* 1996), trees (Buja *et al.* 1996), mosaic plots (Unwin *et al.* 1996), etc. Our particular interest is the application of brushing to spatially referenced data. The relevant works will be considered separately.

The purpose of arrangement of multiple views is to facilitate comparisons. Usually a system for graphical analysis offers some built-in method(s) of arrangement such as the scatterplot matrix rather than any interactive tools for re-arrangement besides ordinary window operations.

Developers of interactive manipulation techniques in statistical graphics have paid significant attention to spatially referenced data. Already in 1982 McDonald had the idea of connecting a map to a scatterplot by painting regions in the map in the same colours as the corresponding points in the scatterplot. Buja *et al.* (1991) consider data referring to geographical locations. They propose to represent locations by a scatterplot of the latitudes and longitudes and link it with other graphics representing thematic information. The same approach is described in MacDougall (1992) for point- and raster-related data. The author also proposes a prototype interface with a map display of geographical objects specified as polygons. Instead of incorporating map displays into a system for interactive graphical analysis, Symanzik *et al.* (1996) achieve linking between maps and statistical graphics by coupling two independent programs: a GIS (ArcView™) displaying maps and a dynamic graphics system XGobi with a variety of statistical graphics (see also Cook *et al.* 1997). An analyst can brush points in either ArcView or XGobi, and the corresponding points are highlighted using the same colour or glyph in another application.

In all the brushing-based approaches, the role of the map is merely to carry the information about spatial location. All thematic information is usually represented and analysed by means of non-cartographic displays. It is reasonable to expect that involvement of the potential of thematic cartography would enhance the analytical capabilities of the tools proposed for exploratory analysis of spatially referenced data.

Cartographers have long adhered to the view of the map as primarily a medium of communication (MacEachren 1995). Recently, influenced by ideas of scientific visualisation and exploratory data analysis, they started to pay more attention to the role of the map as a tool to support visual thinking and decision making (MacEachren 1994b, MacEachren and Kraak 1997, Kraak 1998). To play this role effectively, the map needs two principal additions: interaction and dynamics. Current efforts of cartographers on developing map interactivity and dynamics are reflected in a special issue of the journal *Computers and Geosciences* entitled 'Exploratory Cartographic Visualisation' (1997, 23).

It should be noted that the interactive manipulation techniques considered by cartographers are mostly based on the linking and brushing principles borrowed from the statistical graphics (Monmonier 1989, Dykes 1997). Dykes also implements some additional techniques such as 'dynamic re-expression', that is, replacement of the presentation method used in the map with another method. For example, a choropleth map may be substituted for a map with graduated symbols that presents the same data. Still, very few steps have been made up to now in another direction: advancement of presentation methods developed in thematic cartography, i.e., *making traditional methods interactive and dynamic*. We could find only two relevant works. Egbert and Slocum (1992) on the basis of an earlier work of Yamahira *et al.* (1985) propose an exploratory tool specifically designed for classed choropleth maps. This tool allows the user to change the classification interactively, and the map is repainted to reflect the changes. So, the tool affects the parameters of the presentation method. It is interesting that another manipulation technique for choropleth maps is proposed by researchers in statistics (Unwin and Hofmann 1998). The technique allows transformations of the function mapping from numeric values to shades. Initially the function is linear. Making it non-linear and varying parameters, the user can visually emphasise differences within selected subintervals of the value range.

Development of traditional cartographic presentation methods towards interactivity and dynamics is of primary concern in our system Descartes (though we do not deny the usefulness of linking maps to other types of displays). In general, Descartes is designed as an environment for exploratory cartographic analysis. This implies more than just interactivity of maps and multiple dynamically linked views on data. Kraak (1998) notes that, unlike the traditional cartography where a professional map designer creates a map to communicate spatial information to an audience, exploration usually involves non-cartographers creating maps for their private use. In doing this, the analysts should be able to rely on cartographic expertise provided by the software or some other means. Descartes answers this requirement:

it is a knowledge-based system that incorporates cartographic expertise and on this basis performs automated map design.

3 Visualisation design in Descartes

Descartes greatly facilitates the process of map generation: a user only selects one or more data variables (table columns) to visualise, and the system immediately responds with a list of map presentations that properly correspond to characteristics of the variable(s) and, in the case of several variables, to relationships among them. After selecting a map in this list, a window is created displaying it. The user can now view and study the map and compare it with other ones.

To depict data on maps, Descartes applies presentation methods as developed in cartography. Each presentation method is based on the use of one or more of the *visual variables* introduced by Bertin (1967). These variables, according to their perceptual properties, have different *levels of organisation*: associative, selective, ordered, and quantitative. Presentation techniques for given data should be selected so that each data variable is represented by a visual variable with the corresponding level of organisation.

The cartographic presentation methods used in Descartes can be divided into those based on painting contours (applied only for area objects) and those based on the use of signs. Signs may be either simple or structured. Painting methods exploit the visual variables *colour* and *value* (degree of darkness), the former in cases of nominal data, and the latter for ordered and numeric variables. Some clarification is necessary concerning the use of the visual variable *value* for numeric variables.

According to Bertin, *value* is unsuitable for representing quantitative data because it enables only ordered but not quantitative perception. Nevertheless, encoding numeric data by *value* (e.g., in choropleth maps) has become traditional in cartography. Such usage of this visual variable is productive for analysing spatial distributions and revealing spatial trends and patterns. Therefore Descartes adheres to this cartographic tradition but always complements a choropleth map with another presentation of the same data enabling quantitative perception: it is the general strategy of the system to supply the user with multiple presentations of the same data.

The rules for selecting presentation methods depending on data characteristics and relationships are summarised in Table 18.1.

When several data variables are selected for visualisation, and relationships between them are heterogeneous, or some of them are incomparable, the system divides the variables into groups with homogeneous relationships between their elements; the variables belonging to the same group should be comparable. In particular, a group may contain only one element. Descartes will select presentation methods or visual variables for the groups and then try to combine them. Some possible combinations are listed in Table 18.2.

According to Bertin (1967) and Tufte (1983), relationships between two or more data variables can be efficiently examined by comparing several maps, each presenting spatial distribution of one of the variables. For this reason we have added the presentation by multiple maps to the arsenal of combination techniques employed in Descartes. In this presentation each variable is represented by the area painting

TABLE 18.1
Selection of visualization methods for individual data variables and groups of related variables (N stands for number of variables).

N	Type(s) of variable(s)	Relationship (where appropriate)	Visual variables	Presentation methods	Examples
=1	Nominal (including logical)		Colour or shape	Area colouring, coloured signs, shape signs	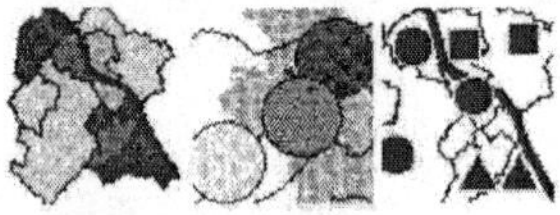
=1	Ordinal		Value	Area shading, shaded signs	
=1	Numeric		Value or size	Area shading, stand-alone bars	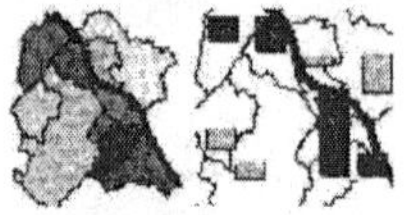
>1	Logical		Colour	Structured signs varying by presence or absence of coloured sectors	
>1	Numeric	Comparable	Size	Parallel bars	
>1	Numeric	Included in a common total	Size	Pies, stand-alone segmented bars, parallel bars	
>1	Numeric	Ordered inclusion (a > b > ...)	Size	Nested squares, parallel bars	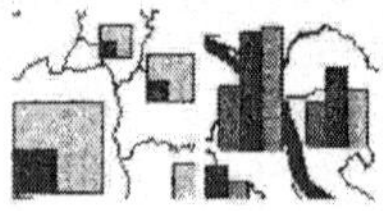

method (choropleth map), which favours perception of the distribution as a single image. Multiple maps are united in one panel. The system supports their simultaneous resizing and zooming so that they always have the same size and show the same territory. This simplifies their joint analysis.

The strategy of Descartes is to apply all presentation methods that are suitable for user-selected data (note that most of the rows in Table 18.1 contain several variants of presentation). All alternative maps are presented to the user. The objective is not only to give the user the freedom to choose the most convenient map. As a rule, such multiple presentations of the same data are *complementary*. Each presentation method suits certain types of analysis tasks better than others. For example,

TABLE 18.2
Selection of combination techniques according to contents of variable groups and relationships among them.

Group descriptions	Combination techniques	Examples
One nominal or ordinal variable + one numeric variable	Coloured or shaded stand-alone bars	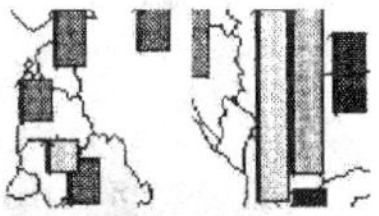
One nominal variable + a group of numeric variables	Area colouring with superimposed diagrams	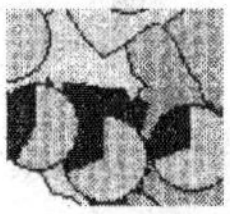
Two groups of comparable numeric variables	Two-sided bar diagrams (2 diverging groups of parallel bars)	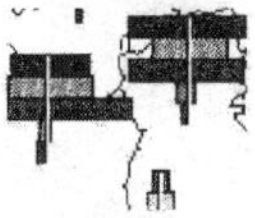
Several groups of numeric variables with inclusion in a common total in each group	Parallel segmented bars if the groups are comparable, radial segmented bars otherwise	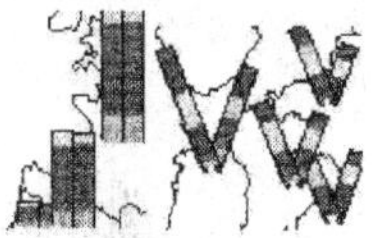
Several incomparable numeric variables	Radial bars	

pie charts are good for estimating proportions but worse for assessing differences between values. In the course of data exploration analysts perform different kinds of tasks, and there hardly exists one 'optimal' presentation method (MacEachren 1994a). To the contrary, the availability of several complementary presentations of the same data can greatly support the variety of activities comprising data analysis, especially exploratory analysis.

4 The interface to view and manipulate maps

Each map built by Descartes appears in a separate window containing the map itself, the legend, tools for zooming and handling geographical layers, and, what is most important, controls affecting visual properties of the data presentation (a map in Descartes may not only contain presentation of attribute data and spatial objects they refer to, but also other geographical features, e.g., rivers or forest areas, forming map layers, or themes, in GIS terminology). In many cases, additional nongeographical presentations of the same data set are provided. Some of them are parts of manipulation devices, like the dot plots in the upper left section of the window in Figure 18.1. For all screenshots given in the paper there are Java applets that show

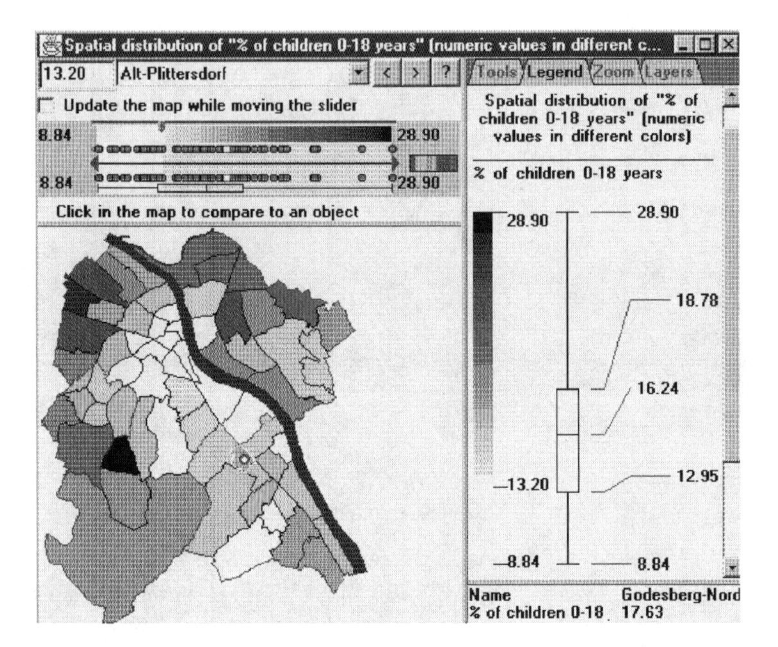

FIGURE 18.1 An example map window with dynamic manipulation tools.

corresponding maps and enable all interactive operations. The applets are available at the URL http://www.ais.fraunhoffer.de/descartes/IcaVisApplet/. The maps have been produced and stored by Descartes. The code supporting map viewing and manipulation is an extract from Descartes. This code does not communicate with the Descartes server and therefore is not disabled by firewalls.

The map is sensitive to mouse movement. When the cursor is positioned on some geographical object for which associated data exist, this object is highlighted in the map, and its name and the corresponding values are shown in a special section of the map window. If the map window contains a dot plot or a scatterplot as an additional non-geographical presentation of the data, the dot representing this object is also highlighted in the graphic. Linking between the map and the additional graphic works also in the opposite direction: when the cursor points at some dot in the graphic, both this dot and the corresponding object in the map are highlighted.

Furthermore, the map may be linked with other currently visible maps. This means that the highlighted object is also highlighted in the other maps, irrespective of their current scales, presented variables, and visualisation methods applied. So, one can observe that we propose our specialised, map-centred interactive manipulation techniques as a complement rather than a substitute to the widely recognised focussing and linking principles.

Manipulation techniques designed for different methods of presenting numeric information exploit similar interactive widgets that are organised in a construct further referred to as **slider unit**, see Figure 18.2.

A slider unit represents the value range of a numeric variable presented in a map. It contains one or more sliders that can be moved using the mouse. Position(s)

FIGURE 18.2 An instance of a slider unit.

of the slider(s) affect visual properties of the map in a way that depends on the visualisation method applied. The dot plots show the distribution of values of the variable relative to its minimum and the maximum values. Each dot corresponds to one geographical object. The box-and-whiskers plot (Tukey 1977) offers a generalised presentation of the value distribution.

The small triangles at both ends of the slider can be used to limit the value interval to be visually examined. Like sliders, these triangles (further called **delimiters**) can be moved using the mouse. The use of delimiters is described in the next section together with the other techniques dynamically affecting the appearance of maps.

5 How to make use of the maps: Exploration by map manipulation

Our approach is to design **specialised** manipulation tools addressing particular presentation methods. In so doing, we take into account (1) what data features and relationships the target method potentially helps to uncover; and (2) what graphical expressive means the method employs. The technique being designed is intended to reinforce visibility of these features and to facilitate their disclosure. It operates with the same graphic means and, possibly, adds redundant visual variables to increase the legibility and expressiveness of the display.

5.1 Dynamic choropleth maps

In this subsection we describe a technique designed for *unclassed choropleth maps* representing numeric data. The 'choropleth map' visualisation method consists in encoding values of some attribute referring to units of territory division by colours or shades. 'Unclassed' means that values of a numeric attribute are converted into degrees of darkness directly, without previous classification: the degrees of darkness are (roughly) proportional to the numbers they represent.

A merit of a choropleth map is that it is capable of producing an integral image of the spatial distribution of data, and thus enables the highest, overall level of map reading (Bertin 1967). Since the whole distribution can be grasped as one image, it is possible to compare two or more distributions presented by choropleth maps. By such comparison one can reveal relationships between several variables: relatedness will manifest itself in similar spatial patterns.

'Dynamic visual comparison' is an interactive tool intended to enhance inherent capabilities of the given presentation method. With this tool some number *N* within the value range of the shown variable is interactively selected, and the map is immediately redrawn using a *diverging*, or *double-ended, colour scheme* (Brewer

1994): values higher than N are encoded by shades of one colour (hue), and those lower than N are shown by shades of another colour. The greater the difference is between some value and N, the darker is the shade used to represent it. The values exactly equal to N are encoded by white. When the user changes N, the map is dynamically repainted.

Descartes offers a number of ways to control the reference value N (see the upper-left section of the window in Figure 18.1):

(*a*) moving the slider;
(*b*) clicking on an object in the map;
(*c*) selecting an object by name in the choice widget. In both cases (**b** and **c**), the value associated with the selected object becomes the reference value for the comparison;
(*d*) automatically locating the object with the previous or with the next value by pressing the buttons "<" and ">";
(*e*) entering an exact number in the edit field.

The sequence of screenshots shown in Figure 18.3 demonstrates how a map dynamically changes as the slider moves from left to right (the reference value N grows). Shades of brown are used for values higher than N, and shades of cyan for those below N.

The map of this example shows the variation of mean household sizes through the districts of Bonn. (The source data were kindly provided by the Statistical Department of the Bonn City Administration). In the first map (A), the reader can observe a certain spatial trend: the values on the outskirts are higher than in the centre. Applying the visual comparison tool, one can investigate the spatial variation more comprehensively.

In the following maps (B)–(D), the cyan areas representing low values are very well distinguished from the brown areas encoding higher values. Map (B) shows that the lowest values are located in the historical centre and in some districts on the south-east. Moving the slider farther to the right, one can observe how the cyan area spreads to the south-east and north-west. At a certain moment the extension in

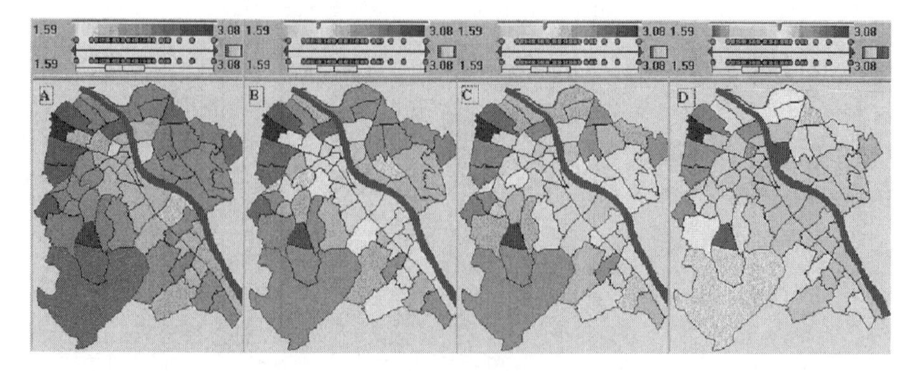

FIGURE 18.3 Several states of a map in the process of 'dynamic visual comparison'. (Figure originally published in colour and in the e-version.)

this direction stops, and the cyan area starts spreading to the west and the east. So, in the process of visual comparison we have found several interesting spatial patterns: centre-periphery, northwest–southeast stretching, and an 'arc' formed by the north and west outskirts. It would be much more difficult to reveal this information only from one static picture like (A). This phenomenon can be explained as follows.

In an unclassed choropleth map different numbers are encoded by (ideally) different shades. This, to some extent, inhibits visual grouping and division into regions since, according to Bertin (1967), *value* is a selective visual variable but not associative. Dynamic comparison adds the variable *hue* to the expressive means used in the map. This variable is not only selective but also associative, and therefore encourages visual grouping of objects: neighbouring objects painted in the same colour tone tend, despite differences in shades, to be associated into a single figure. In this way, dynamic comparison favours revealing spatial patterns. It obviously helps in the task that is considered to be the most important application of the choropleth map presentation method, that is, looking for spatial patterns in value distributions.

A disadvantage of the unclassed choropleth map method is the difficulties humans meet in distinguishing close shades (Dobson 1973). It is especially hard to figure out the right order if compared objects are spatially disjoint and have surroundings with different degrees of darkness. With the dynamic visual comparison technique these difficulties are easy to overcome: to compare values for two objects, it is sufficient to click on one of them and observe in what hue the other is repainted.

Diverging colour schemes are used in cartography for emphasising progressions outward from a critical midpoint of data (Brewer 1994). In exploratory data analysis, however, such critical points are often previously unknown. The 'dynamic comparison' technique can help to empirically reveal them, i.e., to find such reference values that lead to meaningful spatial patterns.

The technique described can be applied as well in a joint study of several choropleth maps presenting different variables. Descartes provides this opportunity whenever a user selects two or more numeric variables to be visualised. One of the presentations the user receives in such cases is a multi-map presentation: several choropleth maps are shown in one window that contains controls for manipulating all the maps. The user can handle each map independently of the others or all maps in parallel.

5.2 Dynamic bars

Another method to depict values of a numeric variable is to encode them by proportional sizes of graphical symbols, e.g. by heights of bars: the greater the value, the higher is the bar. Unlike the choropleth map method, this method does not favour forming integral images. Instead, it is more beneficial for estimating absolute values and differences in values. According to our strategy, the interactive manipulation tool designed for this method should also work in this direction.

It is obvious that the values of a variable for two geographical objects can be more easily compared if the user does not have to mentally subtract the height of one of the bars from the height of the other bar. The system can automatically do this

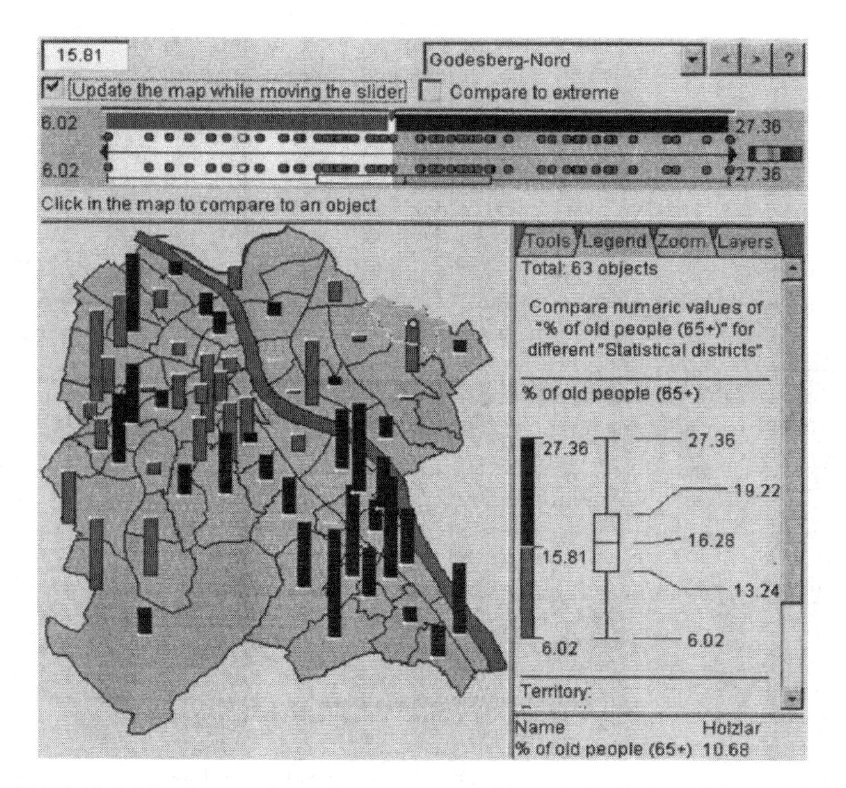

FIGURE 18.4 Visual comparison of percentages of old people in districts of Bonn with that for a selected district (Godesberg-Nord) on a bar map.

for the user: after s/he clicks on an object in a bar map, the system redraws the map so that corresponding bar heights for all objects become proportional to the differences between the associated values and the value for the selected object. The orientations of bars (upward or downward) show the signs of the differences. For better perception, orientation is reinforced by colour, i.e., two different hues are used to represent positive and negative differences.

To control the reference value for visual comparison on a bar map, the same facilities are available as for dynamic choropleth maps (cf. the map windows in Figure 18.1 and Figure 18.4). However, the graphical elements being manipulated are different: colours and shades of area painting in one case and bar heights in the other.

5.3 Dynamic focussing on a value subrange of a numeric variable

To match numeric values with bar heights, as in bar maps, or degrees of darkness, as in choropleth maps, Descartes builds a *linear* encoding function: $f(x) = (v_{max}-v_{min}) \times (x - x_{min})/(x_{max}-x_{min})$, where x is a value to be encoded, x_{max} and x_{min} are the highest and lowest values of the represented data variable in the given sample, and *vmax* and *vmin* are the adopted highest and lowest values of the visual variable (*size* or *value*). If the sample contains both positive and negative values, $x_{min} = 0$, and $x_{max} = max(-s_{min}, s_{max})$, where s_{min} is the minimum (negative) value for the sample, s_{max} is the maximum value.

Sometimes the mapped value set contains, relative to the bulk of it, a few extremely high or extremely low values ('outliers'). In this case, the generated map will not be distinctive or expressive enough with regard to the mainstream values. Look, for example, at the left map in Figure 18.5 (the source data have been taken from *The World Fact Book '95* published by the CIA). Albania has a very high birth rate when compared to all other European countries. The dot plots in the slider unit clearly illustrate the distance from Albania's maximum value of 21.70 to all other values, which are all within the interval from 10.56 to 15.93. As the maximum degree of darkness is assigned to 21.5, the maximum of the remaining values, which is 15.93, is encoded by the degree approximately equal to 48% of the total maximum. As an effect, countries with rather different values (e.g., Switzerland 12.04, France 13.00, Ireland 14.04) are painted in very similar shades. The same problem arises with a bar map: all the bars except that for Albania appear to be of almost the same height.

To handle such data, Descartes allows the user to cut the represented interval, i.e. remove outliers, and subsequently redefine the encoding function. Referring to our example, if Albania is excluded from the presentation, the full scale of darkness will be matched with a significantly shorter interval. The value 15.93 will be encoded by 100% darkness, and differences among the mainstream values will become apparent in the map, as is clearly seen in Figure 18.5 (right).

Outlier removal (and, in general, focussing on any subinterval of the whole value range for studying more in detail) is done interactively by using *delimiter widgets* as introduced in §4. In a slider unit (see upper sections of the map windows in

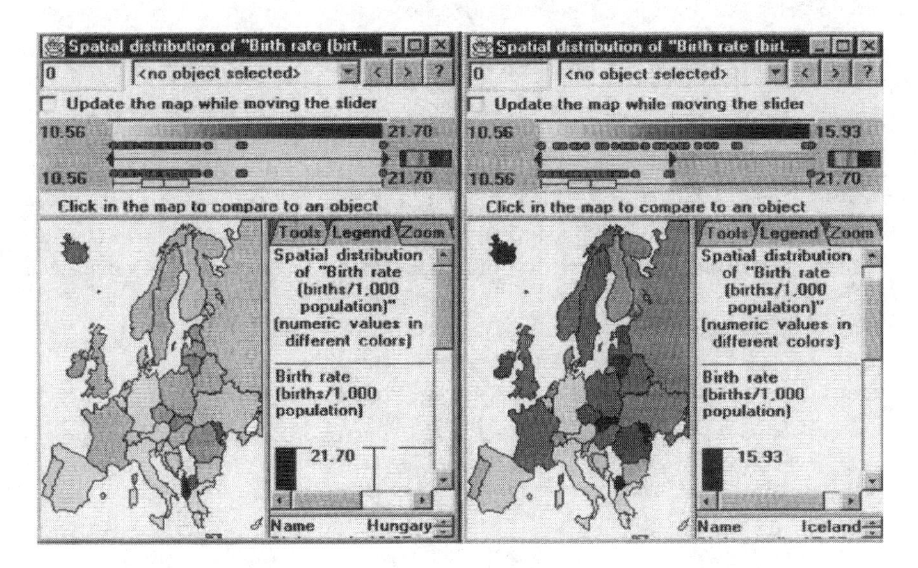

FIGURE 18.5 The effect of dynamic focussing.

Figure 18.5), the line between two dot plots represents the whole value range of the visualised variable. The vertical edges of the delimiters (triangles) mark the boundaries of the currently depicted subinterval. Initially, the delimiters are located at the ends of the line. Moving the delimiters along the line, the user may select a subinterval for further study. The upper dot plot in the slider unit also contains a pair of delimiters (smaller and lighter triangles). They can be used for more exact adjustment of the boundaries of the selected interval. After any movement of a delimiter the encoding function is redefined, and the map redrawn. Geographical objects with values lying outside the focussed interval are marked in the map without painting or drawing signs.

The appearance of the slider unit also changes. The lower dot plot and the box-and-whiskers plot represent, as before, the whole data sample. Light background painting indicates which part of the range is currently presented on the map. The upper dot plot only shows the focussed subinterval stretched to the whole width of the plot area. This zooming makes it easier to distinguish dots representing close values. The dynamic linking between the dot plots and the map is preserved.

The technique of interactive removal of outliers (in application to scatterplots) was proposed by Cleveland (1993). Removing objects standing far apart of the rest allows the scale of the scatterplot to be enlarged and makes the mainstream objects better distinguished.

The outlier problem can also be solved by applying non-linear encoding functions, e.g. logarithmic ones. This expedient is frequently used in statistical graphics (see, for example, Carr *et al.* 1986). Unwin and Hofmann (1998) propose a tool for interactive non-linear transformations of mapping from numbers to shades in a choropleth map. In our opinion, however, such transformations are undesirable for maps, at least those intended for use in exploratory analysis. Unlike scatterplots and other statistical graphics containing axes and labelled ticks, a map itself does not provide any information on how numbers are encoded by shades or sizes. This information can only be retrieved from the legend. At the same time exploratory analysis greatly depends on the immediate impression (MacEachren 1995, pp. 70–71) that comes before the map reader starts learning the legend or applying any knowledge schemata. Besides, in a scatterplot one usually looks for functional dependencies between two variables. Transformation of axes does not impede noticing of such dependencies. In a map one looks for spatial patterns made by neighbouring signs (symbols or painted areas) with similar visual appearance. With a non-linear encoding function, similar appearance may not necessarily reflect closeness of values, and vice versa, close values may be encoded rather differently. So, visual grouping of geographical objects (that usually happens on the pre-attentive stage of image processing by a human) may be inconsistent with actual relationships. Therefore, maps intended for visual exploration should allow intuitive interpretation (see also Tufte 1983, p. 153) that is hardly supported by non-linear encoding functions.

One can observe some similarity between the dynamic focussing technique and the well-known 'Dynamic query' technique (Ahlberg *et al.* 1992). The real similarity, however, mainly refers to the interface rather than function. In 'Dynamic query', the goal is to allow the user to conveniently specify constraints and to see which objects satisfy them. Accordingly, the display is changed by removing the objects not satisfying the query. In our case, the goal is to present values lying within some user selected interval with maximum possible expressiveness. Therefore, the main effect consists in a transformation of the encoding function. Objects with associated values lying outside the focus interval are actually not removed from the view, rather their values are not visually expressed.

5.4 Choropleth maps with dynamic classification

The dynamic classification technique is designed for classed choropleth maps. Unlike the unclassed choropleth maps considered earlier, values of a numeric variable are not directly encoded by proportional degrees of darkness. Instead, the value range of the variable is first divided into two or more intervals, and a specific colour is assigned to each interval. Objects with values fitting in the same interval will thus have identical colour in the map. Due to the classification, which is a kind of generalisation, a part of the initial information concerning differences in values is lost. However, this method is very suitable for grouping and observing spatial distributions of groups of objects.

A choropleth map with a fixed classification, which is typical for 'paper' cartography, inevitably imposes a subjective view on the data. In Descartes, the classification is mobile since the user can interactively move the boundaries of the intervals, easily change the number of intervals, and select various kinds of automatic division such as equal frequency or equal length intervals. All these operations result in immediate repainting of the map according to the new classification. The ease of applying these operations and the immediate feedback greatly encourage users to search and find interesting, geographically meaningful groupings of objects.

Figure 18.6 shows the interface of the dynamic classification technique. The main manipulation widget is the slider unit. Unlike in visual comparison, it may contain several sliders. Positions of the sliders reflect the values of class boundaries.

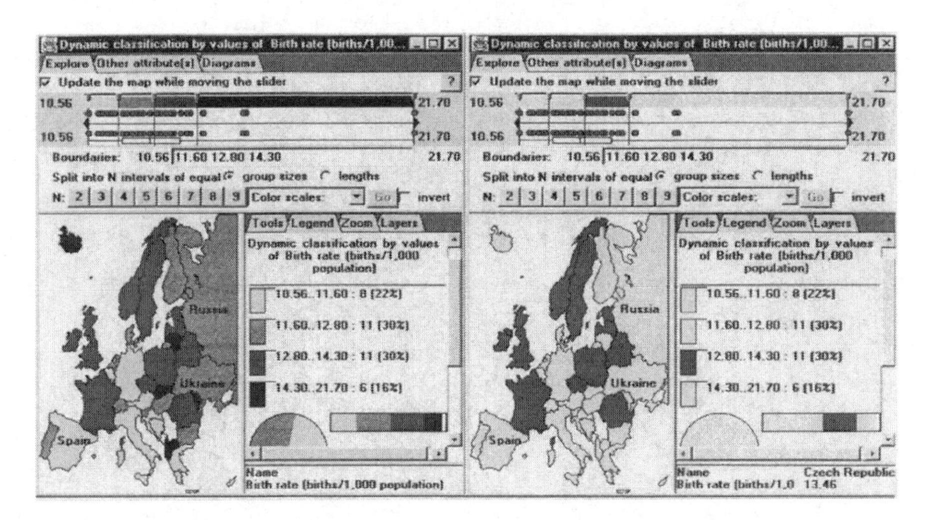

FIGURE 18.6 A classed choropleth map and the tool to manipulate classes.

When the user moves any of the sliders, the corresponding boundary changes. A mouse click between two sliders splits the corresponding interval into two parts, and a new class appears. When a slider closely approaches one of its neighbouring sliders, it is merged with the neighbour, and the number of classes decreases by one.

Besides the slider unit, a number of other widgets are available for interactive classification. They allow the user to manually enter the values of class boundaries, to see results of automatically dividing the interval into 2 to 9 equal length or equal frequency intervals, or to select different variants of colouring.

The dynamic classification tool also allows users to visually isolate a particular class in order to observe its spatial distribution. For this purpose, the user simply clicks on the corresponding colour on the colour band in the slider unit. In response, only the objects belonging to this class will be coloured, whereas the other objects will be shown in neutral grey (see the right map window in Figure 18.6). Another click on the same segment will result in colouring objects from the original class and its two adjacent classes. One more click makes the whole map coloured again. This type of visual isolation also occurs during slider movement: only the two classes affected by the corresponding boundary change are visualised.

In what we have described so far, our dynamic classification tool can be regarded as an upgrade of the work of Egbert and Slocum (1992) by utilising recent progress in graphical hardware, software, and user interface design. The essential novel feature is the opportunity to investigate relationships of the base variable (the one used for classing) with other variables. In Descartes, the user can select one or more variables to study, and the system will then calculate the statistics of value distribution of these variables for each class. Statistical results are presented to the user in the form of averaged 'portraits' of the classes and by 'box-and-whiskers' plots (see Figure 18.7). All statistical results are recalculated when the classes change.

In the example of Figure 18.7, some relationship between birth rates and death rates can be observed across the countries of Europe. The countries are classified according to birth rates. The supplementary 'class statistics' display (Figure 18.7) represents the statistics of distribution of the death rates across the classes. It can be noticed that death rates are lower in the countries either with low or with high birth rates. The dynamic classification tool allows the user to investigate whether this trend preserves when the classification changes.

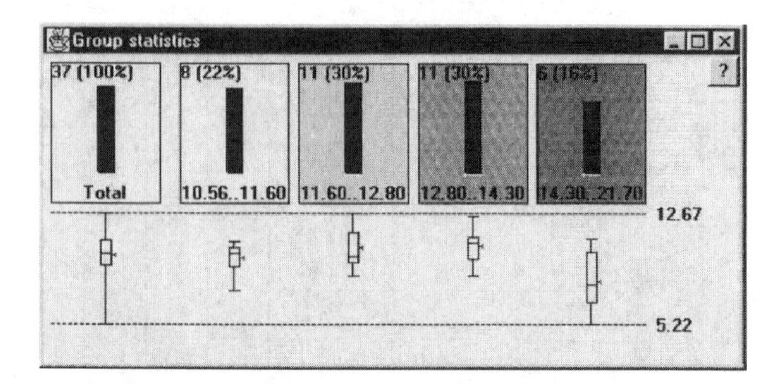

FIGURE 18.7 A display of class statistics.

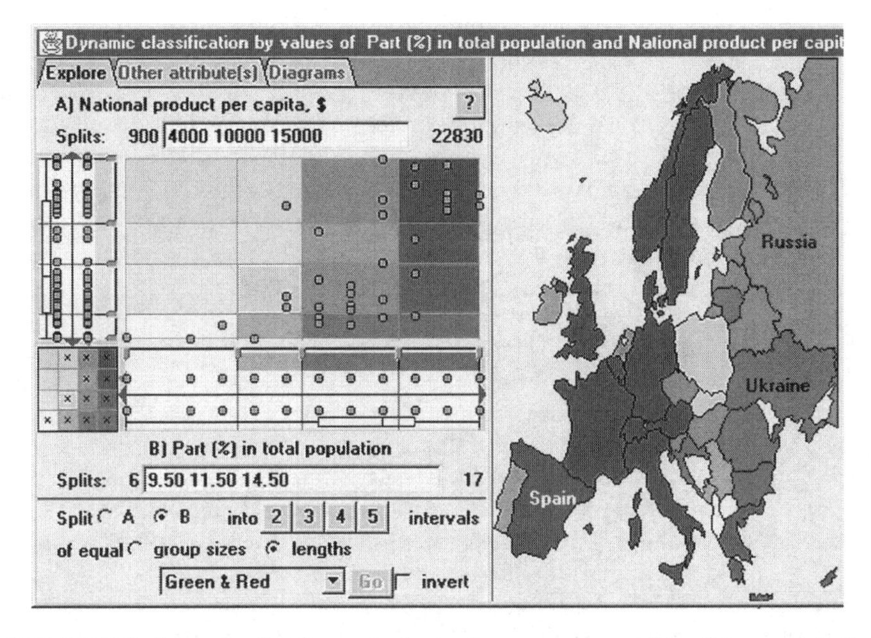

FIGURE 18.8 Cross-classification on the basis of two numeric variables. (Figure originally published in colour and in the e-version.)

A more detailed consideration of the dynamic classification technique as well as an example of data exploration with its use can be found in Andrienko and Andrienko (1998b).

Similar to the dynamic classification that is based on a single numeric variable, we have developed a tool supporting cross-classification with two base variables. The map in Figure 18.8 demonstrates this technique by an example: a classification of European countries on the basis of the variables 'National product per capita, $', and 'Part (%) of old people (65 and more years) in total population'.

Analogous to the previous variant of classification, the value ranges of both variables are divided into intervals. This potentially defines $M \times N$ classes, where M and N are the numbers of intervals for the first and for the second variable, respectively. To assign colours to classes, first a colour scale is separately selected for each variable. In our example, a scale from yellow to green was selected for the first variable, and a scale from yellow to red for the second one. Now, the system generates a colour matrix, like the one seen in Figure 18.8 (left), by mixing colours from the two scales. A geographical object in the map is painted according to the colour of the cell of its associated values. The distribution of colours across the map can manifest interesting relationships between variables. Presence of a relationship is signified by a clear division of the territory into regions formed by similarly painted neighbouring objects.

The investigation of relationships is also supported by a scatterplot **linked** with the map (see the left side of Figure 18.8). This linking and the coloured background of the scatterplot allows the user to establish a correspondence between objects in the scatterplot and those in the map. One may argue that the scatterplot alone is sufficient to find relationships. However, relationships revealed using a map have a different quality than those revealed in scatterplots. In particular, a map manifests relationships inherently based on the spatial distribution, while a scatterplot can show a kind of functional dependency. In the extreme, the map can visually form cohesive regions while the corresponding scatterplot only shows a chaotic diffusion of dots. In any case, a scatterplot does not show any spatial information, e.g., whether objects with close values are also geographically close. For example, the map in Figure 18.8 demonstrates an apparent spatial pattern with brown west and red east that cannot be derived from the scatterplot.

All operations provided for one-dimensional classification are also supported in the two-dimensional case. In addition, the scatterplot can be used as an interactive device to manipulate the map. Clicking in a coloured cell results in visual isolation of objects belonging to the corresponding class: only these are coloured in the map, the others are shown in neutral grey. It is also possible to visually isolate classes belonging to one row or one column of the matrix. The dynamic focussing technique described in the previous section works also for classed choropleth maps. As the focussed interval of one of the base variables is narrowed, the scatterplot is zoomed.

In general, grouping is very important in exploratory data analysis. However, to be really useful, it should necessarily be dynamic. In particular, any *static* division of the value range of a numeric variable into steps is counter-productive for exploration: according to Bertin (1967), 'determining the steps is, in fact, the goal of the graphic operation, not its means'.

5.5 Dynamic isolation of qualitative values

This tool is designed for maps that encode values of qualitative variables by colours or shapes. It allows the user to temporarily switch off the depiction of some values. Corresponding colours or shapes are removed from the map. This helps to concentrate the user's attention on the remaining values and more easily observe their distribution. As the user changes the selection of values, those changes are instantaneously reflected in the display, thus helping the user to concentrate her/his attention on another group of values. The work of the dynamic isolation tool is illustrated by Figure 18.9.

Dynamic isolation is designed for presentations based on the use of the associative visual variables *colour* and *shape*. It is especially useful for analysing presentations based on *shape*, which is, according to Bertin (1967), only associative but not selective. When all values are shown in a map, it is difficult to concentrate on some of them and disregard the others. Dynamic isolation effectively compensates for this lack of selectivity. After masking some of the values, the associative potential of shape or colour promotes seeing the geographical distribution of the remaining values as a single image.

The work of dynamic isolation in Descartes resembles that of 'Dynamic query' (Ahlberg *et al.* 1992): removal of some symbols from the display in response to specifying constraints on values of a variable. However, like in the case of dynamic focussing on value subintervals of a numeric variable, the goal of our tool is different: it is intended primarily for studying spatial distributions rather than search for objects satisfying some constraints.

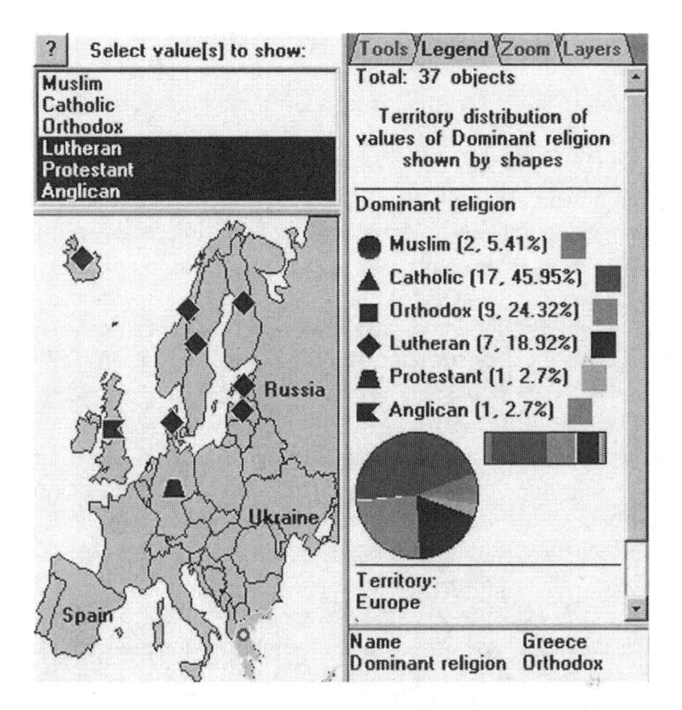

FIGURE 18.9 A map presenting a result of dynamic isolation.

6 Conclusions

Descartes fits in the paradigm of cartographic visualisation defined by MacEachren (1994b) and further elaborated by MacEachren and Kraak (1997). This paradigm includes three key elements: private map use (an individual generates a map for his or her own needs), direction towards revealing unknowns rather than presenting known facts, and high human-map interaction. Accordingly, in developing Descartes we address two objectives:

- allow people in various disciplines (not professional cartographers) to easily present their data by maps (*private map use*).
- support *high interaction* of the analysts with the maps to facilitate *revealing unknowns*.

To achieve the first objective, Descartes incorporates cartographic expertise and on this basis automates map generation with the application of most widely known presentation methods from thematic cartography. The second goal is approached by advancing these presentation methods towards interactivity and dynamics. This approach is intended to complement the popular way of adding interactivity to a map by linking it with other types of data displays. Usually these non-cartographic displays are given the leading role in the analysis of attribute information while the map is used mainly to show the spatial location of objects. In our research we tried to bring the potential of traditional cartographic means of visualisation to exploratory data analysis.

The exploratory tools described here were developed according to our strategy of designing specialised techniques based on particular features of each presentation method, and directly affecting its inherent symbol system. With these tools we try to create more favourable conditions for data analysis by compensating for weaknesses and enhancing potential capabilities of the target presentation methods.

Map authors have always been struggling to increase map expressiveness. For this purpose, such expedients as various colour schemes, classification strategies, non-linear encoding functions, etc., were developed. However, in static presentations the use of these practical aids often results in partial loss of information, distorted perception, or/and imposing some pre-interpretation of source data. Such 'side effects' are counter-productive in exploratory data analysis. Dynamic, interactive displays allow users to achieve necessary expressiveness without significant sacrifices, and may neutralise undesirable consequences. For example, flexible midpoints or class boundaries allow users to reject any pre-set view on data, and to empirically find a view illustrating important, non-trivial properties of the spatial distribution.

Our work on enhancing map interactivity goes on. At this stage, only a part of the cartographic presentation methods implemented in Descartes have been made as flexible as we think is possible and desirable. We hope that the paper has demonstrated the usefulness of such a development.

References

Ahlberg, C., Williamson, C., and Schneiderman, B., 1992, Dynamic queries for information exploration: an implementation and evaluation. In *Proceedings ACM CHI '92* (ACM Press), pp. 619–626.

Andrienko, G. L., and Andrienko, N. V., 1997, Intelligent cartographic visualisation for supporting data exploration in the IRIS system. *Programming and Computer Software*, **23**, 268–282.

Andrienko, G. L., and Andrienko, N. V., 1998a, Intelligent visualisation and dynamic manipulation: Two complementary instruments to support data exploration with GIS. In *Proceedings Advanced Visual Interfaces '98* (New York: ACM Press), pp. 66–75.

Andrienko, G. L., and Andrienko, N. V., 1998b, Dynamic categorization for visual exploration of spatial information. *Programming and Computer Software*, **24**, 108–115.

Bertin, J., 1967–1983, *Semiology of Graphics. Diagrams, Networks, Maps* (Madison: The University of Wisconsin Press).

Brewer, C. A., 1994, Colour use guidelines for mapping and visualisation. In *Visualisation in Modern Cartography* (New York: Elsevier Science Inc.), pp. 123–147.

Buja, A., McDonald, J. A., Michalak, J., and Stuetzle, W., 1991, Interactive data visualisation using focusing and linking. In *Proceedings IEEE Visualization '91* (Washington: IEEE Computer Society Press), pp. 156–163.

Buja, A., Cook, D., and Swayne, D. F., 1996, Interactive high-dimensional data visualization. *Journal of Computational and Graphical Statistics*, **5**, 78–99.

Carr, D. B., Nicholson, W. L., Littlefield, R. J., and Hall, D. L., 1986, Interactive color display methods for multivariate data. In *Statistical Image Processing and Graphics* (New York: Marcel Dekker), pp. 215–250.

Cleveland, W. S., and McGill, M. E., 1988, *Dynamic Graphics for Statistics* (Belmont: Wadsworth and Brooks).

Cleveland, W. S., 1993, *The Elements of Graphing Data* (Murray Hill: AT&T Bell Laboratories).

Cook, D., Symanzik, J., Majure, J. J., and Cressie, N., 1997, Dynamic graphics in a GIS: more examples using linked software. *Computers and Geosciences*, **23**, 371–385.

Dawkes, H., Tweedie, L., and Spence, B., 1996, VICKI — The visualisation construction kit. In *Proceedings of Advanced Visual Interface '96* (New York: ACM Press).

DiBiase, D., 1990, Visualization in the earth sciences. *Earth and Mineral Sciences, Bulletin of the College of Earth and Mineral Sciences, Penn State University*, **59**, 13–18.

Dobson, M. W., 1973, Choropleth maps without class intervals: A comment. *Geographical Analysis*, **5**, 358–360.

Dykes, J. A., 1997, Exploring spatial data representation with dynamic graphics. *Computers and Geosciences*, **23**, 345–370.

Egbert, S. L., and Slocum, T. A., 1992, EXPLOREMAP: an exploration system for choropleth maps. *Annals of the Association of American Geographers*, **82**, 275–288.

Hummel, J., 1996, Linked bar charts: analysing categorical data graphically. *Computational Statistics*, **11**, 23–33.

Hurley, C., 1993, The plot-data interface in statistical graphics. *Journal of Computational and Graphical Statistics*, **2**, 365–379.

Kraak, M.-J., 1998, The cartographic visualization process: from presentation to exploration. *The Cartographic Journal*, **35**, 11–15.

MacEachren, A. M., 1994a, *Some Truth with Maps: A Primer on Symbolization and Design* (Washington, DC: Association of American Geographers).

MacEachren, A. M., 1994b, Visualization in modern cartography: setting the agenda. In *Visualisation in Modern Cartography* (New York: Elsevier Science Inc.), pp. 1–12.

MacEachren, A. M., 1995, *How Maps Work: Representation, Visualization, and Design* (New York: The Guilford Press).

MacEachren, A. M., and Kraak, M.-J., 1997, Exploratory cartographic visualization: advancing the agenda. *Computers and Geosciences*, **23**, 335–344.

McDonald, J. A., 1982, *Interactive Graphics for Data Analysis.* Project Orion, **11** (Stanford: Stanford University).

MacDougall, E. B., 1992, Exploratory analysis, dynamic statistical visualization, and geographic information systems. *Cartography and Geographic Information Systems*, **19**, 237–246.

Monmonier, M., 1989, Geographic brushing: enhancing exploratory analysis of the scatterplot matrix. *Cartographycal Analysis*, **21**, 81–84.

Newton, C. M., 1978, Graphics: from alpha to omega in data analysis. In *Graphical Representation of Multivariate Data* (New York: Academic Press), pp. 59–92.

Symanzik, J., Majure, J., and Cook, D., 1996, Dynamic graphics in a GIS: a bidirectional link between ArcView 2.0 and XGobi. *Computing Science and Statistics*, **27**, 299–303.

Tufte, E. R., 1983, *The Visual Display of Quantitative Information* (Cheshire: Graphics Press).

Tukey, J. W., 1977, *Exploratory Data Analysis* (Reading: Addison-Wesley).

Unwin, A., Hawkins, G., Hofmann, H., and Siegl, B., 1996, Interactive graphics for data sets with missing values — MANET. *Journal of Computational and Graphical Statistics*, **5**, 113–122.

Unwin, A., and Hofmann, H., 1998, New interactive graphic tools for exploratory analysis of spatial data. In *Innovations in GIS*, **5** (London: Taylor & Francis), pp. 46–55.

Wills, G., 1995, Visual exploration of large structured data sets. In *New Techniques and Trends in Statistics* (Amsterdam: IOS Press), pp. 237–246.

Wills, G., 1996, Selection: 524,288 ways to say this is interesting. In *Proceedings IEEE Information Visualization '96* (Washington: IEEE Computer Society Press), pp. 54–60.

Yamahira, T., Kasahara, Y., and Tsurutani, T., 1985, How map designers can represent their ideas in thematic maps. *The Visual Computer*, **1**, 174–184.

Visual Data Exploration: Tools, Principles, and Problems

Gennady L. Andrienko and Natalia V. Andrienko

Our 1999 *IJGIS* paper (Andrienko and Andrienko 1999) was about maps as tools for exploratory analysis of spatially referenced data. After the paper was published, we continued working on tools and techniques for exploratory analysis of spatial, and, more recently, spatio-temporal data. We have written many other papers concerning this topic, and have just finished writing a book, in which we try to approach the topic in a systematic way and to lay theoretical and methodological foundations for exploratory data analysis.

Exploratory data analysis (EDA) means looking at data with an open mind and the aim of detecting and describing patterns, trends, and relationships, and generating hypotheses, which can then be tested using mathematical methods. The very nature of EDA implies that data visualisation plays a crucial role. A dictionary definition of the word *visualise* means "to make perceptible to the mind or imagination." In EDA, it is the mind of a human explorer that is the primary tool of analysis. It is the task of the mind to detect and describe patterns and to generate hypotheses. However, the mind can fulfil its mission only if the data to be explored are made perceptible to it. No thinking is possible without prior perception. Hence, data visualisation is the most important supporting tool (i.e., supplementary to the human mind) in EDA.

For the perception and further reasoning to be valid, a visual representation of data must be structurally similar (isomorphic) to the pertinent features of the underlying phenomenon (Arnheim, 1997). A specific implication of this requirement is that geographically related data can be properly explored only with the use of maps as isomorphic representations of the geographical space. It was therefore natural that we started our research into tools for exploratory analysis of spatial data with maps as the primary means of visualisation of such data. In this way, our research interests became very close to those of the Commission on Visualisation of the International Cartographic Association (see http://kartoweb.itc.nl/icavis/).

In fact, what we did was fully in line with the research agenda of that Commission, with its "emphasis on the role of highly interactive maps ... [in] hypothesis generation, data analysis, and decision-support" (MacEachren and Kraak, 1997). Therefore, the community of researchers in geovisualisation clustering around the Commission enthusiastically accepted our work and adopted us as an integral part. It is interesting that this community does not consist purely of cartographers, but also of scientists from other disciplines including geography, computer science, psychology, and statistics. This creates a fertile ground for a synergy of expertise and experience, ideas, and approaches.

Like many other geovisualisers, we spent a certain amount of time in experimenting with the possibilities provided by the computer screen, such a plastic and responsive representation medium, and thereby contributed to the establishment of the concept of the *interactive map*. In particular, our paper shows that an interactive map allows, in addition to zooming, querying, and brushing, also "playing" with the parameters of the currently applied cartographic representation. Such "playing" can increase map expressiveness and expose initially hidden features, patterns, trends, and relationships.

Our paper differed from contemporaneous works of other researchers concerned with the exploratory use of maps. Our primary focus was the investigation of the properties and capabilities of maps as such, while the others mostly considered maps in combination with other tools, for example, statistical graphics or computational techniques. This does not mean, however, that we denied the utility of tool integration for the exploratory data analysis. We just wanted to gain a deep understanding of maps before extending the scope of our research to other tools.

It is generally recognised that maps alone can seldom be sufficient for a comprehensive exploration of spatial data. In most real cases, data complexity requires interactive maps to be used in combination with other exploratory and analytical tools. Therefore, it is natural that the geovisualisers keep an eye on what is going on in related disciplines such as information visualisation, statistics, and data mining, and adopt techniques and approaches from these disciplines. After publication of the paper our work also developed in this direction.

We have designed and implemented quite a number of exploratory tools and investigated their capabilities. Our particular interest was to discover synergistic effects from joint uses of several tools, especially in combination with interactive maps. The development and further investigation of the tools was often actuated by practical needs, i.e., encountering datasets that could not be properly explored with the available tools. In this manner, helped by our colleagues, we have gradually built a quite powerful toolkit for exploratory analysis of spatial data, with a variety of tightly integrated tools that can be used in diverse combinations.

Among others, we have developed a range of tools supporting the exploration of spatio-temporal data, i.e. data with both spatial and temporal components. There are different kinds of spatio-temporal data requiring different sets of exploratory tools. Thus, discrete events such as earthquakes have to be analysed in a different way than movements of objects or temporal sequences of measurements made in a set of spatial locations. Our focus on spatio-temporal data was quite in line with the general trends of the research in geovisualisation and geoinformation science.

Hence, after publishing the paper about interactive maps, our research developed mostly in breadth, by embracing other types of exploratory tools and extending to other types of space-related data. The same can be said about the geovisualisation research in general. In a recent edited book presenting the state of the art in this area, most of the contributions deal with new visualisation techniques, new tool combinations, or new technologies (Dykes et al., 2005).

Our concern is the absence of a theoretical or methodological framework both for doing comprehensive EDA and for building tools to support it. Is data exploration an arbitrary sequence of trials, with some of them bringing serendipitous discoveries, or is it (or should it be) a purposeful and systematic course of actions? If the latter is true, what is (or should be) the underlying system? What actions are involved and how are they organised? If performing the actions requires specific tools, how can an analyst know which tool is suitable for what action? On the other hand, how can a tool designer anticipate the needs of an analyst and build a tool combination that covers those needs?

Since our methodical and rationalistic minds could not accept the view of EDA as a haphazard activity, we decided to uncover the underlying structure, principles, and driving forces. We switched from developing in breadth to moving in depth. However, the previous in-breadth development was a necessary prerequisite for this movement. We have studied a wide range of tools and tool combinations in applications to various datasets, and in this way acquired a great deal of valuable knowledge and experience. We have proceeded to reflect on this knowledge and experience and generalise them. The resulting theory is presented in our book, *Exploratory Analysis of Spatial and Temporal Data: A Systematic Approach* (Andrienko and Andrienko 2006).

The approach we apply to define the principles of EDA can be characterised as task-centred. We view EDA as consisting of tasks, i.e., finding answers to various questions about data. To find the answers, an analyst needs to apply appropriate tools. The character of a task determines what kind of tool (from a range of tools potentially applicable to the given data) can appropriately support it. Hence, a tool designer who knows the potential tasks of the explorer can create a set of tools capable of covering the explorer's needs. Accordingly, the keystone of our theory is the definition of the tasks emerging in EDA. More precisely, we introduce a system that allows one to define the set of tasks pertinent to a specific dataset or a class of datasets with similar structures (because the tasks are not defined in an abstract way but in terms of data components).

Besides designers of tools for EDA and researchers in geovisualisation, we want the theory to be useful for guiding data analysts in choosing and applying exploratory tools and in doing EDA in a systematic, comprehensive way. Therefore, we have not stopped after introducing our task framework, but made two further steps. First, we have catalogued the existing tools suitable for EDA, described their properties and capabilities (in particular, what tasks they are capable of supporting), and provided examples of their use. Second, we portrayed EDA as a systematic procedure composed of analytic and synthetic activities. We described how high-level tasks are decomposed into lower-level ones, and how the results of the lower-level tasks are integrated into comprehensive answers to the high-level questions. We believe that we have uncovered general principles directing the process of EDA. We do not present these principles as our original invention, but we firmly believe that any experienced data analyst follows these principles, at least intuitively.

Because the principles have been explicitly described, they can be used to educate or guide novices. Tool developers can use the principles as an organising basis for building EDA toolkits and designing their user interface. The results may be beneficial not only for novices but also for experts, who will almost certainly find such systematically designed toolkits more convenient and easy to use. Moreover, a novel exploratory technique appearing as a part of such a system may cause fewer problems for an analyst because its purpose is clear from its particular place in the system.

We recognise that the route from the formulation of the principles to the practical application of them for designing toolkits and guiding users may not be short or obvious. Still, we think that the route is worth trying because it promises at least a partial solution to one of the most serious problems of exploratory tools, namely, the Usability Problem. This problem, which is pertinent not only to geovisualisation but also to other types of visualisation, means that data analysts, both novices and experts in general, may not be willing to use the tools and techniques created for them.

One reason is the lack of knowledge of the EDA concept and principles, at least among novices, and the unconventional nature of the EDA tools and techniques. Another reason is the complexity of the EDA process, which is very demanding on the analyst because it is the human mind that is the primary instrument of analysis. The variety of tasks and corresponding tools, the complexity of real data necessitating the exploration by pieces and slices followed by a laboured synthesis of an overall picture from multiple tiles, make EDA a difficult job. Intelligence embedded in a toolkit for EDA could be an adequate response to these challenges. Intelligent software could "know" the principles of EDA and help and even prompt the users to act in accord with them. It might "know" the tools and assist the users in choosing and utilising them. It could implement the analysis workflow step by step, and relieve the users from the cognitive complexity of the EDA process. It could automatically adapt itself to the particular user, data, tasks, and hardware.

However, there are also other causes for the Usability Problem, and tool designers are, to a large extent, responsible for them. Thus, many researchers tend to implement software prototypes demonstrating particular ingenious approaches rather than toolkits intended to cover the needs of people exploring particular data. The tools created by different researchers are incompatible, and, hence, the needs cannot be accommodated by combining several tools. Furthermore, data analysts do not only need tools that allow them to make discoveries and generate hypotheses but they also need the means to verify these discoveries and to test the hypotheses. Hence, exploratory tools should be linked with appropriate confirmatory techniques, and, for example, implemented as extensions to statistical packages.

One more challenge, which does not yet seem to be properly realised in the visualisation research community, is related to the fact that observations and discoveries that people make in the course of visual data exploration cannot be conveniently captured for later recall and for communication to others. Primary results of using visual data displays are visual impressions, or mental images, which are hard to verbalize or express in any other form without referring to the displays from which they originate. The difficulty of recording and reporting the findings is a serious obstacle to wide recognition and use of visualisation tools.

Hence, the Usability Problem is complex and multifaceted and requires systematic joint efforts of many researchers and tool designers to fully solve. In the future, we will certainly try to contribute to solving the problem and hope to see substantial progress.

Returning to the 1999 *IJGIS* paper and our subsequent work, we would like to note that the Web supplement to the paper with a number of interactive maps is still available at a new address (http://www.ais.fraunhofer.de/and/IcaVisApplet/). In addition, various demonstrators of EDA tools and tutorials explaining their purposes and usage can be accessed from our homepage http://www.ais.fraunhofer.de/and/.

References

ANDRIENKO, G. AND ANDRIENKO, N., 1999, Interactive maps for visual data exploration, *International Journal of Geographical Information Science,* **13**(4), 355–374.

ANDRIENKO, N. AND ANDRIENKO, G., 2006, *Exploratory Analysis of Spatial and Temporal Data: A Systematic Approach* (Berlin: Springer-Verlag).

ARNHEIM, R., 1969, *Visual Thinking,* (University of California Press: Berkeley), reprinted, 1997.

DYKES, J., MACEACHREN, A.M., AND KRAAK, M.-J., (Eds.), 2005, *Exploring Geovisualization* (Amsterdam: Elsevier).

MACEACHREN, A.M. AND KRAAK, M.-J., 1997, Exploratory cartographic visualization: Advancing the agenda, *Computers and Geosciences,* **23**, 335–344.

International Journal of Geographical Information Science,
2001, Vol. 15, No. 7, 591–612.

19 Geographical Categories: An Ontological Investigation

Barry Smith and David M. Mark

Abstract. This paper reports the results of a series of experiments designed to establish how non-expert subjects conceptualize geospatial phenomena. Subjects were asked to give examples of geographical categories in response to a series of differently phrased elicitations. The results yield an ontology of geographical categories — a catalogue of the prime geospatial concepts and categories shared in common by human subjects independently of their exposure to scientific geography. When combined with nouns such as *feature* and *object*, the adjective *geographic* elicited almost exclusively elements of the physical environment of geographical scale or size, such as mountain, lake, and river. The phrase *thing that could be portrayed on a map*, on the other hand, produced many geographical scale artefacts (roads, cities, etc.) and fiat objects (states, countries, etc.), as well as some physical feature types. These data reveal considerable mismatch as between the meanings assigned to the terms 'geography' and 'geographic' by scientific geographers and by ordinary subjects, so that scientific geographers are not in fact studying geographical phenomena as such phenomena are conceptualized by naïve subjects. The data suggest, rather, a special role in determining the subject-matter of scientific geography for the concept of *what can be portrayed on a map*. This work has implications for work on usability and interoperability in geographical information science, and it throws light also on subtle and hitherto unexplored ways in which ontological terms such as 'object', 'entity', and 'feature' interact with geographical concepts.

1 Introduction

What is the field of geographical information science about? What is 'geographical' about geographical information? What makes an object a geographical *object*? Geographers and others have debated these questions for many decades, and we do not propose to resolve the underlying issues here. Instead, we offer a preliminary overview of ontology as this term is currently used by philosophers, information scientists and psychologists. We then use this overview as background to the presentation of detailed evidence as to how the geographical is distinguished from the non-geographical by non-experts. Thus we report on how the domain of geography is defined and conceptualized from the outside.

We believe that this work is of more than simple curiosity value, telling geographers and geographical information scientists how the subject-matter of their work is conceived by others (although already in this respect it contains a number of results which many will find surprising) Our results are also of theoretical importance: they throw light on one common type of human cognition that has seldom been studied (whether by geographers, cognitive scientists, or philosophers). And our results are also of practical importance, in that they can help us to understand how different groups of people exchange (or fail to exchange) geographical information, both when communicating with each other and also when communicating with computers. With the proliferation of geographic information systems (GIS) and GIS-related applications on the World Wide Web, there is an ever-increasing need to know how non-experts conceptualize the geographical domain. A sound, empirically supported ontology of geospatial phenomena will thus form a central part of the foundation for geographical information system design in the future.

2 Categories and ontology

2.1 *Cognition and pre-scientific theories*

With many scientific disciplines we can associate what we might call a *pre-scientific* or *'folk'* counterpart. Psychologists and cognitive anthropologists have in recent years studied theories of naïve physics, of folk psychology, and of folk biology, and the importance of such studies is now widely accepted. Naïve physics, for example, is of practical importance to those confronting the problems involved in the design of mobile robots which would possess sensorimotor capacities equivalent to those of human beings. Folk psychology is important to evolutionary psychologists, whose studies of the biological roots of our inherited psychological capacities are beginning to transform our understanding of human cultural and psychological universals. The importance of folk biology is revealed in fields such as ethnobotany, a scientific discipline devoted to the study of folk theories of plants and of their medicinal properties: it studies the objects and processes identified in the corresponding folk theories from a scientific point of view.

Each such folk discipline is like its properly scientific counterpart in having its own determinate subject-matter. But our pre-scientific thinking taken as a whole, too, has its own subject-matter, which we might think of as 'common-sense reality': this is the niche, or the environment, which we all share in our everyday perceiving and acting. It is the world of affordances, in J. J. Gibson's sense (Gibson 1979).

2.2 *Eliciting ontologies*

Is it legitimate to put these various folk and scientific domains side by side with each other in this fashion? Are we not lending too much credence to folklore in supposing that it, too, can have its own peculiar objects, analogous to the objects of genuine sciences such as physics or chemistry? These are difficult questions, which have been studied for some two thousand years by the branch of philosophy known as ontology. Ontology is distinguished from all the special sciences, and from all the branches of folk theory, in that it seeks to study in a rational, neutral way all of the various types of entities and to establish how they hang together to form a single whole ("reality"). It studies, in other words, the totality of objects, properties, processes and relations which make up the world on different levels of focus and granularity, and whose different parts and aspects are studied by the different folk and scientific disciplines.

In the work of Quine (1953) ontological theorizing seeks to elicit ontologies from scientific disciplines. Ontology thus takes the form of the study, using logical methods, of the ontological commitments or presuppositions embodied in different scientific theories. These methods were extended to the domain of common-sense theories in the papers collected in Hobbs and Moore (1985). Casati et al. (1998) and Smith and Varzi (2000) attempted to apply similar formal methods to the domain of folk or naïve geography; that is, they attempted to elicit the folk ontologies of common-sense geospatial reality to which ordinary human subjects are committed in their everyday cognition in much the same way in which philosophers of science, earlier, had attempted to elicit the ontological commitments of scientific theories by examining the logical structure of such theories.

In relation to the folk domains, however, we have not only such abstract logical methods at our disposal, but also the empirical methods of psychology. It is these empirical methods which have been applied, in the studies of Keil (1979, 1987, 1989, 1994), Medin (Murphy and Medin 1985, Medin and Atran, 1999), Atran (1994) and others, to the task of eliciting the folk ontologies of naïve subjects in such areas as folk physics, folk biology and folk psychology. And it is these same empirical methods which were used in the experiments reported on below.

2.3 Ontology and information systems

The term 'ontology' has another use, however, which arose in recent years within the domain of computer and information science to describe the results of eliciting ontologies (in Quine's sense) not from scientific theories or from human subjects but rather from information systems, database specifications, and the like. To understand the nature of this ontological (data) engineering, it will be useful to introduce first of all the technical notion of a 'conceptualization'.

We engage with the world from day to day in a variety of different ways: we use maps, specialized languages, and scientific instruments; we engage in rituals and we tell stories; we use information systems, databases, ATM machines and other software-driven devices of various types. Each of these ways of engaging with the world, we shall now say, involves a certain *conceptualization*. What this means is that it involves a system of concepts and categories which divide up the corresponding universe of discourse into objects, processes and relations in different sorts of ways. Thus in a religious ritual setting we might use concepts such as *God*, *salvation* and *sin*; in a scientific setting we might use concepts such as *micron*, *force* and *nitrous oxide*; in a story-telling setting we might use concepts such as *magic spell*, *dungeon* and *witch*. These conceptualizations are often tacit, that is, they are often invisible components of our cognitive apparatus, which are not specified or thematized in any systematic way. But tools can be developed to render them explicit (to specify and to clarify the concepts involved and to establish their logical structure). Tom Gruber, one of the pioneers of the use of ontological methods in information science, defines an ontology precisely as 'a specification of a conceptualization' (Gruber 1993).

Gruber proposed this definition in the context of his work on the Knowledge Interchange Project at Stanford, an attempt to address what we might call the Tower of Babel problem in the field of information systems. Different groups of data-gatherers each have their own idiosyncratic terms and concepts which they use to represent the information they receive. When the attempt is made to put this information together, methods must be found to resolve terminological and conceptual incompatibilities. Initially, such incompatibilities were resolved on a case-by-case basis. Gradually, however, it was realized that the provision, once and for all, of a 'concise and unambiguous description of principal, relevant entities of an application domain and their potential relations to each other' (Schulze-Kremer 1997) would provide significant advantages over the case-by-case resolution of successive incompatibilities as they arise in the interactions of specific groups of users or collectors of data. This is because each such group would then need to perform the task of making its terms and concepts compatible with those of other such groups only once: by calibrating its results in the terms of the single canonical description. The term 'upper-level ontology' is used by information scientists to refer to a canonical description of this sort or to an associated classificatory theory, and the IEEE has a study group to define a 'Standard Upper-Level Ontology' (IEEE 2000), which would specify the general ontological concepts — such as part, whole, number and so on — used across all domains.

An ontology, in the information science sense, is thus a neutral and computationally tractable description or theory of a given domain which can be accepted and reused by all information gatherers in that domain. From the ontological engineering perspective, ontology is a strictly pragmatic enterprise: it concerns itself not at all with the question of ontological realism, that is with the question whether its conceptualizations are true of some independently existing reality. Rather, it starts with conceptualizations, and goes from there to a description of corresponding domains of objects or 'closed world data models'. Note that the ontological method so conceived is not restricted to the information systems domain. Indeed it can be used to deal indiscriminately with the generated correlates of conceptualizations of all sorts, and it ends up treating each of these as 'universes of discourse', as 'posits' or 'models', as surrogate created worlds, all of which are treated as being on an equal footing.

2.4 Good and bad ontologies

The relevance of this ontological engineering background to our concerns here is as follows. The project of a common ontology which would be accepted by different information communities in different domains has (thus far at least) failed. We hold that one reason for this failure has been precisely that the attempt at ontology integration was carried out on the basis of a methodology which ignored the real world of flesh-and-blood objects in which we all live, and focused instead on closed world models. Ontological engineering has been, in fact, an exercise not in ontology at all, but rather in model-theoretic (set-theoretic) semantics.

This choice of semantic methodology was understandable for pragmatic reasons: closed world models are much simpler targets, from a mathematical point of view, than are their real-world counterparts. If, however, the real world itself plays a significant role in ensuring the unifiability of our separate ontologies, then it may well be that the project of unification on the basis of a model-based methodology is doomed to failure.

To see what the alternative methodology might be, we need to recognize that not all conceptualizations are equal. Bad conceptualizations (rooted in error, myth-making, astrological prophecy, or in antiquated information systems based on dubious foundations) deal only with created (pseudo-)domains, and not with any transcendent reality beyond. Good conceptualizations, in contrast, are transparent to some corresponding independent domain of reality. Only ontologies based on good conceptualizations, we suggest (Smith 1995b), have a chance of being integrated in a robust fashion into a single unitary ontological system.

Of course to zero in on good conceptualizations is no easy matter: there is nothing like a Geiger counter which we can use to test for truth. Rather, we have to rely at any given stage on the best endeavours of our fellow human beings and proceed, in critical and fallibilistic fashion, from there. Our best candidates for good conceptualizations are then illustrated by those conceptualizations of the developed sciences which have undergone rigorous empirical testing. But there are almost equally good candidates also in the realm of naïve cognition: for many of our folk category systems, too, have undergone rigorous empirical tests, sometimes stretching over many thousands of years, and we can assume that they, too, are to a large degree transparent to the reality beyond: that mothers, apples, milk, and dogs truly do exist, and they have the properties we common-sensically suppose them to have.

We will focus in what follows on good conceptualizations in the folk domain, a notion which we will make more precise in terms of Robin Horton's doctrine of 'primary theory', to be discussed below. We are concerned primarily with the ontology underlying such conceptualizations. However the study of folk conceptualizations along the lines here presented may also be of interest in helping us to provide better theories of common-sense reasoning, for if common-sense reasoning takes place against a background of common-sense beliefs and theories, then we cannot understand the former unless we also develop good theories of the latter. The study of the ways non-experts conceptualize given domains of reality might then help us also in our efforts to maximize the usability of corresponding information systems. The work of ontological engineers such as Nicola Guarino and his co-workers (Guarino 1998, Guarino and Welty 2000) shows further that, for reasons related to those presented here, the study of good conceptualizations can have advantages also in eliminating certain kinds of errors in data-collection and data-representation.

2.5 Ontology in the geographical domain

We will henceforth take it for granted that the geographical concepts shared in common by non-experts represent a good conceptualization in the sense proposed above. They too are transparent to reality: mountains, lakes, islands, roads truly do exist, and they have the properties we commonsensically suppose them to have. The task of eliciting this folk ontology of the geographic domain will turn out to be by no means trivial, but we believe that the effort invested in focusing on good conceptualizations in the geographical domain will bring the advantage that it is more likely to render the results of work in geospatial ontology compatible with the results of ontological investigations of neighbouring domains. It will have advantages also in more immediate ways, above all in yielding robust and tractable standardizations of geographical terms and concepts.

If many of the common-sense concepts of our folk disciplines are transparent to reality, then clearly they are often transparent to different aspects and dimensions of reality than are the concepts illuminated by science. Different conceptualizations may accordingly represent cuts through the same reality which are in different ways skew to each other. The opposition between naïve physics and scientific or sophisticated physics reflects above all a difference in levels of *granularity*. Naïve physics reflects a mesoscopic partioning (that is to say: it is concerned with objects at the human scale), whereas scientific physics involves also microscopic and macroscopic partitions (corresponding to atomic and sub-atomic physics on the one hand and to astronomy and cosmology on the other).

As J. J. Gibson expressed it:

> The world can be analysed at many levels, from atomic through terrestrial to cosmic. There is physical structure on the scale of millimicrons at one extreme and on the scale of light years at another. But surely the appropriate scale for animals is the intermediate one of millimeters to kilometers, and it is appropriate because the world and the animal are then comparable. (Gibson 1979, p. 22)

Note that there are in addition disciplines that are not naïve but yet relate to the very same mesoscopic objects as are associated with our naive-level partitions of reality. As the already mentioned case of ethnobotany shows, there is trained as well as untrained (critical as well as uncritical) knowledge of the common-sense world. Thus there are various specialist extensions of folk disciplines, including: law, economics, land surveying, planning, engineering, paleontology, cookery — and geography. The design of Geographical Information Systems, too, involves a specialist extension of common sense along these lines (Egenhofer and Mark 1995). One not inconsiderable reason why we need to get the ontology of common-sense reality right is that such systems are engineering products in which common-sense reality is embedded.

One task of geographical ontology will be to study the mesoscopic world of geographical partitionings in order to enable the construction of mappings between these mesoscopic partitions and the partitions of associated scientific domains such as geology and meteorology. Mesoscopic geography deals mostly with qualitative phenomena which can be expressed in the qualitative terms of natural language; the corresponding scientific disciplines, in contrast, deal with the same domain but consider features which are quantitative and measurable. GIS thus requires methods that will allow the transformation of quantitative geospatial data into the sorts of qualitative representations of geospatial phenomena that are tractable to non-expert users — and for this purpose, once again, we need a sound theory of the ontology of geospatial common sense.

In earlier work we have claimed that the geographical domain is ontologically distinct from non-geographical domains (Smith and Mark 1998). One of the most important characteristics of the geographical domain is the way in which geographical objects are not merely *located* in space, but are typically *parts* of the Earth's surface, and inherit mereological properties from that surface. At the same time, however, empirical evidence suggests that geographical objects are organized into categories in much the same way as are detached, manipulable objects (Mark *et al.* (1999). Against the background of our remarks on good conceptualizations above, we shall attempt to do justice to these two aspects of geographical common sense in what follows.

3 Primary theory

3.1 The limits of common sense

What are the limits of the common-sense world? Does the belief that the Earth is flat belong to common sense? Did it ever belong to common sense? Does common sense evolve over time? Does the belief that babies are brought by storks belong to common sense, even if everyone in some culture at some given time believes it? To answer these questions, it is useful to divide the messy and problematic totality of naïve or untutored beliefs into two groups, following the anthropologist Robin Horton's distinction between 'primary' and 'secondary' theory.

For Horton (1982) 'primary theory' is that part of common sense which we find in all cultures and in all human beings at all stages of development. 'Secondary theories', in contrast, are those collections of folk beliefs which are characteristic of different economic and social settings. Primary theory consists of basic (naïve) physics, basic (folk or 'rational') psychology, the total stock of basic theoretical beliefs which all humans need in order to perceive and act in ordinary everyday situations. (Forguson 1989, Smith 1995c) Secondary theory consists of folk beliefs which relate to gods and evil spirits, heaven and hell, molecules and microbes.

Our primary naïve beliefs relate to mesoscopic phenomena in the realm that is immediately accessible to perception and action: beliefs about tables and boats, table-tops and snow, neighbourhoods and streets. (This is, once again, the realm of affordances in Gibson's terms.) Our secondary naïve beliefs relate to phenomena which are either too large or too small to be immediately accessible to human beings in their everyday perception and action, or to objects and processes which are otherwise hidden. Primary theory is, as Horton points out, developed to different degrees by different peoples in its coverage of different areas (the primary theory of snow, for example, may be underdeveloped in tropical climates). In other respects, however, it differs hardly at all from culture to culture. In the case of secondary or 'constructive' theory, in contrast, differences of emphasis and degree give way to startling differences in kind as between community and community, culture and culture. For example, the Western anthropologist brought up with a purely mechanistic view of the world may find the spiritualistic world-view of an African community alien in the extreme (Horton 1982, p. 228).

Agreement in primary theory has evolutionary roots: there is a sense in which the theory about basic features must correspond to the reality which it purports to represent, for if it did not do so, its users down the ages could scarcely have survived. At the same time, its structure has a fairly obvious functional relationship to specific human aims and to the specific human equipment available for achieving these aims. In particular, it is well tailored to the specific kind of hand-eye coordination characteristic of the human species and to the associated manual technology which has formed the main support of human life from the birth of the species down to the present day (Horton 1982, p. 232).

From the perspective of survival, we can believe what we like concerning microspirits and macro-devils residing on levels above or below the levels of everyday concern, but we have been constrained, as far as the broad physical structures of everyday reality are concerned, to believe the truth — *otherwise we would not be here*. The commonsensical world as the world that is apprehended in primary theory is thus to a large degree universal. It is apprehended in all cultures as embracing a plurality of enduring substances possessing sensible qualities and undergoing changes (events and processes) of various regular sorts, all existing independently of our knowledge and awareness and all such as to constitute a single whole that is extended in space and time. This body of belief about general regularities in the mesoscopic domain is put to the test of constant use, and survives and flourishes in very many different environments. Thus no matter what sorts of changes might occur in their surroundings, human beings seem to have the ability to carve out for themselves, immediately and spontaneously, a haven of commonsensical reality. Moreover, our common-sense beliefs are readily translated from language to language, and judgments expressing such beliefs are marked by a widespread unforced agreement.

Folk disciplines insofar as they are of concern to us here are based exclusively on beliefs in the domain of primary theory. Such beliefs are also of maximal scientific interest, since they satisfy the constraint of universality (as scientists we are interested primarily in what humans share in common, not in the particular beliefs of this or that culture or community).

3.2 How is primary theory organized?

Primary theory is to a large degree organized qualitatively, and in terms of *objects falling under categories* (such as *dog*, *table*, *hand*). Such categories, like all common-sense categories, are marked systematically by the feature of prototypicality. This means that, as Rosch (1973, 1978), Keil (1979) and others have shown, for most such categories, some members are better examples of the class than others and they are cognized as such. That is to say, humans can distinguish easily between the prototypical instances at the core of common-sense categories and the fringe instances in the penumbra. Furthermore, there is a great degree of agreement among human subjects as to what constitute good and bad examples. For example, robins and sparrows are widely considered to be good examples of *bird*, whereas ostriches and penguins and even ducks are considered bad examples.

Each family of common-sense categories is organized hierarchically in the form of a tree, with more general categories at the top and successively more specific categories appearing as we move down each of the various branches. (Deviations from the tree structure, for example kinds having multiple superordinates, are occasionally proposed. Guarino and Welty (2000), however, provide methods to resurrect the tree structure in such cases, for example via elimination of terminological ambiguities.)

One special level of generality within each such tree is distinguished as consisting of categories which play a special role in learning and memory and in common-sense reasoning. Why do children so readily learn category-terms such as *duck*, *zebra, clock*, and *fork*, while they experience difficulties learning terms like *mammal* or *utensil*? This is because the former belong to what Rosch (1978) called the 'basic level' of cognitive classification, while the latter belong to a level that is superordinate to this basic level. Basic-level categories represent a compromise in cognitive economy between two opposing goals, that of informativeness, on the one hand, and that of minimizing categories based on irrelevant distinctions, on the other. The basic level (*chair, apple*) falls between the superordinate level (*furniture*, *fruit*), which is in general insufficiently informative, and the subordinate level (*lounge chair, golden delicious*), which adds too little informativeness for its additional cognitive cost. Measures of our perception of stimuli, of our responses to stimuli, and of our communication, all converge on this same basic level.

But naïve categories do not walk alone. Each family of naïve categories is organized in such a way as to participate in a corresponding naïve theory. This insight, which we see as providing an important supplement to the work of Rosch on basic-level categories, was first advanced by Murphy and Medin (1985) and it has been applied above all in the sphere of language acquisition by Keil and many others. When we learn categories, we learn them in such a way that they come organized into theory-like structures. Thus we learn how the things falling under given categories are related to each other and how they interact causally. When we acquire the category *bird*, for example, we do this in such a way that we learn, in part through observation, that birds (typically) have wings, that birds (typically) fly, and that these two features are interrelated.

Associated with each family of naïve categories, therefore, is a certain unified domain, analogous to the subject-matter of a scientific theory. As Dowty (1998) express it:

> A key idea in the 'concepts-in-theories' view is that concepts are grouped into large-scale domains, each of which is organized by significantly different principles. Causation in the domain of the physical world is governed by laws of (naïve) physics, whereas in the domain of human individuals, an individual's actions are caused by the

> desires and beliefs of the individuals and so are predictable to an extent from these. Such a view is made more plausible by results of cognitive psychological research not involving language: from a very early age, long before the onset of language acquisition, children have been shown to perceive causation differently, depending on whether human figures or inanimate figures are used to simulate causation.

Children later differentiate further domains: the purposes a manufactured thing can serve for its user yield criteria for distinguishing *artefacts* of different kinds: what makes a chair is that it is something made to sit in — not its colour, its material composition, or its precise shape or size.

In Keil's version of the concepts-in-theories view, there is a two-way interaction between (*i*) understanding (having a theory of) the causal and other properties associated with the categories in a given domain, and (*ii*) identifying the perceptual or other attributes which can be used to identify instances of the corresponding categories in experience. As Rosch has shown, when the child acquires the concept associated with some new word, for example in the domain of animals or artefacts, he or she will attend only to certain kinds of attributes as potentially diagnostic for the concept and ignore other attributes as irrelevant. And then, as Keil notes, in learning concepts:

> People do not simply note feature frequencies and feature correlations; they have strong intuitions about which frequencies and correlations are reasonable ones to link together in larger structures and which are not. Without these intuitions, people would make no progress in learning and talking about common categories given the indefinitely large number of possible correlations and frequencies that can be tabulated from any natural scene. These intuitions seem much like intuitive theories of how things in a domain work and why they have the structure they do. (Keil 1994)

Some examples of domains that have been isolated in developmental studies are listed in Table 19.1a (taken from Dowty 1998). We note that in each domain specific kinds of causal and explanatory principles are at work.

3.3 The primary theory of the geographical domain

Studies of wayfinding and navigation abilities of pre-literate children have shown that by the age of 4, young children can find their ways around familiar neighbourhoods and interpret some aspects of maps and aerial photographs (Hazen et al. 1978, Spencer and Blades 1985, Freundschuh 1990, Blaut 1997, Blades et al. 1998). However, we know of no literature on other aspects of geographical concepts that young children have mastered, especially the development of geospatial object concepts. We hypothesize that the child conceptualizes the geospatial world as a large unitary background of *what does not move*, and that, early on (long before 4 years), she or he has learned to appreciate that there is a difference between things that move, whether by themselves or because caused to move by another object, and the framework within which things move, and which allows him or her to get from place to place. One issue is the point at which infants begin to distinguish between the fixed background spatial world at the level of individual locations (the immediate perceptual environment of single rooms and dwellings) and the larger world of geospatial forms in the strict sense. When do infants first apprehend the difference between location in stationary objects such as rooms and buildings, and location in moving objects such as cars? Is there a distinction, in the spatial background domain, between what is natural and what is constructed or built by human beings (analogous to the distinction, on the table-top scale, between artefacts and non-artefacts)? At what point do young children acquire the capacity to distinguish between natural geographical features such as mountains and lakes on the one hand and places and other fiat geographical objects (Smith 1995a) on the other? How do place concepts relate to children's (and adults') conceptualizations of notions of environment and surroundings (Smith and Varzi 1999, Smith 2000). Leaving such issues to one side, we can conjecture that the relevant extension to Table 19.1(a) for geographical categories is given in Table 19.1(b).

This addendum is hypothetical only, since supporting data pertaining to cognitive categories is based almost entirely on *studies of categorizations of entities at surveyable scales*. Rosch and her associates studied first of all categorization of pets,

TABLE 19.1(a)
Major theory domains.

Domain	Nature of theories for domain	Characteristic causal relations and explanations	Age acquired
Physical	Naive mechanics (physics	Does not move except when caused to by another object. An object is animate if and only if it moves by itself. Actions can be caused by beliefs and desires	By 6 months
'Alive'	Self-initiated action, goal-directed	An object is animate if and only if it moves by itself	By 6–11 months
Human	Capable of actions, perceptions, beliefs, intentions etc.	Actions can be caused by beliefs and desires	By 3 years (?)
Biological	Notion of species	Properties are explained by their utility to the individuals themselves	3–6 years
Artefacts	Manufacture, use	Properties are explained by their utility to others	By 3–4 years

TABLE 19.1(b)
An additional major theory domain.

Geographical	What things move in and through	Properties are explained in relation to systems of landmarks and paths which do not change or move	Before 4 years

tools, and other manipulable artefacts. Work has also been done on more abstract categories such as colours, emotions, events, and on social categories (personal relations, social roles, crimes, ethnic groups). Even when account is taken of the results of the experimental work described below, however, the question has still not been resolved *whether the structure and organizing principles governing our cognitive categories remain the same as we move beyond these families of examples to objects at geographical scales.*

We said that primary theory is to a large degree organized qualitatively, and in terms of objects falling under categories. This holds, too, in regard to the primary theory of phenomena in the geographic domain, which is organized around categories such as *mountain, lake, island.* The primary axis of a folk ontology is its system of objects. The attributes (properties, aspects, features) and relations within the relevant domain form a secondary axis of the ontology, as also do events, processes, actions, states, forces and the like. The system of objects remains primary because attributes are always attributes *of* objects, relations always relations *between* objects, events *involving* objects, and so forth, in ways which, as already Aristotle saw, imply a dependence of entities in these latter categories upon their hosts or bearers in the primary category of objects.

Among properly scientific disciplines we can draw an opposition between those, such as particle physics, molecular chemistry, cell biology, human anatomy which employ an ontology based centrally on the category of *objects*, and those, such as quantum field theory, the physics of electromagnetism and hydrodynamics, which are based centrally on the category of *fields.* We hypothesize that there is no parallel opposition in the realm of folk disciplines. The latter work exclusively (or at least overwhelmingly) with object-based categories. This holds, too, in the realm of geospatial folk categories. Places, for example, are conceptualized by non-experts as objects, and this holds too of the whole of space, which is conceptualized as the totality of places. Since almost all of the experimental data reported in what follows relates to object-categories, we do not claim to have confirmed this hypothesis here. We will see, however, that such relevant data as we have does seem to lend it support.

We are less confident in relation to the claim of Millikan (1998) to the effect that the category of *stuffs* (such as 'gold' or 'milk') is as deeply rooted in our cognitive architecture as are individuals (such as 'Mama', 'Bill Clinton', 'the Empire State Building') and kinds (such as 'cat' and 'chair'). Here again, however, such data as we have seems not to lend it support, at least in relation to geospatial categories of the sort here under review.

3.4 The ontology of geospatial objects

The ontology of objects is itself organized on two conceptual levels: the level of individuals (token, particulars) and the level of kinds (types, universals). Our cognition of individuals is marked by our use of proper names (such as 'Fido', 'Mary' or 'Boston') and of indexical expressions (such as 'this' or 'that' or 'here'). Our cognition of kinds is marked by the use of common nouns such as 'dog' or 'mother' or 'lake'.

As we noted already above, kinds or categories are organized hierarchically in the form of a tree. The lower nodes in such a tree were called 'species' by Aristotle, and the upper nodes were called 'genera', although biologists since Linnaeus and the eighteenth century have used 'genus' and 'species' to refer to two particular levels in the taxonomic tree of organisms (and we note in passing that Aristotle's ideas on hierarchical classification were not only exploited by Linnaeus in his system of biological classification and naming, but also remain alive today, for example in hierarchical database organization, and in the organization of your hard-drive into folders and sub-folders).

Aristotle himself reserved the term 'category' for the topmost node in such a species-genus tree. Here, however, we use 'category' to refer indiscriminately to all such nodes, including nodes corresponding to basic-level categories in the sense of Rosch. Aristotle himself provided various lists of top-level categories, of which the most important for our purposes are the categories of *object* (or 'substance' in Aristotle's own terminology) and various attribute and event categories (referred to by Aristotle as the 'accidents' of substances, because they pertain to what holds of the substance *per accidens*). Accidental categories listed by Aristotle include: quantity, quality, action, relation and place. Both substances and accidents have instances. The prototypical instances of substances in Aristotle's eyes were biological organisms. The prototypical instances of accidents were *whiteness, running, sitting*, and *in the agora* (which must be taken as referring to particular instances of whiteness or to particular runnings or sittings, or to particular cases of being in the agora). Weather phenomena such as storms would have been categorized by Aristotle as accidents of the Earth.

In what follows we are interested, not in instances (tokens, individuals), but rather in types. And we are interested not in accidents (processes, attributes) but rather in substances, or objects, and more precisely still we are interested in the hierarchical classification of object categories within the geospatial realm. Aristotle himself thought that it was possible to give a *definition* (for example *man is a rational animal*) for each category, from which it would then be possible to infer the necessary and sufficient conditions for any given individual's being an instance of that category. At the same time however he recognized that species and genera are organized in such a way that we can distinguish a central core of focal (or typical, or standard) instances and a surrounding penumbra of non-typical or non-standard instances (for example an albino whale or a six-toed man). It is this idea of prototypicality which underlies the empirical work on cognitive categories by Rosch and her associates, and which is presupposed also in our present work.

4 Category norms

If the primary axis of a folk ontology is its system of objects, then our study of the folk ontology of the geospatial realm must begin with an elicitation of the object categories used by non-expert subjects (Smith and Mark 1999). To this end we replicated an experiment carried out by Battig and Montague (1968) to elicit what they called *category norms*.

The norms for a given category are those instances of that category most commonly offered by subjects as exemplifying the category itself. They may be prototypical examples of the category, although this is not necessarily the case. Battig and Montague themselves used an elicitation-of-examples procedure to determine norms for 56 categories. A total of 442 undergraduate subjects in Maryland and Illinois were given category titles, and asked to write down in 30 seconds as many 'items included in that category as you can, in whatever order they happen to occur to you'. Each subject went through all 56 categories in this manner.

Typical of the variety of non-geographical categories tested in the Battig-Montague experiment are: *precious stones*, *birds*, and *crimes*.

Most frequent *precious stones* (442 subjects) were: diamond (435 responses), ruby (419), and emerald (329).
Most frequently mentioned *birds* were: robin (377), sparrow (237), cardinal (208) and blue jay (180).
Most frequently mentioned *crimes* were murder (387), rape (271), and robbery (189).

It is important to note that some perfectly good members of a category may be given infrequently; for example, perjury, which almost all would agree is a crime, was listed by only 22 subjects, about 5% of the total. The number of examples per subject appears to reflect some combination of the familiarity of the category itself and the richness and diversity of familiar category members. Among all 56 categories, the greatest number of examples per subject were recorded for 'parts of the human body' (11.34) and the fewest were observed for 'member of the clergy' (3.82). Subjects listed an average of 5.16 examples of precious stones, 7.35 examples of birds, and 4.97 examples of crimes, numbers which give some measure of the richness and familiarity of the corresponding categories.

Of the 55 categories that Battig and Montague tested, 7 were at least somewhat geographic in nature: *a unit of distance*; *a type of human dwelling*; *a country*; *a natural earth formation; a weather phenomenon*; *a city, and a (US) state*. Some of these (*country, city, state*) produced examples that were specific instances (tokens) rather than type or kinds. Of the remainder, it is significant that *a unit of distance* had a geographical-scale — mile — as the most frequent example (438), closely followed by some common non-geographic units such as foot (417) and inch (411). *A type of human dwelling* showed much lower consensus, with only two examples, house (396) and apartment (316), being listed by at least half of Battig and Montague's subjects. The next most frequent, tent, was listed by just 198 of the subjects. In Battig and Montague's study, *a weather phenomenon* elicited 318 instances of hurricane, 303 of tornado, 297 of rain, and 266 of snow. Higher consensus among the lists of examples from one subject to another is an indicator that the category in question is a natural category in the sense of a category that is rooted more firmly in our cognitive architecture than are categories offered for elicitation which produce lower consensus or no consensus at all.

Most relevant to the geographical domain among Battig and Montague's categories was *a natural Earth formation.* A total of 34 different Earth formations were

TABLE 19.2
Most frequent responses to a natural Earth formation.

A natural Earth formation	N
Mountain	401
Hill	227
Valley	227
River	147
Rock	105
Lake	98
Canyon	81
Cliff	77
Ocean	77
Cave	69

listed by at least 10 of the subjects. Here, the ten most frequently listed terms, with their frequencies among 442 subjects, are listed in Table 19.2 (where *N* is the number of subjects who listed the given feature).

Despite the fact that the category-phrasing offered for elicitation was not prefixed by *a kind of* or *a type of*, only one particular named token was listed: the Grand Canyon, which was mentioned by only 14 subjects. All other terms given five or more times were names of categories, and all but five were at a geographical scale. Nothing movable was on the list, except glacier (very slow moving; 23 subjects) and iceberg (3 subjects).

Given these encouraging and intriguing results, we decided to replicate Battig and Montague's experiment using additional geographical categories. Results of these experiments are reported in the next section.

5 Experiment design and subjects

In our partial replication of Battig and Montague's (1968) study, subjects were tested simultaneously in a large classroom, at the beginning or end of a lecture. Subjects were students in two large sections of a first-year university course called 'World Civilization'. Versions 1 to 5 of the experiment were administered in one classroom, and versions 6 to 10 in the other. Versions 6–10 differed from versions 1–5 only in the order of presentation of stimuli, so that we could test for intercategory priming effects. Within each class, the five versions were printed on different colours of paper, and handed out from piles interleaving the five versions, in order to maximize the chance that the subject pools for the five versions were as similar as possible. Subjects were given a series of nine category names, each printed at the top of an otherwise blank page. They were asked to wait before turning to the first category, and then to write as many items included in that category as they could in 30 seconds, in whatever order the items happened to occur to them. After each 30-second period, they were told to stop, turn the page, and start the next category. A total of 263 subjects completed the first geographic category, with between 51 and 56 subjects responding to each version of the survey. Chi-squared tests showed no significant differences between responses from the two classrooms for any of the questions.

Following a pre-test reported by Mark *et al.* (1999), we chose nine categories to test with larger numbers of subjects. The first category tested was a non-geographic category (*a chemical*), which we hoped would provide a neutral, unprimed basis for the remaining questions. This was followed by a somewhat neutral phrase, *a type of human dwelling*. The third stimulus given to the subjects presented one of five variations on the phrase *a kind of geographic feature.* In this paper, we will focus on the results of our testing of the basic geographic domain, as explored in the third phrase tested; results for the remaining items will be presented in a companion paper.

In reporting on our pre-test (*loc. cit.*), we had observed that the compound noun *geographic feature* elicited solely natural and not artificial geographical features. That was surprising in light of the fact that academic geography — and the school geography curriculum to which most of our subjects had been exposed — is much more strongly a social rather than a natural science, and thus has a greater emphasis on cultural and economic geospatial phenomena than on physical ones. The presence of exclusively natural phenomena in elicited examples of *geographic feature* is thereby in and of itself *prima facie* evidence of a geographical component in our non-expert cognitive architecture that is independent of what subjects learn about the corresponding phenomena in academic settings.

However, as suggested by some commentators on our study, the predominance of physical or natural examples under *geographic feature* may have resulted from effects of the term 'feature', rather than reflecting the subjects' ideas of the meaning of the adjective 'geographic' to which it was attached. The geographical use of the word 'feature' appears to be less familiar among non-experts than, for example, among cartographers and among those accustomed to working with spatial data. We therefore formulated five different wordings of our target phrase, and presented these alternative wordings to five different groups of subjects, in effect changing the base noun of the superordinate category. The five variations we selected were:

- a kind of geographic **feature**
- a kind of geographic **object**
- a geographic **concept**
- **something** geographic
- something that could be **portrayed on a map**

6 Results

6.1 *General geographical things*

Although we had taken the trouble to give five different phrasings of our basic geographical question, we nonetheless expected little difference in subject responses to these different phrasings. As it turned out, however, the responses showed sharp divergences. Evidently, the base nouns in the stimulus phrases placed the geographical categories into different superordinate categories in the minds of the subjects we tested. Our selected superordinates: *feature*, *object*, *thing* and *concept* appear to

TABLE 19.3
Mean numbers of examples per subject.

Category label	Mean responses per subject
Something that could be portrayed on a map	8.21
A kind of geographic feature	7.15
Something geographic	6.17
A kind of geographic object	5.48
A geographic concept	5.15

interact with the adjective 'geographic' in distinct ways. Moreover there is a significant displacement among non-experts as between the extensions of 'geographic' and of 'what can be portrayed on a map'. We shall discuss the implications of these interactions below.

When singular and plural versions of terms were combined and misspellings merged with correctly spelled words, the subjects together gave a grand total of 308 words and phrases as examples of these basic geographical categories. A Chi-square test confirmed that the frequencies of terms in the responses to the different versions of the basic geographical question differed significantly. Also, as shown in Table 19.3, the mean number of examples listed per subject varied considerably across the five phrasings. For reasons already noted, we assume that the mean number of responses per subject within the 30-second time period reflects some combination of the familiarity and richness of the corresponding category. The results suggest that our subjects were very familiar with maps and with the sorts of things that appear on them. But they were also (somewhat counter-intuitively) thoroughly comfortable with the category *geographic feature*, and *less* comfortable with the other phrasings tested. Only 12 of Battig and Montague's original 56 categories produced more examples per subject than did our *something that could be portrayed on a map*. On the other hand all of the phrasings yielded numbers of responses large enough to establish that the categories in question were no less familiar to our subjects than were the categories used by Battig and Montague.

6.2 Frequent terms: Differences among the phrasings

Many of the 308 terms on our list — for example 'soil', 'fjord', 'state park' — were mentioned by very few subjects. Because the relative concentrations of such infrequently mentioned terms across the five phrasings would be heavily influenced by chance, we decided to concentrate our analysis on terms mentioned with a statistically more significant frequency, and arbitrarily chose to study only terms that were listed by at least 10% of the subjects for at least one of the five phrasings.[1] Thirty-five terms met this criterion, and are presented in Table 19.4 through Table 19.7. These terms give an illuminatory overview of the geographical ontology of our

[1] Complete data resulting from our elicitation of examples task for the five basic geographical categories can be found on the Web at http://geog.buffalo.edu/ncgia/ontology/BuffaloGeographicNorms.html.

TABLE 19.4
Terms most frequent for a kind of geographic feature.

Term	Feature	Object	Something	Concept	Map	Total
Number of subjects	54	56	51	51	51	263
Mountain	48	23	32	23	25	151
River	35	18	26	19	31	129
Lake	33	13	25	10	21	102
Ocean	27	16	18	16	18	95
Valley	21	7	4	7	0	39
Hill	20	9	11	3	0	43
Plain	19	6	5	4	1	35
Plateau	17	4	6	8	0	35
Desert	14	6	6	4	0	30
Volcano	10	4	5	3	0	22
Island	8	7	7	7	3	32
Forest	6	4	5	1	3	19
Stream	6	2	2	3	1	14

TABLE 19.5
Terms most frequent for a kind of geographic object.

Term	Feature	Object	Something	Concept	Map	Total
Number of subjects	54	56	51	51	51	263
Map	0	17	11	7	0	35
Globe	0	11	4	0	0	15
Peninsula	8	10	5	6	1	30
Compass	0	8	0	1	2	11
Rock	1	6	3	2	0	12
Atlas	0	6	2	2	0	10

TABLE 19.6
Terms most frequent for something geographic.

Term	Feature	Object	Something	Concept	Map	Total
Number of subjects	54	56	51	51	51	263
Land	2	6	6	5	0	19
The World	0	0	5	1	3	9

TABLE 19.7
Terms most frequent for a geographic concept.

Term	Feature	Object	Something	Concept	Map	Total
Number of subjects	54	56	51	51	51	263
Sea	9	8	9	11	5	42
Delta	4	1	0	6	0	11

subjects, but they also reveal how difficult it is to extract this ontology in the form of a single hierarchy of kinds or types of the sort envisaged by Aristotle or Linnaeus and presupposed also in much contemporary work in folk biology and related fields.

Our geographic ontology is a sinewy thing: as our data shows, it breaks down into categories in significantly different ways according to the terms we use in elicitation. What is noteworthy, however, is the degree to which physical geography predominates. For even when five different elicitation terms are employed, this does not affect in any significant way the predominance of items within the domain of physical geography (and the correspondingly low profile of human geographical items) that we had observed already in our much more limited pilot study. Only five terms reached the 10% threshold on all five versions of this question and all of these are physical: mountain, river, lake, ocean, and sea. This suggests that, for this population of subjects at least, it is the physical environment that provides the most basic examples of geographical phenomena. This predominance of physical geography lends support to the view that concepts for (some) types of geographical objects are very deeply rooted in our primary-theoretic cognitive architecture, namely those — like mountain, river, lake, ocean, and sea — referring to objects of a kind which were (surely) strongly relevant to the survival of our predecessors in primeval environments.

The variation in elicited responses for the five different phrasings also has philosophical import. Philosophical ontologists have long been aware of the problematic character of ontological terminology. What term, for example, should be used for the ontological supercategory within which all beings (things, entities, items, existents, realities, objects, somethings, tokens, instances, particulars, individuals) would be comprehended? Each of these alternatives has its adherents, yet each also brings problems. Thus some of the terms suggested can be held to narrow the scope of ontology illegitimately to some one particular *kind* of being, for example to beings which *exist*, or are *real*, or come ready-demarcated into *items*. Similar arguments have also been seen in the international spatial data standards community. Given the particular meanings of the terms *object*, *entity*, and *feature* in the US Spatial Data Transfer Standard, for example, how should these terms be translated into other natural languages?

Our experiment — which is we believe the first of its kind to address differences in the ways non-expert subjects use general terms of ontology — shows that some counterpart of these problems is present already in the uses of such terms by non-experts (in a way which has posed difficulties for us also in reporting the results of our experiments here).

6.2.1 A kind of geographic feature

Of all the five phrasings, the responses to *a kind of geographic feature* still stand out as most strongly dominated by aspects of the physical environment. In fact, the most frequently listed potentially non-natural item under 'a kind of geographic feature' was 'country', which was listed by only two out of 54 subjects. In other respects however the responses under the geographic feature heading are relatively heterogeneous. Subjects listed shape-based landforms such as mountain, hill, and valley; water bodies such as lake and ocean; water-courses such as river and stream; shore-bounded land features such as island, and other geophysical features such as plain, plateau, desert, and forest. Table 19.4 gives the most frequent examples listed under this heading, with the corresponding frequencies for these items insofar as they were listed under the other headings.

6.2.2 A kind of geographic object

Geographic object stands out from the other phrasings listed in the degree to which it elicits examples of small, portable items. *Map* is the most common term among all of those listed more frequently under this than under any other heading. This heading also elicited a somewhat low mean frequency of responses per subject, suggesting that the English term *object* so strongly connotes a portable, detached thing that many subjects could not readily imagine objects existing at geographical scales. (It is true that *mountain* and *river* were listed as examples of geographic objects even more often than was *map*, but they had much higher frequencies in the geographic features column.) The predominance, here, of *map* is a clear indication that our subjects were thinking of small objects with some geography-related purpose — every subject who listed *globe*, *compass*, or *atlas* also listed *map*, and of the 17 map-listing subjects, only one mentioned *mountain* and only two mentioned *lake*. Other results for geographic object are shown in Table 19.5.

6.2.3 Something geographic

Something geographic is perhaps the most general way to describe in English the domain under review. Not surprisingly, then, this phrasing of the category label picked up a mixture of the responses typical of the other phrasings. Terms predominating also under *geographic feature* — such as mountain, lake, river, ocean — are here most frequent, but then (albeit with a markedly lower frequency) comes map (which is listed by 11 subjects as against 32 for mountain). The only term that was more frequent here than for any of the other phrasings was *the world*, while another term, land, was listed equally often under *something geographic* and under *geographic object*. Both land and the world are very general kinds of geographical phenomena.

6.2.4 A geographic concept

The category elicited by the phrasing a *geographic concept* manifests in our subjects' responses the lowest degree of internal coherence for all the five phrasings. The analysis of mean numbers of examples of categories under the different phrasings suggests that subjects had more difficulty determining what we meant by a *geographic concept*, and thus more difficulty in coming up with examples, than they did for any other phrasing. In everyday English, the term 'concept' refers to something rather abstract. We have no hypothesis to account for *sea* and *delta* appearing here more frequently than elsewhere.

The data under this heading are of some general significance, however, since, of all the five phrasings tested, this was the one least tilted in the direction of eliciting examples of geographic *objects*. In light of our discussion, above, of the object-field dichotomy and of Millikan's (1998) proposal concerning *stuffs*, it is thus significant that this phrasing did not yield significant numbers of examples under headings which could be classified as *field-based* or *stuff-based* geospatial concepts. Thus the field-based term 'elevation' was elicited from only one subject under this heading, and no other field-based term occurred here at greater frequency than under other headings. Our data is less revealing as concerns the issue of stuff-based concepts. Terms such as 'land', 'desert', 'rock' and 'tundra' did indeed occur with a certain frequency (though with no higher frequency here than under object-phrasings), and the data is in any case difficult to interpret in virtue of the fact that all of these terms have both an object- (count) and a stuff-based (mass) reading.

6.2.5 Something that could be portrayed on a map

Prior to running the experiment, we thought that maps generally showed all and only geographical things (phenomena, features, items), and thus we expected 'something that could be portrayed on a map' would turn out to be roughly synonymous with 'something geographic'. But such was not the case (Table 19.8). Things from the domain of human geography — geographical things produced by people, either through construction or by at — appeared far more often in response to this wording than to any other. The subjects apparently were well aware that maps tend to portray cities, states, and counties, roads and streets, yet few listed them under the other categories of geographical things, and especially not under *features*. *Being geographical*, and *being portrayable on a map* are definitely different concepts, at least in terms of the priorities of terms included under them according to our subjects. Moreover, it seems that — again surprisingly — it is *being portrayable on a map* which comes closest to capturing the meaning of 'geographic' as this term is employed in scientific contexts. Geographers, it seems, are not studying geographical things as such things are conceptualized by naïve subjects. Rather, they are studying the domain of what can be portrayed on maps.

7 Summary, conclusions, and further work

Evidence presented in this paper has shown that geospatial concepts together form a coherent knowledge domain in the minds of non-experts in the United States. Although we had a very large sample of subjects, it is important to note that all subjects were native speakers of one language, English. Also, all subjects were from one institution, and most were educated in one region, the State of New York. However, preliminary data from parallel experiments carried out in Finland, Croatia,

TABLE 19.8
Terms most frequent for something that could be portrayed on a map.

Term	Feature	Object	Some-thing	Concept	Map	Total
Number of subjects	54	56	51	51	51	263
City	1	4	5	0	30	40
Road	1	2	3	1	27	34
Country	2	6	8	4	23	43
State	0	5	3	1	15	24
Continent	1	10	8	9	12	40
Street	0	1	1	1	8	11
Town	0	5	2	0	8	15
Highway	1	0	0	0	7	8
Park	0	0	0	0	6	6
Building	0	1	0	0	5	6
County	0	2	0	0	5	7
Elevation	0	0	0	1	5	6

and the United Kingdom produced very similar trends, suggesting that the effects reported here are not an artefact of our particular pool of subjects or of American English.

We believe that our results are of significance both to geographers in general — in throwing light on how the geospatial domain is integrated into the primary cognitive architecture of human beings — and also to those working in the field of geographic information science, in giving a first overview of the geospatial ontologies shared by the users of GISystems. But the results are also of broader significance, and they have implications not only for ontology but also for linguistics and for other cognitive sciences. They show that the interface between language and ontology is not as simple as has hitherto been held. Our data have demonstrated that this is true for ontological terms in the specific realm of geography, but they give strong reason to believe that it will be true in general.

One of the most surprising and potentially significant results of this empirical study is that the base term for the superordinate category to 'geographic' made a considerable difference to subjects' opinions of class members. Depending on whether we asked for *geographic features*, *geographic objects*, or *something geographic*, we observed significant differences in frequencies of terms listed. *Feature* elicited almost exclusively natural geographical things, to the near exclusion of constructed or fiat entities. *Object* apparently triggered on the part of many subjects a mindset wherein they felt they were called upon to provide examples of manipulable, detached objects and this, when combined with *geographic*, caused them to list artefacts with a geographical purpose or meaning such as *map*, *atlas*, *globe*, and *compass*. We also expected that the phrase *can be portrayed on a map* amounted to just another way of saying *geographical*, but it was exclusively under the *mappable* heading that fiat objects such as geopolitical subdivisions and geographical-scale artefacts such as roads and cities were listed with any frequency. In spite of all of this, however, all of the terms produced under any of these questions appear to be terms which to a large degree denote geographical things (items, entities, beings). Thus the results summarized above provide a first approximation to the basic noun lexicon for geographical ontologies, even while pointing out unexpected difficulties in the way of completing an ontology of the geographical (folk) domain.

How then should we express the relations between human conceptualizations for geographic *object, feature*, for *mappable*, and so forth? We suggest that these conceptualizations represent not different *ontologies* that we might ascribe to the subjects in the groups we tested. Rather, they are a matter of different superordinate categories — objects, features, things — that intersect to varying degrees in virtue of the fact that they share a common domain — the domain of geography. Particular kinds of phenomena, such as mountains or maps or buildings, have different relative prominence or salience under these different superordinate categories. We propose, therefore, that there is just one (folk) ontology of the geospatial realm, but that this ontology gets pulled in different directions by contextually determined salience conditions. To appreciate the pervasive effect of such salience conditions, compare the way in which an ornithologist would give a single unified ontology of birds, but would give different examples, or the same examples in different order, in providing a list of birds he likes, or birds he saw today, or birds he likes to eat, and so forth. What we have shown is that analogous differences are triggered by the use of distinct ontological terms. This outcome is significant not least because the distinctions captured by ontological terms are commonly held to be of low or zero practical significance.

Acknowledgments

This paper is based upon work supported by the National Science Foundation, Geography and Regional Science program, under Grant No. BCS-9975557. Support of the National Science Foundation is gratefully acknowledged. Larry Torcello coded the data and assisted with the human subjects' testing. We also wish to thank the student subjects and their instructors for participating in the study.

References

Atran, S., 1993, *Cognitive Foundations of Natural History: Towards an Anthropology of Science* (Cambridge: Cambridge University Press).

Battig, W. F., and Montague, W. E., 1968, Category norms for verbal items in 56 categories: a replication and extension of the Connecticut Norms. *Journal of Experimental Psychology Monograph,* **80**, Part 2, 1–46.

Blades, M., Blaut, J., Darvizeh, Z., Elguea, D., Soni, S., Sowden, D., Stea, D., Surajpaul, R., and Uttal, D., 1998, A cross-cultural study of young children's mapping abilities. *Transactions of the Institute of British Geographers,* **23**, 269–277.

Blaut, J. M., 1997. The mapping abilities of young children — children can. *Annals of the Association of American Geographers,* **87**, 152–158.

Casati, R., Smith B., and Varzi, A. C., 1998, Ontological tools for geographic representation. In *Formal Ontology in Information Systems,* edited by N. Guarino (Amsterdam, Oxford, Tokyo, Washington, DC: IOS Press) (Frontiers in Artificial Intelligence and Applications), pp. 77–85.

Dowty, D., 1998, On the Origin of Thematic Roles Types, Invited lecture, Lexicon in Focus Conference, Wuppertal, August 1998. See http://www.ling.ohio-state.edu/dowty/where-do-roles.abstract, accessed 13 April 2001.

Egenhofer, M. J., and Mark, D. M., 1995, Naive geography. In *Spatial Information Theory: A Theoretical Basis for GIS,* edited by A. U. Frank and W. Kuhn, Lecture Notes in Computer Science 988 (Berlin: Springer-Verlag), pp. 1–15.

Forguson, L., 1989, *Common Sense* (London and New York: Routledge).

Egenhofer, S. M., 1990, Can young children use maps to navigate? *Cartographica,* **27**, 54–66.

Gibson, J. J., 1979, *The Ecological Approach to Visual Perception* (Boston: Houghton-Mifflin).

Gruber, T. R., 1993, A translation approach to portable ontology specifications. *Knowledge Acquisition,* **5**, 199–220.

Guarino, N., 1998, Formal ontology and information systems. In *Formal Ontology in Information Systems. Proceedings of the 1st International Conference,* edited by N. Guarino (Amsterdam: IOS Press), pp. 3–15.

Guarino, N., and Welty, C., 2000, Ontological analysis of taxonomic relationships. In *Proceedings of ER-2000: The 19th International Conference on Conceptual Modeling,* edited by A. Laender and V. Storey, Lecture Notes in Computer Science (Berlin/New York: Springer-Verlag), pp. 210–224.

Hazen, N. L., Lockman, J. J., and Pick, H. L., 1978, The development of children's representation of large-scale environments. *Child Development,* **49**, 623–636.

Hobbs, J. R., and Moore, R. C., (editors), 1985, *Formal Theories of the Commonsense World* (Norwood: Ablex).

Horton, R., 1982, Tradition and modernity revisited. In *Rationality and Relativism,* edited by M. Hollis and S. Luke (Oxford: Blackwell), pp. 201–260.

IEEE, 2000, Standard Upper Ontology (SUO) Working Group http://suo.ieee.org/, accessed 13 April 2001.

Keil, F. C., 1979, *Semantic and Conceptual Development: An Ontological Perspective* (Cambridge, MA: Harvard University Press).

Keil, F. C., 1997, Conceptual development and category structure. In *Concepts and Conceptual Development: Ecological and Intellectual Factors in Categorization,* edited by U. Neisser (Cambridge: Cambridge University Press), pp. 175–200.

Keil, F. C., 1989, *Concept, Kinds and Cognitive Development* (Cambridge, MA: MIT Press).

Keil, F. C., 1994, Explanation based constraints on the acquisition of word meaning. In *The Acquisition of the Lexicon,* edited by L. Gleitman and B. Landau (Cambridge, MA: MIT Press), pp. 169–196.

Mark, D. M., Smith, B., and Tversky, B., 1999, Ontology and geographic objects: an empirical study of cognitive categorization. In *Spatial Information Theory: A Theoretical Basis for GIS,* edited by C. Freksa and D. M. Mark, Lecture Notes in Computer Science (Berlin: Springer-Verlag), pp. 283–298.

Medin, D. L., and Atran, S., editors, 1999, *Folkbiology* (Cambridge, MA: MIT Press).

Millikan, R. G., 1998, A common structure for concepts of individuals, stuffs, and real kinds: more Mama, more milk, and more mouse. *Behavioral and Brain Sciences,* **9**, 55–100.

Murphy, G. L., and Medin, D., 1985, The role of theories in conceptual coherence. *Psychological Review,* **92**, 289–316.

Quine, W. V., 1953, 'On What There Is', as reprinted in *From a Logical Point of View* (New York: Harper & Row).

Rosche, E., 1973, On the internal structure of perceptual and semantic categories. In *Cognitive Development and the Acquisition of Language,* edited by T. E. Moore (New York: Academic Press).

Rosche, E., 1978, Principles of categorization. In *Cognition and Categorization,* edited by E. Rosche and B. B. Lloyd (Hillsdale, NJ: Erlbaum).

Schulke-Kremer, S., 1997, Adding Semantics to Genome Databases: Towards an Ontology for Molecular Biology. In *Proceedings of the Fifth International Conference on Intelligent Systems for Molecular Biology,* edited by T. Gaasterland et al. (Palo Alto: AAAI Press), pp. 272–275.

Smith, B., 1995a, On drawing lines on a map. In *Spatial Information Theory: A Theoretical Basis for GIS,* edited by A. U. Frank and W. Kuhn, Lecture Notes in Computer Science 988 (Berlin/Heidelberg/New York: Springer), pp. 475–484.

Smith, B., 1995b, Formal ontology, common sense, and cognitive science. *International Journal of Human-Computer Studies,* **43**, 641–667.

Smith, B., 1995c, The structures of the commonsense world. *Acta Philosophica Fennica,* **58**, 290–317.

Smith, B., 2000, Objects and their environments: From Aristotle to ecological psychology. In *The Life and Motion of Socioeconomic Units,* edited by A. U. Frank (London: Taylor & Francis), pp. 79–97.

Smith, B., and Mark, D. M., 1998, Ontology and geographic kinds. In *Proceedings, 8th International Symposium on Spatial Data Handling* (SDH98), edited by T. K. Poiker and N. Chrisman (Vancouver International Geographical Union), pp. 308–320.

Smith, B., and Mark, D. M., 1999, Ontology with human subjects testing. *American Journal of Economics and Sociology,* **58**, 245–272.

Smith, B., and Varzi, A. C., 1999, The formal structure of ecological contexts. In *CONTEXT '99: Modeling and Using Context. Proceedings of the Second International and Interdisciplinary Conference,* edited by P. Bouquet, P. Brezillon, L. Serafini, M. Benecereti, and F. Castellani, Lecture Notes in Artificial Intelligence, 1688 (Berlin and Heidelberg: Springer-Verlag), pp. 329–350.

Smith, B., and Varzi, A. C., 2000, Fiat and bona-fide boundaries. *Philosophy and Phenomenological Research,* **60**, 401–420.

Spencer, C., and Blades, M., 1985, How children navigate. *Journal of Geography,* 445–453.

Geographic Categories: An Ontological Retrospective

Barry Smith and David M. Mark

Since it is only five years since the publication of our paper, "Geographical categories: An ontological investigation" (Smith and Mark 2001), it seems somewhat strange to be making retrospective comments on the piece. Nevertheless, the field is moving quickly, and much has happened since the article appeared. A large number of papers have already cited the work, which suggests that there is a seam here that people find worthy of being mined.

In this short essay, we first review the paper and attempt to assess its significance from the perspective of our current work. We then put the paper in the context of our individual and joint works, which led up to it, and summarize our research trajectories since the paper appeared, pointing out what some of this reveals about spatial ontology in general. We conclude with some remarks on the future of ontological research in geographical information science.

Brief overview of our 2001 paper

The paper reported some of the main results of a series of experiments carried out in Buffalo and elsewhere between 1998 and 2001, and the inferences we were able to draw from those results concerning the ways normal human beings conceptualize their geographical environment. The idea for these experiments grew out of our general curiosity concerning the development of a theory of naïve or folk geography, itself reflecting our separate and collective interests in the work of Patrick Hayes and others on the topic of naïve or commonsense physics (Hayes, 1985; see also Smith and Casati, 1994 and Egenhofer and Mark, 1995).

How do nonexpert subjects conceptualize geospatial phenomena? To find a way of answering this question, we developed a series of simple questionnaire-style experiments in which we asked many hundreds of subjects to provide examples of geographical categories in response to a series of differently phrased elicitations. The results, we hypothesized, would yield an ontology of geographical categories — a catalog of the prime geospatial categories shared in common by human subjects independently of their exposure to scientific geography. To some extent this hypothesis was confirmed (for summaries, see our 1999 and 2001 papers presented at the Conference on Spatial Information Theory (COSIT) meetings (Mark et al., 1999, 2001).

Unfortunately, however, we very quickly discovered that the precise formulation of the elicitation question yielded significantly different catalogs of prime geospatial categories. Thus, if we combined the adjective *geographic* with the nouns *feature* or *object*, this yielded almost exclusively elements of the physical environment of geographical scale or size, such as *mountain, lake, river*. Even the words object and feature led systematically to somewhat different lists of examples. The phrase "thing that could be portrayed on a map," which a priori we had assumed would yield a roughly equivalent list of categories, produced instead examples of many geographical scale artefacts (*road, city*, etc.) and fiat objects (*state, country*, etc.: see Smith 2001), alongside the physical feature types elicited overwhelmingly by the *geographic feature* and *geographic object* triggers.

Interestingly, our data also suggested that there is considerable mismatch between the meanings assigned to the terms *geography* and *geographic* by geographic scholars and by ordinary subjects, so that there is a sense in which geographic scholars are not in fact studying geographical phenomenas, as such phenomena are conceptualized by naïve subjects. The data suggest, rather, a special role in determining the subject matter of scientific geography precisely for the concept of *thing can be portrayed on a map* — a result we believe to be worthy of further investigation.

Where we came from

Before writing the *IJGIS* article, we had been collaborating on GIScience research for several years. We met through a meeting of the Buffalo Cognitive Science Center, where Darren Longo, one of Smith's students, presented a paper on Mark's work on cognitive topology of spatial relations such as *across*, itself based on experimental work on human subjects' judgments concerning interrelations between roads intersecting parks in simple sketch maps. The COSIT meetings, and our common association with Andrew Frank in Vienna, played an important role in our research convergence. Mark's first published work on geographic categories was presented at the inaugural COSIT meeting on Elba, Italy, in 1993 (Mark, 1993), and Smith's first publication in the GIS-related literature was at the second COSIT meeting in 1995 in Semmering, Austria. Both papers have been moderately influential — according to Science Citations Index, Mark-1993 has been cited 16 times, and Smith-1995 almost double that (31 citations). Smith's work on fiat objects presented in Semmering was also incorporated into the IEEE Standard Upper Ontology, and formed one starting point of the theory of granular partitions, which Smith then developed in collaboration with Berit Brogaard, Thomas Bittner, and Pierre Grenon (Smith and Brogaard, 2002; Bittner and Smith, 2003; Grenon and Smith, 2004).

Our paper at the 1998 Spatial Data Handling meeting (Smith and Mark, 1998, 1999) laid out some of the basic principles and issues for ontology of the geographic domain, and led to U.S. National Science Foundation (NSF) support for the work beginning in July 1999 (Mark and Smith, 1999). The goal of the NSF project was to develop a formal ontology for geographic entities and categories, based on rigorous empirical research using human subjects. Parallel studies were conducted in several languages and regions, so the resulting ontology is at least to some degree multilingual. We still have much data collected during the original experiments, including data on how beliefs about geographical categories are expressed in a variety of different languages, which we would be happy to make available to researchers who are interested in going further along this trajectory.[2]

Developments since 2001: Smith

In the period immediately following the publication of our *IJGIS* paper, we co-authored additional papers on the more specific issue of the relationship between fieldlike structures captured, for example, in digital elevation models, and those geographic features captured in the lexicon of normal human subjects — the issue of the quantitative-qualitative divide (Smith and Mark, 2003; Mark and Smith, 2004). These papers highlighted the subtle and complicated relationship between the objective reality of the shape of the Earth's crust, and the features that people reason and communicate about in natural language. The World Wide Web is still highly oriented toward content presented as words in natural language, making field-to-feature conversion an important link in connecting geographic information to the Web. Analogous work at the interface between qualitative and quantitative geospatial data and information is also illustrated in other domains, for example in the field of military and intelligence-related information fusion. We both participated in a successful effort to have ontology adopted by the University Consortium for Geographic Information Science (UCGIS) as one of about 14 high-priority research topics for geographic information science in the United States (Mark et al., 2004). Ontology remains a hot topic in information science in general, and geographical information science in particular, and the 2001 special issue of *IJGIS* that included our paper was a key step in the promotion and legitimization of the topic within GIScience.

Since the publication of the paper, Smith has broadened his ontological purview to encompass spatiotemporal entities in all domains, presenting in particular the SNAP and SPAN ontology (Grenon and Smith, 2004), which is an attempt to do justice both to the process-oriented view, which sees the world as a constellation of four-dimensional entities, and the object-oriented view, which sees the world as comprised of continuant entities that endure identically through time. Smith has also directed much attention to the medical domain. In his 2005 COSIT paper, written with colleagues from the domain of biomedical informatics (Smith et al., 2005), he compares the achievements in qualitative and quantitative spatial ontology achieved in the domain of human anatomy with those achieved to date in GIScience. The most impressive achievement in spatial information science on the anatomical side thus far is the Foundational Model of Anatomy (FMA), a map of the human body conceived in ontological terms. Like maps of other sorts, including the maplike representations we find in familiar anatomical atlases, it is a representation of a certain portion of spatial reality as it exists at a certain (idealized) instant of time.

[2] One of the largest data sets, for English, is available on the Web at http://www.geog.buffalo.edu/ncgia/ontology/.

But unlike other maps, the FMA comes in the form of a sophisticated ontology of its object-domain, comprising some 1.5 million statements of anatomical relations among some 70,000 anatomical kinds. It is further distinguished from other maps in that it represents not some specific concrete portion of spatial reality (say, the Bay of Biscay), but rather a generalized or idealized spatial reality associated with a generalized or idealized human being at some generalized or idealized instant in time. Biomedicine provides a rich domain for such idealized qualitative representations of spatial structures, but it offers much more impoverished resources for describing individual instances. This is because your heart, for example, is constantly changing its shape, size, and location (Pilgram et al., 2004). The surface of the earth, on the other hand, provides a relatively impoverished domain for qualitative ontological representations, but much richer possibilities for the gathering of precise, quantitative instance data, by virtue of the fact that changes of shape, size, and location of the objects at or on the surface of the earth are, at least so long as we restrict ourselves to objects of geographic scale and to changes detectable through perception, relatively limited. Another difference between the anatomical and the geographical domain turns on the different role of fiat objects within each. Thus, for example, regions on the surface of the body delimited by fiat play a relatively insignificant role in Western anatomical science, but a central role in traditional Chinese medicine. Fiat demarcations on the surface of the Earth play a central role in the Western understanding of nations and sovereignty in the era since the Treaty of Westphalia, but a relatively insignificant role in the geopolitical ontology of Islam.

Developments since 2001: Mark

Mark's follow-up to the 2001 paper has gone in a quite different direction, focusing on in-depth examinations of definitions of geospatial feature types in other cultures and languages. Mark had been intrigued by the issue of whether spatial cognition was universal to all people, or whether there were significant cultural differences that should influence GIS design, but he had failed to come up with firm evidence either way. A sabbatical in 2002 was spent partly at the Max Planck Institute for Psycholinguistics in Nijmegen, and partly in western Australia. Through Andrew Turk's relationships with an indigenous community in Australia, Mark was finally able to dig into cultural differences in geospatial concepts deeply enough to reveal actual differences (Mark and Turk, 2003). This led in turn to research collaboration with David Stea, including another NSF grant (Mark and Stea, 2004) to compare landscape categories among several arid-lands peoples. Work in collaboration with Stea, Turk, and indigenous collaborators is already underway with the Yindjibarndi in Australia and with the Navajo in New Mexico and Arizona. Why have these studies found cultural differences in geospatial categories, when earlier researchers did not find them for spatial relations? At this point answers to this question are still in the realm of speculation, but it is plausible to hypothesize that spatial relations are more robustly hardwired into human perception, whereas categories for geographic entities are much less determined by basic cognitive factors. Most languages have a relatively small number of spatial relation terms, represented by closed-class grammatical elements such as prepositions in English. Entity categories, on the other hand, are typically encoded in languages by nouns, the most open, extensible class of words. More research is needed, including research into cross-cultural differences in conceptualizations of spatial entity and relation categories, before definitive answers will be available. But if the proposed hypothesis turns out to be correct, the implication is that GIS software may be relatively easy to adapt to other languages and cultures, whereas spatial data infrastructures will need to pay specific attention to multilingual aspects of semantics and categorization.

Concluding comments

Ontology remains an important topic in GIScience, and can be expected to continue to be so for some time. Considerable interest is currently being exhibited in ontology above all by a number of federal government organizations, for example in the context of the development of the Federal Enterprise Architecture Reference Model Ontology. At a meeting in Buffalo in October 2005 there was inaugurated the National Center for Ontological Research, a consortium of government, industry, and academic partners dedicated to raising the standards of ontological research through application of the empirical scientific method. Informal comparisons of spatial ontologies for the geographical and anatomical domains suggest that even for the single domain of spatially extended entities, major differences among ontologies will have to be accepted as the order of the day, and this all the more so if cultural variance must be taken into account.

References

BITTNER, T. AND SMITH, B., 2003, A theory of granular partitions. In *Foundations of Geographic Information Science*, Duckham, M., Goodchild, M.F., and Worboys, M. F., Eds., Taylor & Francis, London, 117–151.

EGENHOFER, M.J. AND MARK, D.M., 1995, Naive geography. In *Spatial Information Theory: A Theoretical Basis for GIS*, Lecture Notes in Computer Sciences 988, Frank, A.U. and Kuhn, W., Eds., Springer-Verlag, Berlin, 1–15.

GRENON, P. AND SMITH, B., 2004, SNAP and SPAN: Towards dynamic spatial ontology, *Spatial Cognition and Computation*, 4(1), 69–103.

HAYES, P.J., 1985, The second naive physics manifesto. In *Formal Theories of the Commonsense World*, Hobbs, J.R. and Moore, R.C., Eds., Ablex, Norwood, 1–36.

MARK, D.M., 1993, Toward a theoretical framework for geographic entity types. In *Spatial Information Theory: A Theoretical Basis for GIS*, Lecture Notes in Computer Sciences 716, Frank, A.U. and Campari, I., Eds., Springer-Verlag, Berlin, 270–283.

MARK, D.M., SKUPIN, A., AND SMITH, B., 2001, Features, objects, and other things: Ontological distinctions in the geographic domain. In *Spatial Information Theory: Foundations of Geographic Information Science*, Lecture Notes in Computer Science 2205, Montello, D., Ed., Springer-Verlag, Berlin, 489–502.

MARK, D.M. AND SMITH, B., 1999, Geographic categories: An ontological investigation. Research Grant BCS-9975557, Geography and Regional Science Program, National Science Foundation.

Mark D.M. and Smith, B., 2004, A science of topography: From qualitative ontology to digital representations. In *Geographic Information Science and Mountain Geomorphology*, Bishop, M.P. and Shroder, J.F., Eds., Springer-Praxis, Chichester, England, 75–100.

Mark, D.M., Smith, B., Egenhofer, M.J., and Hirtle, S.C., 2004, Ontological foundations for geographic information science. In *A Research Agenda for Geographic Information Science*, McMaster, R.B. and Usery, E.L., Eds., CRC Press, Boca Raton, FL, 335–350.

Mark, D.M., Smith, B., and Tversky, B., 1999, Ontology and geographic objects: An empirical study of cognitive categorization. In *Spatial Information Theory: A Theoretical Basis for GIS*, Lecture Notes in Computer Science 1661, Freksa, C. and Mark, D.M., Eds., Springer-Verlag, Berlin, 283–298.

Mark, D.M. and Stea, D., 2004, Collaborative research: Landscape, image, and language among some indigenous people of the American Southwest and Northwest Australia. Research Grants BCS-0423023 and BCS-0423075, Geography and Regional Science Program.

Mark, D.M. and Turk, A.G., 2003, Landscape categories in Yindjibarndi: Ontology, environment, and language. In *Spatial Information Theory: Foundations of Geographic Information Science*, Lecture Notes in Computer Science 2825, Kuhn, W., Worboys, M. and Timpf, S., Eds., Springer-Verlag, Berlin, 31–49.

Pilgram, R., Fritscher, K.D., Fletcher, P.T., and Schubert, R., 2004. Shape modelling of the multiobject organ heart. In *IASTED: International Conference on Biomedical Enigineering — BioMED 2004*, Acta Press, 157–160.

Smith, B., 1995, On drawing lines on a map. In *Spatial Information Theory. A Theoretical Basis for GIS*, Lecture Notes in Computer Science 988, Frank, A.U. and Kuhn, W., Eds., Springer-Verlag, Berlin, 475–484.

Smith, B., 2001, Fiat objects, *Topoi*, 20(2), 131–148.

Smith, B. and Brogaard, B., 2002. Quantum mereotopology. *Annals of Mathematics and Artificial Intelligence,* 7, 591–612.

Smith, B. and Casati, R., 1994, Naive physics: An essay in ontology, *Philosophical Psychology*, 7(2), 225–244.

Smith, B. and Mark, D.M., 1998, Ontology and geographic kinds. In *Proceedings of the 8th International Symposium on Spatial Data Handling (SDH'98)*, Poiker, T.K. and Chrisman, N., Eds., International Geographical Union, Vancouver, BC, 308–320.

Smith, B. and Mark, D.M., 1999, Ontology with human subjects testing: An empirical investigation of geographic categories, *American Journal of Economics and Sociology*, 58 (2), 245–272.

Smith, B. and Mark, D.M., 2001. Geographic categories: An ontological investigation, *International Journal of Geographical Information Science*, 15 (7), 591–612.

Smith, B. and Mark, D.M., 2003. Do mountains exist? Towards an ontology of landforms, *Environment and Planning B: Planning and Design*, 30 (3), 411–427.

Smith, B., Mejino Jr., J.L.V., Schulz, S., Kumar, A., and Rosse, C. 2005, Anatomical information science. In *Spatial Information Theory (COSIT 2005)* Lecture Notes in Computer Science 3693, Cohn, A.G. and Mark, D.M., Eds., Springer-Verlag, Berlin, 149–164.

International Journal of Geographical Information Science,
2003, Vol. 17, No. 1, 25–48.

20 Extending GIS-Based Visual Analysis: The Concept of Visualscapes

Marcos Llobera

Abstract. A Geographical Information System (GIS) is used to retrieve and explore the spatial properties of the visual structure inherent in space. The first section of the article aims to gather, compare and contrast existing approaches used to study visual space and found in disciplines such as landscape architecture, urbanism, geography and landscape archaeology. The concept of a *visualscape* is introduced in the following section as a tentative unifying concept to describe all possible ways in which the structure of visual space may be defined, broken down and represented within GIS independently of the context in which it is applied. Previous visibility studies in GIS are reviewed and further explored under this new concept. The last section presents the derivation of new visual parameters and introduces a new data structure (i.e., a vector field) to describe the *visual exposure* of a terrain.

1 Introduction

This paper describes the use of GIS to study human visual space. To date, the use of GIS to explore human space, i.e., as encountered by an individual, has been very limited. This is partly due to the fact that most GIS operations are based on a traditional geographical view of space which is essentially two-dimensional with a fixed and external frame of reference. The absence of GIS procedures that consider terrain and built environment representations together is a clear indication, among others, of these limitations. Hence, traditional GIS operations are inadequate for developing models of human–space interaction, particularly human perception, whenever a mobile frame of reference is considered. Though some attempts exist to relate GIS with cognition and perception, these have mostly concentrated on landscape preference (Baldwin *et al.* 1996, Germino *et al.* 2001). Ultimately, the design of new GIS routines, and/or the development of new spatial tools that will accommodate human and other factors, will become necessary if cognitive and perceptual factors are to be linked with spatial information. In the meantime, existing GIS can be used to illustrate the necessity and potential of these types of analyses.

The idea that any spatial configuration structures human visual space by virtue of its distribution and geometry, and that such structure can be described spatially using different parameters, underlies the entire paper. Studies that have sought to explore these properties have been developed for the most part within the areas of urbanism and architecture, largely because they have been based on the application of a 'watered down' version of the notion of isovist which permits descriptive parameters to be calculated easily. While visibility studies in 'natural' environments, mostly based on the application of GIS, have not emphasized the structural aspect of visual space, many of the concepts found in these studies, e.g., cumulative viewshed, can still be interpreted as providing a simple description of such structure. The concept of *visualscapes* is introduced here to describe all possible ways in which the structure of visual space may be defined, broken down and represented within GISc independent of the context where it is being applied. Previous visibility studies in GIS are reviewed under the notion of visualscape, and further explored under this new concept. The last section presents the derivation of new visual parameters and introduces a new data structure (i.e., a vector field) to describe the *visual exposure* of a terrain.

The nature of all of the examples used in this paper is purposely generic. Although this limits the possibilities of exploring 'real' implications, it also guarantees the applicability of new concepts to any context.

2 Background

Formal approaches to the study of visual space can be found in various fields, such as urbanism (Batty 2001, Turner *et al.* 2001), architecture (Benedikt 1976), geography (Fisher 1995) or archaeology (Wheatley 1995, Fisher *et al.* 1997, Lake *et al.* 1998). These studies tend to fall into two categories: the built environment and 'natural' landscape.

In the following sections, some of these studies will be reviewed and compared with others focussing on 'natural' landscapes. The following discussions are centered around the basic units of analysis used in both approaches, i.e., isovists and viewsheds.

2.1 Urban landscape: The study of isovists

Recently, several works have appeared that focus on the properties of urban or architectural visual patterns, which ultimately seek to elicit and derive possible social implications (Turner *et al.* 2001). Most approaches are based on Benedikt's inspiring work on isovist and isovist fields (1979). In this work, Benedikt defined and explored the concept of an isovist in detail, a notion first introduced by Tandy (1967). This constitutes the basic element of analysis used in recent research on visibility within an urban context. An isovist is defined as a subset of points in space — all of those points in a visible surface D that are visible from a 'vantage' [view]point (x) (Benedikt 1979). However, it is most often thought of in relation to its geometry, i.e., usually as a two-dimensional polygon representing the area of visibility associated with a specific viewpoint. It is vital to note that while isovists have been calculated as two-dimensional entities ever since their inception, Benedikt defined them originally as being three- and four-dimensional (3D + time). Benedikt derived and explored several numerical properties of isovists, such as *area, perimeter, occlusivity, variance, skewness* and *circularity* of isovists, and 'mapped' them in order to generate some sort of mathematical scalar field, or isovist field.

Isovists (Figure 20.1) are usually the result of ad hoc programming (the exception being CASA's *DepthMap*). Generally, they are derived from urban and architectural plans by disregarding any information on the height variability of urban elements (e.g., buildings, fences). The possibility of being able to look beyond an obstacle, once a line-of-sight (LoS) has reached it, is usually never considered. As a result, isovists do not present 'holes', which means that they can be represented easily by simple polygons (de Berg *et al.* 1997). An individual positioned at any location in the isovist can walk straight up to the original viewpoint without ever losing sight of it. The calculation of isovists has traditionally been carried out in continuous space, without any need for sampling (isovists are still conceived as continuous). Recently, the possibility of mapping numerical characteristics of isovists back into space has precipitated the adoption of discrete representations (Batty 2001). Discussions on the effects of distance over visibility are not present in these studies because the range of isovists tends to be short within an urban context (Batty 2001). Because of the above factors the computation and description of isovists is a relatively quick and unproblematic process. It can be argued, however, that the restrictions imposed on their calculation may ultimately reduce their usefulness.

Batty (2001) and Turner *et al.* (2001) have recently extended Benedikt's work by representing isovists as a subgraph of a visibility graph (De Floriani *et al.* 1994) from which several properties, such as *average distance, minimum distance, maximum distance*, area, perimeter compactness and cluster ratio, could be calculated

FIGURE 20.1 Typical example of an isovist.

and mapped back into space. When such properties are computed for each point within a sample space, a scalar field is created similar to that found in Benedikt (1979). Both articles discuss the possibility of deriving social information whenever an isovist is treated as a graph.

The process of creating a scalar field, by mapping numerical properties derived from an 'isovist graph' back into space, and the subsequent interpretation of these patterns, is quite critical. Properties of an 'isovist graph' not only describe the interrelationship between a viewpoint and its visible points, but also the interrelationship between each visible point within the isovist (O'Sullivan and Turner 2001). This means that, occasionally, the value at a particular location, or viewpoint, may be due predominantly to the interrelationships among each of the other points in the isovist, rather than to the relationship between the viewpoint itself and its visible points within the isovist. Hence the coupling of a value (describing a certain property) to a location may be very loose and difficult to interpret at times.

An example of the difficulties of interpreting the spatial aspect of visual space is found in Turner *et al.* (2001). In this study, the authors make the claim that the clustering coefficient 'indicates how much of an observer's visual field will be retained or lost as they move away from that point' (Turner *et al.* 2001). Here it is assumed that visual field refers to the visible area associated to a specific viewpoint. If so, this cannot be read, at least not directly, from the definition of the clustering coefficient of an 'isovist graph.' The clustering coefficient of an isovist graph describes the interrelationship between all locations within the visibility graph but does not, at least directly, describe the relationship between a specific location and its neighbors. The next example (Figure 20.2) clearly demonstrates this point. It shows the plan of a room onto which a regular grid has been laid out as a way of sampling the space. In this case, the room could be 3 × 4m and the sampling rate 1m. The clustering coefficient is defined as:

> the number of edges between all the vertices in the neighbourhood of the generating vertex (i.e., the number of lines of sight between all the locations comprising the isovist) divided by the total number of possible connections with that neighbourhood size [neighborhood being all the vertices that are visible from a location] (Turner *et al.* 2001)

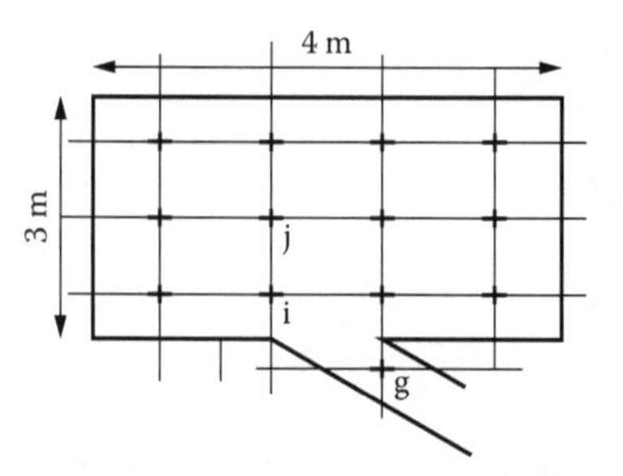

FIGURE 20.2 This example illustrates how locations *i*, *j* may share the same cluster coefficients but not necessarily the same visual change when moving from them to any of their neighbouring locations.

Formally the clustering coefficient is defined as

$$C_i = \frac{\left|\bigcup\{e_{jk} : v_j, v_k \in N_i\}\right|}{k_i(k_i - 1)} \tag{20.1}$$

where the numerator describes the number of LoS in an isovist/ viewpoint (i) and the denominator represents the number of possible LoS for that same isovist/viewpoint (i).

The clustering coefficient is the same at locations i and j (i.e., $C_i = C_j$), the neighbourhood around both locations is 13, and almost identical in composition (the only difference being that j is part of N_i while i is part of N_j). The number of LoS within this neighbourhood is less than the maximum number of possible LoS for a neighbourhood of similar size, given that 13 out of 7 locations do not have visual contact with g and vice versa. Hence while both locations have the same cluster coefficient it does not follow that the effect of moving from each location to its immediate neighbours is necessarily the same. Moving from i to g represents a big visual change, as opposed to moving from j to any of its immediate neighbours. Strictly speaking, how much of an observer's visual field is retained does not follow from the definition of clustering coefficient of an isovist graph.

Looking at the clustering coefficients of points across an area gives an indication of the nature of the change in the visual environment but cannot be easily applied to a specific case. While it may be true that locations with high clustering coefficients tend to indicate that there will be small visual variations when moving to neighbouring locations, as it is in the case of location j, this implication does not follow directly from having a high clustering coefficient. The identification of a location, such as i, in space is significant, as it marks a location (within our sample space) in where some aspect of the visual field changes dramatically (e.g., the shape or area). Although, strictly speaking, the interpretation of Turner *et al.* (2001) does not follow from the definition of the clustering coefficient, the possibility of increasing the number of locations in the sample space does support their original observation.

Outside the context of urban and architectural studies, references on visual analyses tend to use the term viewshed for isovist. Currently, viewsheds are closely linked to the use of commercial GIS and although they are very well established, they have not been explored in the same manner as isovists.

2.2 Visual patterns in a 'natural' landscape: The viewshed

The viewshed procedure (used to generate viewsheds) is a standard procedure among most GIS packages today. It is used, essentially, to calculate which locations (i.e., grid cells) in a digital elevation model (DEM) can be connected by means of an uninterrupted straight line (i.e., LoS) to a viewpoint location within any specified distance. Effectively, it calculates which locations or objects are not obstructed by topography and therefore may be visible from the specified viewpoint location. While it is true that for any location to be visible it must be connected by at least one uninterrupted LoS to the viewpoint location, this does not guarantee that it is visible from that viewpoint, i.e., atmospheric conditions may render an unobstructed object invisible. Whether a location, or an object on it, can be distinguished or identified is never considered.

Any interpretation based on the results of the viewshed calculation, particularly when a human component is present, is subject to the limitations of the DEM (e.g., altitude errors, curvature of the Earth); the absence of detailed coverage (e.g., vegetation, built environment); the effect of atmospheric conditions and the ability of the observer to resolve features. Unfortunately a large proportion of examples on the application of viewsheds do not address these restrictions directly (Gaffney and Stani 1991, Fels 1992, Miller *et al.* 1994, Gaffney *et al.* 1996, Lee and Stucky 1998). While the inclusion of the term view in the term viewshed can be seen as partly contributing to misleading interpretations (Tomlin 1990 for a more neutral term), it is the lack of better algorithms which is ultimately responsible for setting the limits for further interpretations. Gillings and Wheatley (2000) have provided a useful nontechnical synthesis on the problems and risks surrounding the use and interpretations of viewshed results.

Viewsheds are most often calculated and represented using a raster data model (though algorithms using triangular-irregular-networks or TINs also exist, see De Floriani *et al.* 1994, De Floriani and Magillo 1999). They are derived by means of algorithms which require terrain heights to be checked along each LoS calculated (Fisher 1991, 1992, 1993). Their computation is therefore far more intensive than the one needed to compute isovists as these are currently found. The nature of the results is also different. Viewsheds are usually irregular and fragmented; often comprising of discrete patches, rather than a single continuous bounded area or polygon. While discussion on the appropriateness and variability of methods for calculating viewsheds has been an important issue (Fisher 1993), it has not been so with isovists. All of these reasons have contributed to viewsheds resisting the type of parameterization that can be found with isovists. Parameters that describe, for instance, geometrical properties such as compactness (Batty 2001) or shape. Instead it is possible, in contrast with isovists, to find several studies in which certain parameters refer to the content in the viewsheds (that is characteristics of the terrain found within the area delimited by the viewshed, Miller *et al.* 1994, Bishop *et al.* 2000, Germino *et al.* 2001). These parameters, however, are seldom mapped back into space in order to generate new surfaces.

In an attempt to overcome these limitations, Llobera (1999) distinguished the *near viewshed* as the smallest continuous area immediately surrounding a viewpoint and tried to describe some of its characteristics, such as *maximum* and *minimum axis* (maximum or minimum distance), *orientation of maximum* and *minimum axis, angle between maximum* and *minimum axis*, as the subset of a viewshed equivalent to an isovist. Given the raster nature of the representation, these parameters proved to be exceedingly coarse and revealed the limitations of using two-dimensional data to explore what is essentially a three-dimensional construct.

3 Visualscapes

In the following sections, visualscape is introduced as a concept within GISc that is analogous to Benedikt's isovist fields (1970), and that may help to unify, under one term, the scope and ideas found in current analyses on 'human' visual space, independently of their scale or context. While the definition provided here is purposely abstract and generic in character, and not all possible combinations of what may constitute a visualscape are necessarily explored, it is hoped that the definition will be seen as an extension to Benedikt's initial ideas.

At a theoretical level, the concept of visualscape (as with isovist fields) finds its source of inspiration in Gibson's (1986) *ambient optic array* insofar as it relates to the visual structure inherent in an environment, although, strictly speaking, a visualscape could only be equated with Gibson's optic array if a light source was also included.

3.1 Visualscape defined

A visualscape is defined here as the **spatial representation** of any **visual property** generated by, or associated with, a **spatial configuration**.

To expand:

Spatial representation refers to the way in which a visual property (see below) at a location is stored and represented. This representation is related to a sample space (a discrete space within which observations are taken or calculated at a certain rate) that has a resolution as fine as necessary for the analytical purpose. It is at one, or various, locations in this sample space that an imaginary model of a human individual is situated in order to capture a visual property. Such a model can admittedly be very simple, e.g., the height and orientation of the body, but it is hoped that knowledge in ergonomics and/or human physiology will be incorporated to improve future models. Traditionally, this representation has taken the form of a scalar field (Bendedikt 1970, Batty 2001). Scalar fields are easily created and show the spatial pattern of properties in a familiar way (rasters being GIS analogous to them). This form of spatial representation, however, is not exclusive or the best one for certain purposes. Llobera (in press) demonstrates how the visual structure of simulated landscape, and that associated with a series of antennae, can be represented using a vector field, and that such a representation is far better suited to describe visual changes related to changes in body orientation during movement along a path, than a scalar field.

Visual property refers to the measure of any 'visual characteristic' associated with a location in the sample space. This property may be the description of some aspect linked to the viewshed/isovist generated at that or some other location. An example of the former, for instance, would be the mapping of the average distance of the isovist at each location in a sample space (Batty 2001), an example of the latter may be a cumulative viewshed, where locations store the number of times that they are visible from other locations.

Finally, the notion behind what constitutes a **spatial configuration** lies at the heart of the visualscape concept. The idea here is that by varying the selection of what spatial components make up a spatial configuration, we can vary the scope, scale and intent of the visual analysis. The analytical potential of the visualscape is not only linked to the choice of what constitutes the spatial configuration but also to the way in which it is represented and stored, whether by means of traditional spatial primitives, e.g., a single point to represent a building, or more complex data structures, like 3D solid models. So far the former way of representing spatial structure has prevailed in GIS, but the possibilities of using more complex spatial data structures are near and likely to precipitate the generation of new data. For instance, to understand the visual impact that a building with some symbolic relevance, such as a temple (e.g., *il duomo de Firenze*), has on its surroundings, and to answer questions such as: where can it be seen from? How much of it can be seen at each location? How does its visual presence change as we walk to and from, or around it? As important is the fact that the visual structure of specific spatial components can be targeted, since this allows us to incorporate and explore another of Gibson's important contributions, the idea of perception as the education of attention. As a landscape evolves through time, so does the relevance of features in it (whether natural or built). During some periods certain features become more salient than others ('anchorpoint theory' by Golledge 1978). Understanding what is the nature of the visual structuring during a certain period, and how it transforms through time, can be achieved by generating and studying the various visualscapes associated with these salient features (Thomas 1993, Tilley 1994). The relevance of varying the spatial composition to study visual structure is illustrated in the following figure (Figure 20.3). In this case the visual areas associated with an entire house, the façade and the interior are broken down one by one to create a set of nested spaces. The ability to recognize such spaces may be used to understand how distinct spatial patterns of social behaviour surrounding a house are generated.

To summarize, the notion of a visualscape is put forward here in the hope that it will be a useful GISc term used to describe the fact that any spatial configuration creates its own visual structure which:

- Can be studied in its entirety or with respect to any relevant subset of the spatial configuration.
- Can be spatially represented in various ways (e.g., scalar or vector fields).
- Is essentially three-dimensional. They may be explored using any of the standard concepts that apply to 3D surfaces (Figure 20.4).
- Can be described by means of a multitude of parameters, as shown in some urban studies, and eventually mapped back into space for further research. Besides a few exceptions (Bishop *et al.* 2000, Germino *et al.* 2001), this possibility has seldom been explored in the context of current GIS visibility applications. There are even fewer examples (most notably O'Sullivan and Turner 2001) of visual parameters being mapped back into space and their spatial properties further studied.
- Occurs both in 'natural' and urban landscapes. The adoption of new spatial structures and the development of analytical procedures to explore the

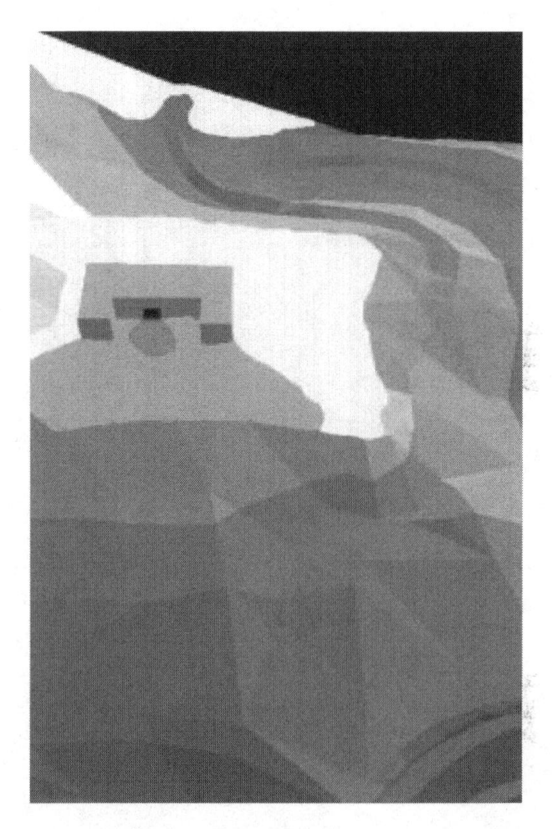

FIGURE 20.3 Breaking down the visual structure of a single monument into a set of nested visualscapes.

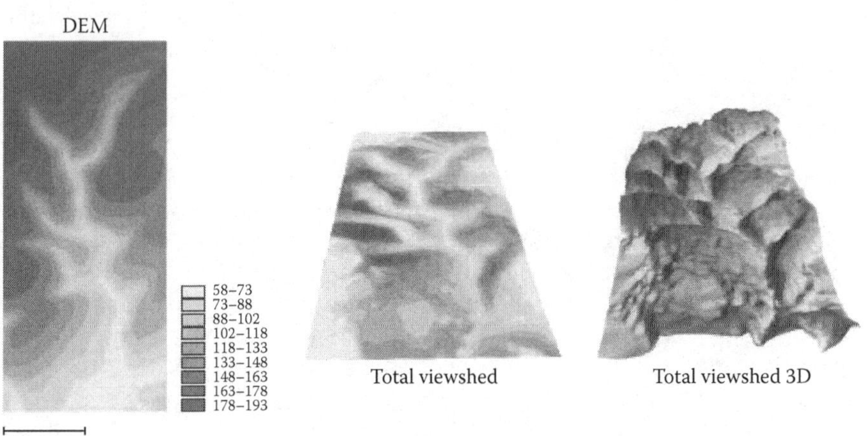

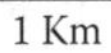

FIGURE 20.4 Within GISc visual patterns have generally been approached as two-dimensional patterns, however, the possibility of new analytical possibilities becomes clearer when visual information is better understood as a 3D surface.

benefits of such representations will, in this author's opinion, unite efforts to study visual space.

3.2 Cumulative and total viewsheds

Perhaps the most popular concept used to explore visual space in a 'natural' landscape has been the *cumulative viewshed* (Wheatley 1995), sometimes called *times seen* (Fisher *et al.* 1997). Cumulative viewsheds in general are created by calculating repeatedly the viewshed from various viewpoint locations, and then adding them up one at a time using map algebra (Tomlin 1990), in order to produce a single image. The result tells us how many of the viewpoints can be seen from, or are seen at each location, i.e., their *visual magnitude*. Fels (1992) distinguished two types, depending on whether the offset (e.g., representing the height of an individual) was included at the viewpoint (*projective viewshed*) or at the target location of the LoS (*reflective viewshed*). *Total viewsheds* are created in the same way as a cumulative one except that all locations are employed. Lee and Stucky (1998) distinguished two types of total viewshed based on the location of the offset on the viewpoint (*viewgrid*) or the target (*dominance viewgrid*).

Cumulative and total viewsheds are subject to the same limitations as single viewsheds (Gillings and Wheatley 2000). In both, locations near to the boundaries of the study area suffer from an edge effect, i.e., fewer locations are available, which in the case of a total viewshed can be calculated as a function of the position of each viewpoint location and the radius used for the viewshed, and represented as an additional raster or scalar field. Given the limitations of the viewshed (i.e., no distance attenuation), this edge effect decreases once the radius in the viewshed exceeds the radius of the largest possible circle that can be fully contained within the study area. From that radius onwards, the number of locations within the study area surrounding any location tends to converge towards the maximum number of locations as the value of radius reaches the maximum (Euclidean) distance separating any two locations. At that point, and given the lack of visual attenuation with distance, any location is theoretically visible for any other, no matter where it might be located.

Both cumulative and total viewsheds are examples of visualscapes that can be calculated using standard GIS. They use a set of points to describe the spatial configuration (differing only in the number), as a visual property, they record the number of locations that may be visible, to or from, each point in the spatial configuration and present this information as a scalar field.

In landscape planning, cumulative viewsheds have been used to determine visual impacts (Fels 1992). In archaeological landscape research, they have been used primarily to discuss the intervisibility level among monuments, in order to determine social cohesion and the importance of visible awareness, as a way of establishing territorial rights (Wheatley 1995), or to assess the level of cross-visibility (Llobera 1999) or visual continuity among monuments belonging to different periods (Gaffney *et al.* 1996).

The sole use of cumulative viewsheds as a possible measure for inter- and cross-visibility capitalizes on a static and 'pointillist' view of space. A view where the focus of the analysis is on understanding the relationship between points in isolation, and where concern about space in-between is lost and deemed meaningless and inert. The cumulative viewsheds calculated for a set of features provides a simple description of the visual structuring that these generate. At a very basic level, it can be used to identify where the visual presence of these features may be greatest, providing a series of 'anchorpoints' (Golledge 1978) integral for building a sense of place. In order to establish the significance of the visual patterns found in a cumulative viewshed, whether they conform to our expectations given the existing terrain, or to some other spatial configuration, it is necessary to establish some sort of comparison. Such a comparison may be obtained by calculating the cumulative viewshed of locations obtained through some sampling strategy and comparing them with the original cumulative viewshed (Lake *et al.* 1998) or by calculating the total viewshed for the entire terrain (Figure 20.5). Both strategies provide the information necessary for implementing, if required, statistical measures of comparison (not shown here). While the sampling strategy is computationally less intensive, the scope of the results is generally bounded to the questions at hand (e.g., are these patterns significant?) and cannot be incorporated as easily as total viewsheds in future research (see below).

While cumulative viewsheds may be used to describe the visual structure generated with respect to certain locations, or features on them, the total viewshed provides a first description of the visual structure for an entire terrain. Figure 20.6 shows the total viewshed for different types of landscapes. These have been calculated using the same offset of 1.74m at both the viewpoint and target location.

Simple histograms, and/or more sophisticated methods (e.g., kernel density estimates, image analysis methods) can be employed to examine the values obtained

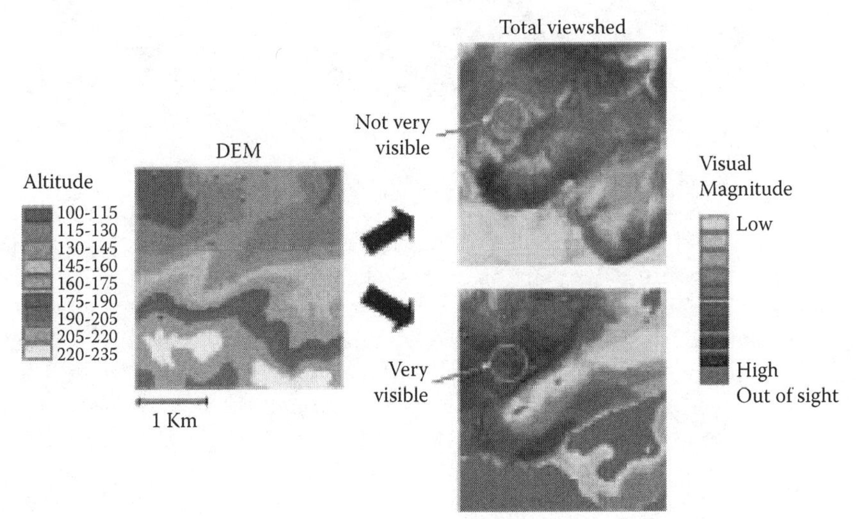

FIGURE 20.5 Comparison of a cumulative viewshed and a total viewshed. Darker areas represent higher visual magnitude.

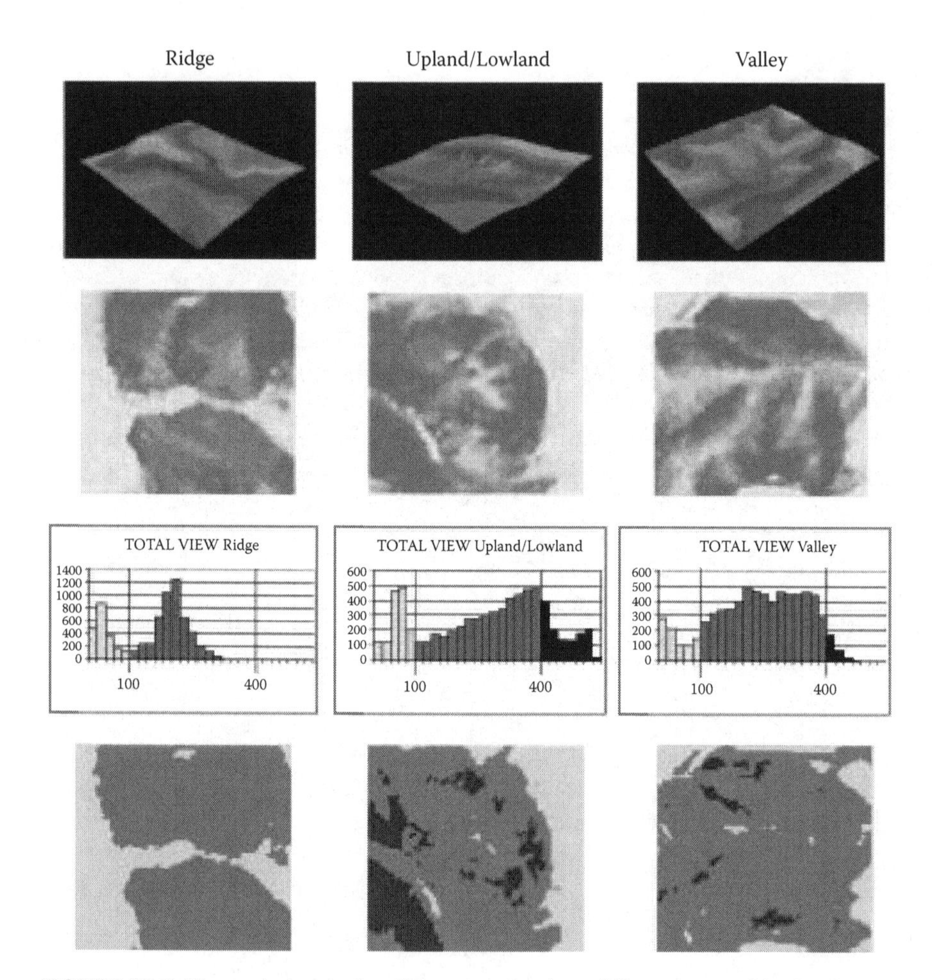

FIGURE 20.6 The total viewshed and histogram for three different types of landscapes. All histograms have the same bin-width and terminal points (darker areas represent higher visual magnitude). The last row shows all three total viewsheds reclassified using the same criteria points.

in the total viewshed of a terrain, in order to identify locations with similar visual magnitudes. Figure 20.6 shows, for instance, the total viewshed for three different terrains, and their respective histograms (all have been standardized to the same bin-width and terminal points). In addition, each total view has been reclassified into three different categories to facilitate easy comparison. The following discussion results from the visual examination of the histograms and re-classified images.

By comparing the histograms alone it is clear that unlike other terrains, the ridge landscape does not contain locations from which the entire terrain is visible, in fact the maximum number of visible locations corresponds to half of the total area. This is because the ridge is wide, thus impeding views of both sides simultaneously. The ridge is effectively acting as a visual barrier. This landscape is visually well defined and very homogeneous, given the large number of locations sharing the same level of visual magnitude. Its histogram seems to point out the existence of two distinct clusters centered at a low visibility level (bin 400–600), and the other at a medium low level (≈2500). However, the sharp peaks also indicate the possibility of acute visual changes (further analyses may be achieved by calculating the spatial variogram of the total viewshed). The upland/lowland landscape is the only one of the three examples that contains locations from where the entire terrain is visible. The histogram of this landscape shows the possibility of well differentiated visual clusters. The same applies to the ridge landscape, the histogram shows that the transition from some clusters to others can at times be quite sharp indicating the existence of abrupt visual changes. The largest classification (dark grey) shows a trend in which the number of locations with greater visual magnitude increases steadily, spatially this seems to correlate with an increase in visibility from right to left (the right being dominated by locations with low visibility). Finally, the valley landscape shows an even spread of locations with different visual magnitudes, i.e., less abrupt changes. As might be expected given the terrain geometry, the valley landscape is visually diverse but compact at the same time, i.e., there are many locations with different but related visual magnitudes.

Given that, in this case, the same offset was used at both the viewpoint and target location during the viewshed calculations, it is possible to read the histograms in relation to intervisibility. Landscapes with high intervisibility levels are characterized by a greater accumulation of locations at the higher end of the histogram. The ridge landscape is characterized by two very well defined groups of locations with low and medium-low intervisibility levels, while the intervisibility for the highland/lowland landscape is much more evenly spread although two groups of locations, with low and relatively high intervisibility, can also be distinguished. Of all the landscapes, the valley type offers the most even spread of intervisibility, mostly concentrated within the medium-low to the medium-high (for similar discussions see also O'Sullivan and Turner 2001).

The creation of total views can be manipulated so that the viewshed at each viewpoint is calculated only for those locations within a certain distance band. Rather than calculating what is visible from the viewpoint location to the maximum radius, say r_{max}, the visibility can be calculated, for instance, for any distance r_1 away from the viewpoint, where $0 < r_1 < r_{max}$, to another arbitrary distance r_2, where $0 < r_1 < r_2 < r_{max}$. This allows us to study the visual impact that each location has at various ranges (Figure 20.7). That is, it shows how important each location is when it is part of the foreground, middleground or background. In this case, as opposed to the previous example, these 'partial' total viewsheds are normalized by the maximum number of locations that are potentially visible within each distance band. This is done to compensate for the fact that in some occasions the number of locations, contained within a particular distance band, can be severely reduced at certain positions in the study area.

The intersection of linear features, such as walking paths, with total viewsheds provides an approximate indication of the nature of visual changes along a path (Fisher 1995, Batty 2001). This approach can be extended to other visualscapes, e.g., generated by a different spatial configuration such as a cumulative viewshed or by mapping out different visual properties (Batty 2001). In addition, Lee and

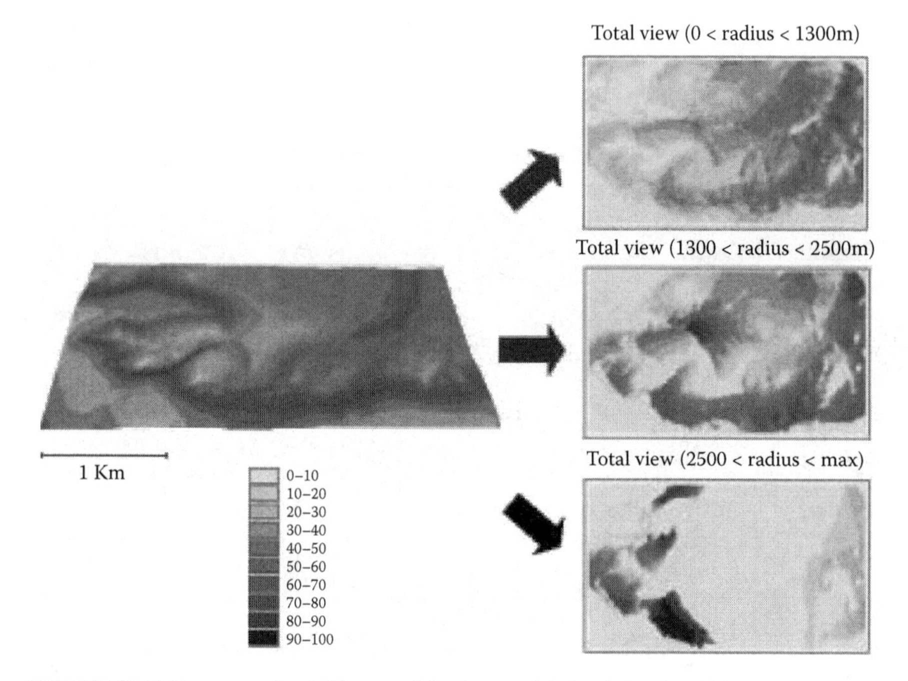

FIGURE 20.7 Foreground, middleground, background defined, for simplicity, using arbitrary Euclidean distances. Darker areas represent higher visual magnitudes.

Stucky (1998) have already shown how total viewsheds may be reclassified and used in combination with cost surface analyses to generate paths of different visual qualities. The following example discusses some simple possibilities.

Figure 20.8a shows a valley-like landscape which, at first glance, seems relatively smooth. The total viewshed for the same landscape, shown right beside it (Figure 20.8b), has been given further relief using standard hillshade functionality, as found in most GIS. In spite of its simplicity, the image provides immediately a good indication of the amount of visual complexity inherent in it and an appreciation of the visual changes that someone moving in the landscape may encounter. These changes become more apparent when examined along a track or path (Figure 20.8c,d). In Figure 20.8e, the cross section of the total viewshed along path A shows how the visual magnitude increases quite steadily until it reaches a maximum towards the end of the route, after which it decreases very rapidly. The scenario is totally different for path B (Figure 20.8f) where changes in the visual magnitude occur in a roller-coaster fashion. Initially, the path is characterized by a steady descent into what appears to be a large visual enclosure, as seen in Figure 20.8b and Figure 20.8d. This is followed by a plateau and a pronounced increase in visual magnitude. The path then drops sharply into a second possible enclosure, not as deep as the first one, with pronounced boundaries. After an extremely steep ascent the visual magnitude plunges down dramatically towards the end of the path.

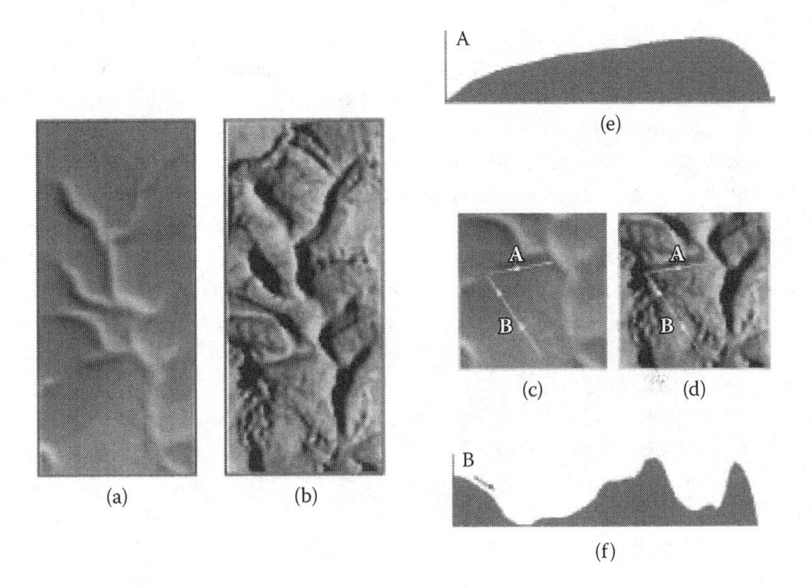

FIGURE 20.8 Profile of linear features (e.g., a path) on a total viewshed.

3.3 Visual prominence

An (intuitive) understanding of something being **prominent in space** arises after some sort of comparison has been established between the level, or amount, of some property (p) at a location (l), with those found at adjacent locations. Thus it can be said that some idea of neighbourhood, i.e., what constitutes neighbourhood, is implied in the comparison. Many criteria may be used to define what is a neighbourhood or what locations are part of it, here a neighbourhood will be defined by reference to an arbitrary Euclidean distance around any location. The area, comprising all locations within this radius, will define the neighbourhood at any location. Once a neighbourhood is defined, it is possible to produce a simple definition of prominence,

Prominence of a property p, at any location i, Prom$(t\,)_p$ is defined as the average difference between the property at that location $p(i)$, and that found at each of the other locations j_n, $p(j_n)$, within an arbitrary neighbourhood of i, N_i, such that given an arbitrary distance where N_i is the neighbourhood of i, $N = \text{Card}(N_i)$ and $n \in N$ (Natural numbers).

$$i\ \text{Prom}(i)_p = \frac{\sum_{j_n \in N_i} p(i) - p(j_n)}{N} \tag{20.2}$$

The values for the prominence at any location can have any positive or negative value. The upper and lower bounds will change with each neighbourhood size. The value and sign of the prominence reflects the morphological character of the location. For example, when calculated using a DEM, i.e., altitude being the property that is being compared, higher positive values tend to indicate a sharper hilltop, while more moderate values point towards a more rounded hilltop; values close to zero indicate flat locations and negative values channel- or pit-like locations. It is important to note that the definition at this stage does not include any sort of normalization. This is because the index may be normalized in various ways; the normalization of the entire raster by the maximum prominence value produces a result that is informative, if the analysis is restricted to one image and to its entirety. However, if the analysis represents the comparison of indexes throughout various neighbourhood sizes (Figure 20.10) then it is preferable to normalize each value by the local maximum found within each neighbourhood.

It is fairly obvious from the definition that a prominence index can be derived for different properties or magnitudes. Here, two different, though related, types of prominence are generated, a topographic and a visual prominence. The former is defined by reference to the altitude (derived from DEM) while the latter uses the visual magnitude as described by the total viewshed. Both prominences tend to be easily interchanged, i.e., more prominent locations are thought of as being visually prominent as well, but as Figure 20.9 shows the relationship between topographic and visual prominence is not always straightforward. Such a comparison helps investigate the interplay between physical and visual aspects in a landscape. In this case, we can detect locations that have a high visual magnitude (from where we can see a lot of terrain or from where one can be seen easily, if we accept intervisibility) but are not themselves physically prominent.

Figure 20.10 shows the visual prominence for different neighbourhood sizes. In this case, initial prominence values were transformed and normalized using local maxima, i.e., the maximum prominence value found within the neighbourhood at each location. Calculating visual prominence for various neighbourhoods allows determining at which scale a location can start to be considered as being visually prominent and/or to identify locations that for instance, maintain a high level of prominence independently of any scale. Such locations are very significant as they tend to be important territorial landmarks and provide useful navigational information about the physical structure of the landscape. Elsewhere Llobera (2001) has shown how prominence at various scales may be combined together and mapped to create a single raster that describes how it changes at each location. It points towards the possibility of describing prominence as a signature (throughout different scales) whose characteristics may be mapped back into space.

To summarize, cumulative and total viewsheds can be used to describe and explore, at a basic level, the visual structure that a spatial configuration, such as the physical topography of a landscape, generates. Properties of this structure can be studied using traditional GIS capabilities (e.g., histograms, reclassification, etc.) and/or by further manipulating them as shown through the visual prominence example. However, the description they offer remains relatively coarse due to factors such as: the sampling interval (i.e., raster resolution), choice of radius, sensitivity of the LoS algorithm and in particular, the choice of visual property that is being recorded, in this case using a Boolean value to describe in-sight or out-of-sight.

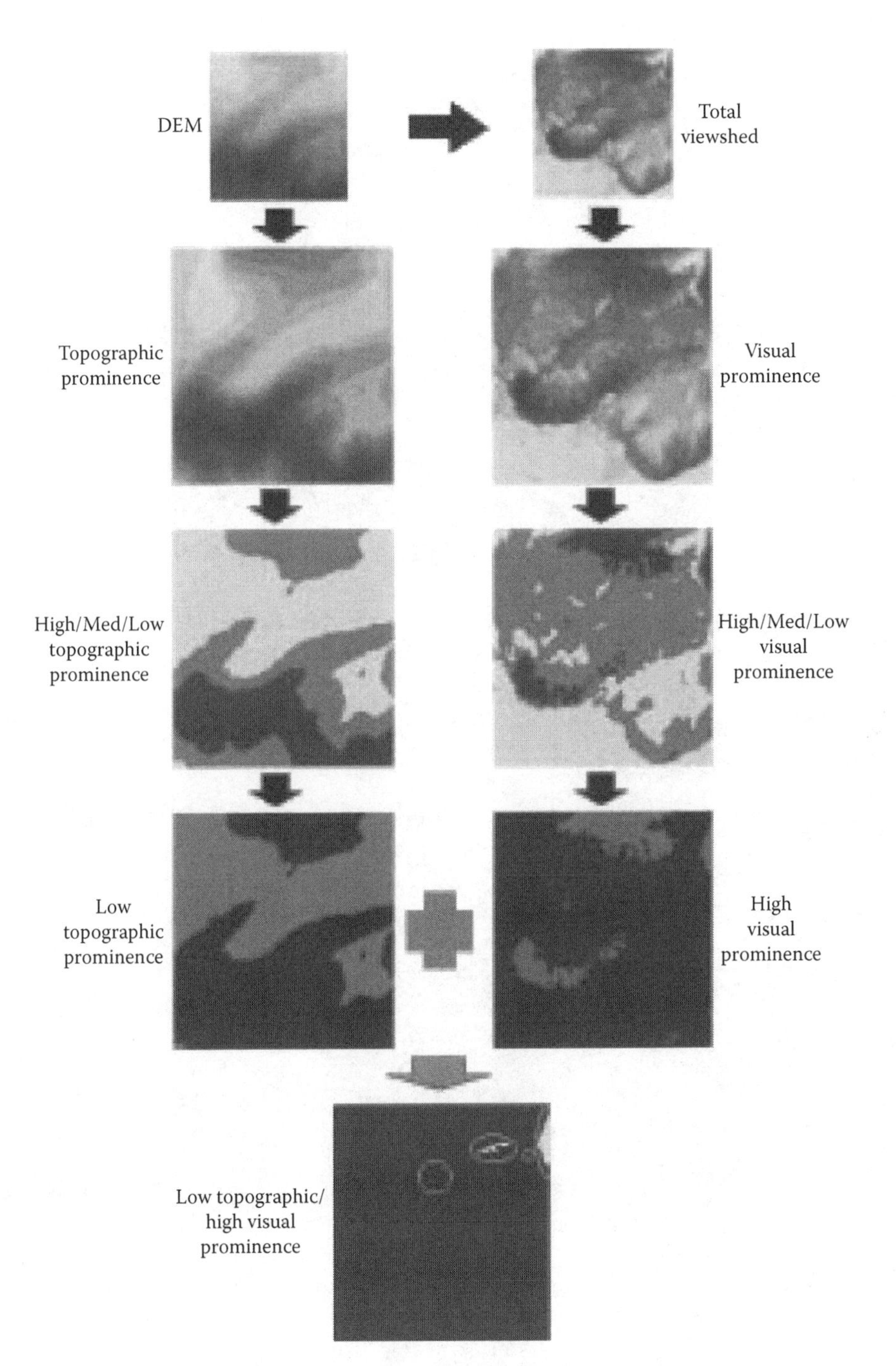

FIGURE 20.9 An example comparing topographic prominence with visual prominence. Here both prominences have been calculated for the maximum radius. Both topographic and visual prominence images are reclassified into high, medium and low values. These are further reclassified and combined in order to obtain locations with low topographic and high visual magnitude.

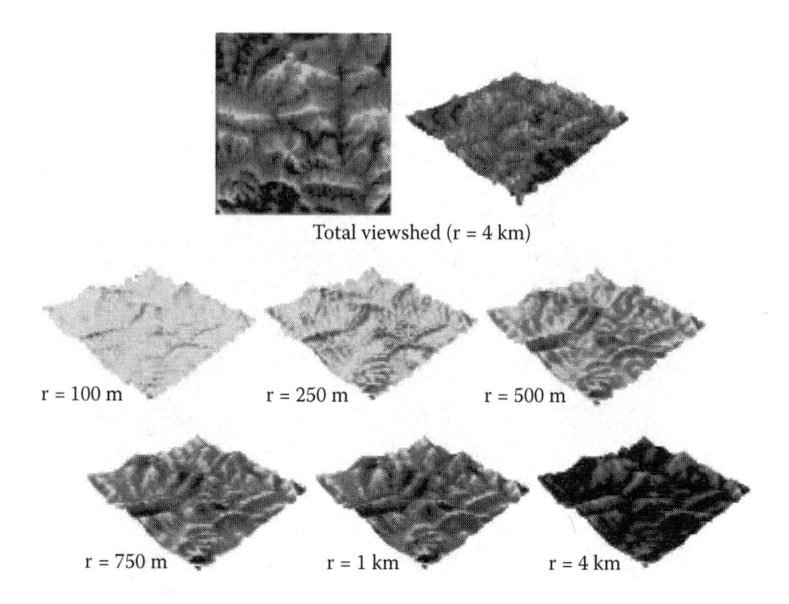

FIGURE 20.10 Visual prominence image. The top two images show the total view for a landscape. These are followed by the visual prominence at various scales, the more visually prominent a location, the lighter it appears.

3.4 Visual exposure

Reference to the more dynamic aspect surrounding visibility was present in the previous discussion on total viewsheds (Batty 2001, Lee and Stucky 1998) but was not really dealt with directly. Such absence among GIS and other spatial studies may be partly the consequence of relying heavily on two-dimensional spatial representations, given their obvious limitations and the fact that such representations may lead researchers to think in a 'static way'.

The limitations of current approaches are important as they hinder our understanding of visual space, especially when movement, and therefore change, has been identified as a key element in visual perception (Gibson 1986). Works as comprehensive as those cited above made only limited reference to this aspect, concentrating instead on the various visual parameters as **static** attributes of space, without further exploring their change in space.

In the following examples the dynamic aspect of visualscapes is considered by concentrating on the study of *visual exposure*, another type of visualscape. Visual exposure is created by assigning to each location in our sample space a measure of the visible portion of whatever is the focus of the investigation, whether the entire landscape or a set of features. Previous work by Travis *et al.* (1975) and Iverson (1985) showed the importance of mapping how visible each location in a landscape (i.e., DEM) (i.e., cell) from a viewpoint was and pointed out the possibility of doing the same with respect to specific landscape features. When calculated in relation to some feature it can be thought of as a description of how much the feature occupies the field of view of an individual at any location. The stress is on the visual patterns created by the physical presence of a feature (i.e., its visible portions). Cumulative and total viewsheds provide a crude depiction of these patterns but their description is not sensitive enough and requires, in the case of a cumulative viewshed, the presence of at least two or more features. The interest here is in determining **how much** of a feature or a terrain is visible at each location, rather than finding out whether a location is visible or not (or how many times it is visible).

On this occasion, *visual angles* are used to generate visual exposure, i.e., visual property. They describe the visible span (both horizontal and vertical) of a feature or terrain facet that can be seen at any viewpoint location, and are well suited to describe the visual exposure of a feature as the field of view of an individual is generally described in terms of angular ranges. Intuitively, the closer we are to the feature the more we expect to notice it, as its presence occupies more of our field of view. Figure 20.11 through Figure 20.15 illustrate the possibilities for a very simple feature (ideally this can be extended to consider full three-dimensional objects).

Figure 20.11 shows feature A, a vertical pole with a height of 10 m, which could represent a communications antenna. The visual exposure for feature A, after calculating 'vertical' visual angle with a sampling frequency of 5m, is shown in Figure 20.12. Because values at locations near the pole are very high in comparison with those further away, it is necessary to use an adequate colour palette, where higher outliers are grouped into one single category.

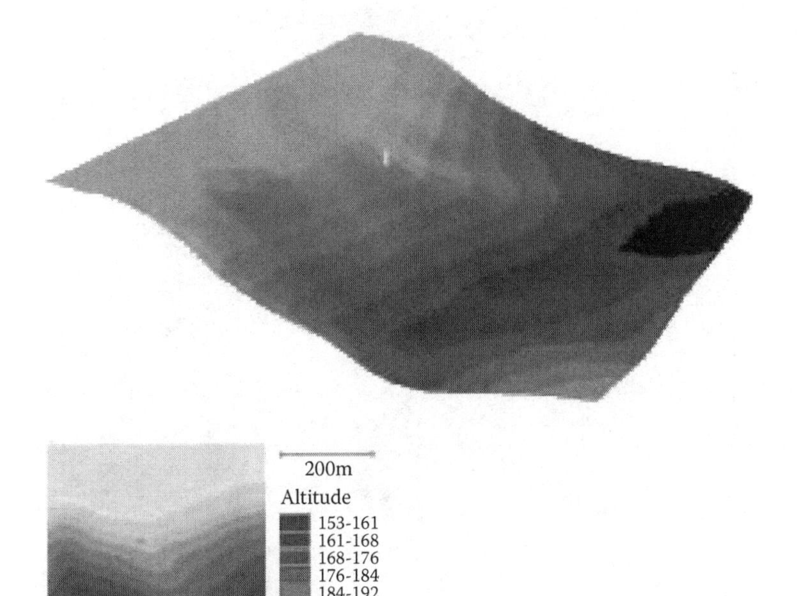

FIGURE 20.11 Feature A (pole) on example DEM.

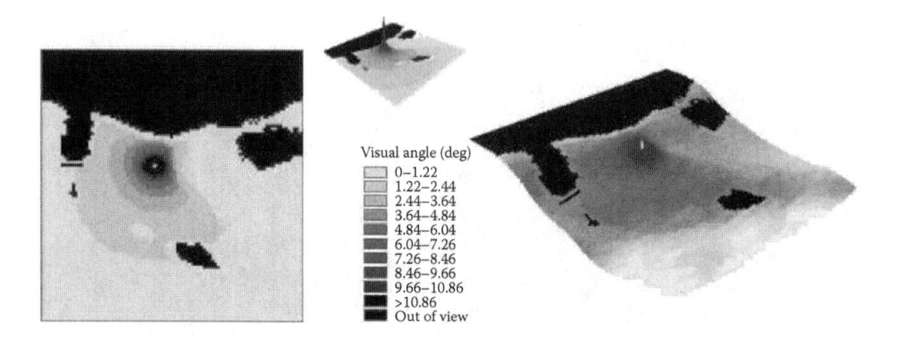

FIGURE 20.12 Magnitude of visual angle calculated for feature A. In plan view, as a 3D surface (small inset) and on the DEM.

One of the benefits of mapping out the visual exposure is that we can use real numbers (floating-point), which translate into smoother surfaces than those obtained for cumulative or total viewsheds. This, in turn, allows us the possibility of further processing the visual exposure using standard mathematical techniques. By calculating the local gradient of the visual exposure, i.e., the first surface derivative (slope), we can identify where (local) visual changes occur, their magnitude (darker areas representing higher change) and the direction of change, e.g., direction in which maximum visual exposure is obtained (see Figure 20.13). From this it follows, that

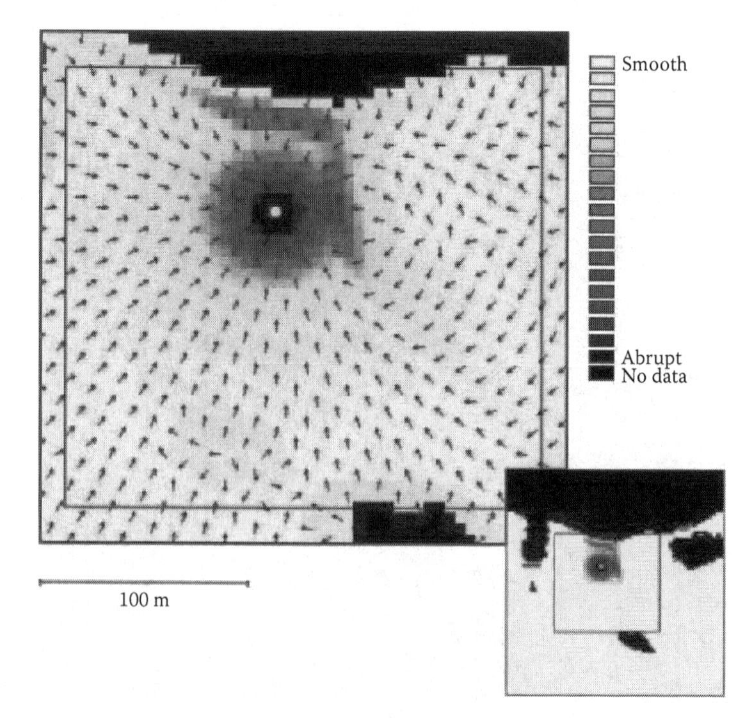

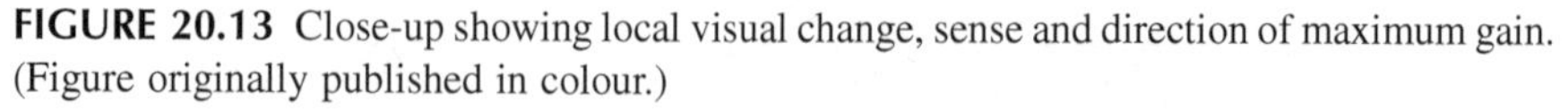

FIGURE 20.13 Close-up showing local visual change, sense and direction of maximum gain. (Figure originally published in colour.)

the opposite direction represents the quickest way to get out of sight from the feature, and that by moving perpendicular to this direction will not incur any visual change.

Figure 20.14 to Figure 20.15 describe the shape of the visual change. These images were obtained by calculating the second (discrete) derivative of the visual exposure in the direction of maximum change, and in the orthogonal direction (Figure 20.16).

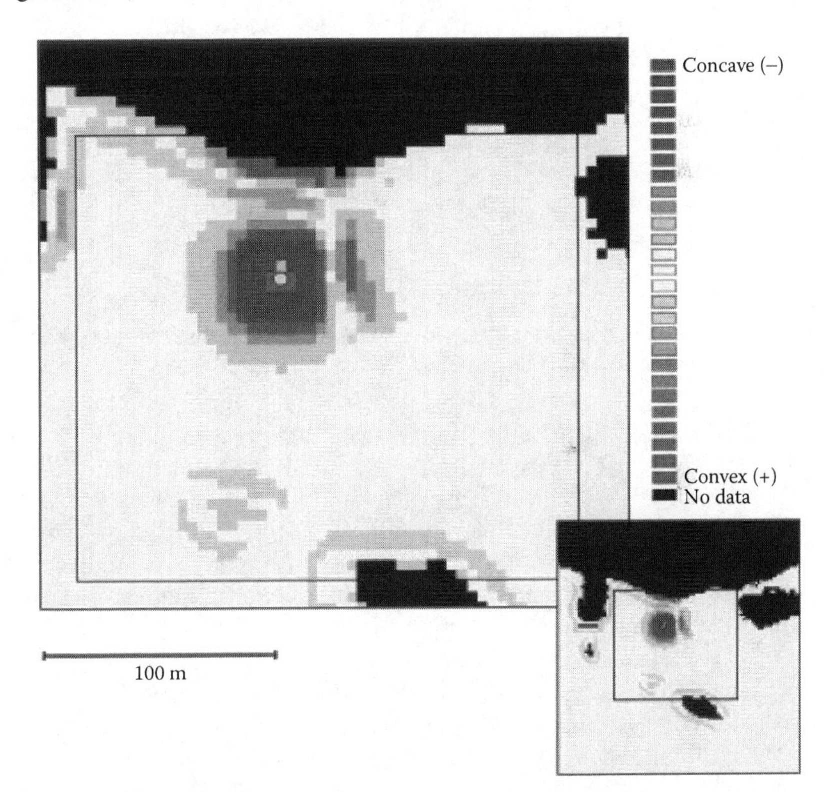

FIGURE 20.14 Local curvature of the visual exposure **in the direction of maximum change**. The shading indicates the intensity of the curvature. (Figur eoriginally published in colour.)

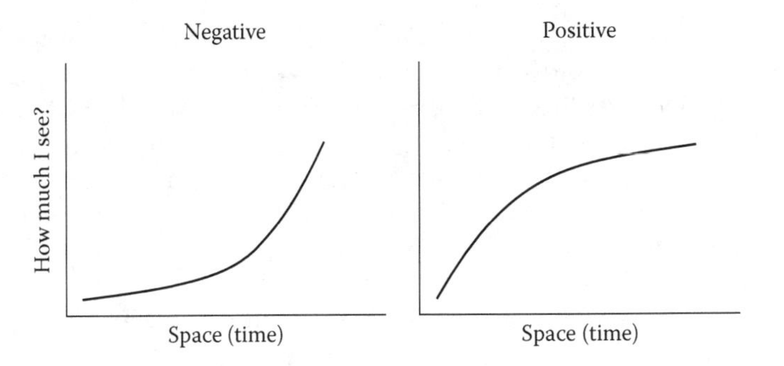

FIGURE 20.15 Profile cut in the **direction of maximum change** showing the concavity and convexity of the visual exposure for feature *a* in Figure 20.14.

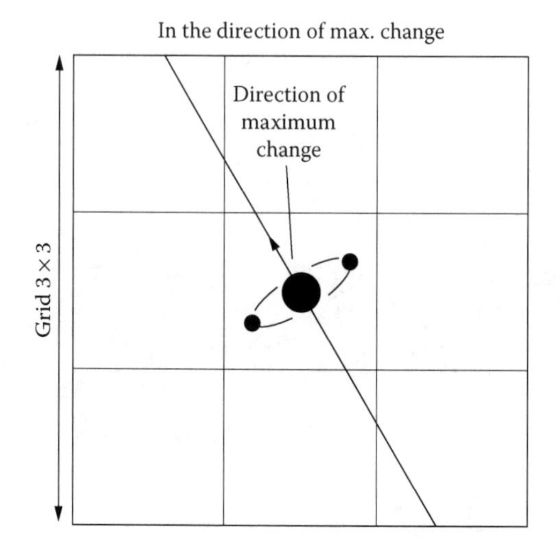

FIGURE 20.16 In this occasion the concavity and convexity are measured locally (i.e., within a 3 × 3 window) in the direction of maximum change in Figure 20.14.

The idea is better understood by imagining the visual exposure as a three-dimensional representation of a landscape where instead of altitude values (as we would have with a DEM) these are magnitudes of solid angles that describe how much we see of feature *A*. Moving up or down this 'landscape' along contours represents changes in visual exposure while movement preserves the same level of visual exposure. If the amount or rate of visibility associated with the feature increases each time more along a certain direction we get a concavity (see Figure 20.15), while if it decreases we get a convexity.

Figure 20.17 and Figure 20.18 are based on the same principles as Figure 20.14 except that it shows the shape of the visual exposure, the concavity or convexity, in the orthogonal direction of maximum change (see Figure 20.19 and Figure 20.20).

Potentially it can be used to identify locations where we would get 'visual corridors' (i.e., concavities, or more visibility towards the sides than towards the middle) and 'visual ridges' (i.e., convexities, or more visibility towards the middle than towards the sides).

So far, mapping where changes in visual exposure occur and the nature of those changes has only been explored in relation to the direction of maximum change (given by the local gradient). While not shown here, these calculations could be easily adapted to allow mapping similar information for any direction of movement by applying the general definition of a *surface directional derivative* (from where the gradient derives). Similarly, methods of morphometric characterization (Wood 1996) as found for DEMs, could be extended to the visual exposure (and other visualscapes) to describe their morphological characteristics, i.e., shape, properties.

Several interesting insights can be drawn from the previous examples. The sampling rate that was used (i.e., 5m intervals) is likely to be inadequate for modelling the visual impact that features have on people, at least, if these are meant to be calculated at a close range. The closer an observer is to the feature or part of the

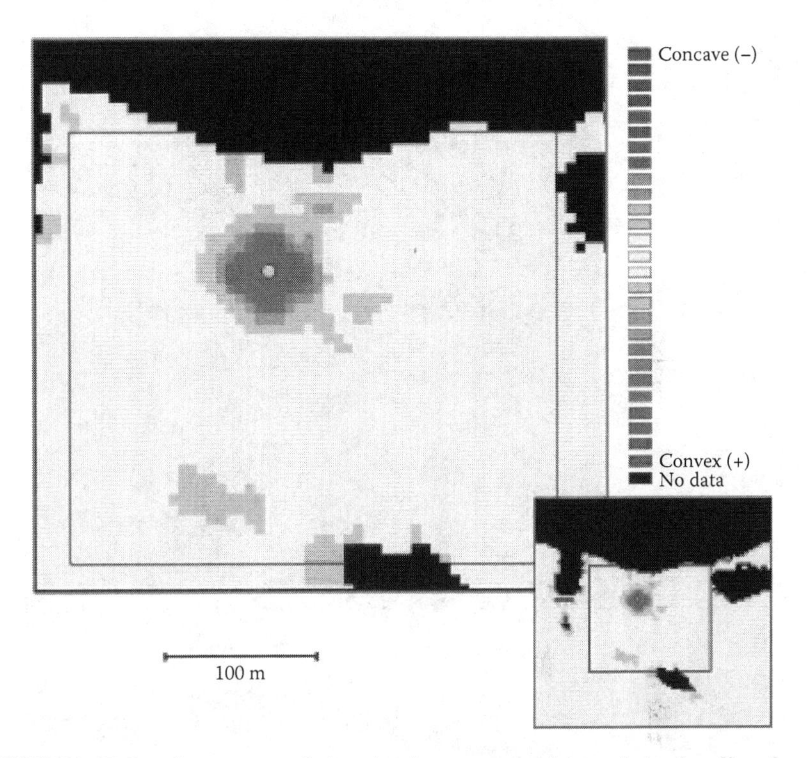

FIGURE 20.17 Local curvature of the visual exposure **orthogonal to the direction of maximum change**. The shading indicates the intensity of the curvature. (Figure originally published in colour.)

landscape of interest, the smaller the distance the viewer has to move in order to appreciate a substantial change, and vice versa. This observation points towards an important finding, the use of a scalar field or raster to represent visual information limits the possibility of detecting changes to those we would observe at a constant speed of movement.

To conclude this section, a final example of the visual exposure as a vector field is provided. Figure 20.21 shows a landscape represented by a TIN data structure. To calculate a vector version of the visual exposure for an entire landscape, the Digital Terrain Model (DTM) is sampled at a certain fixed rate (20m), and the following procedure is repeated at every target location (Figure 20.22).

1. A viewpoint location is selected.
2. An orthonormal vector (i.e., perpendicular vector with a magnitude of one) to the terrain is calculated at the target.
3. A normal LoS vector is also calculated with its origin on the target and direction pointing towards the current viewpoint.
4. The visual exposure is the vector obtained by projecting the surface orthonormal onto the LoS. In this case, the vector is multiplied by an additional factor derived from the distance between the viewpoint and the target point (the further away, the less you see).

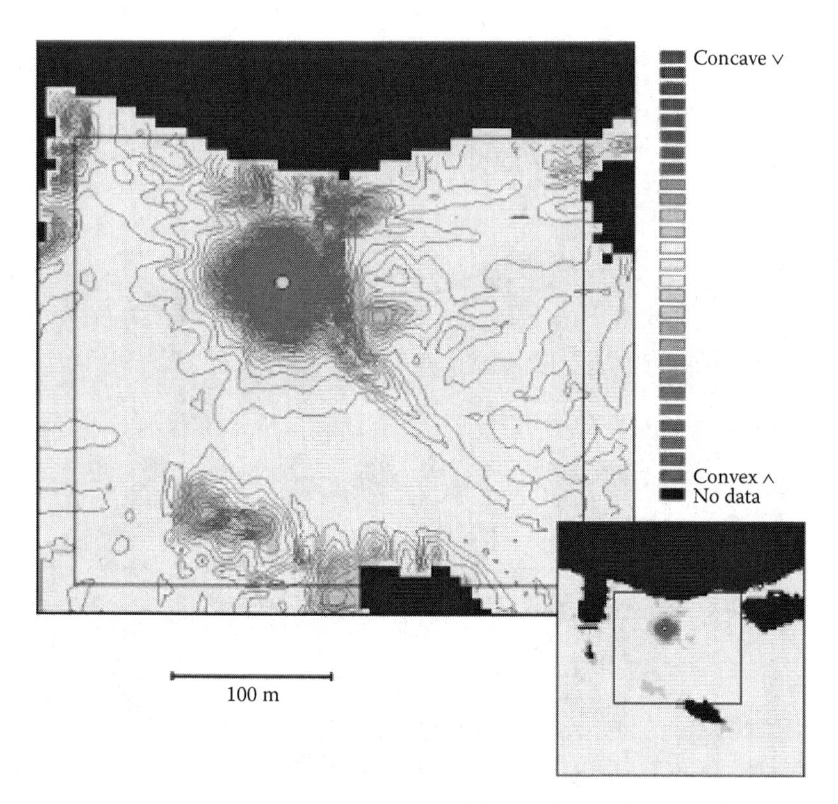

FIGURE 20.18 Local curvature of the visual exposure **orthogonal to the direction of maximum change**. The shading indicates the intensity of the curvature. Contour lines are added to give a better horizontal sense. (Figure originally published in colour.)

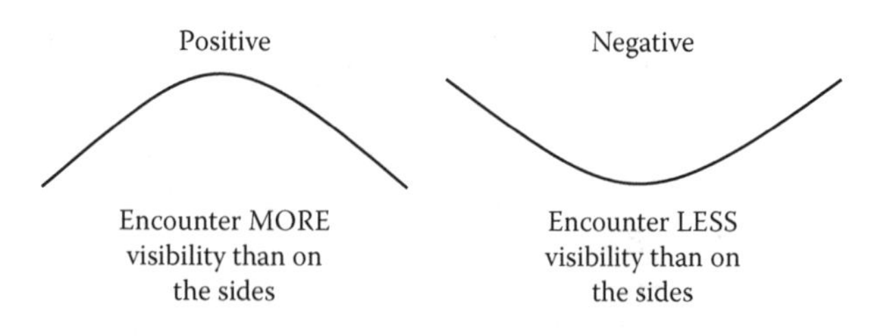

FIGURE 20.19 Profile cut in the **PERPENDICULAR direction of maximum change** showing the concavity and convexity of the visual exposure for feature *A*.

The sum of all vectors obtained after paring the target location with every viewpoint location represents the total visual exposure for that target location. Figure 20.23 shows the total viewshed and the magnitudes of the vectors describing the total visual exposure for the same landscape side by side. This is similar to using a vector field, in this case to represent the total visual exposure of a landscape, is an important improvement over other ways of representation for it not only provides a

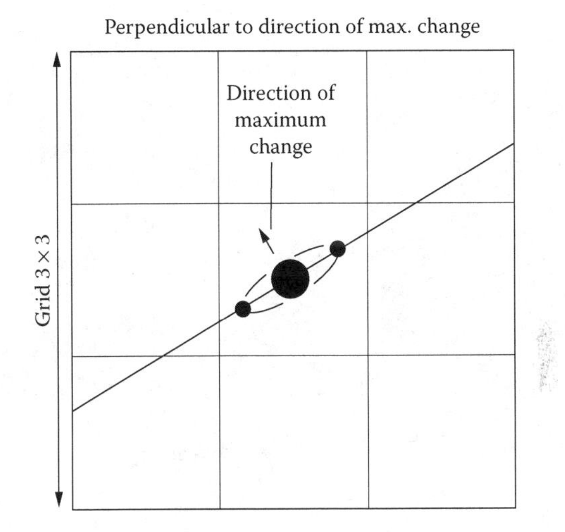

FIGURE 20.20 The concavity and convexity are measured locally (i.e., within a 3 × 3 window) perpendicular to the direction of maximum change.

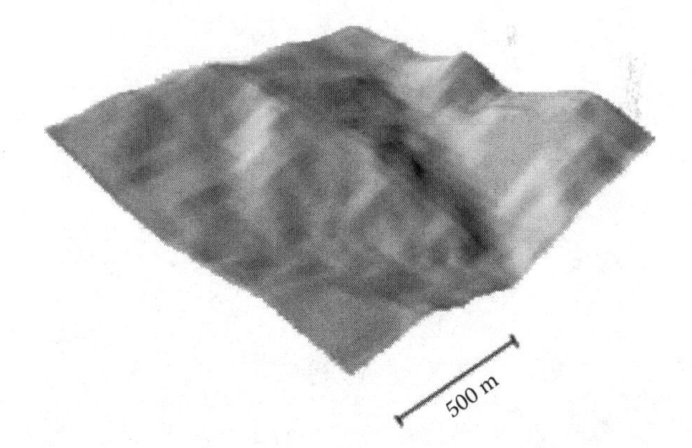

FIGURE 20.21 TIN representation of a landscape.

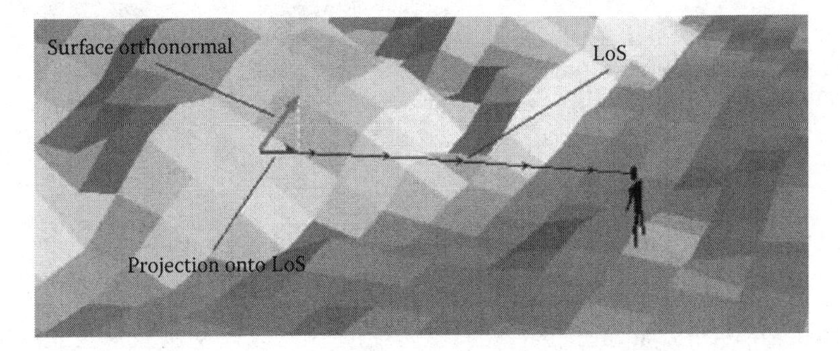

FIGURE 20.22 Generating a vector field to describe visual exposure.

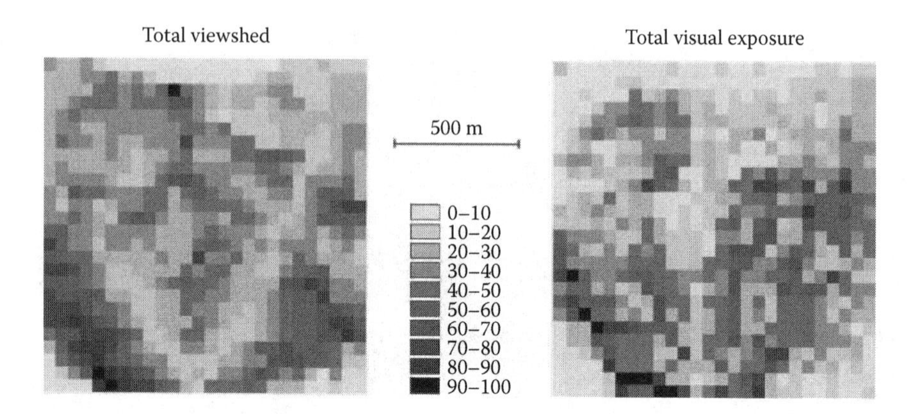

FIGURE 20.23 Comparison of the total viewshed with the total visual exposure for the previous landscape.

measure of a visual property but also its direction, information that is indispensable in order to model and map more efficiently visual changes due to movement (Llobera in press).

4 Conclusions

The exploration of visual space crosses various disciplines. This paper provided a critical synthesis of some of the issues associated with the analysis of visual space that are relevant to many disciplines, and attempted to unite them under a single common notion: visualscape.

Any spatial configuration, whether it is the physical topography of a 'natural' landscape or the façade of several buildings, structures space visually. This paper showed how isovists, isovist fields, cumulative viewsheds and total viewsheds may be interpreted as providing simple descriptions of such structure, i.e., visualscapes, that represent just as a tiny fraction of the possible ways in which this structure can be described and represented. With such interpretation in mind, the characteristics of total viewsheds were further explored and the idea of visual prominence was put forward. Visual exposure was introduced as a new visualscape that describes how much of a feature, or an entire landscape, can be seen at any location. Properties of the visual impact were illustrated using several examples, and the possibility and benefits of using a vector field to represent the visual exposure were discussed.

While it is true that factors such as landscape or environmental perception cannot be reduced to a mere physical measure of some structuring property of space, no matter how complex, their role, nevertheless, is an important one which is only now beginning to be explored.

Underlying the content of this paper is the realization that in order to understand how people experience and associate meaning to space we need to develop new analytical tools that help us retrieve the structure of space as it is encountered and unfolds in relation to a mobile subject. Studies of visual space fall within this orientation.

Acknowledgments

I would like to thank James McGlade and Aaron Shugard (Institute of Archaeology, UCL), Pamela Wace and Helen Walkington (Oxford University), Mike Batty (CASA, UCL) and particularly two anonymous referees for useful comments on earlier drafts of this paper.

References

Baldwin, J., Fisher, P., Wood, J., and Langford, M., 1996, Modelling environmental cognition of the view with GIS. In *Proceedings of the Third International Conference/Workshop on Integrated GIS and Environmental Modeling.*

Batty, M., 2001, Exploring isovist fields: space and shape in architectural and urban morphology. *Environment and Planning B*, **28**, 123–150.

Benedikt, M. L., 1979, To take hold of space: isovists and isovist fields. *Environment and Planning B*, **6**, 47–65.

Bishop, I. D., Wherrett, J. R., and Miller, D., 2000, Using image depth variables as predictors of visual quality. *Environment and Planning B*, **27**, 865–875.

de Berg, M., van Kreveld, M., Overmars, M., and Scharwrzkopf, O., 1997, *Computational Geometry* (Berlin: Springer-Verlag).

De Floriani, L., and Magillo, P., 1999, Intervisibility on terrains. In *Geographic Information Systems: Principles, Techniques, Management and Applications*, edited by P. A. Longley, M. F. Goodchild, D. J. Maguire, and D. W. Rhind (London: John Wiley & Sons), pp. 543–556.

De Floriani, L., Manzano, P., and Puppo, E., 1994, Line of sight communication on terrain models. *International Journal of Geographic Information Systems*, **8**, 329–342.

Fels, J. E., 1992, Viewshed simulation and analysis: an interactive approach. *GIS World*, **Special Issue July**, 54–59.

Fisher, P., 1991, First experiments in viewshed uncertainty: the accuracy of the viewshed area. *Photogrammetric Engineering and Remote Sensing*, **57**, 1321–1327.

Fisher, P., 1992, First experiments in viewshed uncertainty: simulating fuzzy viewsheds. *Photogrammetric Engineering and Remote Sensing*, **58**, 345–327.

Fisher, P., 1993, Algorithm and implementation uncertainty in viewshed analysis. *International Journal of Geographic Information Systems*, **7**, 331–347.

Fisher, P., 1995, An exploration of probable viewsheds in landscape planning. *Environment and Planning B*, **22**, 527–546.

Fisher, P., Farrelly, C., Maddocks, A., and Ruggles, C., 1997, Spatial analysis of visible areas from the bronze age cairns of Mull. *Journal of Archaeological Science*, **25**, 581–592.

Gaffney, V., and Stani, Z., 1991, *GIS Approaches to Regional Analysis: a Case Study of the Island of Hvar* (Ljubljana: Znanstveni institute Filozofske Fakultete, University of Ljubljana).

Gaffney, V., and Stani, Z., and Watson, H., 1996, Moving from catchments to cognition: tentative steps towards a larger archaeological context for GIS. *Anthropology, Space and Geographic Information Systems*, edited by M. Aldendefer and H. Maschner (Oxford: Oxford University Press), pp. 132–154.

Germino, M. J., Reiners, W. A., Blasko, B. J., McLeod, D., and Bastian, C. T., 2001, Estimating visual properties of Rocky Mountain landscapes using GIS. *Landscape and Urban Planning*, **53**, 71–84.

Gibson, J. J., 1986, *The Ecological Approach to Visual Perception* (New Jersey: Lawrence Erlbaum Associates Inc.).
Gillings, M., and Wheatley, D., 2000, Vision, perception and GIS: developing enriched approaches to the study of archaeological visibility. *Beyond the Map*, edited by G. Lock (Amsterdam: IOS Press), pp. 1–28.
Golledge, R., 1978, Learning about urban environments. In *Timing Space and Spacing Time I: Making Sense of Time*, edited by T. Carlstein, D. N. Parkes and N. J. Thrift (London: Edward Arnold).
Iverson, W. D., 1985, And that's about the size of it: visual magnitude as a measure of the physical landscape. *Landscape Journal*, **4**, 14–22.
Lake, M. W., Woodman, P. E., and Mithen, S. J., 1998, Tailoring GIS software for archaeological applications: an example concerning viewshed analysis. *Journal of Archaeological Science*, **25**, 27–38.
Lee, J., and Stucky, D., 1998, On applying viewshed analysis for determining least-cost-paths on Digital Elevation Models. *International Journal of Geographic Information Systems*, **12**, 891–905.
Llobera, M., 1996, Exploring the topography of mind: GIS, landscape archaeology and social theory. *Antiquity*, **70**, 612–622.
Llobera, M., 1999, Landscapes of Experiences in Stone: Notes on a Humanistic Use of a GIS to Study Ancient Landscapes. Unpublished D.Phil. Thesis, Oxford: Oxford University.
Llobera, M., 2001, Building past landscape perception. Understanding topographic prominence. *Journal of Archaeological Science*, **25**, 1005–1014.
Llobera, M., in press, The nature of everyday experience: examples on the study of visual space. *In Re-Presenting GIS*, edited by D. Unwin (London: John Wiley & Sons).
Miller, R. D., Morrice, J., Horne, P., and Aspinall, R., 1994, Characterization of landscape views. *EGIS 1994 Proceedings*, 223–232.
O'Sullivan, D., and Turner, A., 2001, Visibility graphs and landscape visibility analysis. *International Journal of Geographic Information Science*, **15**, 221–237.
Tandy, C. R. V., 1967, The isovist method of landscape survey. In *Methods of Landscape Analysis*, edited H. C. Murray (London: Landscape Research Group).
Thomas, J., 1993, The politics of vision and the archaeologies of landscape. In *Landscape: Politics and Perspectives*, edited B. Bender (Oxford: Berg), pp. 19–48.
Tilly, C., 1994, The *Phenomenology of Landscape* (Oxford: Berg).
Tomlin, D. C., 1990, *Geographic Information Systems and Cartographic Modeling* (New Jersey: Prentice-Hall).
Travis, M .R., Elsner, G. H., Iverson, W. D., and Johnson, C. G., 1975, VIEWIT: Computation of seen areas, slope, and aspect for land use planning, General Technical Report PSW 11/1975, Forest Service, US Department of Agriculture.
Turner, A., Doxa, M., O'Sullivan, D., and Penn, A., 2001, From isovists to visibility graphs: a methodology for the analysis of architectural space. *Environment and Planning B*, **28**, 103–121.
Wheatley, D., 1995, Cumulative viewshed analysis: a GIS-based method for investigating intervisibility, and its archaeological application. *Archaeology and Geographic Information Systems: A European Perspective*, edited by G. Lock and Z. Stani (London: Taylor & Francis), pp. 171–186.
Wood, J. D., 1996, The Geomorphological Characterisation of Digital Elevation Models. Unpublished Ph.D. Thesis. Leicester: University of Leicester.

Seeing Is Believing: GISc and the Visual

Marcos Llobera

The volume and diversity of specialist studies comprising what is known as geographic information science (GISc) is constantly increasing. And while many of these studies are built on abstract concepts and constructs that escape our direct experience (for example, properties of cyberspace), there are others that remain anchored in our daily encounter with the world. The notion of a viewshed is one such concept. Once introduced to its definition, most Western people have no difficulty associating the idea of a viewshed with their visual perception of the world (in spite of the difficulties this association entails). To talk about viewsheds without reference to the world makes little sense.

There is little doubt that the possibility of calculating viewsheds, as offered by most GISc, has had an important impact on various disciplines, such as forestry, landscape architecture, landscape archaeology, communications, and even real estate (Lake et al., 1998). At times, this impact may amount to the simple act of deriving a new layer of information that was not considered previously (for example, in real estate). In other instances, it has triggered academic discussions about the role of vision in society (Gillings and Wheatley, 2000).

Conceptually, viewsheds were never meant to capture the complexities of visual space. Yet putting aside any uncertainties caused by errors in the Digital Elevation Model (DEM), they can be seen as defining a necessary condition for visibility: while it is not always true that one may be able to see an object within the viewshed area, it is almost certain that an object will not be visible outside the viewshed.

In the early 1990s Fisher generated a series of landmark publications that addressed important aspects surrounding the calculation, reliability, and application of viewsheds (Fisher, 1991, 1992, 1993 [Chapter 10 in this volume], 1995.). He also showed that other types of information related to visibility could be mapped as an attribute of space (Fisher, 1996). Most of the studies that have followed have been aimed toward improving the viewshed algorithm (more specifically the line-of-sight or *los* procedure, but see Wang et al., 2000, for an important exception) in various ways. These improvements have sought to increase the efficiency and speed when calculating heights along an *los* (Franklin and Ray, 1994; Wang et al., 1996; Kidner et al., 2001; Izraelevitz 2003; Kidner, 2003,), reduce the number of candidate viewpoints (Kim et al., 2004) and/or use cluster computing techniques to distribute the *los* calculation among various computers (Mineter et al., 2003; Llobera et al., in press). But for the most part, they have all aimed at producing the same or similar end result: the classification of a location as being visible (1) or not (0).

It is within the context of urban and architectural studies that in recent years a wider array of visual characteristics have been considered and mapped as spatial attributes (Batty, 2001; O'Sullivan and Turner, 2001; Turner, 2003; Ratti, 2005). As already mentioned (Llobera, 2003) this was possible in part because of the well-defined nature of the spatial boundaries (that is, it is easier to establish what is visible or not in a city) and their reliance on the *isovist* (Benedikt, 1979), a simpler 2D version of the viewshed. A question that begs to be asked of these investigations is whether 2D representations are the right approximations when describing the visual experience in a city. Currently there are studies that fully consider the 3D nature of visual space in an urban context (Fisher-Gewirtzman and Wagner, 2003; Teller, 2003) but these remain a minority (possibly because of computational overhead).

Effort toward resolving any of the difficulties surrounding the application of viewsheds to real-world scenarios has not been very forthcoming either. Whereas studies on meteorological conditions (e.g., refraction) have been around for quite some time (Middleton, 1952), these have hardly been incorporated into viewshed calculations except in a very simplified form (for example, ArcGIS version 9.01 documentation). The same can be said about the effect that a source of light may have on the actual calculation of visibility (Fisher, 1995). The absence of surface vegetation, and its effect on visibility and/or radiowave propagation, is a major limiting factor that has gone mostly unchecked. In spite of its importance, interest in resolving this limitation has drawn little attention. Dean's work (1997) represents a first attempt at solving this problem.

Given the large GISc community, the limited effort shown to solve some of these questions or to expand the range of visual descriptors is surprising. This may well be due to the user's lack of interest (is there any commercial value in this work?); to reluctance to increase the complexity of what is already considered a computer-intensive routine; or to the weight of traditional, and often narrow, ways of thinking about visual space. The notion of *visualscape* originated as a response by this author to the latter. It was conceived, among various other reasons, to unite studies done with GISc and other geo-, or spatio-modeling methods under a single term. But most importantly, it sought, or rather seeks, to point out the extensive array of opportunities that are available for research once a wider selection of visual qualities of space are explored, and novel representations and data structures for this information are adopted. Some of these, and their implications, are discussed further in the following paragraphs.

The study of visual space would potentially benefit from the development of new procedures aimed at computing visual indexes (even the standard viewshed output) that integrated both topography and the built environment. Currently the calculation of viewsheds is still based on the traditional spatial primitives: point, lines, and polygons. The use of other spatial representations, like a plain solid model such as a cube to represent spatial features or objects, are simply not used. If we wanted to describe the visual experience of someone approaching a house, or for that matter a Greek temple located within the landscape, we would have serious difficulties given the range of tools we have at present (for a very simple attempt, see Llobera, 2003).

It is not to say that exploring this scenario is not already feasible within the capabilities of some GISc. The process and results that we would obtain from using such systems are, however, far from optimal. GISc systems are not designed with these possibilities in mind. This is true, in spite of the fact that our ability to integrate this information is well within our technological reach. One has just to contemplate the constant stream of advances in computer graphics that dominate the personal computing market, both in algorithms and hardware (for example, graphic cards). But there is not the commercial incentive in doing so. Industries such as the games industry are definitely interested in speed when it comes to determining what is visible and what is not, but have no interest in disassembling the scene, in probing it, calculating its area, or describing its geometry numerically. Yet the possibility of using slightly more realistic spatial representations and of extending the range of descriptions used at present is vital to developing new, more formal, approaches in areas such as landscape archeology and architecture.

Interestingly, the idea of developing such new approaches requires reconciling implicitly the use of spatial objects (that is, the cube) with other spatial representations such as a raster, which brings to the frontline the old discussion about objects versus fields (Couclelis, 1992). Today, there is no question that when it comes to representing information that is considered to vary continuously in space, scalar (i.e., raster) and vector (though less so) mathematical representations are the ones favored by modelers in most disciplines (that is, geography, physics, and ecology). Breaking down space into regular bits eases the computation of new indexes. In this sense, the repetitive use of an *los*, the traditional way of calculating a viewshed, may prove to be ultimately a good analogy to the way in which the human eye scans the environment. But how will we map out these new indexes when using a 3D model of the world? How will we query this world constituted by objects and field-like information? These are questions we need to ask if we are going to introduce more complicated (that is, realistic) spatial structures.

Change in visual space is another area that needs further consideration. At the moment the computation of visibility in GISc is reduced to a *focal function* (borrowing Tomlin's terminology, Tomlin, 1990). This relegates the description of visual space to a discrete scalar field, where only one value is stored per unit of space. While useful, this representation is somewhat incomplete as directionality is ultimately lost. The structure of visual space changes as we move. Moreover, ecological psychologists would maintain that it is only through movement that we are able to retrieve or perceive such structure in its entirety, a belief encapsulated in the notion of *optic flow* (Gibson, 1986). But the complexity of visual space may be broken down into components, later quantified and reassembled, the same way that complex terrain calculations are broken into pieces when using *map algebra* (Tomlin, 1990). A simple illustration of this comes from Lee and Stucky's (1998) visibility path delineations, where paths of maximum visibility are generated after the total viewshed is first calculated. Once strategies and methods like these are set in place, we will still have to face the difficulties of assessing the correspondence between any measures of change, and change as experienced by a real individual. Such mapping may ultimately prove to be less a matter of degree and more a matter of a qualitative change (Galton, 2000).

If ultimately the aim is to generate convincing descriptions of properties in the visual world, we will need to bridge the gap between our digital representations and the physical world. Such an endeavor calls for an expansion of the GISc agenda into the realm of experimentation proper. Experiments of this sort (that is, in the physical world) are common practice in disciplines like psychophysics and ecological psychology. Examples can also be found in the GISc literature (Dean, 1997; Conroy, 2003) though they are less common and may lack the same level of rigor. The benefits of constructing such experiments are undeniable (and necessary if we consider some of GISc to be an aspect of modeling). But effort in this direction is likely to be quite taxing. It is very doubtful that any developments will occur unless the implications of following this route (that is, in the way of time, resources, and expertise) are understood and fully accepted by the research community at large.

All of the above highlight the benefits of widening the current scope of visibility studies within GISc. A case in point is the work done by Koenderink and his colleagues on the nature of visual space (Koenderink et al., 2000, Koenderink et al., 2002, Todd et al., 2001). Through their findings, many of which are based on experiments in the "real" world, Koenderink and his colleagues have exposed our limited understanding of the geometrical properties of visual space. These results, while difficult to integrate because of their tentative character, are important to keep in mind. How the inability to fully comprehend and define these properties affects current, and the design of future, applications, is a question that will require close attention.

The concept of *visualscape* refers to all these possibilities. Some of the ideas and suggestions presented are well in line with the current efforts in GISc; others call for opening up the field to work done in other disciplines that have similar or complementary goals — the description and modeling of visual space.

References

BATTY, M., 2001, Exploring isovist fields: Space and shape in architectural and urban morphology, *Environment and Planning B*, 28, 123–150.

BENEDIKT, M.L., 1979, To take hold of space: Isovists and isovist fields, *Environment and Planning B*, 6, 47–65.

CONROY, D.R., 2003, The secret is to follow your nose: Route path selection and angularity, *Environment and Behavior*, 35, 132–160.

COUCLELIS, H., 1992, People manipulate objects (but cultivate fields): Beyond the raster vector debate in GIS. In *Theories and Methods of Spatio-Temporal Reasoning in Geographic Space*, Lecture Notes in Computer Science 639, Frank, A.U. and Campari, I., Eds., Springer-Verlag, Berlin, 65–77.

DEAN, D.J., 1997, Improving the accuracy of forest viewsheds using triangulated networks and the visual permeability method, *Canadian Journal of Forest Research*, 27, 969–977.

FISHER, P.F., 1991, First experiments in viewshed uncertainty: The accuracy of the viewshed area, *Photogrammetric Engineering and Remote Sensing*, 57(10), 1321–1327.

FISHER, P.F., 1992, First experiments in viewshed uncertainty: Simulating fuzzy viewsheds, *Photogrammetric Engineering and Remote Sensing*, 58(3), 345–327.

FISHER, P.F., 1993, Algorithm and implementation uncertainty in viewshed analysis, *International Journal of Geographic Information Systems*, 7(4), 331–347.

FISHER, P.F., 1995, An exploration of probable viewsheds in landscape planning, *Environment and Planning B* 22, 527–546.

FISHER, P.F., 1996. Reconsideration of the viewshed function in terrain modelling, *Geographical Systems*, 3(1), 33–58.

FISHER-GEWIRTZMAN, D. AND WAGNER, I.A., 2003, Spatial openness as a practical metric for evaluating built-up environments, *Environment and Planning B*, 30 (1), 37–49.

FRANKLIN, W.R. AND RAY, C., 1994. Higher isn't necessarily better: Visibility algorithms and experiments. In *Advances in GIS Research: Sixth International Symposium on Spatial Data Handling*, Waugh, T.C. and Healey, R.G., Eds., Taylor & Francis, London, 751–770.

GALTON, A., 2000, *Qualitative Spatial Change*, Oxford University Press, Oxford.

GIBSON, J.J., 1986, *The Ecological Approach to Visual Perception*, Lawrence Erlbaum Associates, Inc., New Jersey.

GILLINGS, M. AND WHEATLEY, D., 2000, Vision, perception and GIS: Developing enriched approaches to the study of archaeological visibility. In *Beyond the Map*, Lock, G., Ed., IOS Press, Amsterdam, 1–28.

IZRAELEVITZ, D., 2003, A fast algorithm for approximate viewshed computation, *Photogrammetric Engineering & Remote Sensing*, 69(7), 767–774.

KIDNER, D.B., 2003, Higher-order interpolation of digital grid digital elevation models, *International Journal of Remote Sensing*, 24(14), 2981–2987.

KIDNER, D.B., SPARKES, A.J., DOREY, M.I., WARE, J.M., AND JONES, C.B., 2001, Visibility analysis with the multiscale implicit TIN, *Transactions in GIS*, **5**(1), 19–37.

KIM, Y.-H., RANA, S., AND WISE, S., 2004, Exploring multiple viewshed analysis using terrain features and optimization techniques, *Computers & Geosciences*, 30(9–10), 1019–1032.

KOENDERINK, J.J., VAN DOORN, A.J., KAPPERS, A.M.L., AND TODD, J.T., 2002, Pappus in optical space, *Perception & Psychophysics,* 64(3), 380–391.

KOENDERINK, J.J., VAN DOORN, A.J., AND LAPPIN, J.S., 2000, Direct measurement of curvature of visual space, *Perception,* 29(1), 69–79.

LAKE, I.R., LOVETT, A.A., BATEMAN, I.J., AND LANGFORD, I.H., 1998, Modelling environmental influences on property prices in an urban environment, *Computers, Environment and Urban Systems*, 22, 121–136.

LEE, J. AND STUCKY, D., 1998, On applying viewshed analysis for determining least-cost-paths on digital elevation models, *International Journal of Geographic Information Systems*, 12, 891–905.

LLOBERA, M., 2003, Extending GIS based analysis: The concept of visualscape, *International Journal of Geographic Information Science*, 17(1), 25–49.

LLOBERA, M., WHEATLEY, D., STEELE, J., COX, S. AND PARCHMENT, O., in press, Calculating the inherent visual structure of a landscape (total viewshed) using high-throughput computing. In *Beyond the Artifact: Digital Interpretation of the Past.* Computer Applications in Archaeology CAA'04, Prato, Italy, BAR International Series, Oxford.

MIDDLETON, W.E.K., 1952, *Vision through the Atmosphere*, University of Toronto Press, Toronto.

Mineter, M.J., Dowers, S., Caldwell, D.R., and Gittings, B.M., 2003, High-throughput computing to enhance visibility analysis. In *Proceedings of the 7th International Conference on GeoComputation 2003*, CD ROM, University of Southampton, United Kingdom, 8–10 September.

O'Sullivan, D. and Turner, A., 2001, Visibility graphs and landscape visibility analysis,, *International Journal of Geographic Information Science*, 15, 221–237.

Ratti, C., 2005, The lineage of the line: Space syntax parameters for the analysis of urban DEMs, *Environment and Planning B*, 32(4), 547–566.

Teller, J., 2003, A spherical metric for the field-oriented analysis of complex urban open spaces, *Environment and Planning B*, 30(3), 339–356.

Todd, J.T., Oomes, A.H.J., Koenderink, J.J., and Kappers, A.M.L., 2001, On the affine structure of perceptual space, *Psychological Science*, 12(3), 191–196.

Tomlin, D.C., 1990, *Geographic Information Systems and Cartographic Modeling*, Prentice-Hall, Upper Saddle River, NJ.

Turner, A., 2003, Analysing the visual dynamics of spatial morphology, *Environment and Planning B*, 30(5), 657–676.

Turner, A., Doxa, M., O'Sullivan, D., and Penn, A., 2001, From isovists to visibility graphs: A methodology for the analysis of architectural space, *Environment and Planning B*, 28(1), 103–121.

Wang, J., Robinson, G.J., and White, K., 1996, A fast solution to local viewshed computation using grid-based digital elevation models, *Photogrammetric Engineering & Remote Sensing*, 62(10), 1157–1164.

Wang, J., Robinson, G.J., and White, K., 2000, Generating viewsheds without using sight-lines, *Photogrammetric Engineering & Remote Sensing*, 66(1), 87–90.

21 Collaboration Networks Revealed by *IJGIS* Authors

Cristina Arciniegas and Jo Wood

A reevaluation of classic papers in geographical information (GI) Science allows us to reflect on the impact that individuals and their scientific contributions have had on the discipline. In doing so, we are able to consider in greater depth, the nature of the discipline, the way in which it has evolved, and the impact it has had on the wider scientific community. Yet GI Science includes complex, changing, and vaguely defined research domains (see Goodchild, Chapter 9 in this volume) with contributions from individuals, many of whom may not choose to place themselves within an agreed domain boundary of GI Science. In this chapter, we offer an insight into the structure of GI Science by considering how those who have published in *IJGIS* have collaborated, or indeed, have chosen not to. We borrow some of the techniques from bibliometrics (Broadus, 1987) and social network theory (Newman, 2004) to consider whether the structure of co-authorship in *IJGIS* can tell us something about the nature of GI Science and the role *IJGIS* has played in it.

A definition of GI Science that avoids having to draw boundaries around a number of allied disciplines is to consider it as

work carried out by those who publish it in *IJGIS* (21.1)

Clearly there is some circularity in this definition which is only resolved by passing the responsibility of domain definition to the editors and reviewers of the journal (work considered out of scope by reviewers will not be published in that journal). It does, however, provide us with a measurable quantity with which we can define the core of the discipline. We can further quantify distance from that core in relation to the amount of work published in *IJGIS* (someone who has published more than 10 papers in IJGIS could be regarded as more centrally involved in GI Science than someone who has published 1 paper in *IJGIS* and 9 elsewhere). We might enlarge our definition by considering GI science also to include

work carried out in collaboration with those involved in (21.1) defined above (21.2)

And to some extent, there may be an involvement in GI Science by those who collaborate in work defined in (21.2) above, and so on. Thus we can conceive of a network of collaborations among researchers with a core somewhere firmly in the GI Science domain and a periphery based on collaborative research with those nearer the core.

To investigate this structure, we have considered all published output from *IJGIS* between 1991 and 2003 and built a network of contributors based upon authorship and coauthorship of those publications. While this represents a form of bibliographic domain analysis (Melin and Persson, 1996), it does not suffer from the same levels and types of bias inherent in analysis of citation networks (Hjørland, 2002).

The network, when represented as a graph, comprises nodes, where each node is an author or coauthor of a paper, and edges, where each edge represents a collaboration between two authors in the same paper. From this graph (see Table 21.1), a number of quantifiable characteristics can be measured such as mean node degree (average number of coauthors), mean geodesic distance (average path length between authors), and diameter (the longest shortest path between any two authors).

To build a network that summarizes GI Science collaboration, all publications from *IJGIS* between 1991 and 2003 were extracted from the Institute for Scientific Information's Web of Knowledge database (WoK). From this list, the frequency distribution of papers per author was calculated (see Figure 21.1). If we regard authors with higher numbers of papers published in *IJGIS* as being more centrally involved in the subject domain, we can expand the publication boundaries of GI Science by considering the other journals in which these more prolific authors have published. Empirical evaluation led to the selection of all publications in journals contributed to by at least 4 authors who have themselves published 5 or more articles in *IJGIS* (See Table 21.2 and Table 21.3). The result is a network of 12,457 papers written by a total of 13,950 authors, upon which the analysis in this chapter was performed. It is worth noting that a small number of journals that may be considered central to GI Science are not stored in the WoK (for example, *Transactions in GIS*), and as such will result in slightly more fragmented networks than would otherwise be expected.

TABLE 21.1
Global summary statistics for simple networks.

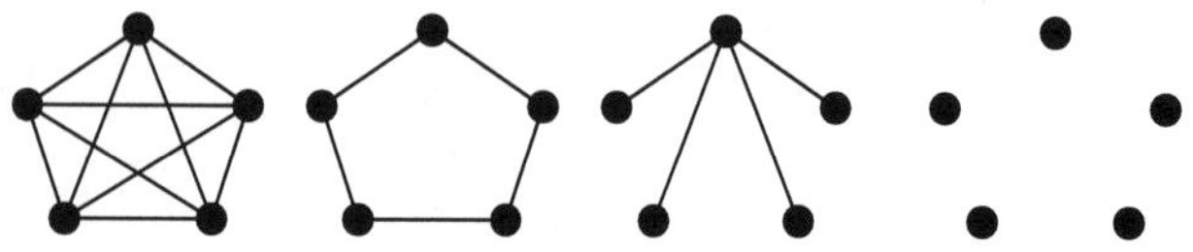

Mean node degree	4	2	1.6	0
Mean geodesic distance	1	1.5	1.6	0
Diameter	1	2	2	0

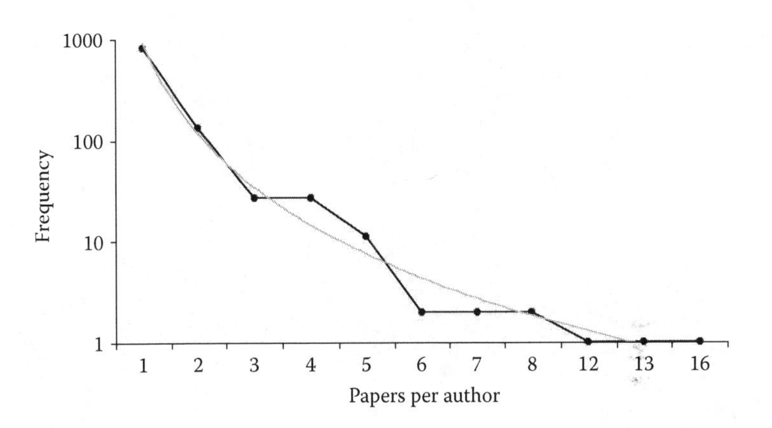

FIGURE 21.1 Frequency distribution of author publication in *IJGIS* 1991–2003. The modelled power function is shown in grey and is consistent with Lotka's Law (Lotka, 1926).

TABLE 21.2
All authors who have published 5 or more papers in IJGIS *between 1991 and 2003.*

Papers published in *IJGIS*	Author name
16	Goodchild, M.
13	Burrough, P.
12	Egenhofer, M.
8	Aangeenbrug, R.
8	Fisher, P.
7	Martin, D.
7	Stuart, N.
6	Frank, A.
6	Heuvelink, G.
5	Carver, S.
5	Govers, G.
5	Jankowski, P.
5	Jones, C.
5	Lee, J.
5	Mark, D.
5	Mather, P.
5	Muller, J.
5	Shi, W.
5	Wise, S.
5	Worboys, M.

TABLE 21.3
All journals in the ISI Web of Knowledge Database who have had at least 4 authors from Table 21.2 publish within them.

Journal name	Number of authors from Table 21.2 publishing in a journal
Environment and Planning B-Planning & Design	13
Computers & Geosciences	7
Photogrammetric Engineering and Remote Sensing	6
International Journal of Remote Sensing	5
Spatial Information Theory Lecture Notes in Computer Science	5
Computers Environment and Urban Systems	4
Earth Surface Processes and Landforms	4
Environment and Planning A	4
Geoinformatica	4
Progress in Human Geography	4

TABLE 21.4
Summary statistics for GI Science paper co-authorship and other subject domains.

Measure	GI Science	Computer Science	Human Computer Interaction	Biology (Biomedical)	Physics	Mathematics
Database source	ISI WoK	NCSTRL	ACM	Medline	Physicse-print archive	Mathematics Review
No. authors	13,950	11,994	23,624	1,520,251	52,909	253,339
No. papers	12,457	13,169	22,887	2,163,923	98,502	—
Papers per author	1.7	2.6	2.2	6.4	5.1	6.9
Authors per paper	2.0	2.2	2.3	3.8	2.5	1.5
Ave no. collaborators	2.9	3.6	3.7	18.1	9.7	3.9
Largest component	31%	57%	51%	92%	85%	82%
Ave distance	10.0	9.7	6.8	4.6	5.9	7.6
Diameter	28	31	27	24	20	27
Time period	1992–2002	1995–1999	1998–2002	1995–1999	1995–1999	1995–1999

Global statistics for all GI Science papers selected were first compared with the results of equivalent studies in other subject domains (see Table 21.4). Within the sample selected, authors in the GI Science domain would appear to be less prolific and less collaborative in paper writing than those in other disciplines (mean of 1.7

papers per author and 2.0 authors per paper). Of the network as a whole, the largest connected part of the coauthorship graph comprises only 31% of all authors. This reflects a more fragmented community of coauthored publications than is found in all other disciplines considered here. This is reinforced by a larger average distance between any two connected authors in the network (10.0) than in all other disciplines, despite having an equivalent maximum distance (28). Analysis of trends over the time period 1992 to 2002 has suggested that while there has been a gradual increase in connectivity over time, this is no more than would be expected from a growing random network.

These observations may be explained by a number of possible causes. GI Science as a coherent discipline is still relatively young, and as such, there is not a consensus as to the "normal" publication outlets. While *IJGIS* may be considered central, its multidisciplinary nature means that there is a wide and poorly defined set of allied publication outlets that contribute to advances in the subject. We may also hypothesise that collaboration within GI Science may involve, to a greater extent than other disciplines, coauthorship in other types of publication excluded from this analysis, such as edited books and teaching-oriented texts. Finally, we may conclude that the nature and publication culture of GI Science is associated with a greater proportion of contributions of individuals than other domains considered here.

To investigate the nature of GI Science collaboration, we can consider two important measures of centrality associated with individual authors. *Closeness* represents distance of each node from all others in the network (see Table 21.5). While it is influenced by publication productivity, higher values of closeness also reflect a diversity of collaborations with those in different parts of the GI Science network. An author with high paper output, but who worked alone or collaborated with the

TABLE 21.5
Top 25 authors ordered by network closeness. Higher closeness indicates co-authorship with many collaborators who are connected to a large proportion of the network.

	Author	Closeness		Author	Closeness
1	Justice, C.	0.051	14	Cihlar, J.	0.048
2	Belward, A.	0.050	15	Muller, J.	0.047
3	Townsend, J.	0.050	16	Defries, R.	0.047
4	Estes, J	0.049	17	Lewis, P.	0.047
5	Goodchild, M.	0.049	18	Scepan, J.	0.047
6	Openshaw, S.	0.049	19	Strahler, A.	0.047
7	Teillet, P.	0.049	20	Vermote, E.	0.047
8	Malingreau, J.	0.049	21	Gong, P.	0.047
9	Chen, J.	0.048	22	Shimabukuro, Y.	0.047
10	Tucker, C.	0.048	23	Hansen, M.	0.047
11	Li, Z.	0.048	24	Wise, S.	0.047
12	Goward, S.	0.048	25	Loveland, T.	0.047
13	Krug, T.	0.048			

TABLE 21.6
Top 25 authors ordered by network betweenness. Higher betweenness indicates a pivotal role in linking different sub-networks within the discipline.

	Author	Betweenness		Author	Betweenness
1	Goodchild, M.	0.016	14	Atkinson, P.	0.007
2	Gong, P.	0.014	15	Cracknell, A.	0.007
3	Justice, C.	0.014	16	Richards, K.	0.007
4	Openshaw, S.	0.013	17	Shokr, M.	0.007
5	Estes, J.	0.012	18	White, K.	0.007
6	Li, Z.	0.011	19	Krug, T.	0.007
7	Chen, J.	0.010	20	Wang, J.	0.007
8	Belward, A.	0.009	21	Mattikalli, N.	0.007
9	Townsend, J.	0.008	22	Ramsay, B	0.007
10	Foody, G.	0.008	23	Martin, D.	0.006
11	Jackson, T.	0.008	24	Curran, P.	0.006
12	Ledrew, E.	0.008	25	Jensen, J.	0.006
13	Mutlow, C.	0.007			

same small set of authors would not have a high closeness value. Those with higher closeness measures tend to combine productivity with diversity. This may be due to a longer publication career that has seen an evolution of research direction, or research that has been applied in a wider range of application areas.

Betweenness allows us to refine our notion of centrality by representing the number of shortest (geodesic) paths that run through a given node. It can be used to identify "pivotal" authors who provide important links among different parts of the network. Again, this measure is partly a function of publication productivity, but further identifies authors who provide unique or rare links among significant sub-networks in the graph (see Table 21.6).

Because the WoK database includes author affiliations, it is possible to gain some insight into the spatial distribution of collaborating authors. Figure 21.2 shows international collaboration between coauthors. The width of each line is proportional to the *collaboration exclusivity* $e_{ij} = n_{ij} / (n_i+n_j)$ where n_{ij} is the number of papers with coauthors from countries *i* and *j*, n_i is the total number of papers from country *i*, and n_j, the total from country *j*. The higher the exclusivity, the greater the proportion of published output involving collaboration among the given countries.

Clearly geographical proximity plays a role in the way in which international collaboration effort is concentrated (strong links between the United States and Canada, between Australia and New Zealand, and within Europe). Language and cultural similarities also appear to play a part (for example, the United Kingdom, United States, and Australia maintain strong collaborative links). Countries with strong international publishing patterns, often encouraged by research funding opportunities, can also be identified (for example, Greece, Portugal, Russia, and China all have high collaboration exclusivity values with a small set of collaborators).

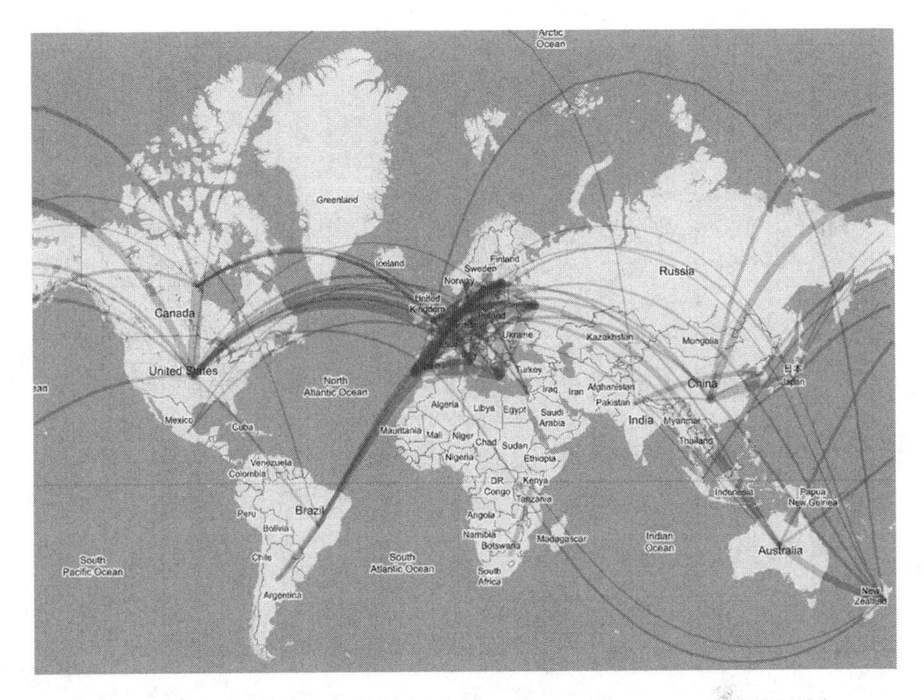

FIGURE 21.2 International collaboration in GI Science journal papers, 2002. All international collaborations where n_i and n_j are both greater than 10 and e_{ij} is at least 2%.

Together with topological analysis of coauthorship, we are able to describe and explain how those individuals who have published papers in *IJGIS* have become embedded in an international network of GI Science.

References

Broadus, R.M., 1987, Toward a definition of "bibliometric," *Scientometrics*, 12(5), 373–379.

Hjørland, B., 2002, Domain analysis in information science: Eleven approaches—traditional as well as innovative, *Journal of Documentation*, 58(4) 422–462.

Lotka, A.J., 1926, The frequency distribution of scientific productivity, *Journal of the Washington Academy of Sciences*, 16(12), 317–323.

Melin, G. and Persson, O., 1996, Studying research collaboration using co-authorships, *Scientometrics*, 36(3), 363–377.

Newman, M.E.J., 2004, Co-authorship networks and patterns of scientific collaboration, *Proceedings of the National Academy of Sciences U.S.A (PNAS)*, 101(Suppl. 1), 5200–5205.

Contributors

Gennady L. Andrienko
Fraunhofer AIS — Institute for Autonomous Intelligent Systems
Schloss Birlinghoven
Sankt-Augustin, Germany
gennady.andrienko@ais.fraunhofer.de

Natalia V. Andrienko
Fraunhofer AIS — Institute for Autonomous Intelligent Systems
Schloss Birlinghoven
Sankt-Augustin, Germany
natalia.andrienko@ais.fraunhofer.de

Cristina Arciniegas
Department of Information Science
City University
Northampton Square, London, United Kingdom
at705@soi.city.ac.uk

Yaser Bishr
Image Matters, LLC
Leesburg, Virginia, USA
yaserb@imagem.cc

Kurt E. Brassel
Department of Geography
University of Zürich
Zürich, Switzerland

Chris Brunsdon
Department of Geography
University of Leicester
Leicester, United Kingdom
cb179@leicester.ac.uk

Peter A. Burrough
Department of Physical Geography
Utrecht University
Utrecht, The Netherlands
p.burrough@geo.uu.nl

Martin E. Charlton
National Centre for Geocomputation
National University of Ireland, Maynooth
County Kildare, Ireland
Martin.Charlton@nuim.ie

Keith C. Clarke
Department of Geography
University of California, Santa Barbara
Santa Barbara, California, USA
kclarke@geog.ucsb.edu

Charles K. Dietzel
Department of Geography
University of California, Santa Barbara
Santa Barbara, California, USA

Max J. Egenhofer
Department of Spatial Information Science and Engineering
University of Maine
Orono, Maine, USA
max@spatial.maine.edu

Peter F. Fisher
Department of Information Science
City University
Northampton Square, London, United Kingdom
pff1@soi.city.ac.uk

A. Stewart Fotheringham
National Centre for Geocomputation
National University of Ireland, Maynooth
County Kildare, Ireland
stewart.fotheringham@nuim.ie

Andrew U. Frank
Department of Geoinformation and Cartography
Technical University Vienna
Vienna, Austria
frank@geoinfo.tuwien.ac.at

Nicholas Gazulis
Department of Geography
University of California, Santa Barbara
Santa Barbara, California, USA

Noah C. Goldstein
Department of Geography
University of California, Santa Barbara
Santa Barbara, California, USA

Michael F. Goodchild
National Center for Geographic Information and Analysis and
Department of Geography
University of California
Santa Barbara, California, USA
good@geog.ucsb.edu

Gerard B.M. Heuvelink
Soil Science Centre
Wageningen University and Research Centre
Wageningen, The Netherlands
gerard.heuvelink@wur.nl

Piotr Jankowski
Department of Geography
San Diego State University
San Diego, California, USA
piotr@geography.sdsu.edu

Harri T. Kiiveri
CSIRO Mathematical and Information Sciences
The Leeuwin Centre
Wembley, Australia
harri.kiiveri@csiro.au

David E. Livingstone
Faculty of Computing, Information Systems and Mathematics
Kingston University
Kingston Upon Thames, United Kingdom
d.livingstone@kingston.ac.uk

Marcos Llobera
Department of Anthropology
University of Washington
Seattle, Washington, USA
mllobera@u.washington.edu

David J. Maguire
Environmental Systems Research Institute, Inc.
Redlands, California, USA
dmaguire@esri.com

David M. Mark
Department of Geography
University of Buffalo
Amherst, New York, USA
dmark@geog.buffalo.edu

Harvey J. Miller
Department of Geography
University of Utah
Salt Lake City, Utah, USA
harvey.miller@geog.utah.edu

Jonathan F. Raper
Department of Information Science
City University
Northampton Square, London, United Kingdom
raper@soi.city.ac.uk

Andrew K. Skidmore
ITC International Institute for Geo-Information Science and Earth Observation
Enschede, The Netherlands
skidmore@itc.nl

Barry Smith
Department of Philosophy
University of Buffalo
Amherst, New York, USA
phismith@buffalo.edu

Alfred Stein
ITC International Institute for Geo-Information Science and Earth Observation
Enschede, The Netherlands
stein@itc.nl

Robert Weibel
Department of Geography
University of Zürich
Zürich, Switzerland
weibel@geo.unizh.ch

Jo Wood
Department of Information Science
City University
Northampton Square, London, United Kingdom
jwo@soi.city.ac.uk

Michael F. Worboys
Department of Spatial Information Science and Engineering
University of Maine
Orono, Maine, USA
worboys@spatial.maine.edu

Contact details for the following contributors to original papers are not given here: Alan Craft, Robert D. Franzosa, Hilary Hearnshaw, Leonard J. Gaydos, Stan Openshaw, and Colin Wymer. If need be they can be contacted via the authors of the related commentary.

Index

Page numbers in italics indicate tables and/or figures.

C

D

H

I

N

Q

R

S

T

U

V

W

Z